UBC BOTANI
READING

Volume 24

Rhamnales
Rhamnaceae
Leeaceae
Vitaceae

Linales
Erythroxylaceae
Linaceae

Polygalales
Malpighiaceae
Tremandraceae
Polygalaceae
Xanthophyllaceae

Volume 25

Sapindales
Melianthaceae
Akaniaceae
Sapindaceae
Aceraceae
Burseraceae
Anacardiaceae
Simaroubaceae

Volume 26

Meliaceae
Rutaceae
Zygophyllaceae

Volume 27

Geraniales
Oxalidaceae
Geraniaceae
Tropaeolaceae

Apiales
Araliaceae
Apiaceae

Volume 28

Gentianales
Loganiaceae
Gentianaceae
Apocynaceae
Asclepiadaceae

Volume 29

Solanales
Solanaceae

Volume 30

Convolvulaceae
Cuscutaceae
Menyanthaceae
Polemoniaceae
Hydrophyllaceae

Lamiales
Boraginaceae
Verbenaceae

Volume 31

Lamiaceae

Volume 32

Callitrichales
Callitrichaceae

Plantaginales
Plantaginaceae

Scrophulariales
Oleaceae
Scrophulariaceae
Globulariaceae

Volume 33

Myoporaceae
Orobanchaceae
Gesneriaceae
Acanthaceae
Pedaliaceae
Bignoniaceae
Lentibulariaceae

Volume 34

Campanulales
Sphenocleaceae
Campanulaceae
Stylidiaceae
Donatiaceae

Volume 35

Brunoniaceae
Goodeniaceae

Volume 36

Rubiales
Rubiaceae

Dipsacales
Caprifoliaceae
Valerianaceae
Dipsacaceae

Volume 37,38

Asterales
Asteraceae

Volume 39

Alismatales
Limnocharitaceae
Alismataceae

Hydrocharitales
Hydrocharitaceae

Najadales
Aponogetonaceae
Juncaginaceae
Potamogetonaceae
Ruppiaceae
Najadaceae
Zannichelliaceae
Posidoniaceae
Cymodoceaceae
Zosteraceae

Triuridales
Triuridaceae

Arecales
Arecaceae

Pandanales
Pandanaceae

Arales
Araceae
Lemnaceae

Volume 40

Commelinales
Xyridaceae
Commelinaceae

Eriocaulales
Eriocaulaceae

Restionales
Flagellariaceae
Restionaceae
Centrolepidaceae

Juncales
Juncaceae

Volume 41,42

Cyperales
Cyperaceae

Volumes 43,44

Poaceae

Volume 45

Hydatellales
Hydatellaceae

Typhales
Sparganiaceae
Typhaceae
Bromeliales
Bromeliaceae

Zingiberales
Musaceae
Zingiberaceae
Costaceae
Cannaceae

Liliales
Philydraceae
Pontederiaceae
Haemodoraceae
Liliaceae

Volume 46

Iridaceae
Aloeaceae
Agavaceae
Xanthorrhoeaceae
Hanguanaceae
Taccaceae
Stemonaceae
Smilacaceae
Dioscoreaceae

Volume 47

Orchidales
Burmanniaceae
Corsiaceae
Orchidaceae

Volume 48

Gymnospermae
Pteridophyta

Volume 49,50

Oceanic Islands

Volume 51 et seq.

Non-vascular plants

FLORA OF AUSTRALIA

Dampiera adpressa A.Cunn. ex DC., a species of south-eastern Queensland and eastern New South Wales. Painting by Diana Boyer.

AUSTRALIAN BIOLOGICAL RESOURCES STUDY

FLORA OF AUSTRALIA

Volume 35

Brunoniaceae, Goodeniaceae

An AGPS Press publication
Australian Government Publishing Service Canberra

© Commonwealth of Australia 1992

This work is copyright. Apart from any use as permitted under the *Copyright Act 1968*, no part may be reproduced by any process without prior written permission from the Australian Government Publishing Service. Requests and inquiries concerning reproduction and rights should be addressed to the Manager, Commonwealth Information Services, Australian Government Publishing Service, GPO Box 84, Canberra ACT 2601.

EXECUTIVE EDITOR
Alexander S. George

EDITORIAL ASSISTANCE
Helen S. Thompson (1987–)
Barbara Randell (1989–1990) under ABRS grants

Arthur D. Chapman (–1991)
(Bibliography)

This work may be cited as

Flora of Australia Volume 35, Brunoniaceae, Goodeniaceae, Australian Government Publishing Service, Canberra (1992)

Individual contributions may be cited thus:

R.C.Carolin, *Goodenia*, *Flora of Australia* 35: 147–281 (1992).

National Library of Australia
Cataloguing-in-Publication entry

Flora of Australia. Volume 35, Brunoniaceae, Goodeniaceae.

Includes bibliographical references and index.

ISBN 0 644 14553 6.

ISBN 0 644 14454 8 (pbk.).

ISBN 0 642 07013 X (set).

ISBN 0 642 07016 4 (set:pbk.).

1. Botany – Australia – Classification. 2. Plants – Identification. I. George, Alexander S., 1939– . II. Australian Biological Resources Study. III. Title : Brunoniaceae, Goodeniaceae.

581.994

This volume was produced using Word for Windows 1.1a.
Printed by State Print Netley South Australia

CONTENTS

Endpapers

Front: Contents of volumes in the *Flora of Australia*, the families arranged according to the system of A.Cronquist, *An Integrated System of Classification of Flowering Plants* (1981).

Back: *Flora of Australia*: Index to families of flowering plants.

CONTRIBUTORS TO VOLUME 35

Dr Roger C. Carolin, 30 Pullman St, Berry, New South Wales 2535.

Dr David A. Morrison, Department of Applied Biology, University of Technology, Sydney, PO Box 123, Broadway, New South Wales 2007.

Dr Tahir Rajput, 281-C Block D, Unit 6, Latifabad, Hyderabad, Sind, Pakistan.

ILLUSTRATORS

Mrs Diana Boyer, Bobbara Creek, Binalong, New South Wales 2584 (cover, frontispiece).

Mr David Mackay, c/- National Herbarium of New South Wales, Royal Botanic Gardens, Mrs Macquarie's Road, Sydney, New South Wales 2000

PHOTOGRAPHERS

Mr Murray Fagg, Australian National Botanic Gardens, G.P.O. Box 1777, Canberra, Australian Capital Territory 2601.

Mr Alexander S. George, Australian Biological Resources Study, Australian National Parks & Wildlife Service, GPO Box 636, Canberra, Australian Capital Territory 2601.

Mr Tim Low, 48 Orchard Tce, St Lucia, Queensland 4067.

Ms Belinda Pellow, School of Biological Sciences, University of Sydney, Sydney, New South Wales 2006.

Mr Malcolm Ricketts, School of Biological Sciences, Wollongong University, Wollongong, New South Wales 2500.

Mr John Wrigley, c/- Australian National Botanic Gardens, G.P.O. Box 1777, Canberra, Australian Capital Territory 2601.

FLORA EDITORIAL COMMITTEE

The Committee was established in 1980.

COMMITTEE MEMBERS

Bryan A. Barlow (1982-84, 1986-1988)

Peter Bridgewater (ex officio, 1983–1988)

Barbara G. Briggs (1980) (Chairman, 1982–1987)

Roger C. Carolin (1985–1987)

Robert J. Chinnock (1985–)

Hansjoerg Eichler (1980–1985)

Gordon P. Guymer (1989–)

Leslie Pedley (1980–1988)

Sir Rutherford Robertson (Chairman, 1980–1981)

James H. Ross (1980–) (Chairman, 1988–)

George A. M. Scott (1988–)

David E. Symon (1980–1984)

Judy G. West (1989–)

Paul G. Wilson (1980–)

EXECUTIVE OFFICERS

Alexander S. George (Secretary, 1989–)

Roger J. Hnatiuk (Executive Officer, 1984–1989)

Alison McCusker (Secretary, 1980–1984)

INTRODUCTION

Volume 35 of the *Flora of Australia* contains two families – Brunoniaceae and Goodeniaceae. The former is sometimes included within the latter, but is maintained in the Cronquist system of family classification followed in this *Flora*.

The Brunoniaceae is a monotypic family, although its single species is widespread mainly in drier regions of Australia and is quite variable.

The Goodeniaceae is a moderately sized family with c. 400 species (377 Australian) in 11 genera. Only the genus *Scaevola* is well-represented outside Australia (mainly on islands of the Pacific but also in Africa and the Caribbean), but *Lechenaultia*, *Goodenia* and *Velleia* occur in New Guinea, *Goodenia* in Java and South-east Asia, and *Selliera* in New Zealand and Chile. The bulk of the family is thus endemic in Australia. It is represented in most habitats except rainforest and mangrove forests.

The work published here is the result of research over the past 30 years by Roger Carolin and his students at the University of Sydney. All genera have been revised during this period resulting in a modern classification including a review of the status of each genus and the description of many new taxa.

Scope and Presentation of the *Flora*

The geographical area covered by the *Flora* includes the six Australian States, the Northern Territory, the Australian Capital Territory and immediate offshore islands. Other Australian and State-administered territories such as Christmas Is. and Lord Howe Is. are excluded, but the occurrence in those territories of species included in the *Flora* is added to the notes on distribution. Complete Floras of the oceanic islands are in preparation as Volumes 49 and 50.

Descriptions and discussion in the *Flora* are concise and supplemented by important references, synonymy, and information on type collections, chromosome numbers, distribution, habitat, and published illustrations. Descriptions are based on Australian material except for some taxa not confined to Australia for which the collections in Australian herbaria are inadequate. Synonymy is restricted to names based on Australian types or used in Australian literature. Misapplied names are given in square brackets together with an example of the misapplication. Alien taxa established in one or more localities, other than under cultivation, are considered naturalised and are included and asterisked (*).

Families are arranged in the system of A.Cronquist, *An Integrated System of Classification of Flowering Plants* (Columbia University Press, New York, 1981). Within families, genera and species are arranged to show natural relationships as interpreted by contributors. Although relationships cannot be shown adequately in a linear sequence, such an arrangement in a *Flora* assists comparison of related taxa. Infraspecific taxa are keyed out under relevant species. Up to seven collections are cited for each species and infraspecific taxon.

Maps showing distribution in Australia are arranged in the same sequence as the descriptions and are grouped together at the end of the main text (pp. 301–328). The term 'Malesia' is sometimes used in the notes on geographical distribution for species that occur widely in the region covered by *Flora Malesiana*, i.e. Malaysia, Singapore, Indonesia, the Philippines, New Guinea and adjacent islands. New taxa and lectotypifications are included in an Appendix where they are formally published in accordance with the

International Code of Botanical Nomenclature (Bohn, Scheltema & Holkema, Utrecht, 1988). Abbreviations, contractions and references to the format for author and bibliographic citations are listed after the Appendix.

A key to families of flowering plants in Australia and a glossary of technical terms are provided in Volume 1 of the *Flora*. Supplementary glossaries are included in each volume as necessary.

Nomenclatural data have been checked against the *Australian Plant Name Index*; where there is a difference, the data in the *Flora* may be taken as correct.

Acknowledgments

There are 11 contributors, illustrators and photographers to Volume 35. Their co-operation is gratefully acknowledged.

The directors and staff of Australian and overseas institutions have assisted preparation of the Volume with loans of specimens to contributors and illustrators as well as with additional information and reviews of manuscripts during the editorial stages.

The Australian National Botanic Gardens slide collection provided a number of the colour photographs used in this volume.

Several illustrations are reproduced by permission from previously published works. These are acknowledged in the appropriate figure captions.

Many people have helped Roger Carolin with his research over the years. John Ford and Belinda Pellow gave considerable assistance with *Goodenia* and *Dampiera*. Sue Cook (née Dyer) and David Salt did a great deal of the groundwork for the *Scaevola* treatment. The University of Sydney has consistently funded this research over the years, and grants from the Australian Biological Resources Study assisted the completion of treatments of *Scaevola*, *Lechenaultia* and *Anthotium*.

The initial editing of the Goodeniaceae was carried out by Barbara Randell, Adelaide, under grants from the Australian Biological Resources Study. David Symon assisted in testing the keys to species.

The Executive Editor acknowledges with pleasure the assistance of the staff of the Australian Biological Resources Study in preparing this Volume. Roger Hnatiuk provided managerial support in the early stages, and Helen Hewson over the final months. Computing assistance was provided by Wayne Murray. Initial adaptation of the typesetting system to Microsoft 'Word for Windows' was carried out by Lyn Randall. Data entry was performed by Savita Meek. In the final stages of preparation support was provided by Barbara Barnsley and Jane Mowatt.

The co-operation of the Australian Government Publishing Service, Canberra, and State Print, Adelaide, is gratefully acknowledged.

Figure 1. *Brunonia australis*
Photograph — A.George

Figure 2. *Anthotium rubriflorum*
Photograph — J.Wrigley (ANBG)

Figure 3. *Lechenaultia tubiflora*
Photograph — A.George

Figure 4. *Lechenaultia acutiloba*
Photograph — M.Fagg

Figure 5. *Lechenaultia formosa*
Photograph — J.Wrigley (ANBG)

Figure 6. *Lechenaultia biloba*
Photograph — A.George

Figure 7. *Lechenaultia macrantha*
Photograph — A.George

Figure 8. *Lechenaultia lutescens*
Photograph — A.George

Figure 9. *Dampiera linearis*
Photograph — A.George

Figure 10. *Dampiera rosmarinifolia*
Photograph — M.Fagg (ANBG)

Figure 11. *Dampiera stricta*
Photograph — M.Fagg

Figure 12. *Dampiera luteiflora*
Photograph — M.Fagg (ANBG)

Figure 13. *Dampiera diversifolia*
Photograph — M.Fagg

Figure 14. *Dampiera wellsiana*
Photograph — M.Fagg

Figure 15. *Dampiera purpurea*
Photograph — M.Fagg (ANBG)

Figure 16. *Coopernookia barbata*
Photograph — M.Fagg (ANBG)

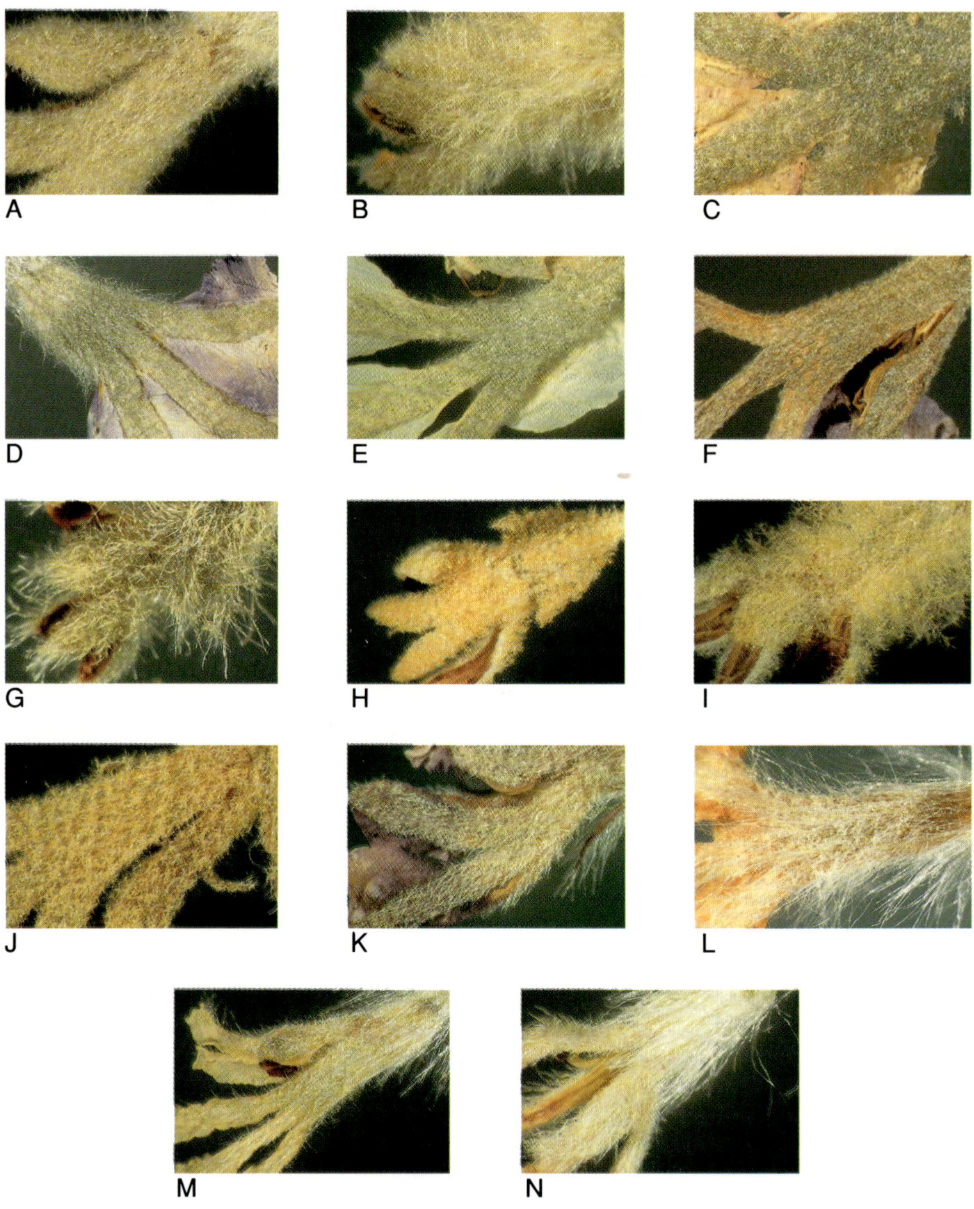

Figure 17. Hairs on the outside of flowers of *Dampiera*. **A**, *D. discolor* X20 (W.Peacock 6111.36.1, SYD). **B**, *D. atriplicina* X16 (A.George 9066, PERTH). **C**, *D. krauseana* X16 (A.Ashby 2272, PERTH). **D**, *D. spicigera* X10 (A.Ashby 723, PERTH). **E**, *D. stenostachya* X20 (D.Perry, Oct. 1961, CANB). **F**, *D. teres* X20 (M.Phillips 973, SYD). **G**, *D. cinerea* X20 (A.George 8715, SYD). **H**, *D. conospermoides* X20 (N.Byrnes D2457, DNA). **I**, *D. candicans* X20 (E.Jackson 2972, AD). **J**, *D. ramosa* X20 (A.George 8122, SYD). **K**, *D. plumosa* X20 (R.Royce 1047, PERTH). **L**, *D. eriocephala* X20 (E.Bennett, 1531 SYD). **M**, *D. wellsiana* X20 (P.Weston 124, SYD). **N**, *D. dentata* X20 (A.Beauglehole 60785 & E.Errey 4485, SYD). Photographs by B.Pellow.

Figure 18. Hairs on the outside of flowers of *Dampiera*. **A**, *D. purpurea* X10 (C.Burgess, 19 Nov. 1969, SYD). **B**, *D. ferruginea* X16 (P.Sharpe, 28 Nov. 1971, NSW). **C**, *D. altissima* X20 (R.Carolin 10647, SYD). **D**, *D. salahae* X20 (W.Peacock 6044.1, SYD). **E**, *D. tephrea* X20 (R.Carolin 10628, SYD). **F**, *D. marifolia* X20 (M.Phillips, 8 Sept. 1961, SYD). **G**, *D. incana* var. *fuscescens* X20 (R.Carolin 10685, SYD). **H**, *D. orchardii* X20 (B.Benn, 7 Oct. 1967, SYD). **I**, *D. tenuicaulis* var. *curvula* X20 (M.Phillips 1664A, SYD). **J**, *D. scaevolina* X20 (K.Newbey 1533, PERTH). **K**, *D. tenuicaulis* var. *tenuicaulis* X20 (B.Briggs, 30 Sept. 1960, SYD). **L**, *D. diversifolia* X20 (A.Ashby 2330B, SYD). **M**, *D. roycei* X10 (A.Beauglehole 60643 & E.Errey 4343, SYD). Photographs by B.Pellow.

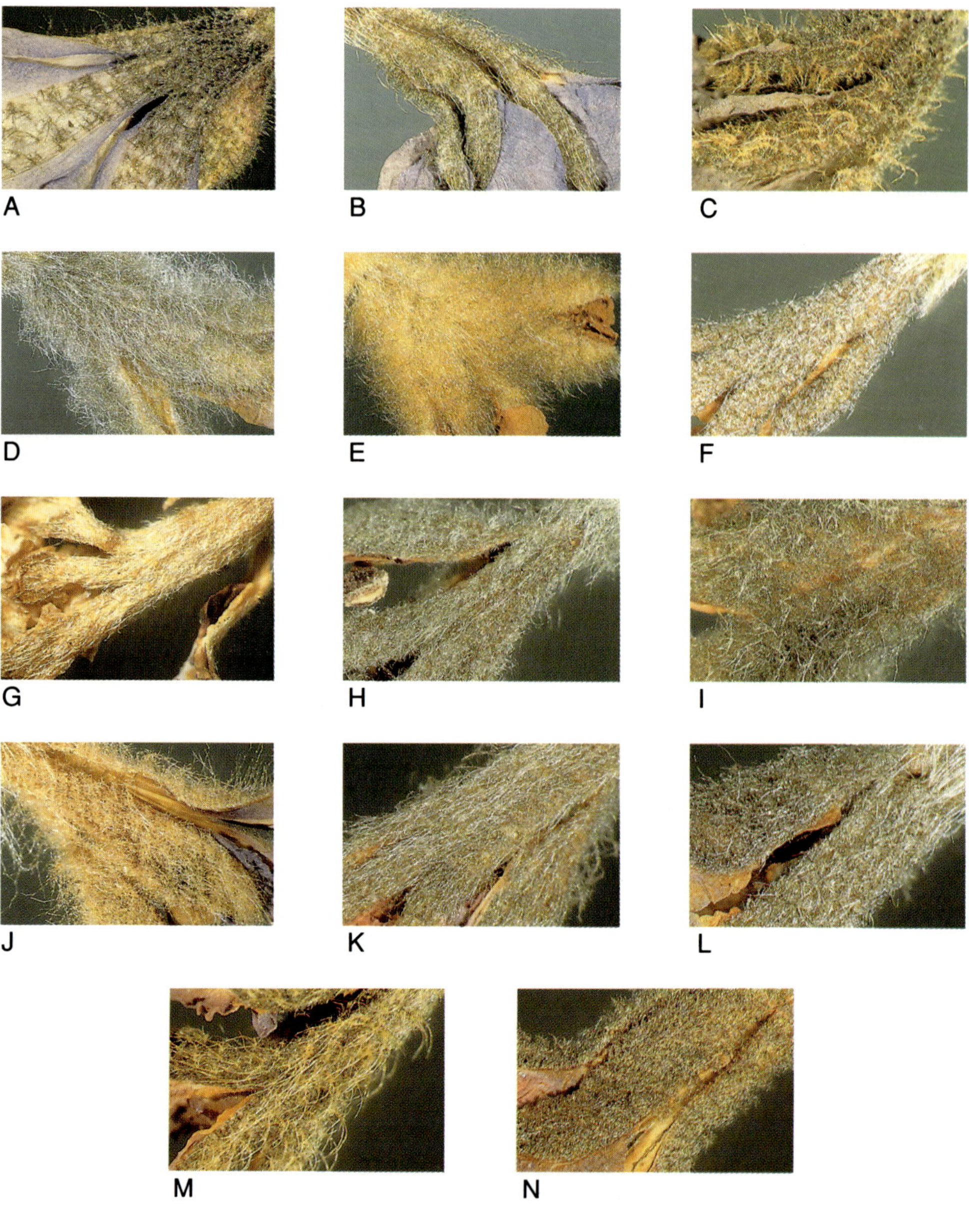

Figure 19. Hairs on the outside of flowers of *Dampiera*. **A**, *D. haematotricha* subsp. *haematotricha* X20 (P.Weston 197, SYD). **B**, *D. haematotricha* subsp. *dura* X10 (A.Strid 20023, SYD). **C**, *D. hederacea* X20 (D.Whibley 3298, SYD). **D**, *D. tomentosa* X10 (M.Crisp 1092, SYD). **E**, *D. luteiflora* X10 (B.Briggs, 30 Sept. 1960, SYD). **F**, *D. stenophylla* X20 (P.Wilson 7626, NSW). **G**, *D. fitzgeraldensis* X20 (R.Carolin 3564, NSW). **H**, *D. adpressa* X20 (R.Carolin 13154, SYD). **I**, *D. lanceolata* X20 (M.Phillips, 30 Aug. 1964, SYD). **J**, *D. eriantha* X16 (R.Helms, 17 Sept. 1891, NSW). **K**, *D. scottiana* X20 (W.Peacock 6110.17.1, SYD). **L**, *D. rosmarinifolia* X20 (A.Orchard 2165, SYD). **M**, *D. dysantha* X20 (D.Symon 10708, AD). **N**, *D. lavandulacea* X20 (R.Carolin 3539, SYD). Photographs by B.Pellow.

Figure 20. Hairs on the outside of flowers of *Dampiera*. **A**, *D. oligophylla* X10 (R.Carolin 10659, SYD). **B**, *D. juncea* X10 (B.Pellow 57, SYD). **C**, *D. linearis* X20 (L.Jackson 3299, SYD). **D**, *D. pedunculata* X20 (M.Corrick 8857, SYD). **E**, *D. latealata* X20 (A.Beauglehole 49381, SYD). **F**, *D. decurrens* X20 (A.Strid 21175, SYD). **G**, *D. trigona* X20 (A.George 9696, SYD). **H**, *D. loranthifolia* X20 (B.Benn, 10 Oct. 1963, SYD). **I**, *D. leptoclada* X20 (A.Strid 20435, SYD). **J**, *D. galbraithiana* X20 (A.Beauglehole 43382, SYD). **K**, *D. triloba* X20 (W.Fitzgerald, Aug. 1903, NSW). **L**, *D. parvifolia* X20 (A.Strid 21226, SYD). **M**, *D. sericantha* X20 (A.Beauglehole 49311, SYD). Photographs by B.Pellow.

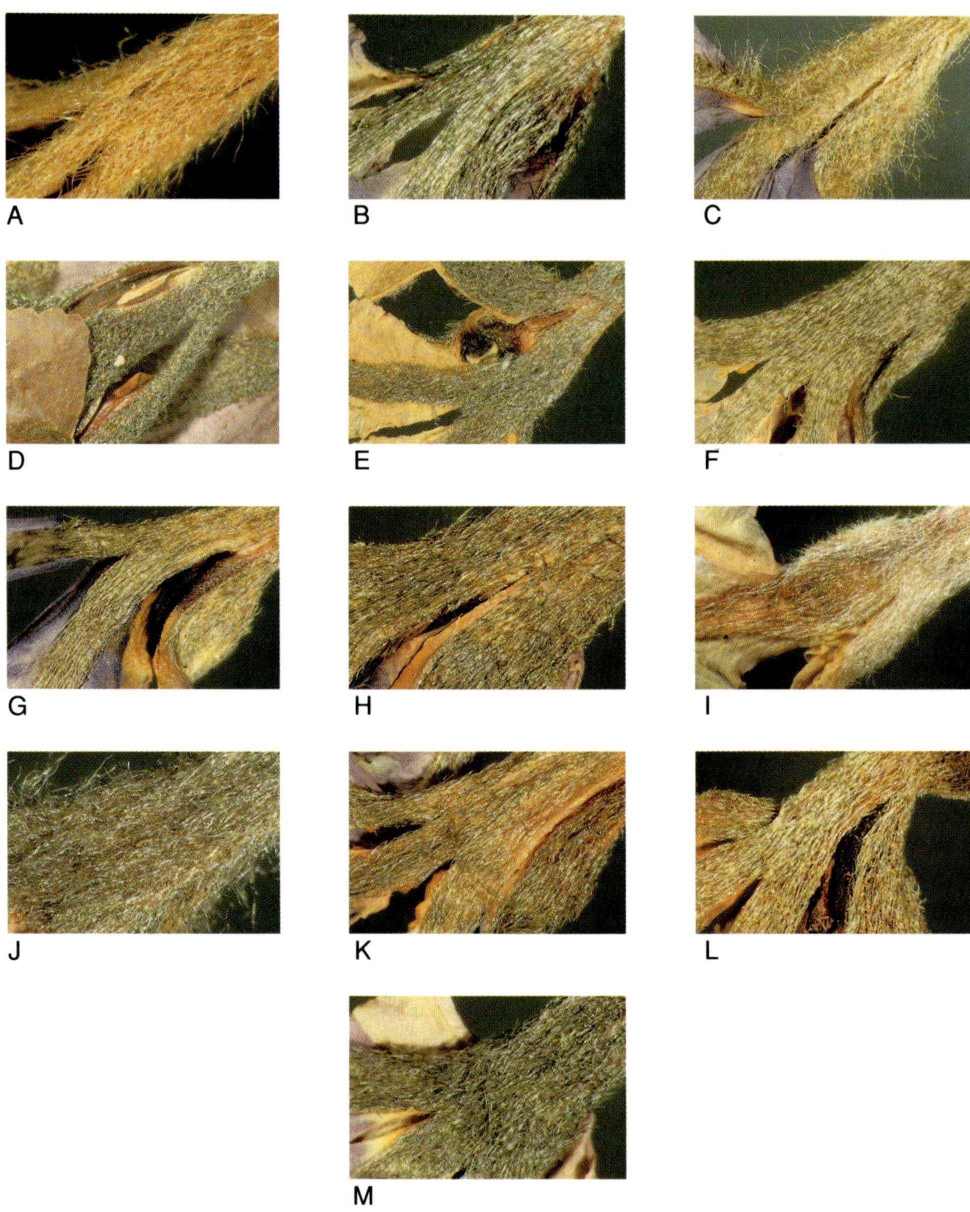

Figure 21. Hairs on the outside of flowers of *Dampiera*. **A**, *D. stricta* X20 (J.Grieve A33, SYD). **B**, *D. fasciculata* X20 (A.Ashby 3598, SYD). **C**, *D. sylvestris* X10 (P.Weston & J.Ford 30, SYD). **D**, *D. glabrescens* X20 (K.Newbey 2002, SYD). **E**, *D. obliqua* X20 (B.Benn, 5 Oct. 1963, SYD). **F**, *D. angulata* X20 (A.Beauglehole 49287, SYD). **G**, *D. heteroptera* X20 (R.Chinnock 5326, SYD). **H**, *D. coronata* X20 (A.Strid 20632, SYD). **I**, *D. carinata* X20 (A.Morrison 16064, NSW). **J**, *D. alata* X20 (W.Peacock 60822.1, SYD). **K**, *D. deltoidea* X20 (A.George 7167, SYD). **L**, *D. lindleyi* X20 (A.George 7924, SYD). **M**, *D. sacculata* X20 (A.Beauglehole 49195, SYD). Photographs by B.Pellow.

BRUNONIACEAE

R.C.Carolin

Perennial or annual herbs; hairs multicellular with terminal cell papillate; roots primary. Leaves mostly basal, on a stock, villous in axils. Scapes usually numerous each bearing a single terminal head surrounded by an involucre of green bracts. Flowers bisexual, regular, each with several membranous hyaline bracts. Sepals 5, connate, usually densely hairy. Petals 5, connate. Stamens 5, opposite sepals, epipetalous; anthers connate forming a tube, 2-locular, dehiscing through longitudinal slits. Ovary superior, 1-locular; ovule solitary, attached basally on septum or ridge; style entire with hollow indusium enclosing ±bifid tip. Fruit a nut enclosed in persistent calyx. Seed without endosperm; testa membranous; embryo straight. $x = 9$, W.J.Peacock, *Proc. Linn. Soc. New South Wales* 88: 8–27 (1963), W.J.Peacock & S.Smith-White, *Brunonia* 1: 33 (1977).

A family of 1 genus endemic in Australia.

G.Bentham, *Brunonia* in Goodenovieae, *Fl. Austral.* 4: 120–121 (1868); K.Krause, Brunoniaceae, *Pflanzenr.* 54, IV, 277a: 1–6 (1912); R.C.Carolin, Floral Structure and Anatomy in the family Goodeniaceae Dumort., *Proc. Linn. Soc. New South Wales* 84: 252–255 (1959); R.C.Carolin, The concept of the inflorescence in the order Campanulales, *Proc. Linn. Soc. New South Wales* 92: 7–26 (1967); R.C.Carolin, The trichomes of the Goodeniaceae, *Proc. Linn. Soc. New South Wales* 96: 8–22 (1970); R.C.Carolin, The systematic relationships of *Brunonia*, *Brunonia* 1: 9–29 (1978); W.J.Peacock & S.Smith-White, Cytogeography of *Brunonia australis* Sm. ex R.Br., *Brunonia* 1: 31–43 (1978).

Investigation of the structure of the flower (R.C.Carolin, *loc. cit.,* 1959) indicates that before anthesis pollen is shed into the indusium while it is terminal on the style. Two lobes of the style tip develop later; the stigmatic surface is located in the cleft and at the ends of these lobes.

R.C.Carolin (*loc. cit.,* 1978) considered that *Brunonia* should be placed in the family Goodeniaceae. Both families possess the stylar indusium which is unknown elsewhere in the flowering plants. It is therefore doubtful that the family Brunoniaceae should be recognised, but it is maintained in this work in accordance with the classification of A.Cronquist.

There is no record of the single species being used for any purpose although, were it to be brought into cultivation, it could prove to be an attractive ornamental flower with its bright blue composite-like heads.

BRUNONIA

Brunonia Smith ex R.Br., *Prodr.* 589 (1810); named after the Scottish botanist Robert Brown (1773–1858), a leading taxonomist of his day and author of the first flora of Australia, in which the genus was described.

Type: *B. australis* Smith ex R.Br.

A monotypic genus; the species is rather variable.

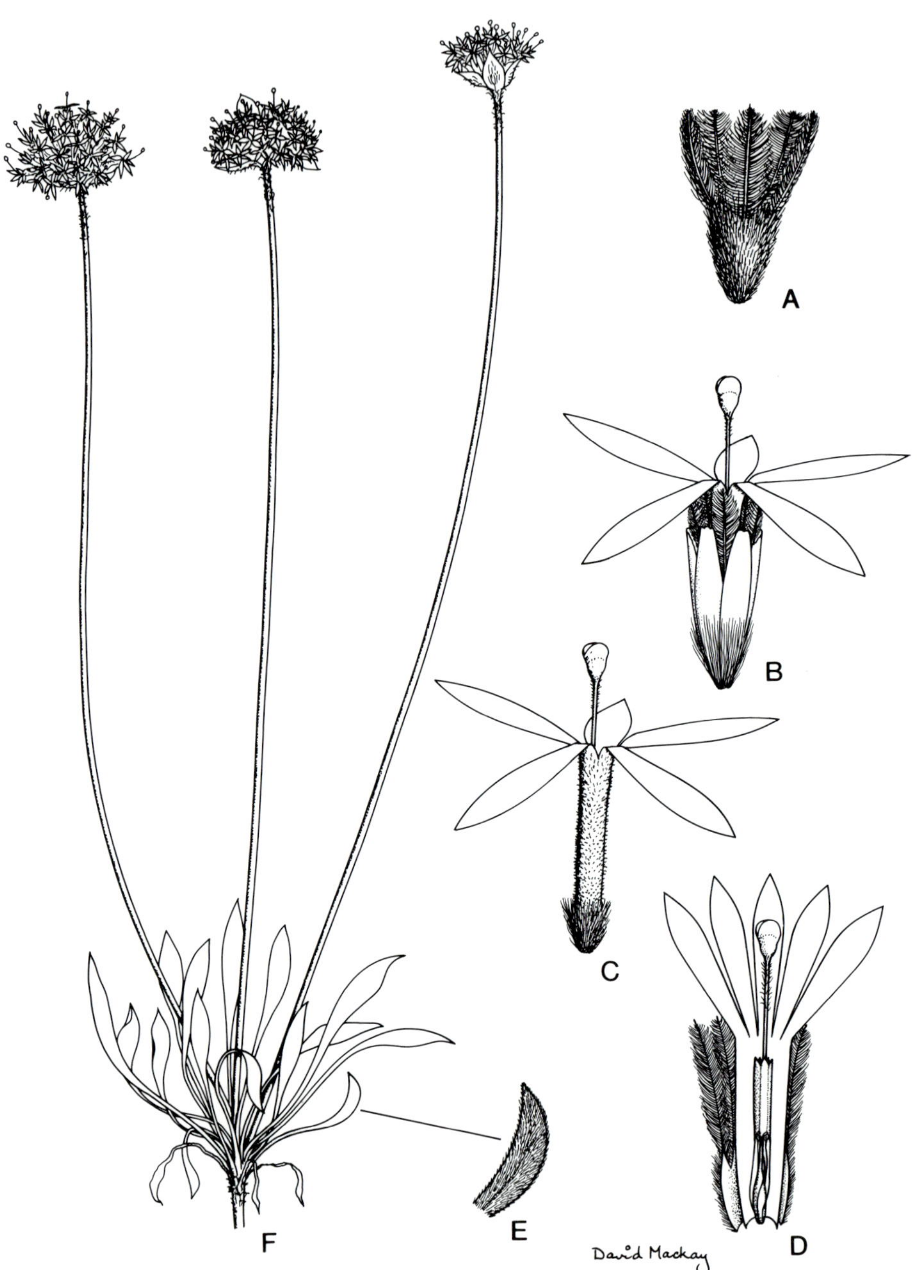

Figure 22. *Brunonia australis*. **A**, fruit surrounded by persistent sepals X5 (W.Bäuerlen, Oct. 1903, NSW). **B**, flower surrounded by bracts X5. **C**, flower with calyx lobes removed X5; **D**, dissected flower X5; **E**, leaf tip X1.5; **F**, habit X0.5 (**B–F**, K.Williams 84217, NSW). Drawn by D.Mackay.

Brunonia australis Smith ex R.Br., *Prodr.* 590 (1810)

T: near the coast, Port Dalrymple, Van Diemens Land [Tas.], Jan. 1804, *R.Brown*; holo: LINN.

B. sericea Smith ex R.Br., *Prodr.* 590 (1810); *B. australis* var. *sericea* (Smith ex R.Br.) Colozza, *Nuovo Giorn. Bot. Ital.* n. ser. 14: 302 (1907). T: Pine Point, east coast of New Holland, [Qld], Aug. 1802, *R.Brown*; holo: LINN.

B. simplex Lindley in T.L.Mitchell, *Exped. Trop. Australia* 82 (1848); *B. australis* var. *simplex* (Lindley) Colozza, *Nuovo Giorn. Bot. Ital.* n. ser. 14: 302 (1907). T: subtropical New Holland, *T.L.Mitchell*; syn: NSW.

B. australis var. *macrocephala* Colozza, *Nuovo Giorn. Bot. Ital.* n. ser. 14: 301 (1907). T: numerous syntypes; none seen.

Illustrations: J.E.Smith, *Trans. Linn. Soc.* 10: t. 28, t. 29 (1810); *Bot. Mag.* t. 1833 (1836); K.Krause, *Pflanzenr.* 54, IV, 277a: 4, fig. 1 (1912); W.T.Stearn, *The Australian Flower Paintings of Ferdinand Bauer* t. 15 (1976).

Herb to 35 cm tall, silky, with silvery or brown hairs, rarely almost glabrous. Leaves oblanceolate to obovate, 3–15 cm long, 3–45 mm wide, contracting gradually towards base, entire. Inflorescence bracts to 8 mm long; flowers to 7 mm long; flower bracts elliptic to oblong, c. 4 mm long. Calyx hairy; tube swollen apically when mature; lobes linear, to 5 mm long, with margins long-ciliate. Corolla tube narrow, to 5 mm long; lobes linear, to 5 mm long, spreading, bright blue. Style straight, exceeding corolla tube. Indusium exserted, erect, orbicular to broadly elliptic, c. 2 mm long, with scattered hairs; lips without bristles. $n = 9$, 18, 36, W.J.Peacock, *Proc. Linn. Soc. New South Wales* 88: 10 (1963). *Pincushion.* Figs 1, 22.

Widespread in Australia in all States south of the 17th parallel from the arid regions to the higher rainfall areas on the mountains of the south-east; mostly on sandy soils. Flowers winter and spring. Map 1.

W.A.: near Glen Cumming, Rawlinson Ra., *A.S.George 8857* (PERTH, SYD). N.T.: c. 8 km N of Aileron, *R.Swinbourne 379* (DNA, NSW). S.A.: 104 km N of Cook on Cook–Vokes road, *D.E.Symon 12202* (AD). Qld: 90 km E of Bollon, *K.A.Williams 84217* (BRI, NSW). N.S.W.: Iona, *C.A.Booth 633* (NSW). Vic.: Narbethong near Healesville, *R.C.Carolin 1109* (SYD). Tas.: c. 3 km W of Powranna on Cressy road, *W.J.Peacock 6111.61.1* (SYD).

A species which inhabits a variety of different environments and is very variable, particularly in the degree of hairiness. The cytological variation reported by W.J.Peacock and S.Smith-White *loc. cit.* does not appear to be reflected in morphologically distinguishable entities.

GOODENIACEAE

R.C.Carolin, M.T.M.Rajput & D.Morrison

Herbs, shrubs or rarely scramblers, glabrous, viscid or variously hairy; roots usually primary, sometimes adventitious. Leaves mostly alternate, often with axillary tuft of hairs. Flowers bisexual, in dichasia, thyrses, racemes, umbels, spikes or heads, or solitary. Sepals 5, 3 reduced to rim or obsolete, free or connate. Corolla tubular but more deeply slit in 1 or 2 positions (i.e. fanlike or two-lipped), sometimes pouched. Petals 5, ±winged, often with auricles around stamens. Stamens 5, opposite sepals, epigynous or rarely apparently hypogynous; anthers free or connate, 2-locular, dehiscing through longitudinal slits. Ovary superior, half-inferior or inferior; locules 2 (usually incomplete), rarely 4 or apparently 1; placentas axile or basal; style entire with hollow pollen cup (indusium) enclosing ±bifid tip, rarely 2- or 3-branched. Fruit a drupe, inferior nut or capsule, rarely separating transversely into 1-seeded woody segments. Seeds with endosperm; testa membranous to thickened; embryo straight. x = 7, 8, 9, *W.*J.Peacock, Chromosome numbers and cytoevolution in the Goodeniaceae, *Proc. Linn. Soc. New South Wales* 88: 8–27 (1963).

A family of 11 genera and c. 400 species almost confined to the southern hemisphere. Almost all the genera are confined to Australia and New Guinea; *Scaevola* alone is an important component of the floras of the islands of the Pacific and the strand lines of the Atlantic and Indian Oceans. *Goodenia pilosa* extends to Indonesia, southern China and the Phillipines and *Selliera radicans* occurs naturally in New Zealand and Chile. The present treatment covers 377 species for Australia.

W.H.de Vriese, Goodenovieae, *Natuurk. Verh. Holl. Maatsch. Wetensch. Haarlem* ser. 2, 10: 1–194 (1854); G.Bentham, Goodenovieae (excluding *Brunonia*), *Fl. Austral.* 4: 37–120 (1868); K.Krause, Goodeniaceae, *Pflanzenr.* 54: 1–207 (1912); R.C.Carolin, Floral Structure and Anatomy in the family Goodeniaceae Dumort., *Proc. Linn. Soc. New South Wales* 84: 252–255 (1959); R.C.Carolin, The structures involved in the presentation of pollen to visiting insects in the order Campanales, *Proc. Linn. Soc. New South Wales* 85: 197–205 (1960); W.J.Peacock, Chromosome numbers and cytoevolution in the Goodeniaceae, *Proc. Linn. Soc. New South Wales* 88: 8–27 (1963); R.C.Carolin, Seeds and fruit of the Goodeniaceae, *Proc. Linn. Soc. New South Wales* 91: 58–83 (1966); R.C.Carolin, The concept of the inflorescence in the order Campanulales, *Proc. Linn. Soc. New South Wales* 92: 7–26 (1967); R.C.Carolin, The trichomes of the Goodeniaceae, *Proc. Linn. Soc. New South Wales* 96: 8–22 (1970).

RELATIONSHIPS WITH OTHER FAMILIES

There has been no systematic analysis of the group of families which may be related to Goodeniaceae. It has been argued (R.C.Carolin, *Brunonia* 1: 9–29, 1977) that the position of the family in the Campanulales should be retained but the relationship with either Campanulaceae or Lobeliaceae is not particularly close. The most notable similarities are concerned with the pollen presentation apparatus but, when analysed, these amount to the possession of hairs on the style which take part in pollen presentation in very different ways in the 3 families (R.C.Carolin, *loc. cit.*). The structure of the inferior ovary and the vasculation of the ovary of Goodeniaceae and Campanulaceae are so different as to preclude a very close relationship.

The Metachlamydae are basically a natural group and the relationships of the

Campanulaceae and Goodeniaceae are themselves to be sought in the Archichlamydeous families. Based upon ovary structure and embryological characters, the best candidate for this appears to be the Saxifragales, particulary the woody groups, such as the Escalloniaceae.

RELATIONSHIPS WITHIN THE FAMILY

Research by R.C.Carolin, *loc. cit.* indicates that there are 2 distinct groups in Goodeniaceae *sens. lat.* One of these, the *Lechenaultia–Anthotium–Dampiera* group, also includes *Brunonia* which indicates that Goodeniaceae, as recognised here, is not a natural family when that genus is excluded.

The *Dampiera* group is characterised by connate anthers and a base chromosome number of 9. Moreover, the vasculation of the inferior ovary of this group shows significant differences from that of the *Goodenia* group (R.C.Carolin, *loc. cit.* 1959) and indicates that this condition is separately derived in the 2 groups. The *Goodenia* group is characterised by free anthers, a base chromosome number of 7 or 8 and the vasculation of the ovary which more clearly indicates a 4-carpellary condition than that of the *Dampiera* group.

HAIR STRUCTURE

There are 2 main types of hair found in the family – glandular and non-glandular.

Thick-walled simple and/or stellate non-secretory hairs (Fig. 23A–F) are found almost throughout the genera *Goodenia*, *Scaevola*, *Coopernookia*, *Diaspasis* etc., but the related T-shaped hair (Fig. 23G) is found only in *Goodenia mueckeana*. Many species of *Scaevola* have simple hairs with small warts on the surface. The multicellular hairs illustrated in Fig. 24A, B, D–F, are the cottony and webby hairs found in a number of *Goodenia* spp. and the axillary hairs found in a large number of species in the family.

Verreauxia and *Pentaptilon* have characteristic branched hairs (Fig. 24C) as well as simple and sometimes glandular ones, and almost all *Dampiera* species have some form of dendritic hair (illustrated under that genus).

The function of these non-secretory hairs is a source of speculation but little has been demonstrated in this regard.

The glandular hairs (Fig. 25) are almost all secretory, or potentially so, except possibly those of some species of *Lechenaultia* (Fig. 25J). They may be large and quite conspicuous (Fig. 25A, B, D, G–I) but sometimes they are flattened and peltate (Fig. 25C, E) making the surface of the plant appear glabrous except under high magnifications. The secretion may make the younger parts of the plant clammy or viscid. In many species of *Goodenia* and *Scaevola* this viscid product develops into a varnish over the surface of the plant as it ages, eventually cracking and fragmenting. The function of these products is not clear although some have speculated that they assist in water conservation. Some of the varnishes appear to contain substances used as drugs by the Aborigines, e.g., the varnish of *Goodenia ovata*.

The secretory hairs of *Scaevola* sect. *Scaevola* appear to be derived from the more conventional type of glandular hair but the cells of the head are elongate giving a stellate appearance to the hair (Fig. 25F). The thin walls of these cells, however, distinguish these hairs from the thick-walled stellate hairs described above.

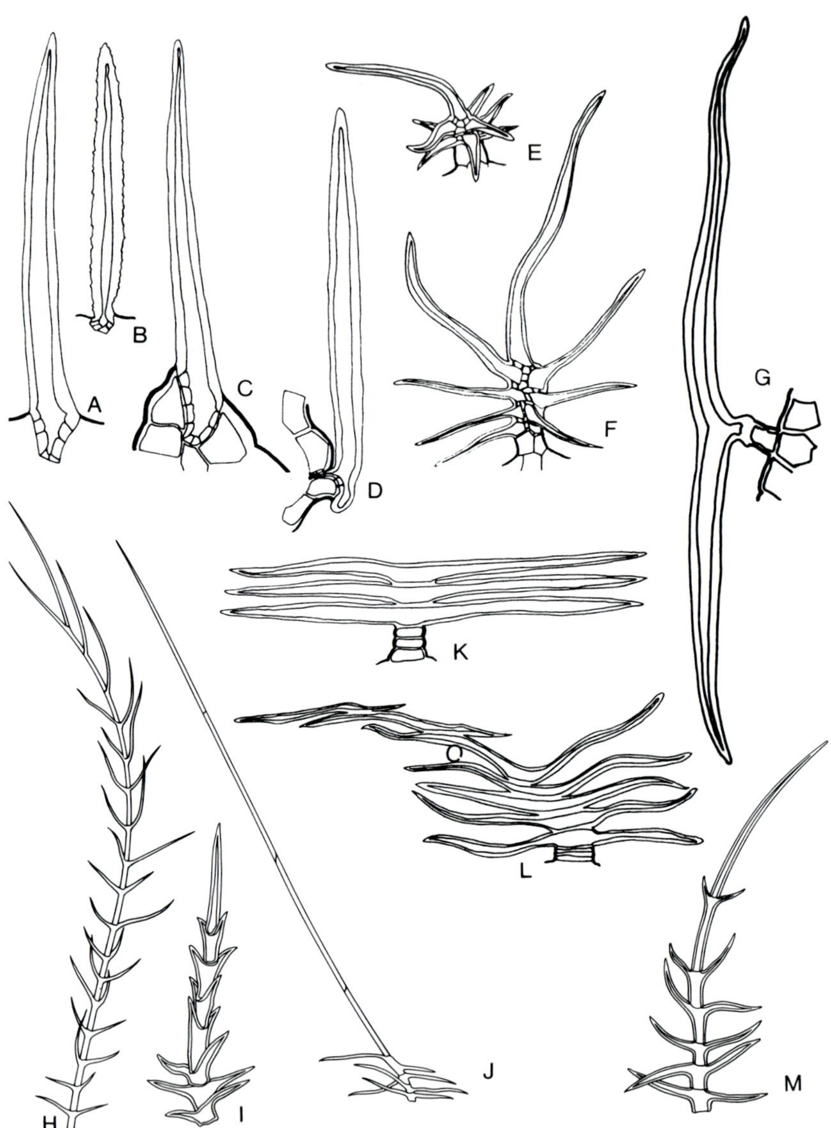

Figure 23. Thick-walled, non-secretory, simple or stellate hairs and T-shaped hairs on outer surfaces of corollas. **A–D**, simple. **A**, *Goodenia ovata*. **B**, *Diaspasis filifolia*. **C**, *Goodenia pinnatifida*. **D**, *Goodenia fascicularis*. **E–F**, stellate. **E**, *Goodenia stelligera*. **F**, *Coopernookia polygalacea*. **G**, T-shaped, *Goodenia mueckeana*. **H–M**, *Dampiera* corolla hairs. **H,** type i, *D. tomentosa* X40 (NSW83455, NSW). **I**, type ia, *D. hederacea* X80 (J.Powell 3156 & M.Hardie, NSW). **J**, type ii, *D. eriocephala* X40 (R.Helms, 13 Nov. 1891, NSW). **K**, type iii, *D. coronata*. **L**, type iv, *D. stricta*. **M**, type v, *D. lavandulacea* X80 (B.Smith 742, NSW). **A–G**, **K–L**, reproduced by permission from *Proc. Linn. Soc. New South Wales* 96: 9–10 (1971). **H–J, M,** drawn by D.Mackay.

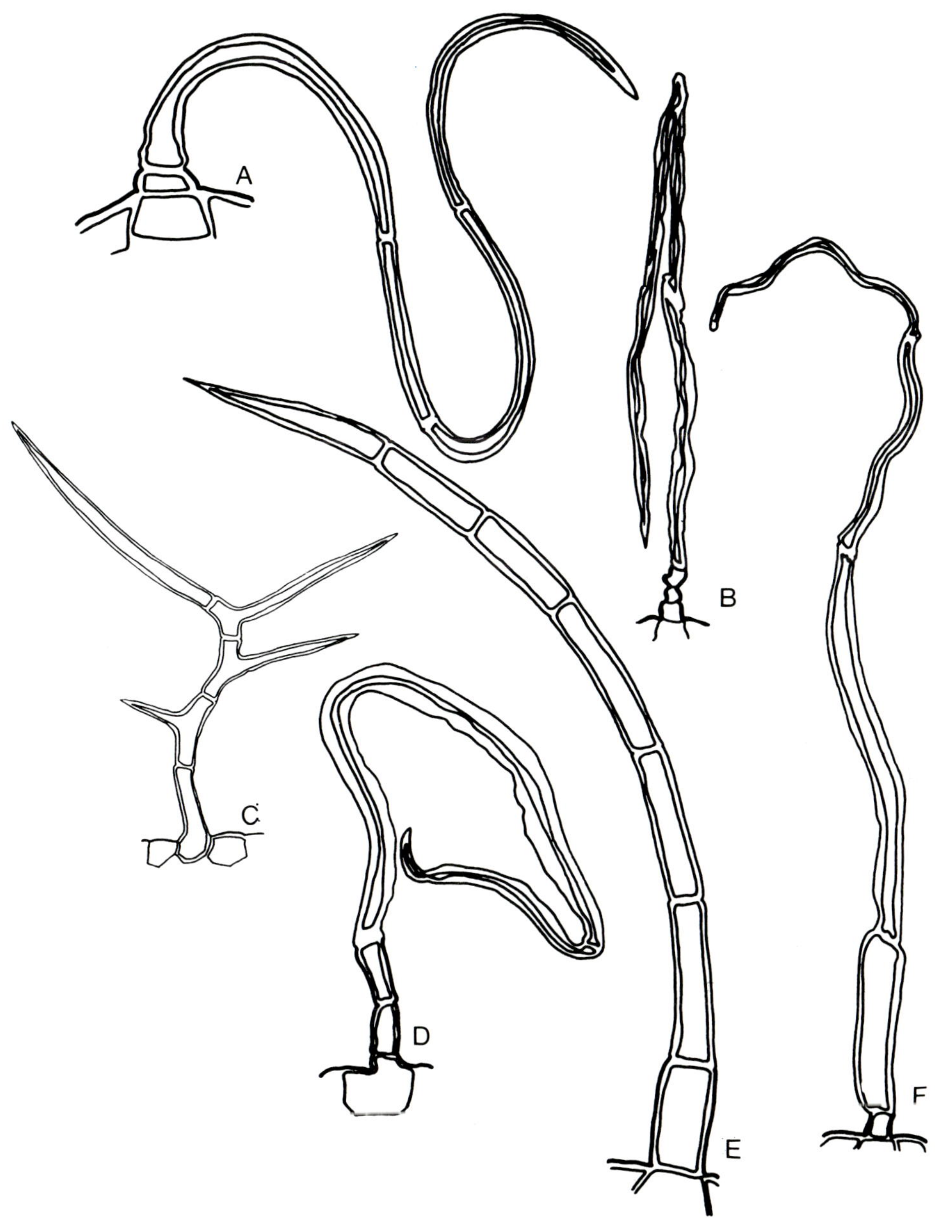

Figure 24. Branched (**C**) and uniseriate multicellular (**A–B**, **D–F**) hairs. **A**, *Goodenia incana*. **B**, *Goodenia hederacea* subsp. *hederacea*. **C**, *Verreauxia reinwardtii*. **D**, *Goodenia blackiana*. **E**, *Pentaptilon careyi*. **F**, *Goodenia heterochila*. Reproduced by permission from *Proc. Linn. Soc. New South Wales* 96: 10–12 (1971).

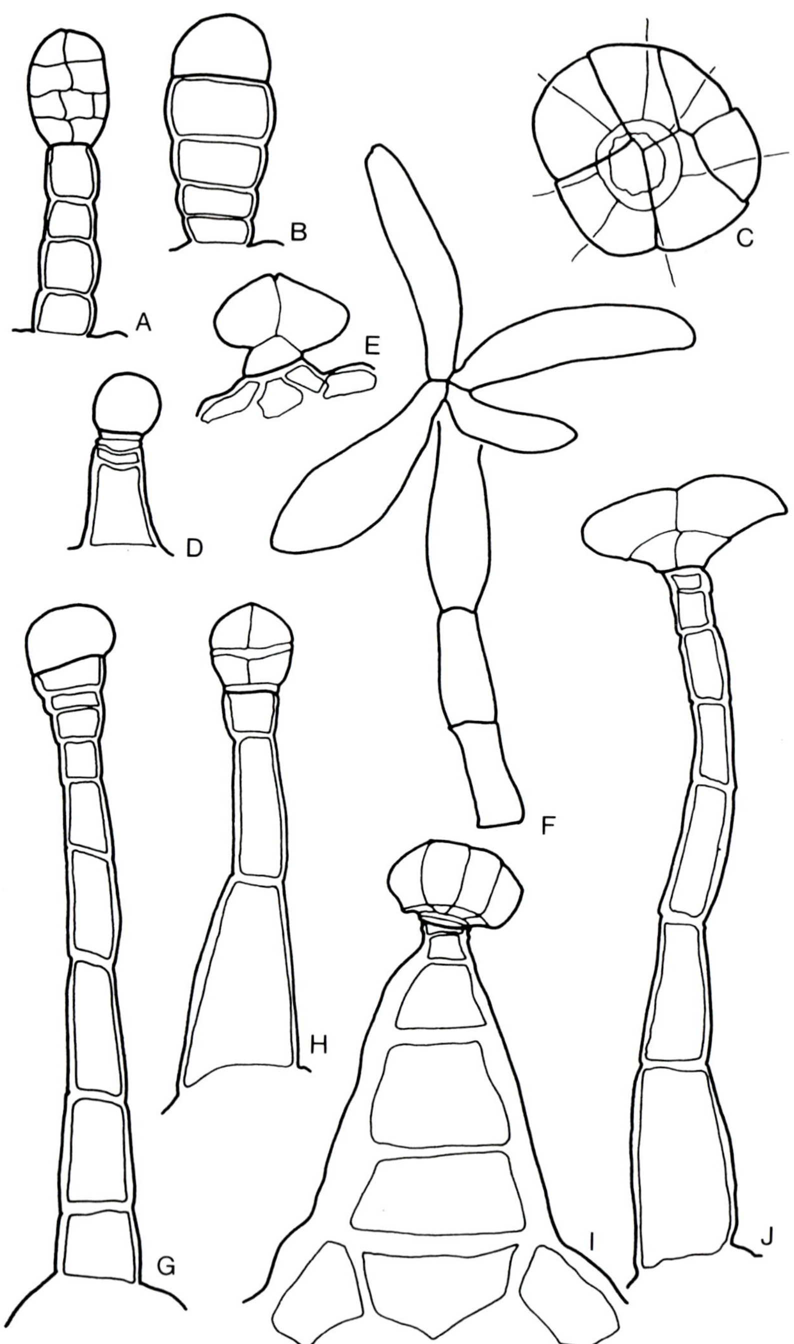

Figure 25. Glandular hairs. **A**, *Goodenia holtzeana*. **B**, *Goodenia decurrens*. **C**, *Goodenia ovata* (peltate). **D**, *Goodenia concinna*. **E**, *Coopernookia polygalacea*. **F**, *Scaevola coriacea* (an Hawai'ian species) (pseudo-stellate). **G**, *Goodenia stephensonii*. **H**, *Pentaptilon careyi*. I, *Goodenia caerulea*. **J**, *Lechenaultia hirsuta*. Reproduced by permission from *Proc. Linn. Soc. New South Wales* 96: 11 (1971).

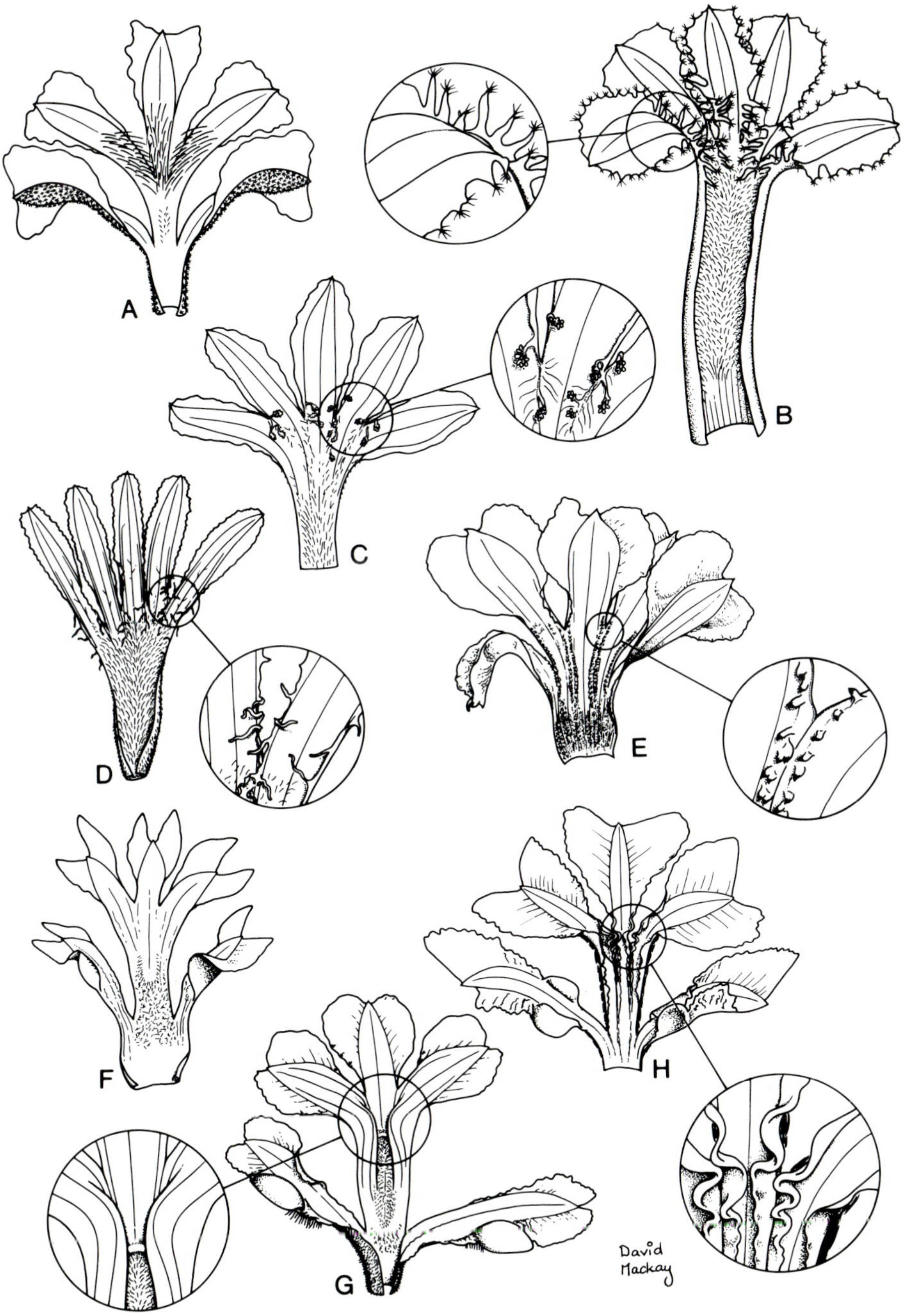

Figure 26. Pollinator tactile guides in the corollas of Goodeniaceae. **A**, *Coopernookia chisholmii*, corolla X2.75 (R.Coveny 12240 & W.Bishop, NSW). **B**, *Scaevola taccada*, corolla X2.75, detail X6 (NSW 83321, NSW). **C**, *Scaevola humifusa*, corolla X4, detail X9 (J.Tepper, Sept. 1892, NSW). **D**, *Scaevola humilis*, corolla X2.75, detail X9 (E.Jackson 2152, NSW). **E**, *Goodenia calcarata*, corolla X2, detail X9 (M.Corrick 10518, NSW). **F**, *Goodenia gibbosa*, corolla X2.75 (M.Lazarides & J.Palmer 381, NSW). **G**, *Goodenia sepalosa*, corolla X2, detail X4 (M.Lazarides 8537, NSW). **H**, *Dampiera leptoclada*, corolla X2.75, detail X9 (F.Rodway 5490, NSW). Drawn by D.Mackay.

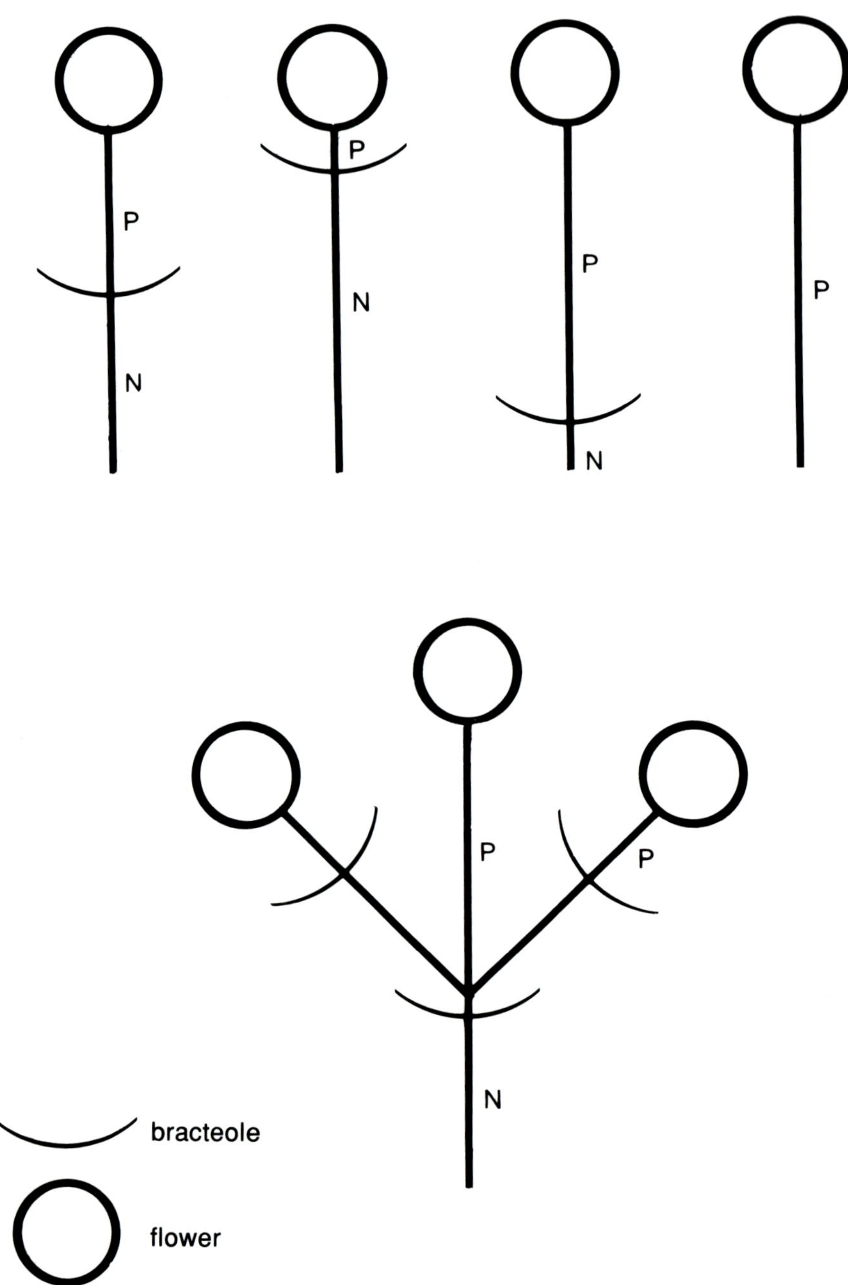

Figure 27. Diagram illustrating the use of the terms pedicel and peduncle as used in this volume in Goodeniaceae. (P = pedicel, N = peduncle).

GOODENIACEAE

INFLORESCENCE

In the Goodeniaceae the inflorescence often forms a significant part of the habit of the plant. For a discussion relating to the inflorescence of Goodeniaceae see R.C.Carolin, *Proc. Linn. Soc. New South Wales* 92: 7–26 (1967). The term *bract* is applied to any appendage (morphologically a leaf) that subtends a flower whether it has the appearance of a foliage leaf, is reduced to a scale, is brightly coloured or appears otherwise. The term *bracteole* is applied to the appendages (2 or more) on the flower stalk. The part of the stalk below the bracteoles is referred to as the *peduncle* whilst that part above the bracteoles is referred to as the *pedicel*. When there are no bracteoles the whole stalk is referred to as the *pedicel*. These concepts are illustrated in Fig. 27. Sometimes the bracteoles themselves subtend flowers, in which case the whole unit is a cyme, and in this treatment only the appendage which subtends the whole cyme is referred to as a bract.

The primitive form of the inflorescence in *Goodenia*, *Scaevola*, *Selliera*, *Verreauxia*, *Pentaptilon* and *Diaspasis* is considered to be a thyrse (e.g. Fig. 47B, *Goodenia scaevolina*), in which the primary bracts are leaf-like or reduced. Reduction of each constituent axillary cyme has resulted in the raceme and spike which are much commoner inflorescence types (e.g. Fig. 48G, *Goodenia ramelii*). A number of *Goodenia* species show a condensation of the main axis to form subumbels (e.g. Fig. 82B, *Goodenia megasepala*). In *Velleia*, the condensation of the main axis of the plant and the expansion of each partial inflorescence into often complex dichasia has produced a very characteristic arrangement (e.g. Fig. 88E, *Velleia spathulata*). In a small group of *Scaevola* species related to *S. canescens* the spikes are arranged laterally along a stem (e.g. Fig. 41F, *Scaevola canescens*) (R.C.Carolin, *Telopea* 3: 477–516, 1990) but in the majority of species the spikes, racemes and thyrses are terminal although there may be subsidiary lateral inflorescences as well.

In the *Dampiera* group of genera the inflorescences are based on the cymo-panicle with a terminal flower (e.g. Fig. 30D, *Dampiera ramosa*). Those of *Dampiera* sect. *Dampiera* are particularly difficult to interpret since the flowers are borne alternately along branches which arise in clusters in the axils of the leaves and are apparently reduced cymo-panicles (M.T.M.Rajput & R.C.Carolin, *Telopea* 3: 183–216, 1988). In this account the term 'cyme-like panicle' is used for this structure.

FLOWER

The flower of the Goodeniaceae has, like that of many of the Lobeliaceae, a tubular corolla which is slit on the adaxial side to varying degrees. The flower of Lobeliaceae, however, is resupinate so the similarity is probably only superficial. In almost all species the corolla lobes have thin but often large wings which are usually the most brightly coloured part of the flower and serve as the main attractant for the pollen vector. In *Scaevola* the corolla is opened along the slit to expose the indusium so that the corolla resembles a fan (e.g. Fig 26D), hence the name fan-flower for these plants. In other genera the adaxial petals are folded around the indusium (e.g. Fig. 35C, *Coopernookia* scabridiuscula) and in some cases the flower is distinctly 2-lipped (e.g. Fig. 33A, *Dampiera decurrens*) even to the extent that the adaxial and abaxial petals are different in colour (*Goodenia bicolor*). In many species of *Goodenia* (e.g. Fig. 80F, *Goodenia pinnatifida*) and throughout *Dampiera* (e.g. Fig. 31E, *Dampiera scaevolina*) the lower wings of the adaxial petals are differentiated in their lower part into auricles which envelope the indusium in the flower so that it is only exposed when the insect vector pushes the corolla lobes apart as it seeks the nectar at the base of the flower.

The pollen indusium is found only in Goodeniaceae and Brunoniaceae. It is a cup at the top of the style and surrounding the stigmatic initial that collects pollen from the stamens which dehisce in the bud. Thus, when the flower opens the pollen is retained in the pollen indusium and is presented in it to the pollen vector (R.C.Carolin, *Proc. Linn. Soc. New South Wales* 85: 197–207, 1960). In some cases, particularly in *Dampiera* species, the collection of pollen into the indusium is rather inefficient and much is actually presented in the auricles and on the outside of the indusium. In *Lechenaultia* the pollen appears to be deposited entirely on the outside surface of the indusium although the homologies of the structures there are not clear. The pollen is protected by the indusium until released in some way by the vector as it probes to the base of the abaxial side of the flower where the nectar is mostly produced either on the top of the ovary, as in *Scaevola*, or in a pouch as in *Coopernookia* and *Goodenia* and its relatives (R.C.Carolin, *Proc. Linn. Soc. New South Wales* 84: 252–255, 1959). In *Dampiera* there is usually a pouch opposite each petal containing nectar (R.C.Carolin, *loc. cit.*). In most species except those of *Dampiera*, a number of bristles arranged along the lip of the indusium are disturbed by the vector while probing for nectar and this results in pollen being released from the indusium. Later the stigmatic initials mature and grow out of the indusium and collect pollen from pollen vectors.

Many genera of Goodeniaceae have various types of guides within the corolla which seem to orientate the pollinator (usually an insect) into the correct position to obtain the reward (mostly nectar) and to pick up pollen or to pollinate the stigma. These guides may be coloured lines (visual guides) and/or various outgrowths (tactile guides, fig. 26) on the corolla. Tactile guides vary between the genera. Thus *Coopernookia* and *Goodenia* subg. *Monochila* (Fig. 46C) have long, flat hairs on the margin of the wings and in the throat and projecting into the throat. Many species in *Scaevola* sect. *Scaevola* (Fig. 40C) have similar hairs but in this case they are attached to the tops of laciniae on the wing margins. In other sections of *Scaevola* the laciniae are narrower and are referred to here as *barbulae* (R.C.Carolin, *Telopea* 3: 477–516, 1990). The hairs on these barbulae may be reduced to papillae at the apex (papillate barbulae) (Fig. 26C) or they may be absent (Fig. 26D), in which case the barbulae are referred to as *simple.* Some simple barbulae are practically indistinguishable from the hairs that are almost always present inside the corolla.

In many species of *Velleia* (Fig. 88B) and *Goodenia* subsect. *Goodenia* and sect. *Caeruleae* (e.g. Fig. 47C), the tactile guides are small outgrowths in the throat, here referred to as *enations* of the corolla. They are probably not homologous with the barbulae of *Scaevola.* In *Goodenia* subsect. *Ebracteolatae*, hairs on the surface of the throat of the corolla (Fig. 86B) appear to be the tactile guides, whilst in many species of subsect. *Borealis* there are folds on the inner surface of the corolla (Fig. 26G) that also appear to be tactile guides. The hairs at the base of the corolla of many species of *Goodenia* and *Lechenaultia* in particular may prevent the entry of non-pollinating insects to the region of the reward.

In *Dampiera* there are ridges leading down into the throat and in many species there are small outgrowths on these ridges, here referred to as *calli*. Both of these features appear to act as tactile guides leading to the nectar.

Almost throughout the family the ovary consists of 3 zones: a basal 2-locular zone, which may be very short, a middle 1-locular zone and an upper very small 2-locular zone (R.C.Carolin, *Proc. Linn. Soc. New South Wales* 84: 252–255, 1959). Ovules may be inserted in either or both the 2 lower zones on axile or parietal placentas respectively. There is, however, some evidence from the vascular structures of the ovary to indicate that the ovary is 4-carpellary, at least in the *Goodenia* group of genera (R.C.Carolin, *loc. cit.*). In 1 species only (*Scaevola porocarya*), there are, in fact, 4 fertile locules, while in other species of *Scaevola* there are 2 fertile locules and 2 sterile cavities (R.C.Carolin, *Proc. Linn. Soc. New South Wales* 91: 58–83, 1966) although the homologies of the sterile cavities are quite unclear.

GOODENIACEAE

FRUIT AND SEED STRUCTURE

Dispersal mechanisms within the family vary from genus to genus, and sometimes within a genus. The morphology of the fruits has been discussed in R.C.Carolin, *loc. cit.* (1966). In *Velleia*, *Verreauxia*, *Selliera*, *Coopernookia*, *Anthotium* and most *Goodenia* spp., the fruit is a capsule from which the seed is shed. The first 3 genera and *Goodenia* all have compressed or flat seeds with a wing (see Fig. 45). This wing may take various forms and functions in different ways.

The wing of *Goodenia cycloptera* (Fig. 45F) and *Goodenia heterochila* is large and thick and apparently acts as an elaiosome since ants carry the seed around apparently to use the wing. More commonly in those species which occur in the moister areas, e.g., species of *Goodenia* subg. *Monochila*, sect. *Porphyranthus* (Fig. 45G), subsect. *Scaevolina* (Fig. 45C) and part of subsect. *Goodenia* (Fig.), the wing is narrow, swells in contact with water and is mucilaginous. It, too, may act as an elaiosome and/or assist in the control of germination, although the evidence for either is meagre. In fact the mucilaginous wing reaches its greatest development in the group of arid region species allied to *Goodenia havilandii* where it overlaps the body of the seed to a considerable extent (Fig. 45E).

In many members of *Goodenia* subsect. *Ebracteolatae* (Fig. 45D) the wing is broad and assists in wind dispersal; at the same time many of these wings are at least slightly mucilaginous and may also have some importance in the control of germination. The seeds of *Goodenia* subsect. *Caeruleae* are also broad and assist in wind dispersal but they do not appear to be mucilaginous. *Goodenia hirsuta* (Fig. 45B) and *G. stellata* are unique in the genus since the body of the seed gradually passes into the wing without a well-demarcated boundary between them. The wing does not appear to be mucilaginous in this case.

In *Goodenia* subsect. *Borealis* (Fig. 45I) and some species of *Goodenia* subsect. *Goodenia* (Fig. 45H), the wing is almost obsolete and presumably serves no function.

The seeds of *Selliera* are particulary interesting. The fruit is a fleshy, tardily dehiscent capsule whilst the seeds have a narrow, but extremely mucilaginous, sticky wing. These features make it ideally suited to animal dispersal. Since this same species is found in salt marshes in Chile, New Zealand and Australia, it seems likely that sea birds have played a part in its dispersal.

Coopernookia has obloid seeds with a very distinct caruncle which is an enlargement of the funicle. The caruncle is an elaiosome and assists in the dispersal of the seeds by ants.

The dispersal units of *Lechenaultia* are unique in the family. The fruit breaks up into a series of single-seeded, angular articles consisting of part of the pericarp surrounding each seed and the seed itself. *Scaevola* (e.g. Fig. 39), *Diaspasis*, *Dampiera* (e.g. Fig. 34E, *D. altissima*), and several species of *Goodenia* have hard indehiscent fruits with thick endocarps. Many of these fruits have more than 1 fertile seed which germinate together, and it is not uncommon to see 2 plants of *Scaevola*, for instance, growing from almost the same position, presumably having germinated from the same fruit. Only in *Scaevola* sect. *Scaevola* (where *S. spinescens* and *S. tomentosa* are certainly dispersed by Emus, *Dromaius novaehollandiae*), *Scaevola* sect. *Enantiophyllum* and the coastal *S. calendulacea,* and its relatives in New Zealand and Tonga, have these fruits developed a conspicuous fleshy mesocarp and thus a more sophisticated animal dispersal system.

It is not clear whether the thin, green mesocarp of other *Scaevola* spp., *Dampiera* spp., some species of *Goodenia* sect. *Monochila* and *Goodenia goodeniacea* has any function in dispersal. The pericarp of *Goodenia neogoodenia*, which surrounds the single seeds, is very thin and quite unlike those species with hard endocarps.

Pentaptilon has an indehiscent fruit, also without a hard endocarp. The pericarp is light and spongy and has 5 longitudinal wings of various sizes. This all suggests a wind dispersal mechanism, although a relatively inefficient one.

KEY TO GENERA

1 Anthers connate

2 Ovules and seeds more than 2; fruit dehiscent or fragmenting into articles, rarely indehiscent and then fruit beaked

3 Leaves all cauline; indusium 2-lipped ..... **2. LECHENAULTIA**

3: Leaves cauline and basal; indusium not 2-lipped ..... **1. ANTHOTIUM**

2: Ovules and seeds 1 or 2; fruit indehiscent, not beaked

4 Corolla without auricles; hairs simple ..... **6. DIASPASIS**

4: Corolla auriculate; hairs branched, rarely absent ..... **3. DAMPIERA**

1: Anthers free

5 Ovules (and usually seeds) more than 2 per locule

6 Seeds ovoid, obloid or irregularly compressed, without a wing, rim or peripheral groove

7 Corolla yellow ..... **9. VELLEIA**

7: Corolla blue to pinkish or white ..... **4. COOPERNOOKIA**

6: Seeds flattened or concave-compressed, with a wing, rim or peripheral groove

8 Corolla lobes without wings; fruit ±fleshy, indehiscent; prostrate succulent salt-marsh plant ..... **8. SELLIERA**

8: Corolla lobes winged (although sometimes very narrowly); fruit a dry dehiscent capsule; rarely salt marsh plants

9 Flowers in thyrses, racemes, spikes or subumbels; ovary inferior or half-inferior (superior in *Goodenia macroplecta* only) ..... **7. GOODENIA**

9: Flowers in axillary dichasia; ovary superior or apparently so ..... **9. VELLEIA**

5: Ovules and seeds 1 or 2 per locule

10 Fruit dehiscent (capsular); seed strophiolate ..... **4. COOPERNOOKIA**

10: Fruit indehiscent (nutlike, soft-shelled with several seeds, or a 1- or 2-seeded drupe); seed not strophiolate

11 Ovary and fruit winged ..... **10. PENTAPTILON**

11: Ovary and fruit not winged

12 Style with stiff purplish hairs ..... **7. GOODENIA**

12: Style with whitish hairs or glabrous

13 Corolla yellow; plants herbaceous

14 Plants glabrous or hairy with unbranched cottony hairs ..... **7. GOODENIA**

14: Plants hairy with branched and simple hairs ..... **11. VERREAUXIA**

13: Corolla blue to white, or rarely yellow and then scrambling shrubs ..... **5. SCAEVOLA**

1. ANTHOTIUM

D.A.Morrison

Anthotium R.Br., *Prodr.* 582 (1810); from the Greek *anthos* (flower) and *otos* (ear), referring to the auriculate adjacent wings of the adaxial petals which enclose the indusium.

Type: *A. humile* R.Br.

Perennial herbs, glabrous except indusium. Stems 1–many; rootstock woody. Leaves radical, linear to spathulate, entire to serrulate. Inflorescence cymose, compound, 1–5 heads each of 1–5 monochasia, each of 1–3 flowers; bracts and bracteoles leaf-like; flowers sessile. Sepals linear to lanceolate. Corolla bilabiate, red, light blue or cream; tube open nearly to base adaxially, obliquely spreading abaxially; abaxial lobes lanceolate, the wings rounded on free parts; shorter adaxial lobes falcate, the wings rounded on opposite margins, auriculate on adjacent margins. Filaments free; anthers cohering in a tube. Ovary inferior, 2-locular; ovules numerous, axial; indusium cupular, glabrous or bearded. Fruit capsular; valves 4, longitudinal. Seeds with hardened epidermis.

A genus of 3 species, endemic in south-western W.A.

G.Bentham, *Anthotium*, *Fl. Austral.* 4: 44–45 (1868); K.Krause, *Anthotium*, *Pflanzenr.* 54: 109–112 (1912); D.A.Morrison, The genus *Anthotium* (Goodeniaceae), *Nuytsia* 7: 49–58 (1989).

1 Leaves lanceolate or spathulate, more than 2 mm wide **2. A. rubriflorum**

1: Leaves linear to terete, 1 mm wide

2 Flowering stems 2–7 cm long; corolla usually creamy white **1. A. humile**

2: Flowering stems 12–40 cm long; corolla pale blue or mauve **3. A. junciforme**

1. Anthotium humile R.Br., *Prodr.* 582 (1810)

Lechenaultia humilis (R.Br.) Sprengel, *Syst. Veg.* 1: 720 (1824). T: Lucky Bay and King George Sound, [W.A.], *R.Brown*; lecto: BM; isolecto: BM, *fide* D.A.Morrison, *Nuytsia* 7: 51 (1989).

Anthotium glabrum R.Br. ex Poiret, *Dict. Sci. Nat.* 2: Suppl. 80 (1816). T: probably based on R.Brown's collections, cf. *A. humile*.

Goodenia pygmaea Vriese in J.G.C.Lehmann, *Pl. Preiss.* 1: 413 (1845). T: near Perth, W.A., *L.Preiss 1492*; lecto: LD; isolecto: G *n.v.* photo seen SYD, *fide* D.A.Morrison, *loc. cit.*

Illustrations: K.Krause, *Pflanzenr.* 54: 111, fig. 21a (1912); D.A.Morrison, *Nuytsia* 7: 53, fig. 1A (1989).

Tufted herb to 10 cm high; up to 5 tufts from rootstock. Leaves somewhat fleshy, usually linear to terete, 4.5–11 cm long, 0.5–1 mm wide, sometimes narrowly lanceolate or narrowly spathulate, then 3–6.5 cm long, 1.5–3 mm wide, entire. Flowering stalks 2.5–7 cm long, usually shorter than leaves; heads compact. Sepals 2.5–3.5 mm long. Petals 6.5–8 mm long, light blue through mauve or pink to cream; wings almost equal, 0.5–1 mm wide. Ovary 2.5–4 mm long; style 2.5–5 mm long; indusium glabrous. Fruit and seeds not seen.

Widespread in near-coastal south-western W.A., from the Stirling Ra. to Cape le Grand, with outliers W to Broke Inlet and N to near Pingelly. Grows in sand and sandy clay, in heath and woodland. Flowers usually Dec.–Jan. Map 2.

W.A.: between West Mt Barren and Point Anne, *T.E.H.Aplin 5696* (PERTH); W. Australia, *J.Drummond 181* (MEL, NSW); c. 82 km E of Ravensthorpe, *A.S.George 2251* (PERTH); 11 km S of Dumberning Siding, *G.J.Keighery 7861* (PERTH); c. 95 km E of Lake King, *E.Wittwer 1498* (PERTH).

2. **Anthotium rubriflorum** F.Muell. ex Benth., *Fl. Austral.* 4: 45 (1868)

T: south-western W.A., 1848, *J.Drummond 5: 180*; syn: K *n.v.*, MEL, photo seen SYD; south-western Australia, *G.Maxwell*; lecto: K *n.v.*, photo seen SYD; isolecto: MEL, *fide* D.A.Morrison, *Nuytsia* 7: 54 (1989).

Illustrations: K.Krause, *Pflanzenr.* 54: 111, fig. 21c–h (1912); D.A.Morrison, *Nuytsia* 7: 53, fig. 1C (1989).

Tufted herb to 15 cm high; up to 5 tufts from rootstock. Leaves fleshy, lanceolate to spathulate, usually 4.5–8 cm long, usually 3–6 mm wide, entire to serrulate. Flowering stalks 9–16 cm long, usually twice as long as leaves; heads compact. Sepals 3.5–4.5 mm long. Petals usually 6.5–8.5 mm long, usually bright scarlet; wings unequal, 0.2–1.5 mm wide. Ovary 4–5 mm long; style 2.5–3.5 mm long; indusium bearded on upper side. Fruit 7–9 mm long; seeds usually 7–9 pairs. Figs 2, 29A–B.

Widespread in inland south-western W.A., between New Norcia and Ravensthorpe. Grows in sand or sandy gravel, usually in heath or scrub but also in woodland and mallee. Flowers Nov.–Dec. Map 3.

W.A.: W of Ravensthorpe, *J.C.Anway 584* (MEL, NSW, PERTH); Uberin Hill, Jan. 1918, *C.A.Fontleroy* (BRI); Tammin, *C.H.Ostenfeld 960* (PERTH); Frank Hann Natl Park, *R.D.Royce 10250* (PERTH); 10 km E of Rabbit Proof Fence on Hyden–Norseman road, *P.Weston 332* (SYD).

Easily distinguishable from the other species by the petal colour.

3. **Anthotium junciforme** (Vriese) D.Morrison, *Nuytsia* 7: 55 (1989)

Goodenia junciformis Vriese in J.G.C.Lehmann, *Pl. Preiss.* 1: 413 (1845); *Anthotium humile* var. *junciforme* (Vriese) E.Pritzel in F.L.E.Diels & E.Pritzel, *Bot. Jahrb. Syst.* 35: 554 (1905). T: near Perth, W.A., 12 Feb. 1840, *L.Preiss 1522*; lecto: LD *n.v.* photo seen SYD; isolecto: G *n.v.* photo seen SYD, K, L *n.v.* photo seen SYD, MEL, P *n.v.*, W *n.v.*, *fide* D.A.Morrison, *Nuytsia* 7: 55 (1989).

Goodenia geniculata Vriese in J.G.C.Lehmann, *Pl. Preiss.* 1: 413 (1845) *non* R.Br. (1810); *G. genuflexa* Vriese in J.G.C.Lehmann, *Pl. Preiss.* 2: 244 (1848). T: near Tobys [Torbay] Inlet, W.A., 27 Dec. 1839, *L.Preiss 1456*; lecto: LD *n.v.*, photo seen SYD; isolecto: G *n.v.*, photo seen SYD, L *n.v.*, photo seen SYD, MEL, W *n.v.*, *fide* D.A.Morrison, *Nuytsia* 7: 55 (1989).

Illustrations: K.Krause, *Pflanzenr.* 54: 111, fig. 21b (1912); D.A.Morrison, *Nuytsia* 7: 53, fig. 1B (1989).

Grass-like herb to 40 cm high; tuft solitary. Leaves linear to terete, 9–14.5 cm long, 0.5–1 mm wide, somewhat fleshy, entire. Flowering stalks 18–40 cm long, usually twice as long as leaves; heads loose. Sepals 3–5.5 mm long. Petals 7.5–9.5 mm long, purple to light blue; wings unequal, 0.2–1.2 mm wide. Ovary usually 3.5–6.5 mm long; style 4–5 mm long; indusium glabrous. Fruit 8–12 mm long; seeds 11–15 pairs. Fig. 29C.

Endemic in south-western W.A. along the coast between Perth and Busselton. Grows in eucalypt woodland. Flowers Dec.–Feb. Map 4.

W.A.: Midland Junction, Dec. 1899, *W.V.Fitzgerald* (NSW, PERTH); Wattle Grove, *A.S.George 627* (PERTH); Waterloo, *G.J.Keighery 3844* (PERTH); Bayswater, *A.Morrison 18031* (BRI, NSW); Busselton, no date, J.*C.Rosselloty* (MEL).

The central adaxial sepal lobe is 0.5–1 mm longer than the remainder. A rare species that occurs in a densely settled area, and is no longer common.

2. LECHENAULTIA

D.A.Morrison

Lechenaultia R.Br., *Prodr.* 581 (1810); named after Jean-Baptiste Louis-Claude-Theodore Leschenault de la Tour (1773–1826), botanist with Nicolas Baudin's expedition to Australia (1800–1804).

Type: *L. formosa* R.Br.

Lechenaultia a. *Latouria* Endl., *Gen. Pl.* 508 (1838); *Latouria* (Endl.) Lindley, *Veg. Kingd.* 2nd edn, 695 (1847); *Lechenaultia* sect. *Latouria* (Endl.) Benth., *Fl. Austral.* 4: 43 (1868). T: *Lechenaultia filiformis* R.Br.

Ericopsis C.Gardner, *J. & Proc. Roy. Soc. W. Australia* 9: 42 (1923). T: *E. formosus* C.Gardner.

Perennial herbs or small shrubs, often suckering. Stems several to many, usually glabrous; rootstock woody. Leaves sessile, linear, narrowly lanceolate to ovate, entire, usually glabrous. Inflorescence cymose; flowers sessile; bracts and bracteoles leaf-like, rarely reduced. Sepals linear to narrowly lanceolate, usually glabrous. Corolla bilabiate, usually glabrous outside, hairy inside, red, blue, mauve, white or yellow; free lobes lanceolate, winged. Filaments free; anthers cohering in a tube; pollen grains in tetrads. Ovary inferior, usually erect, 2-locular; ovules numerous, axial; indusium 2-lipped, ciliate. Fruit capsule-like, elongated, retaining outer floral whorls; valves 4, longitudinal. Seeds in articles, with woody endocarp of inner locule wall. $x = 9$, 18 (6 species) W.J.Peacock, *Proc. Linn. Soc. New South Wales* 88: 8–27 (1963).

A genus of 26 species, all occurring in Australia, with one extending to New Guinea; 20 species endemic in south-western W.A. The scientific name is often incorrectly spelt as *Leschenaultia*, although this spelling is used in vernacular names.

G.Bentham, *Leschenaultia*, *Fl. Austral.* 4: 38–44 (1898); K.Krause, *Leschenaultia*, *Pflanzenr.* 54: 97–109 (1912); D.A.Morrison, Taxonomic and nomenclatural notes on *Lechenaultia* R.Br. (Goodeniaceae), *Brunonia* 9: 1–28 (1986); D.A.Morrison, The phytogeography, ecology and conservation status of *Lechenaultia* R.Br. (Goodeniaceae), *Kingia* 1: 85–133 (1987); D.A.Morrison, Notes on the fruits of *Lechenaultia* R.Br. (Goodeniaceae), with a new species from northern Australia, *Telopea* 3: 159–166 (1988).

KEY TO SECTIONS

1 Corolla tube forming a complete cylinder, erect — sect. 3. **LECHENAULTIA**

1: Corolla tube cleft to base, open on adaxial side

  2 Wings on adaxial petal lobes always present — sect. 1. **PATENTES**

  2: Wings on adaxial petal lobes usually absent — sect. 2. **LATOURIA**

KEY TO SPECIES

1 Inside of corolla hairy throughout tube or only towards top of tube, usually also hairy on free part of lobes

  2 Central adaxial sepal conspicuously longer than other sepals

3 Leaves more than 12 mm long, narrow **15. L. filiformis**

3: Leaves less than 10 mm long, ovate to narrowly ovate **16. L. ovata**

2: All sepals same length

4 Wings on adaxial petals almost as wide as those on abaxial petals; adaxial wings usually more than 1 mm wide

5 Leaves and sepals hairy **4. L. pulvinaris**

5: Leaves and sepals glabrous

6 Style more than 11 mm long

7 Tips of branches curved downwards **24. L. linarioides**

7: Tips of branches straight; flowers yellow and red **2. L. stenosepala**

6: Style less than 9 mm long; flowers blue and white

8 Wings on petals more than 2 mm wide; ovary more than 9 mm long **1. L. biloba**

8: Wings on petals less than 2 mm wide; ovary less than 8 mm long **3. L. expansa**

4: Wings on adaxial petals much narrower than those on abaxial petals; adaxial wings usually less than 1 mm wide and often absent

9 Stems with rough bark except on new growth

10 Leaves, sepals and ovaries papillate; leaves usually less than 3 mm long **6. L. papillata**

10: Leaves, sepals and ovaries not papillate; leaves more than 4 mm long **5. L. floribunda**

9: Stems with rough bark only at base of plant or rarely also on lowest branches

11 Sepals less than 2.5 mm long

12 Leaves more than 6 mm long **7. L. subcymosa**

12: Leaves less than 3 mm long

13 Leaves more than 1 mm long; fruit persistent, deeply constricted between seeds **8. L. divaricata**

13: Leaves less than 1 mm long or absent; fruit deciduous, not constricted **9. L. aphylla**

11: Sepals more than 3 mm long

14 Leaves rare and scattered distally on flowering stems

15 Petals dark blue **13. L. brevifolia**

15: Petals pale blue or sometimes cream

16 Bracts and bracteoles more than 7 mm long; ovules and articles more than 15 pairs **12. L. striata**

16: Bracts and bracteoles less than 7 mm long; ovules and articles less than 8 pairs **14. L. juncea**

14: Leaves common and evenly distributed on flowering stems

17 Leaves less than 9 mm long; petals usually pale blue **10. L. heteromera**

17: Leaves more than 10 mm long; petals usually pale yellow **11. L. lutescens**

1: Inside of corolla hairy only in a tuft at base of tube

18 Stems with rough bark only at base of plant

19 Stems, leaves, sepals and outside of petals hairy **19. L. hirsuta**

19: Stems, leaves, sepals and outside of petals glabrous

20 Petal wings usually less than 4 mm wide; leaves less than 25 mm long **20. L. longiloba**

20: Petal wings more than 4 mm wide; leaves usually more than 25 mm long **21. L. macrantha**

18: Stems with rough bark except on new growth

21 Leaves rigid with a pellucid, acute point; petal wings cohering **23. L. tubiflora**

21: Leaves soft and fleshy; petal wings (if present) free along outside margin

22 Petals pale bluish green or greenish yellow

23 Leaves less than 6 mm long; sepals less than 5 mm long **22. L. acutiloba**

23: Leaves more than 6 mm long; sepals more than 7 mm long **25. L. chlorantha**

22: Petals red, orange or yellow

24 Leaves more than 11 mm long **18. L. superba**

24: Leaves less than 11 mm long

25 Adaxial petals connivent along adjacent margins above tube, and enclosing indusium **26. L. formosa**

25: Adaxial petals not connivent above tube **17. L. laricina**

Sect. 1. Patentes

Lechenaultia sect. **Patentes** D.Morrison, *Brunonia* 9: 6 (1987)

Type: *L. biloba* Lindley

Inflorescence a monochasial or occasionally dichasial cyme, rarely reduced to a single flower; bracts and bracteoles leaf-like. Corolla blue, often with a pale yellow centre; tube open to base on adaxial side, obliquely spreading to abaxial side; indumentum throughout inside of tube, spreading onto lower part of free lobes and sometimes also wings; adaxial lobes lanceolate, as long as abaxial lobes; wings on adaxial lobes always present. Style usually curved.

A section of 6 species endemic in south-western W.A.

1. Lechenaultia biloba Lindley, *Sketch Veg. Swan R.* xxvii (1839)

T: Swan R., W.A., 1839, *J.Drummond s.n.*; lecto: CGE, *fide* D.A.Morrison, *Brunonia* 9: 6 (1987); isolecto: K; Swan R., W.A., 1838, *J.Drummond s.n.*; syn: K.

L. grandiflora Lindley, *Sketch Veg. Swan R.* xxvi (1839). T: Swan R., W.A., *Capt. Mangles*; lecto: CGE, *fide* D.A.Morrison, *Brunonia* 9: 6 (1987); Vasse R., W.A., 1839, *Mrs Molloy*; syn: CGE.

L. grandiflora DC., Prodr. 7: 519 (1839), *nom. illeg.* T: Swan R., W.A., 1839, *J.Drummond s.n.*; lecto: G-DC *n.v.*, microfiche seen, *fide* D.A.Morrison, *Brunonia* 9: 7 (1987).

L. drummondii Vriese, *Natuurk. Verh. Holl. Maatsch. Wetensch. Haarlem* ser. 2, 10: 182 (1854). T: Swan R., W.A., 25 Aug. 1839, *L.Preiss 1463*; lecto: LD, *fide* D.A.Morrison, *Brunonia* 9: 7 (1987); isolecto: L *n.v.*, photo seen SYD, MEL; Swan R., W.A., *J.Drummond 292*; syn: W *n.v.*, photo seen SYD.

Illustrations: K.Krause, *Pflanzenr.* 54: 101, fig. 18A–D (1912); B.J.Grieve & W.E.Blackall, *How to Know W. Austral. Wildfl.* 2nd edn, pt IV, t. v (1975); J.R.Wheeler, *Fl. Perth Region* 2: 633, fig. 234 (1987).

Diffuse, ascending, moderately branched subshrub or shrub to 100 cm high, often suckering. Bark rough. Leaves, sepals and ovary glabrous. Leaves crowded, narrow, usually 6–11 mm long, fleshy. Flowers in compact monochasia or occasionally dichasia. Sepals 6–7 mm long. Corolla usually 14–20 mm long, brilliant dark blue through pale blue to creamy white (usually pale blue or white on lobes); tube pilose inside, white; lobes almost equal, spreading, with sparse hairs often extending onto wing margins; wings almost equal, triangular to deeply bilobed, usually 3–6 mm wide. Ovary usually 11–17.5 mm long; style 5.5–8 mm long, sparsely glandular-hairy; indusium pilose on back. Fruit usually 23–35 mm long; articles 8–15 pairs. n = 9, 18, W.J.Peacock, *Proc. Linn. Soc. New South Wales* 88: 13 (1963). *Blue Leschenaultia.* Fig. 6.

Widespread in south-western W.A. from Three Springs to Cape Leeuwin, and E to Lake King. Occurs in sand, often over laterite, in heath, scrub, woodland and (mainly Jarrah) forest. Flowers Aug.–Nov. Map 5.

W.A.: 3.5 km SE of Armadale, *H.Eichler 15759* (AD, CANB, PERTH); c. 80 km E of Coorow, *C.H.Gittins 1708* (BRI, NSW, PERTH); Wooroloo, *M.Koch 1396* (MEL, NSW, PERTH); 30 km W of Lake King, *R.H.Kuchel 1864* (AD, BRI, NSW); c. 0.8 km from Wubin towards Buntine, 13 Sept. 1968, *M.E.Phillips* (CBG, NSW, SYD).

Varies greatly in flower and leaf size as well as flower colour, but it remains distinctive because of the large corolla wings.

2. Lechenaultia stenosepala E.Pritzel in F.L.E.Diels & E.Pritzel, *Bot. Jahrb. Syst.* 35: 552 (1905)

T: Strathmore Rd Reserve (no. 26248), W.A., 5 Nov. 1975, *A.S.George 14197*; neo: PERTH, *fide* D.A.Morrison, *Brunonia* 9: 8 (1987).

Diffuse, ascending herb or subshrub to 40 cm high, often suckering. Bark rough basally, coarsely striate. Leaves, sepals and ovary glabrous. Leaves crowded on short leafy stems, scattered on flowering stems, narrow, usually 7.5–13 mm long, fleshy or somewhat rigid. Flowers in compact monochasia. Sepals usually 7.5–11 mm long. Corolla usually 13–19 mm long, blue through pale blue to creamy white, midrib and edges often darker blue; tube pubescent inside, the apex sometimes yellow spotted; lobes equal, spreading, with sparse hairs often extending onto wing margins; wings almost equal, rounded to triangular, usually 0.9–2 mm wide. Ovary 9.5–18.5 mm long; style 11.5–18.5 mm long, sparsely glandular-hairy; indusium pilose on back. Fruit 16–22 mm long; articles 8–16 pairs.

Endemic in south-western W.A. between Three Springs and Gingin. Found in sand or sandy gravel, often over clay or laterite, usually in low open heath. Flowers Oct.–Dec. Map 6.

W.A.: 12 km W of Gingin–Dongara road on road to Nambung Natl Park, *T.A.Halliday 163* (AD, PERTH); c. 14 km N of Gingin, *V.Mann & A.S.George 178* (MEL, NSW, PERTH); c. 6 km S of Marchagee, *B.R.Maslin 1429* (PERTH, SYD); 1 km W of junction of Badgingarra West Rd and road S to Badgingarra Research Stn, *R.A.Saffrey 185* (PERTH, SYD); just W of [New] Badgingarra, *D.J.E.Whibley 4834* (AD, PERTH).

Very variable in leaf and flower size. This variation appears to correlate with latitude, with the largest leaves and flowers occurring on plants in the northern part of the distribution.

3. Lechenaultia expansa R.Br., *Prodr.* 581 (1810)

T: King George Sound, [W.A.], Dec. 1801, *R.Brown*; lecto: BM, *fide* D.A.Morrison, *Brunonia* 9: 8 (1987); isolecto: BM.

L. pallescens var. *congesta* Vriese in J.G.C.Lehmann, *Pl. Preiss.* 1: 415 (1845); *L. expansa* var. *congesta* (Vriese) Vriese in J.G.C.Lehmann, *Pl. Preiss.* 2: 402 (1848). T: towards Canning R., W.A., 2 Nov. 1839,

L.Preiss 1468; lecto: LD, *fide* D.A.Morrison, *Brunonia* 9: 9 (1987); isolecto: K, L *n.v.*, photo seen SYD, MEL.

L. parviflora Vriese in J.G.C.Lehmann, *Pl. Preiss.* 1: 416 (1845). T: Princess Royal Harbour, W.A., Oct. 1840, *L.Preiss 1462*; lecto: LD, *fide* D.A.Morrison, *Brunonia* 9: 9 (1987).

L. tenuifolia Vriese in J.G.C.Lehmann, *Pl. Preiss.* 1: 415 (1845). T: near Mt Barrow, W.A., 9 Nov. 1840, *L.Preiss 1461*; lecto: LD, *fide* D.A.Morrison, *Brunonia* 9: 9 (1987); W.A., *L.Preiss 1462*; syn: LD.

Illustration: K.Krause, *Pflanzenr.* 54: 106, fig. 20C–F (1912).

Virgate, ascending, moderately branched subshrub to 60 cm high. Bark rough on lower branches only. Leaves, sepals and ovary glabrous. Leaves crowded, narrow, 5.5–9.5 mm long, fleshy. Flowers in compact groups of axillary monochasia each with terminal flower. Sepals 4.5–5.5 mm long. Corolla 9–11 mm long, pale blue-purple; tube pilose inside, pale yellow-white; lobes equal, spreading, with sparse hairs often extending onto wing margins; wings almost equal, rounded, 1–1.6 mm wide. Ovary 3.5–6.5 mm long; style 5.5–6.5 mm long, sparsely glandular-hairy; indusium hispid on back, ofen extending around side and underneath. Fruit 5–10 mm long; articles 2–5 pairs. Fig. 28A.

Endemic in south-western W.A. near the coast between Perth and Albany. Usually found in swamp heath or around the edge of paperbark swamps or in other permanently damp or seasonally waterlogged areas. Flowers Oct.–Jan. Map 7.

W.A.: CalSil brickworks, c. 5 km S of Jandakot, *J.C.Anway 543* (AD, MEL, PERTH); 30 mile [c. 48 km] mark between Goomalling and Toodyay, *S.Carlquist 3908* (AD, NSW); Long Swamp, 10 km NE of Augusta, *T.A.Halliday 260* (AD, PERTH); Bow R., Dec. 1912, *S.W.Jackson* (NSW, PERTH, SYD); King George Sound, Nov. 1909, *J.H.Maiden* (BRI, NSW).

Closely related to *L. floribunda* and often confused with it. However, *L. expansa* has larger leaves which are minutely pitted, a more densely hairy floral tube, smaller and almost equal petal wings, a more thickly hairy indusium, and smaller fruit with fewer articles.

4. Lechenaultia pulvinaris C.Gardner, *J. Roy. Soc. W. Australia* 47: 63 (1964)

T: c. 40 km S of Corrigin, W.A., 19 Oct. 1961, *C.A.Gardner 13620*; holo: PERTH; iso: PERTH.

Illustration: J.Leigh *et al.*, *Extinct & Endangered Pl. Australia*, between 192 & 193.

Hemispherical, procumbent, much-branched subshrub to 15 cm high. Bark rough. Leaves narrow, 4.5–8.5 mm long, crowded, rigid, tomentose. Flowers solitary and terminal, rarely in compact monochasia. Sepals 3.5–4 mm long, densely tomentose. Corolla 9–11 mm long, pale blue or purple; tube pubescent inside, pale yellow; lobes almost equal, spreading, sparsely pubescent; wings almost equal, rounded 0.8–1 mm wide. Ovary 5–6 mm long, glabrous or often sparsely tomentose on upper half; style 4–4.5 mm long, sparsely hairy; indusium pilose on back. Fruit 5–7 mm long; articles c. 8 pairs.

Endemic in south-western W.A. between Corrigin and Wagin, occurring in open patches in sand in low scrub. Flowers Oct.–Dec. Map 8.

W.A.: 14 km SE of Tincurrin, *D.A.Morrison 224* (SYD); 300 m S of NE corner of Reserve No. 19089, 23 Nov. 1982, *S.J.Patrick* (PERTH); 3 m from NW corner of Reserve No. 19096, 24 Nov. 1982, *S.J.Patrick* (PERTH); c. 16 km E of Stretton, K.Newbey 2489 (PERTH); c. 5 km SW of Harrismith, *E.Wittwer 1513* (PERTH).

Restricted geographically, mainly to the edges of a number of conservation reserves.

5. **Lechenaultia floribunda** Benth. in S.F.L.Endlicher *et. al.*, *Enum. Pl.* 70 (1837)

T: Swan R., W.A., *C.A.A. von Hügel*; holo: W *n.v.*, photo seen SYD; iso: K.

L. glauca Lindley, *Sketch Veg. Swan R.* xxvii (1839). T: Swan R., W.A., 1839, *J.Drummond s.n.*; lecto: CGE, *fide* D.A.Morrison, *Brunonia* 9: 10 (1987); Swan R., [W.A.], *Mr Toward*; syn: CGE.

L. pallescens Vriese in J.G.C.Lehmann, *Pl. Preiss.* 1: 415 (1845). T: near Bull Creek, Perth, W.A., Nov. 1841, *L.Preiss 1464*; lecto: LD, *fide* D.A.Morrison, *Brunonia* 9: 11 (1987); isolecto: MEL, W; near Perth, W.A., 22 Oct. 1839, *L.Preiss 1460*; syn: L *n.v.*, photo seen SYD, LD, MEL, W *n.v.*

L. floribunda var. *borealis* E.Pritzel in F.L.E.Diels & E.Pritzel, *Bot. Jahrb. Syst.* 35: 553 (1905). T: between Dongara and Northampton, W.A., 19 Sept. 1932, *W.E.Blackall 2656*; neo: PERTH, *fide* D.A.Morrison, *Brunonia* 9: 11 (1987).

L. drummondiana Colozza, *Nuovo Giorn. Bot. Ital.* n. ser. 15: 204 (1908). T: Swan R., W.A., 1839, *J.Drummond s.n.*; holo: FI-W *n.v.*, photo seen SYD.

Illustration: A.Colozza, *op. cit.* t. VIII as *Lechenaultia drummondiana*.

Diffuse, ascending, moderately branched subshrub or shrub to 100 cm high. Bark rough. Leaves, sepals and ovary glabrous. Leaves crowded, narrow, 4.5–7.5 mm long, fleshy. Flowers in compact groups of axillary monochasia with terminal flowers. Sepals 3–4.5 mm long. Corolla usually 11–15 mm long, pale blue through pale mauve to almost creamy white, lobes paler; tube sparsely pubescent inside, white or pale yellow; lobes almost equal, spreading, sparse hairs often extending onto wing margins; wings on abaxial lobes triangular, usually 1.3–2.2 mm wide, on adaxial lobes rounded to triangular, usually 0.4–0.8 mm wide. Ovary usally 6–8 mm long; style usually 7.5–8.5 mm long, sparsely glandular-hairy; indusium pilose on back. Fruit usually 11–18 mm long; articles 7–15 pairs. $n = 9$, W.J.Peacock, *Proc. Linn. Soc. New South Wales* 88: 13 (1963).

Occurs in south-western W.A. between Kalbarri and Perth, and inland to Wongan Hills, growing in sand or loamy sand, in heath, scrub or woodland with a heath understorey. Flowers Aug.–Dec. Map 9.

W.A.: c. 38 km N of Murchison R. on North West Coastal Hwy, *J.S.Beard 6742* (NSW, PERTH, SYD); 0.4 km S of Reynoldson Reserve, *S.Carlquist 3905* (AD, NSW); 15 km SE of Mingenew on Geraldton Hwy, *T.A.Halliday 129* (AD, PERTH); c. 11 km N of Gingin, *V.Mann 175 & A.S.George* (MEL, NSW, PERTH); Coorow–Green Head road, 30 km E of Brand Hwy, *J.Taylor 963*, *M.D.Crisp & R.Jackson* (CBG, PERTH).

A very variable species, particularly in leaf length, flower size, and flowering habit, with a morphocline running from north to south.

6. **Lechenaultia papillata** D.Morrison, *Brunonia* 9: 12 (1987)

T: c. 58 km N of mouth of Oldfield R., W.A., 21 Oct. 1968, *H.Eichler 20392*; holo: AD; iso: CANB, PERTH.

Illustration: D.A.Morrison, *Brunonia* 9: fig. 2 (1987).

Diffuse, ascending, moderately branched subshrub or shrub to 45 cm high. Bark rough. Leaves, sepals and ovary glabrous but papillate. Leaves crowded, narrow, 1.5–3 mm long, somewhat fleshy. Flowers in compact monochasia or occasionally dichasia. Sepals 3–5 mm long. Corolla 11–14 mm long, pale blue; tube pubescent inside; lobes almost equal, spreading, sparse hairs often extending onto wing margins; wings on abaxial lobes triangular, 1.9–2.6 mm wide, on adaxial lobes rounded, 0.8–1.6 mm wide. Ovary 5–8 mm long; style 6–8 mm long, sparsely glandular-hairy; indusium pilose on back. Fruit 10–16 mm long, papillate; articles 9–14 pairs.

Occurs in inland south-western W.A. between Newdegate and Mt Ragged, growing in sand, loamy sand, or gravelly loam, usually in heath, low open scrub, or eucalypt mallee scrub.

Flowers Oct.–Nov. Map 10.

W.A.: Lake King–Norseman road, between Rabbit Proof Fence & 100 Mile Tank, *J.S.Beard 3784* (PERTH); c. 92 km E of Lake King crossroads, 14 Nov. 1965, *F.W.Humphreys* (PERTH); c. 45 km E of Cross Roads, Forrestania, *F.Lullfitz 3868* (PERTH); near Mt Madden, *K.Newbey 1835* (PERTH); 15 km W of Lake Cronin, *K.Newbey 6613* (PERTH).

The papillate leaves, sepals, ovary and fruit are unique within the genus.

Sect. 2. Latouria

Lechenaultia sect. **Latouria** (Endl.) Benth., *Fl. Austral.* 4: 39, 43 (1868); *Lechenaultia* a. *Latouria* Endl., *Gen. Pl.* 508 (1838); *Latouria* (Endl.) Lindley, *Veg. Kingd.* 2nd edn 695 (1847)

Type: *L. filiformis* R.Br.

Inflorescence a monochasial or occasionally a dichasial cyme, or reduced to single flower; bracts and bracteoles leaf-like or sometimes reduced. Corolla blue or pale yellow; tube open to base on adaxial side, obliquely spreading to abaxial side; indumentum throughout inside of tube, spreading onto lower part of free lobes and sometimes also wings; adaxial lobes lanceolate-falcate, shorter than abaxial lobes; wings on adaxial lobes usually absent. Style usually curved.

A section of 10 species – 2 in tropical Australia, 5 in arid W.A., N.T. and S.A., and 3 in south-western W.A.

7. Lechenaultia subcymosa C.Gardner & A.S.George, *J. Roy. Soc. Western Australia* 46: 134 (1963)

T: 56 miles [c. 90 km] S of Learmonth, W.A., 2 June 1961, *A.S.George 2433*; holo: PERTH; iso: MEL.

Illustration: C.Gardner & A.S.George, *op. cit.* 136, fig. 5.

Divaricate, ascending, sparsely branched herb or subshrub to 30 cm high. Bark sometimes rough basally, coarsely striate. Leaves, sepals and ovary glabrous. Leaves scattered on flowering stems, crowded on short, leafy stems, narrow, 7–11.5 mm long, rigid. Flowers in loose dichasia. Sepals 1–2.5 mm long. Corolla 14–19 mm long, creamy white to pale mauve, sometimes with reddish brown midribs and edges of lobes; tube pubescent inside; lobes spreading, sparsely pubescent; wings on abaxial lobes rounded to triangular, 1–3 mm wide, on adaxial lobes rounded, 0.2–0.6 mm wide. Ovary 7.5–14.5 mm long; style 8.5–11.5 mm long, sparsely hairy; indusium pubescent on back. Fruit 19–33 mm long; articles 7–11 pairs.

Occurs in W.A. on North West Cape and on the Shark Bay islands and coastline, growing in sand or loam over limestone, usually with scattered low shrubs. Flowers sporadically, perhaps only after recent fire. Map 11.

W.A.: Biddy Giddy Camp to Mt Direction and Steep Point, *T.E.H.Aplin 3414* (PERTH); Shothole Canyon, Cape Ra., *A.S.George 10319* (CANB, PERTH); 2 km S of Hawknest Well, Dirk Hartog Is., *A.S.George 11573* (CANB, PERTH); Dorre Is., *K.F.Kenneally 4653* (CANB, PERTH); c. 5 km W of Giralia Stn homestead, Aug. 1963, *J.Tonkinson* (PERTH).

8. **Lechenaultia divaricata** F.Muell., *Fragm.* 3: 33 (1862); 167 (1863)

T: Coopers Creek, N.T., *Dr Wheeler*; holo: MEL; iso: K.

Illustrations: R.C.Carolin in J.P.Jessop, *Fl. Centr. Australia* 360, fig. 459 (1981); G.M.Cunningham *et al.*, *Pl. W. New South Wales* 636 (1981); D.A.Cooke in J.P.Jessop & H.R.Toelken, *Fl. S. Australia* 3: 1407, fig. 634A (1986).

Divaricate, erect, much-branched subshrub to 100 cm high. Bark rough only basally, coarsely striate. Leaves, sepals and ovary glabrous. Leaves scattered, narrow, 1.5–2 mm long, membranous. Flowers in monochasia. Sepals 1.5–2 mm long. Corolla 13–20 mm long, yellow, through pale yellow to creamy white; tube sparsely pubescent inside; lobes spreading, sparse hairs often extending onto wing margins; wings on abaxial lobes triangular, fimbriate, 1.5–2.5 mm wide, on adaxial lobes (if present) rounded, 0.1–0.4 mm wide. Ovary 4.5–7 mm long; style usually 7–9 mm long, glabrous; indusium pubescent on back. Fruit 9–32 mm long, woody, almost moniliform, articles 1–4 pairs. *Tangled Leschenaultia.* Fig. 28D.

Widespread and common throughout the eastern arid and semi-arid areas of central Australia. Occurs on red sand dunes, sandplains, or alluvial soil, often in floodplains, swales, or other periodically wet depressions, in open grassland or open mulga. Flowers sporadically. Map 12.

N.T.: c. 30 km S of Alice Springs, *D.J.Nelson 1802* (AD, CANB, DNA, MEL, NSW); c. 40 km NE of Ooratippra Stn, *R.A.Perry 3452* (AD, BRI, CANB, DNA, MEL, NSW, PERTH). Qld: c. 90 km WNW of Birdsville, *D.E.Boyland 315* (BRI, MEL). N.S.W.: Fort Grey, *P.L.Milthorpe 883* (NSW). S.A.: Coopers Creek, 17 June 1972, *U.Johnson* (AD, NSW).

The fruit are often persistent for several years. The Aborigines apparently ate the roots, and a gum from the roots was used as an adhesive.

9. **Lechenaultia aphylla** D.Morrison, *Fl. Australia* 35: 332 (1992)

T: 11.0 km NE Mt Finke, S.A., 5 Oct. 1987, *D.E.Symon NPWS 1093*; holo: NSW; iso: AD.

Divaricate, ascending, moderately branched herb or subshrub to 50 cm high. Bark rough only basally, coarsely striate. Leaves, sepals and ovary glabrous. Leaves scattered, narrow, 0.5–1 mm long, caducous. Flowers in monochasia or solitary, terminal. Sepals 1–1.5 mm long. Corolla 15–18 mm long, yellow; tube pubescent inside; lobes spreading, sparsely pubescent; wings on abaxial lobes triangular, fimbriate, 1–1.5 mm wide, on adaxial lobes (if present) rounded, 0.1 mm wide. Ovary 8–9 mm long; style 8–9 mm long, glabrous; indusium sparsely hairy on back. Fruit 15 mm long; articles 6–9 pairs.

Known from only two collections, from mallee patches in swales of hard red dune sand near Yellabina, S.A. Flowers sporadically. Map 13.

S.A.: W of Mt Finke, *NPWS 2602* (AD).

The almost leafless habit of this species is unique within the genus.

10. **Lechenaultia heteromera** Benth., *Fl. Austral.* 4: 43 (1868)

T: East Mt Barren, W.A., *G.Maxwell*; lecto: MEL, *fide* D.A.Morrison, *Brunonia* 9: 14 (1987); isolecto: BM; Swan R., W.A., *Mylne*; syn: K; Phillips R., W.A., *G.Maxwell*; syn: K; Oldfield R. and Moir Inlet, W.A., ?*G.Maxwell*; syn: MEL.

Illustration: K.Krause, *Pflanzenr.* 54: 106, fig. 20G (1912).

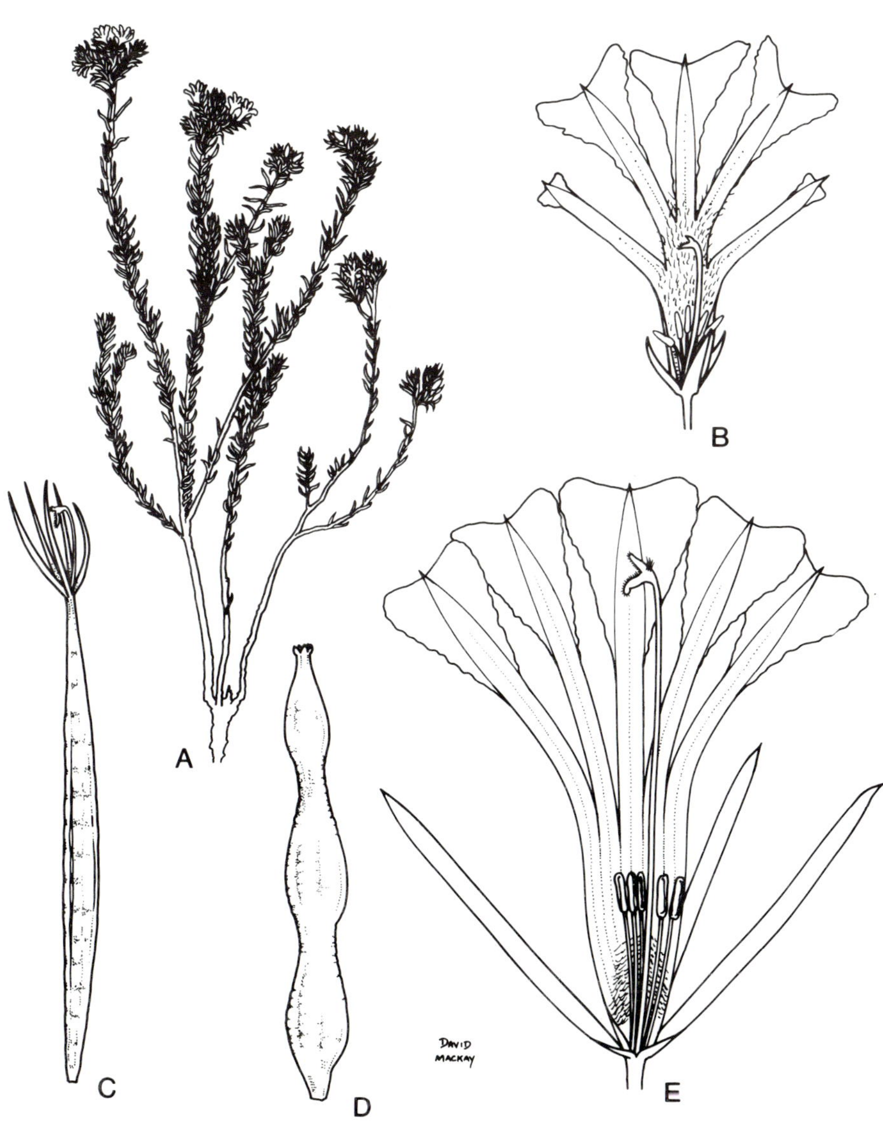

Figure 28. *Lechenaultia*. A, *L. expansa*, habit X0.5 (S.Jackson, Dec. 1912, PERTH). B, *L. lutescens*, flower X2 (A.George 8254, PERTH). C, *L. striata*, fruit X2 (R.Morland 8254, PERTH). D, *L. divaricata*, fruit X2 (G.Chippendale NT135). E, *L. longiloba*, flower X2 (A.Ashby 1894, AD96730168). Drawn by D.Mackay.

Virgate, ascending, moderately branched subshrub to 60 cm high. Bark rough on lower branches only, striate. Leaves, sepals and ovary glabrous. Leaves scattered to crowded, often recurved, 4.5–6.5 mm long, somewhat fleshy. Flowers in monochasia. Sepals 6.5–9 mm long. Corolla 14–19 mm long; tube pilose inside, often white; lobes spreading, sparsely pilose; wings on abaxial lobes triangular, 2–2.2 mm wide, and on adaxial lobes (if present) rounded, 0.1 mm wide; wings and lobes pale blue. Ovary 10.5–14.5 mm long; style 9.5–10.5 mm long, sparsely glandular-hairy; indusium pilose on back. Fruit 20–28 mm long; articles 6–8 pairs. $n = 9$, W.J.Peacock, *Proc. Linn. Soc. New South Wales* 88: 14 (1963).

Endemic in south-western W.A. along the coast between Starvation Boat Harbour and West Mt Barren. Found in deep white sand, in heath, open scrub or woodland. Flowers Aug.–Nov. Map 14.

W.A.: c. 0.8 km N of Hopetoun, *S.Carlquist 3495* (NSW); near Hopetoun, *H.Eichler 21111* (AD, CANB); Daniels Rd, S of Ravensthorpe, *A.S.George 5731* (PERTH, SYD); East Mt Barren, 14 Oct. 1961, *J.H.Willis* (MEL); c. 3 km from Hopetoun towards Ravensthorpe, 27 Oct. 1968, *J.W.Wrigley* (CBG, PERTH).

11. **Lechenaultia lutescens** D.Morrison & Carolin in D.A.Morrison, *Brunonia* 9: 15 (1987)

T: Yuendumu Reserve, N.T., 13 Jan. 1972, *N.Henry 334*; holo: DNA; iso: CANB, K, MEL.

[*Lechenaultia helmsii auct. non* K.Krause: J.Maconochie & N.Byrnes, *Muelleria* 2: 135 (1971); G.Chippendale, *Proc. Linn. Soc. New South Wales* 96: 267 (1972)]

Illustration: D.A.Morrison, *Brunonia* 9: 16, fig. 3 (1987).

Virgate, ascending, moderately branched herb or subshrub to 40 cm high. Bark rough basally, striate. Leaves, sepals and ovary glabrous. Leaves not crowded, narrow, 14.5–20.5 mm long, rigid. Flowers in loose monochasia or dichasia, rarely solitary and terminal. Sepals 3.5–4.5 mm long. Corolla 16–22 mm long; tube pilose inside; lobes spreading, sparsely pilose; wings on abaxial lobes deeply bilobed, usually 2–3 mm wide, on adaxial lobes triangular to deeply bilobed, usually 0.8–1 mm wide; lobes and wings orange-yellow, through pale yellow to creamy white. Ovary 7.5–11 mm long; style 8–9.5 mm long, glabrous or sparsely glandular-hairy; indusium pubescent on back. Fruit 15–25 mm long; articles 10–13 pairs. Figs 8, 28B.

Widespread in the western arid and semi-arid areas of central Australia. Occurs on red sand dunes, sandy loam plains, or around the gravelly edges of lateritic breakaways, usually among mallee, desert oak, open *Triodia* grassland, or spinifex open scrub. Flowers sporadically, perhaps only after recent rain. Map 15.

W.A.: c. 25 km E of Cosmo Newberry, *A.S.George 8105* (PERTH); 53 km SW of Warburton, *A.S.George 12178* (CANB, PERTH); 40 km SW of Giles Meteorological Stn, *P.K.Latz 7661* (BRI, DNA). N.T.: c. 8 km W of Docker River Settlement, *N.M.Henry 415* (BRI, CANB, DNA, NSW, PERTH); 50 km SSW of Vaughan Springs, *P.K.Latz 8668* (CANB, CBG, DNA, NSW).

12. **Lechenaultia striata** F.Muell., *Fragm.* 8: 245 (1874)

T: Mount Olga, N.T., *E.Giles*; holo: MEL.

Lechenaultia helmsii K.Krause, *Pflanzenr.* 54: 105 (1912). T: Victoria Desert, W.A., 7 Sept. 1891, *R.Helms*; holo: K; iso: AD, MEL, NSW.

Illustration: D.A.Cooke in J.P.Jessop & H.R.Toelken, *Fl. S. Australia* 3: 1407, fig. 634B (1986).

Virgate, ascending, sparsely branched herb or subshrub to 60 cm high. Bark rough basally, coarsely striate. Leaves, sepals and ovary glabrous. Leaves crowded on short, leafy stems, scattered on flowering stems, narrow, 10.5–20.5 mm long, fleshy. Flowers in loose monochasia or dichasia. Sepals 4.5–7.5 mm long. Corolla usually 14–18 mm long; tube pubescent inside; lobes spreading, with sparse hairs often extending onto wing margins; wings on abaxial lobes triangular or deeply bilobed, usually 2–3 mm wide, on adaxial lobes (if present) rounded, 0.1–0.3 mm wide; pale blue through pale yellow to creamy white. Ovary usually 12–16.5 mm long; style 7.5–8 mm long, sparsely glandular-hairy; indusium pilose on back. Fruit usually 23–37 mm long; articles 16–20 pairs. Fig. 28C.

Widespread in the western arid and semi-arid areas of central Australia. Occurs on red sand dunes, in open *Triodia* grassland or spinifex open scrub. Flowers sporadically, perhaps only after recent rain. Map 16.

W.A.: Comet Vale, *C.A.Gardner 11117* (PERTH); near Muggun Rockhole, Cosmo–Warburton road, *A.S.George 8141* (MEL, NSW). N.T.: c. 50 km N of Angas Downs on Wallera Ranch road, *A.C.Beauglehole 20253* (AD, CANB, DNA, MEL, NSW); c. 9 km N of Olunga Well, Angas Downs, *G.Chippendale 610* (AD, DNA, MEL, NSW). S.A.: Seismic-2 road, 27 June 1967, *W.S.Reid* (AD).

The fruit is unusual in dehiscing through 4–8 valves.

13. Lechenaultia brevifolia D.Morrison, *Brunonia* 9: 18 (1987)

T: between Red Kangaroo Hill and Yilgarn, W.A., Nov. 1891, *R.Helms*; holo: AD; iso: CANB, MEL, NSW.

Illustration: D.A.Morrison, *Brunonia* 9: 19, fig. 4 (1987).

Tufted, ascending, sparsely branched subshrub to 40 cm high, often suckering. Bark rough basally, striate. Leaves, sepals and ovary glabrous. Leaves crowded on short, leafy stems, scattered on flowering stems, narrow, 2–3 mm long, fleshy. Flowers in loose monochasia. Sepals 3–4.5 mm long. Corolla 13–18 mm long; tube pubescent inside, usually white; lobes spreading, sparsely pubescent; wings on abaxial lobes triangular, 2.5–3.5 mm wide, on adaxial lobes rounded, 0.1–1 mm wide; wings and lobes dark blue. Ovary 7.5–12 mm long; style 6.5–8 mm long, sparsely glandular-hairy; indusium pilose on back. Fruit 22–29 mm long; articles 15–21 pairs.

Endemic in inland south-western W.A. between Southern Cross and Mt Ragged, occurring in sand in low scrub or heath. Flowers Oct.–Dec. Map 17.

W.A.: Scaddan, *R.J.Cranfield 1059* (MEL, PERTH); 242 mile post [c. 390 km] on the Great Eastern Hwy, *H.Demarz 4868* (PERTH); c. 136 km ENE from Esperance Bay, 2 Nov. 1891, *P.A.Gwynne* (AD, NSW); near Gnarlbine, 12 Nov. 1891, *R.Helms* (AD, MEL); c. 11 km from Lake King towards Newdegate, 6 Nov. 1968, *J.W.Wrigley* (CBG, PERTH).

14. Lechenaultia juncea E.Pritzel in F.L.E.Diels & E.Pritzel, *Bot. Jahrb. Syst.* 35: 553 (1905)

T: between Moore and Murchison Rivers, W.A., Nov. 1901, *E.Pritzel 982*; lecto: AD, *fide* D.A.Morrison, *Brunonia* 9: 20 (1987); isolecto: BM, K, L *n.v.*, NSW, W *n v*

Illustration: K.Krause, *Pflanzenr.* 54: 106, fig. 20A–B (1912).

Grass-like, erect, sparsely branched herb or subshrub, to 50 cm high. Bark rough basally, minutely striate. Leaves, sepals and ovary glabrous. Leaves crowded on short, leafy stems, scattered on flowering stems, narrow, 8.5–16 mm long, fleshy. Flowers in loose monochasia. Sepals 5–6 mm long. Corolla 14–18 mm long; tube pubescent inside; lobes spreading, sparsely pubescent; wings on abaxial lobes rounded or triangular, 2.5–3.5 mm wide, on adaxial lobes (if present) rounded, 0.3–0.9 mm wide; pale blue, edges of lobes often outlined in darker blue. Ovary 11–19 mm long; style 7.5–8.5 mm long, glandular-hairy; indusium pubescent on back. Fruit 15–25 mm long; articles 6–8 pairs.

Endemic in south-western W.A. between Three Springs and Gunyidi, occurring in sand or sandy gravel, in heath. Flowers Nov.–Dec. Map 18.

W.A.: c. 5 km S of Three Springs, *S.Carlquist 3941a* (NSW); c. 22 km SW of Three Springs, *A.S.George 3225* (PERTH); Gunyidi, *F.Lullfitz 1885* (PERTH); 1.3 km S of Carnamah–Eneabba road (Rd 1) on Rd 40, *D.A.Morrison 189* (SYD); Carnamah–Eneabba road, 2 km W of Rd 40, *P.H.Weston 158* (SYD).

Restricted geographically and currently known only from a few roadside verges and relatively undisturbed fields in farmland.

15. Lechenaultia filiformis R.Br., *Prodr.* 581 (1810)

Latouria filiformis (R.Br.) Vriese, *Natuurk. Verh. Holl. Maatsch. Wetensch. Haarlem* ser. 2, 10: 187 (1854). T: Carpentaria Island h, [N.T.], 20 Dec. 1802, *R.Brown*; lecto: BM, *fide* D.A.Morrison, *Brunonia* 9: 21 (1987); isolecto: BM, K, MEL.

Lechenaultia agrostophylla F.Muell., *Fragm.* 6: 8 (1867). T: upper Victoria R., [N.T.], *F.Mueller*; lecto: K, *fide* D.A.Morrison, *Brunonia* 9: 21 (1987); Depot Ck, Arnhem Land, [N.T.], Mar. 1856, *F.Mueller*; syn: MEL; Arnhem Land; syn: MEL.

Illustrations: F.Mueller, *Fragm.* 6: t. xlvii (as *L. agrostophylla*) & t. xlviii (1867); Britten, *Illustr. Bot. Cook's Voyage* 2: t. 171 (1901); P.W.Leenhouts, *Fl. Males.* ser. I, 5: 338, fig. 2 (1957); D.A.Morrison, *Telopea* 3: 165, fig. 3d–f (1988).

Grass-like, ascending, sparsely branched herb to 40 cm high. Leaves, sepals and ovary glabrous. Leaves scattered, narrow, 12.5–24 mm long, somewhat fleshy. Flowers in loose monochasia or dichasia. Sepals 3–8.5 mm long. Corolla usually 11–18 mm long; tube tomentose inside, often white or pale yellow; abaxial lobes spreading, adaxial ones erect and strongly coherent by interlocking hairs, otherwise sparsely tomentose; wings on abaxial lobes rounded, usually 1.3–3.5 mm wide, on adaxial lobes (if present) rounded, usually 0.5–2.5 mm wide; deep blue, through purple, pale blue to creamy white, paler on wing edges. Ovary usually 20–30 mm long; style 7.5–12 mm long, glabrous or sparsely glandular-hairy; indusium pubescent on back. Fruit usually 25–50 mm long; articles usually 8–14 pairs.

Widespread in tropical northern Australia from the Kimberley, W.A., to Cape York, Qld, extending to the south-eastern coast of New Guinea. Usually found in *Triodia* grassland in *Eucalyptus* or *Melaleuca* woodland, usually in sand or sandy loam near water-courses or other low-lying areas, but sometimes on sandstone plateaus or granitic pebble hillsides. Flowers sporadically, perhaps only after recent rain. Map 19.

W.A.: c. 9 km N of New Drysdale homestead, *J.R.Maconochie 1226* (AD, BRI, CANB, DNA, MEL, NSW, PERTH). N.T.: c. 40 km W of Borroloola, *G.Chippendale 5551* (AD, DNA, MEL, NSW, PERTH); c. 20 km WSW of Mt Evelyn, *M.Lazarides 7968* (BRI, CANB, DNA, NSW, PERTH). Qld: 27 km NW of old Corinda outstation on the Doomadgee–Borroloola road, *R.Pullen 9136* (CANB, NSW); 19.3 km E of Wollogorang on Qld–N.T. border, *D.E.Symon 5043* (AD, BRI, DNA, SYD).

Probably grows annually from a non-woody rootstock. The central adaxial sepal lobe may be 0.5–2 mm longer than the rest, a character shared with *L. ovata*. Very variable in flower size. This appears to correlate with longitude, plants with larger flowers predominating in the western part of the distribution.

16. Lechenaultia ovata D.Morrison, *Telopea* 3: 164 (1988)

T: 30 km SE of Jabiru, N.T., 27 Feb. 1973, *L.Craven 2438*; holo: CANB; iso: A, BRI, DNA, L.

Illustration: D.A.Morrison, *Telopea* 3: 165, fig. 3a–c (1988).

Grass-like, ascending, sparsely branched herb to 15 cm high. Leaves, sepals and ovary glabrous. Leaves not crowded, ovate to narrowly ovate, usually 7.5–10 mm long, somewhat fleshy. Flowers solitary, terminal. Sepals 3–5 mm long. Corolla 8–10 mm long; tube tomentose inside; abaxial lobes spreading, adaxial ones erect, strongly coherent by interlocking hairs, otherwise glabrous; wings on abaxial lobes rounded, 0.7–0.8 mm wide, on adaxial lobes rounded, 0.1 mm wide; white. Ovary 17–24 mm long; style 5–5.5 mm long, glabrous or sparsely glandular-hairy; indusium pubescent on back. Fruit 22–28 mm long; articles usually 6–9 pairs.

Known from only two collections in northern N.T., from short sedgeland in sandy depressions on sandstone plateaus. Flowers sporadically. Map 20.

N.T.: Kakadu Natl Park, *C.R.Dunlop 8568 & P.F.Munns* (AD, BRI, CANB, DNA, MEL, NSW, PERTH).

Probably growing anually from a non-woody rootstock. Shares with *L. filiformis* the longer central sepal. The ovate leaves of this species are unique within the genus.

Sect. 3. Lechenaultia

Lechenaultia R.Br. sect. **Lechenaultia**

Inflorescence a monochasial cyme or reduced to a solitary flower; bracts and bracteoles leaf-like. Corolla tube erect and cylindrical, red, orange or green, the adaxial lobes often coherent; indumentum only at base on inside of tube; petals with adaxial lobes lanceolate, as long as abaxial lobes; wings on adaxial lobes usually present. Style usually erect.

A section of 10 species endemic in south-western W.A.

17. Lechenaultia laricina Lindley, *Sketch Veg. Swan R.* xxvii (1839)

T: Swan R., W.A., 1839, *J.Drummond s.n.*; lecto: CGE, *fide* D.A.Morrison, *Brunonia* 9: 22 (1987); isolecto: K, MEL.

L. splendens Hook., *Bot. Mag.* 72: t. 4256 (1846). T: cultivated in England from seeds from Western Australia; not found; illustration, W.J.Hooker, *Bot. Mag.* 72: t. 4256 (1846); lecto, *fide* D.A.Morrison, *loc. cit.*

Illustration: B.J.Grieve & W.E.Blackall, *How to Know W. Austral. Wildfl.* 2nd edn, t. v (1975).

Diffuse, ascending, much-branched shrub to 70 cm high, often suckering. Bark rough. Leaves, sepals and ovary glabrous. Leaves narrow, 5.5–11.5 mm long, crowded, somewhat fleshy. Flowers in compact monochasia. Sepals 5–7.5 mm long. Corolla 19–23 mm long, scarlet to orange-red, usually more orange on lobes; tube densely pilose inside, usually orange; lobes equal, abaxial lobes spreading, adaxial ones erect; wings on abaxial lobes triangular, 1.5–2.5 mm wide, on adaxial lobes triangular, 0.9–1.2 mm wide. Ovary 6.5–9.5 mm long; style 13.5–19.5 mm long, sparsely glandular-hairy; indusium pilose on back. Fruit 17–29 mm long; articles 10–20 pairs. *Scarlet Leschenaultia.*

Endemic in south-western W.A. between Meckering and Clackline and S to Kukerin, occurring in sand or occasionally gravelly loam, usually in woodland. Flowers Oct.–Dec. Map 21.

W.A.: c. 8 km W of Kukerin, *J.S.Beard 2131* (PERTH); Mokine, *M.Koch 1868* (NSW, PERTH); 12 km from Clackline towards Spencers Brook, *D.A.Morrison 216* (SYD); N of Great Southern Hwy on Berry Brow Rd, *D.A.Morrison 217* (SYD); near Meenaar, *E.Pritzel 904* (AD, NSW).

Once apparently common in the area between Meenaar, Meckering and Northam, but now known from only a couple of populations in farmland.

18. **Lechenaultia superba** F.Muell., *Fragm.* 6: 10 (1867)

T: Phillips R., W.A., 26 July 1863, *G.Maxwell*; lecto: MEL, *fide* D.A.Morrison, *Brunonia* 9: 23 (1987); isolecto: K, MEL.

Illustration: R.Erickson *et al., Fl. Pl. W. Australia* 92, t. 262 (1973).

Erect, moderately branched shrub to 70 cm high. Bark rough. Leaves, sepals and ovary glabrous. Leaves crowded, narrow, 11–22.5 mm long, fleshy. Flowers solitary and terminal, sometimes in compact monochasia. Sepals 6–7.5 mm long. Corolla 17–23 mm long, yellow, often suffused with orange or red (sometimes completely red); tube densely pilose inside; lobes equal, abaxial lobes spreading, adaxial ones erect and coherent at tip; wings on abaxial lobes rounded to triangular, 1.7–2.8 mm wide, on adaxial lobes rounded, 0.9–1.3 mm wide. Ovary 8–9.5 mm long; style 18–20.5 mm long, sparsely glandular-hairy; indusium pilose on back. Fruit 13–20 mm long; articles 5–8 pairs. *Barrens Leschenaultia.*

Endemic in south-western W.A. at the eastern end of the Barrens, occurring in quartzite soil on rocky hillsides, in open scrub. Flowers Sept.–Oct. Map 22.

W.A.: East Mt Barren, *H.Eichler 21125* (AD); southern side of Mt Barren, *R.Filson 9188* (MEL); Middle Mt Barren, Sept. 1926, *C.A.Gardner* (PERTH); Whoogarup Ra., *A.S.George 7198* (PERTH); NE side of East Mt Barren, *D.A.Morrison 229* (SYD).

Restricted geographically, with the entire known distribution within Fitzgerald River Natl Park.

19. **Lechenaultia hirsuta** F.Muell., *Fragm.* 6: 9 (1867)

T: south-western Australia, 1851–52, *J.Drummond 6: 145*; lecto: MEL, *fide* D.A.Morrison, *Brunonia* 9: 23 (1987); isolecto: BM, K, LD, MEL, NSW.

Illustration: B.J.Grieve & W.E.Blackall, *How to Know W. Austral. Wildfl.* 2nd edn, pt IV, t. vi (1975).

Straggling, procumbent, sparsely branched herb or subshrub to 50 cm high. Stems striate, sparsely hispid. Bark rough basally. Leaves scattered, narrow, usually 17.5–27 mm long, somewhat fleshy, sparsely hispid. Flowers in monochasia. Sepals 7–8.5 mm long, densely glandular-hispid. Corolla 29–36 mm long, glandular-tomentose outside, scarlet, sometimes orange-red on lobes; tube densely pilose inside, sometimes orange-red basally; lobes almost equal, abaxial ones spreading, adaxial ones erect and adjacent but not coherent; wings on abaxial lobes triangular to rounded, 3–5 mm wide, on adaxial lobes rounded, 0.9–1.5 mm wide. Ovary usually 17.5–26 mm long, glandular-hispid; style 21.5–29.5 mm long, glandular-hairy; indusium pilose on back. Fruit 35–42 mm long, glandular-hispid; articles 7–9 pairs. *Hairy Leschenaultia.* Fig. 25J.

Endemic in south-western W.A. on near-coastal sandplains between Kalbarri and Badgingarra, occurring in sand, often over laterite, commonly in low open heath. Flowers Sept.–Dec. Map 23.

W.A.: 42 km N of Murchison R. crossing, *W.R.Barker 2185* (AD, BRI); c. 32 km W of Nerren Nerren, *J.S.Beard 6749* (NSW, PERTH); 40 km ENE of Three Springs, *M.G.Corrick 8296* (MEL, PERTH); c. 13 km S of Eneabba, *R.W.Johnson 3326* (BRI); c. 13 km from Eneabba towards Jurien Bay, 23 Sept. 1968, *M.E.Phillips* (CBG, SYD).

The hairy stems and hairy external floral parts are unique within the genus.

20. **Lechenaultia longiloba** F.Muell., *Fragm.* 6: 10 (1867)

T: W.A., *J.Drummond s.n.* [6th coll.]; lecto: MEL, *fide* D.A.Morrison, *Brunonia* 9: 23 (1987); W.A., *J.Drummond s.n.* [6th coll.]; syn: K, MEL.

Illustration: B.J.Grieve & W.E.Blackall, *How to Know W. Austral. Wildfl.* 2nd edn, pt IV, t. v (1975).

Straggling, procumbent, sparsely branched herb or subshrub to 30 cm high, often suckering. Bark rough basally, striate. Leaves, sepals and ovary glabrous. Leaves scattered or moderately crowded, narrow, 9.5–26 mm long, somewhat fleshy. Flowers in compact monochasia. Sepals 16.5–21 mm long. Corolla 25–35 mm long, pale yellow or green, usually suffused with deep pink or red on wings; tube densely pilose inside; lobes equal, abaxial lobes spreading, adaxial ones erect; wings almost equal, triangular, 1.7–4.2 mm wide. Ovary 15.5–20.5 mm long; style 21–27.5 mm long, sparsely glandular-hairy; indusium pilose on back. Fruit 20–26 mm long; articles 13–16 pairs. *Irwin Leschenaultia* Fig. 28E.

Endemic in south-western W.A. between Mullewa and Dongara, growing in sand, in open heath. Flowers July–Oct. Map 24.

W.A.: Casuarinas Rd, off Burma Rd, *A.M.Ashby 2292* (AD, PERTH); c. 9 km E of Ellendale on Burma Rd, *H.W.Bennetts 12117* (PERTH); 39 km NW of Strawberry on Burma Road, *M.G.Corrick 8286* (MEL, PERTH); 6 km from Strawberry on Walkaway road, *G.Perry 573* (PERTH); c. 22 km from Walkaway towards Strawberry, 15 Sept. 1968, *M.E.Phillips* (CBG, NSW).

Restricted geographically, and now known from only a few populations along roadsides in farmland.

21. **Lechenaultia macrantha** K.Krause, *Pflanzenr.* 54: 100 (1912)

T: Jibberding, W.A., Sept. 1905, *M.Koch 1327*; holo: K; iso: AD, MEL, NSW, PERTH.

Illustration: B.J.Grieve & W.E.Blackall, *How to Know W. Austral. Wildfl.* 2nd edn, pt IV, t. v (1975).

Wreath-like, procumbent, sparsely branched herb or subshrub to 15 cm high. Bark rough basally, striate. Leaves, sepals and ovary glabrous. Leaves crowded, narrow, usually 25–45 mm long, somewhat fleshy. Flowers in compact monochasia. Sepals 20.5–33 mm long. Corolla 25–35 mm long, yellow, often suffused with deep pink or red on wings; tube densely pilose inside; lobes equal, abaxial lobes spreading, adaxial ones erect; wings almost equal, triangular, 4.5–8.5 mm wide. Ovary 17.5–22.5 mm long; style 18.5–29 mm long, sparsely glandular-hairy; indusium pilose on back. Fruit 22–33 mm long; articles 15–20 pairs. *Wreath Leschenaultia.* Fig. 7.

Occurs in inland south-western W.A. between Tallering Peak and Coorow, and near Nerren Nerren and Boolardy pastoral stations. Found in sandy or gravelly soil, in open areas. Flowers Aug.–Oct. Map 25.

W.A.: 390–394 mile peg [c. 625 km] on Carnarvon–Geraldton section of North West Coastal Hwy, *A.C.Burns 1045* (PERTH); c. 0.5 km N of Tardun on the Tardun–Pindar road, *S.Carlquist 2945* (AD, NSW); Boolardy Stn, *A.B.Cashmore 127* (PERTH); c. 24 km E of Mullewa on Mt Magnet road, *R.Filson 8701* (MEL); c. 16 km from Pindar towards Tardun, 1 Oct. 1962, *M.E.Phillips* (CBG, SYD).

The wreath-like flowering habit, growing annually from a persistent rootstock, makes this species unique in the genus.

22. **Lechenaultia acutiloba** Benth., *Fl. Austral.* 4: 41 (1868)

T: Young R., W.A., *G.Maxwell*; holo: K; iso: MEL.

Illustration: K.Krause, *Pflanzenr.* 54: 101, fig. 18g (1912).

Hemispherical, ascending, much-branched shrub to 100 cm high, often suckering. Bark rough. Leaves, sepals and ovary glabrous. Leaves crowded, narrow, 3–5.5 mm long, fleshy. Flowers solitary, terminal. Sepals 3–5 mm long. Corolla 21–25 mm long, greenish yellow, blue towards tips of lobes; tube densely pilose inside; lobes equal, abaxial lobes reflexed, adaxial ones erect and curved with tips crossing behind style; wings (if present) flattened, 0.1–0.2 mm wide. Ovary 5–6.5 mm long; style 17.5–21.5 mm long, sparsely hairy; indusium pubescent on back. Fruit 12–16 mm long; articles 5–7 pairs. Fig. 4.

Endemic in south-western W.A. between Ravensthorpe and Ongerup, occurring in sand or sandy gravel, usually near river banks or swamps. Flowers Sept.–Dec. Map 26.

W.A.: Fitzgerald R. crossing, Ongerup–Ravensthorpe road, *R.C.Carolin 3563* (SYD); 57 km E of Jerramungup, *H.Demarz 9460* (PERTH); Hamersley R., 12 Nov. 1935, *C.A.Gardner* (PERTH); rabbit proof fence, N of Jerramungup, *A.S.George 7001* (PERTH); E bank of West R. at crossing of Ravensthorpe–Jerramungup road, *D.A.Morrison 230* (SYD).

Restricted geographically and habitat specific.

23. **Lechenaultia tubiflora** R.Br., *Prodr.* 581 (1810)

T: King George Sound, [W.A.], Dec. 1801, *R.Brown*; lecto: BM, *fide* D.A.Morrison, *Brunonia* 9: 24 (1987); isolecto: BM, K, MEL.

L. pinastroides Lehm., *Pl. Preiss.* 2: 244 (1848). T: between Mt Manypeak and Cape Riche, W.A., 23 Nov. 1840, *L.Preiss 430*; lecto: S, *fide* D.A.Morrison, *Brunonia* 9: 24 (1987); isolecto: L *n.v.*, photo seen SYD, MEL, W *n.v.*

L. tubiflora var. *purpurea* E.Pritzel, in F.L.E.Diels & E.Pritzel, *Bot. Jahrb. Syst.* 35: 552 (1905). T: near Gibson Soak, *L.Diels 5973*; *n.v.*

Ericopsis formosus C.Gardner, *J. & Proc. Roy. Soc. Western Australia* 9: 42 (1923). T: Hotham R., Popanyinning, W.A., 5 Dec. 1922, *C.A.Gardner 1880*; holo: PERTH; iso: PERTH.

Illustrations: K.Krause, *Pflanzenr.* 54: 103, fig. 19 (1912); B.J.Grieve & W.E.Blackall, *How to Know W. Austral. Wildfl.* 2nd edn, t. iv (1975).

Hemispherical subshrub or somewhat erect perennial to 70 cm high and 60 cm diam. Stems procumbent or ascending, much-branched, often suckering. Bark rough. Leaves, sepals and ovary glabrous. Leaves crowded, narrow, 7.5–11.5 mm long, rigid. Flowers solitary, terminal. Sepals 4–5.5 mm long. Corolla 13–17 mm long, bright red through pale yellow, pale greenish cream to creamy white, often suffused with orange, red or blue on wings and lobes; tube densely pilose inside; lobes equal, usually erect; wings on each lobe connivent along adjacent margins, triangular, 0.3–0.7 mm wide. Ovary 2.5–6 mm long; style 11–13.5 mm long, glabrous; indusium pilose on back. Fruit 5–7 mm long; articles c. 6 pairs. $n = 9$, W.J.Peacock, *Proc. Linn. Soc. New South Wales* 88: 14 (1963). *Heath Leschenaultia.* Fig. 3.

Widespread in inland south-western W.A. between Coorow and Albany, and along the coast to Israelite Bay. Occurs in sand, often over limestone or granite, usually in heath or woodland. Flowers spring and summer. Map 27.

W.A.: N side of Tambellup, 9 Oct. 1960, *B.G.Briggs* (NSW, SYD); c. 13 km S of Pingelly, *S.Carlquist 3959* (AD, NW); Munglinup, *R.J.Cranfield 1081* (MEL, PERTH); Fitzgerald R. Natl Park, *H.Eichler 21088* (AD, CANB); c. 5 km NW of Cape Riche, *K.Newbey 2397* (PERTH).

Varies markedly in stature and flower colour.

24. **Lechenaultia linarioides** DC., *Prodr.* 7(2): 519 (1839)

T: Swan R., W.A., 1839, *J.Drummond s.n.*; holo: G-DC *n.v.*, microfiche seen; iso: K.

Scaevola grandiflora Benth. in S.F.L.Endlicher *et. al.*, *Enum. Pl.* 70 (1837); *Lechenaultia grandiflora* (Benth.) Druce, *Bot. Exch. Club Brit. Isles Rep.* 1916, Suppl. 2: 632 (1917). T: Swan R., [W.A.], *C.A.A. von Hügel*; holo: W *n.v.*, photo seen SYD.

L. arcuata Vriese in J.G.C.Lehmann., *Pl. Preiss.* 1: 416 (1845). T: near Fremantle, W.A., 17 Dec. 1838, *L.Preiss 1465*; lecto: LD, *fide* D.A.Morrison, *Brunonia* 9: 25 (1987); isolecto: L *n.v.*, photo seen SYD, MEL.

Illustrations: K.Krause, *Pflanzenr.* 54: 101, fig. 18e–f (1912); B.J.Grieve & W.E.Blackall, *How to Know W. Austral. Wildfl.* 2nd edn, pt IV, t. v (1975); J.R.Wheeler, *Fl. Perth Region* 2: 633, fig. 235 (1987).

Tangled, much-branched subshrub to 100 cm high, suckering. Bark rough basally, striate. Leaves, sepals and ovary glabrous. Leaves narrow, usually 8–12.5 mm long, scattered on stems and crowded on branchlets, somewhat fleshy. Flowers solitary, terminal, or in compact monochasia. Sepals usually 5–6.5 mm long. Corolla usually 16–22 mm long; tube densely hirsute inside; lobes almost equal, abaxial spreading and reflexed, adaxial erect and enclosing indusium; wings on abaxial lobes rounded, usually 3.5–4.5 mm wide, on adaxial lobes adjacent (not connivent), rounded, usually 1.8–2.5 mm wide; abaxial lobes yellow, adaxial lobes and wings deep pink to purplish red; wings usually suffused with deep pink or red. Ovary usually 9.5–11.5 mm long; style 12.5–15 mm long, sparsely glandular-hairy; indusium pilose on back. Fruit usually 19–33 mm long, woody; articles usually 7–11 pairs. $n = 9$, W.J.Peacock. *Proc. Linn. Soc. New South Wales* 88: 14 (1963). *Yellow Leschenaultia.*

Endemic in south-western W.A. on coastal sandplains between Kalbarri and Perth, and on the Shark Bay islands and coastline. Occurs in sand over limestone, in heath, scrub, or occasionally woodland. Flowers sporadically. Map 28.

W.A.: c. 85 km N of the Murchison R. bridge, *N.T.Burbidge 6449* (AD, BRI, CANB, MEL, PERTH); c. 16 km from Badgingarra towards Moora, 25 Sept. 1968, *E.M.Canning* (CBG, SYD); Dirk Hartog Is., *A.S.George 11554* (PERTH); c. 11 km E of Kojarena, *R.Melville 4198 & J.H.Calaby* (AD, MEL, NSW, PERTH); c. 8 km NNE of Kwinana, *H.Salasoo 3987* (NSW).

In this species, the stems are ascending but curved downwards at the tips, and the corolla tube is gibbous on the abaxial side.

25. **Lechenaultia chlorantha** F.Muell., *Fragm.* 2: 20 (1860)

L. formosa var. *chlorantha* (F.Muell.) K.Krause, *Pflanzenr.* 54: 108 (1912). T: Murchison R., W.A., ? *A.Oldfield*; holo: MEL; iso: K.

Diffuse, ascending, much-branched subshrub or shrub to 30 cm high, often suckering. Bark rough. Leaves, sepals and ovary glabrous. Leaves crowded, narrow, 6.5–13.5 mm long, fleshy. Flowers solitary, terminal. Sepals 7.5–9 mm long. Corolla 21–25 mm long, pale bluish green; tube densely pilose inside, sometimes dull orange apically; lobes almost equal, abaxial spreading and reflexed, adaxial erect and connivent along adjacent margins and enclosing indusium; wings on abaxial lobes rounded to triangular, 4.5–6 mm wide, on adaxial lobes 0.1–1.5 mm wide, usually absent on cohering margin. Ovary 7–10 mm long; style 16–17 mm long, glabrous; indusium pilose on back. Fruit 18–23 mm long; articles 8–10 pairs.

Endemic in south-western W.A. near the mouth of the Murchison R., occurring in rocky sandstone gullies and on ledges. Flowers Aug.–Sept. Map 29.

W.A.: Kalbarri, *D. & B.Bellairs 1440* (PERTH); Murchison R. area, *A.C.Burns 1011* (PERTH); between Northampton and Shark Bay, 1884, *S.Carey* (MEL); Red Bluff, Murchison R., 10 Aug. 1966, *R.Erickson* (MEL, PERTH); near Murchison House, *B.G.Muir 1859* (AD).

In this species, the corolla tube is gibbous on the abaxial side and the ovary is often nodding. Restricted geographically and not locally common.

26. Lechenaultia formosa R.Br., *Prodr.* 581 (1810).

T: Bay I [Lucky Bay], South Coast, [W.A.], Jan. 1802, *R.Brown*; lecto: BM, *fide* D.A.Morrison, *Brunonia* 9: 26 (1987); isolecto: BM, K, MEL.

L. baxteri Don in J.C.Loudon, *Hort. Brit.* 79 (1830), *nom. superfl.*

L. oblata Sweet, *Fl. Australas.* t. 46 (1828); *L. multiflora* G.Lodd. ex DC., *Prodr.* 7: 519 (1839); *L. formosa* var. *oblata* (Sweet) E.Pritzel in F.L.E.Diels & E.Pritzel, *Bot. Jahrb. Syst.* 35: 552 (1905). T: cultivated by Robert Barclay from seeds from Western Australia, not found; illustration in R.Sweet, *Fl. Australas.* t. 46 (1828); lecto: *fide* D.A.Morrison, *Brunonia* 9: 26 (1987).

Illustration: B.J.Grieve & W.E.Blackall, *How to Know W. Austral. Wildfl.* 2nd edn, pt IV, t. v (1975).

Prostrate or erect subshrub or shrub to 40 cm high and 40 cm diam. Stems procumbent or ascending, much-branched, often suckering. Bark rough. Leaves, sepals and ovary glabrous. Leaves crowded, narrow, usually 4.5–8.5 mm long, fleshy. Flowers solitary, terminal. Sepals 5–6 mm long. Corolla usually 15–21 mm long; tube densely pilose inside; lobes almost equal, abaxial ones spreading and reflexed, adaxial ones erect and connivent along adjacent margins and enclosing indusium; wings on abaxial lobes triangular, 2.2–4 mm wide, on adaxial lobes rounded, 0.5–1.5 mm wide, usually absent on cohering margin; scarlet through orange-red to pale orange, often redder on wings. Ovary usually 7.5–11.5 mm long; style 11–17 mm long, sparsely glandular-hairy; indusium pilose on back Fruit 10–29 mm long; articles 10–22 pairs. n = 9, 18, W.J.Peacock, *Proc. Linn. Soc. New South Wales* 88: 13 (1963). *Red Leschenaultia.* Fig. 5

Widespread in inland south-western W.A. between Coorow and Albany, along the coast to Israelite Bay, and at Point Culver, Eyre and Eucla. Occurs in sand over laterite and granite, as well as in clay and gravelly clay, in heath, scrub, mallee and woodland. Flowers sporadically but mainly winter-spring. Map 30.

W.A.: c. 95 km W of Frank Hann Natl Park, *R.J.Cranfield 815* (MEL, PERTH); 6.5 km S of Narrogin, *H.Eichler 1583* (AD, CANB, NSW, PERTH); c. 41 km W of Coorow, *C.H.Gittins 1684* (BRI, NSW, PERTH); c. 6 km E of Kalgan R. on Albany–Mt Manypeaks road, *R.Melville 4411 & R.D.Royce* (AD, BRI, MEL, NSW, PERTH); next to microwave repeater station on road to Eyre Bird Observatory, *D.A.Morrison 168* (SYD).

In this species the corolla tube is gibbous on the abaxial side, and the ovary is often nodding. It varies greatly in flower and leaf size as well as in flower colour and habit.

3. DAMPIERA

M.T.M.Rajput and R.C.Carolin

Dampiera R.Br., *Prodr.* 587 (1810); named for William Dampier (1652–1715), English navigator, pirate and naturalist who collected plants in north-western Australia in 1699.

Type: *D. incana* R.Br.

Linschotenia Vriese in T.L.Mitchell, *J. Exped. Trop. Australia* 345 (1848). T: *L. discolor* Vriese

Subshrubs, many-stemmed perennials, or rosette herbs, sometimes suckering, glabrous or dendritic hairy. Flowers in racemes, thyrses, sometimes condensed into heads, or irregular cymo-panicles; bracteolate. Sepals obsolete or 5, free, adnate to ovary to summit. Corolla bilabiate, often with calli in throat, auriculate, usually blue with yellow throat, rarely red, pink, yellow or whitish; lobes winged. Stamens epigynous; anthers connate. Ovary sometimes gibbous, 1–2-locular, usually hairy as corolla; indusium horizontal, ±globular, sometimes bilabiate; ovule one per locule, erect to horseshoe-shaped. Fruit a nut; sepals, style, stamens and base of corolla often persistent. Seeds ovoid or curved, without wing or rim. Embryo terete. $x = 9$, W.J.Peacock, *Proc. Linn. Soc. New South Wales* 88: 8–27(1963).

A genus of 66 species all endemic in Australia, c. 38 of which are endemic in south-western W.A.

The type of hairs on the outside of the corolla is particularly diagnostic. In this work they are designated as types i–v, illustrated in Fig 23H–M. The colour and appearance of these hairs to the naked eye are also quite characteristic of species.

G.Bentham, *Dampiera*, *Fl. Austral.* 4: 106–120 (1868); K.Krause, *Dampiera*, *Pflanzenr.* 54: 171–200 (1912); M.T.M.Rajput & R.C.Carolin, Stem structure and vascularisation in *Dampiera* (Goodeniaceae), *Proc. Linn. Soc. New South Wales* 107: 479–485 (1984); M.T.M.Rajput & R.C.Carolin, The genus *Dampiera* (Goodeniaceae): systematic arrangement, nomenclatural notes and new taxa, *Telopea* 3: 183–216 (1988).

KEY TO SECTIONS, SUBSECTIONS AND SERIES

1 Flowers arranged in scapose thyrses, racemes or heads; phyllotaxis 2/5; stems terete; subshrubs ..... sect. 1. **LINSCHOTENIA**

1: Flowers arranged in cymo-panicles or panicles; stems ribbed angled or grooved; plants many-stemmed from a rootstock ..... sect. **2. DAMPIERA**

2 Stems 5-ribbed at maturity ..... subsect. 1. **DAMPIERA**

2: Stems 3-angular compressed or flat except at base ..... subsect. 2. **ANGULARES**

3 Ovary not gibbous but sometimes oblique ..... ser. 1 **ANGULARES**

3: Ovary gibbous ..... ser. 2 **CAMPTOSPORA**

KEY TO ARTIFICIAL GROUPS

1 Leaves mostly basal ..... **GROUP 1**

1: Leaves mostly cauline

2 Ovary and fruit gibbous ..... **GROUP 2**

2: Ovary and fruit not gibbous

3 Young stems compressed, flat, triangular or with a narrow groove on each side of stem immediately below uppermost node ..... **GROUP 3**

3: Young stems terete or with five narrow grooves

4 Flowers in terminal racemes spikes or rarely thyrses; leaves subtending flowers much smaller than other cauline leaves ..... **GROUP 4**

4: Flowers in leafy inflorescences, usually with several flowers on each axillary branch

5 Leaves revolute or strongly recurved ..... **GROUP 5**

5: Leaves flat, concave or slightly recurved

6 Mature leaves glabrous ..... **GROUP 6**

6: Mature leaves hairy, at least below

7 Mature leaves hairy below, glabrous or much less hairy above **GROUP 7**

7: Mature leaves hairy on both surfaces **GROUP 8**

GROUP 1

(Leaves mostly basal)

1 Leaves glabrous or glabrescent below

2 Bracts obtuse; leaves mostly entire or crenulate **13. D. wellsiana**

2: Bracts acute; leaves dentate **14. D. dentata**

1: Leaves tomentose at least below

3 Corolla hairs yellowish **10. D. ramosa**

3: Corolla hairs white to grey

4 Sepals 2–4 mm long, but hidden by long hairs; leaves spathulate, sessile **11. D. plumosa**

4: Sepals 0.5 mm long with tuft of long hairs at tip, or entirely replaced by tuft of long hairs; leaves obovate to elliptic, petiolate **12. D. eriocephala**

GROUP 2

(Ovary gibbous)

1 Upper stems triangular or with 3 wings or with two narrow grooves below nodes

2 Stems with 2 narrow grooves below nodes; upper cauline leaves similar to others **62. D. carinata**

2: Upper stems triangular or with 3 wings

3 Leaves entire; branches with obtuse angles **59. D. angulata**

3: Leaves dentate; branches with very acute angles or wings **61. D. coronata**

1: Upper stems compressed or flattened with 2 wings or angles

4 Wing of adaxial petal obsolete above auricle **60. D. heteroptera**

4: Wing of adaxial petal present above auricle although sometimes smaller than others

5 Leaves triangular-ovate **64. D. deltoidea**

5: Leaves not triangular-ovate

6 Stems compressed, biconvex in T.S. **66. D. sacculata**

6: Stems flattened, usually with a midrib

7 Corolla hairs not appressed **63. D. alata**

7: Corolla hairs appressed **65. D. lindleyi**

GROUP 3

(Stems triangular, compressed, flattened or with 2 narrow grooves)

1 Ovary 2-locular; W.A. only

2 Stem with 3 wings each 2.5–5 mm wide **44. D. latealata**

2: Stem angled or with wings to 2 mm wide

3 Weak ascending herb; leaves linear to lanceolate, 2–6 mm wide **46. D. trigona**

3: Stiff, robust plant; leaves ovate to elliptic, decurrent into very acute angles, 5–23 mm wide **45. D. decurrens**

1: Ovary 1-locular; widespread

4 Ovary with an oblique top **58. D. obliqua**

4: Ovary with a flat top

5 Stems with 2 narrow grooves immediately below uppermost nodes

6 Bracteoles numerous below flower; flower sessile **51. D. parvifolia**

6: Bracteoles 1 or 2 below flower; flower usually pedicellate **52. D. sericantha**

5: Stems with distinct angles and no narrow grooves

7 Ovary glabrous or sparsely hairy **48. D. leptoclada**

7: Ovary densely hairy

8 Corolla hairs yellowish grey, yellowish brown or golden

9 Corolla hairs loose, yellowish grey; southern N.S.W. and Vic. **55. D. fusca**

9: Corolla hairs appressed, yellowish brown or golden; widespread

10 Leaves with 3 lobes or teeth towards top; W.A. **50. D. triloba**

10: Leaves irregularly toothed or entire; east coast of Qld, N.S.W., Vic. and Tas. **54. D. stricta**

8: Corolla hairs grey to black

11 Corolla hairs appressed

12 Upper leaves in pseudowhorls; W.A. **53. D. fasciculata**

12: Upper leaves not in pseudo-whorls; widespread

13 Stem without rib between angles; W.A. **57. D. glabrescens**

13: Stem with rib between each pair of angles; Vic. **49. D. galbraithiana**

11: Corolla hairs loose

14 Sepals distinct, more than 1 mm long; N.S.W. and Qld **56. D. sylvestris**

14: Sepals obsolete; W.A. **47. D. loranthifolia**

GROUP 4

(Inflorescence with reduced leaves; flowers in racemes, spikes or thyrses)

1 Mature leaves densely hairy on both surfaces

2 Corolla to 7 mm long **7. D. cinerea**

2: Corolla more than 7 mm long

3 Leaves petiolate, closely tomentose **2. D. atriplicina**

3: Leaves sessile, loosely tomentose often in patches **5. D. stenostachya**

1: Mature leaves glabrous, or with few scattered hairs above, or glabrous above and hairy below

4 Corolla to 7 mm long

5 Leaves oblong to lanceolate **8. D. conospermoides**

5: Leaves ovate to orbicular **9. D. candicans**

4: Corolla more than 7 mm long

6 Leaves petiolate **1. D. discolor**

6: Leaves sessile

7 Leaves linear to narrowly lanceolate, mostly to 2 mm wide, glabrescent on both surfaces **6. D. teres**

7: Leaves mostly more than 2 mm wide, hairy below

8 Hairs on young stems brownish; wing of adaxial petal obsolete above auricle **3. D. krauseana**

8: Hairs on young stems white to grey or yellowish; wing of adaxial petal present above auricle but often smaller **4. D. spicigera**

GROUP 5

(Leaves closely revolute)

1 Mature stems pubescent

2 Corolla hairs yellowish or brownish grey; S.A., Vic. **26. D. marifolia**

2: Corolla hairs silvery or grey

3 Corolla 6–7 mm long, with short white silky hairs outside; W.A. **31. D. stenophylla**

3: Corolla 8–12 mm long, with pale grey or silvery hairs outside; S.A., N.S.W.

4 Inflorescence branches with 1–3 flowers; N.S.W. **36. D. scottiana**

4: Inflorescence branches with 3–9 flowers; S.A. **34. D. lanceolata**

1: Stems glabrous except when very young, or sometimes with hairs in grooves

5 Sepals obsolete; Great Victoria Desert, W.A. **35. D. eriantha**

5: Sepals distinct (although sometimes hidden by hairs), mostly more than 0.5 mm long; widespread

6 Leaves not clustered in axils **34. D. lanceolata**

6: Two or more smaller leaves clustered in axils of main stem leaves

7 Corolla hairs yellowish or brownish grey **38. D. dysantha**

7: Corolla hairs pale grey to silvery

8 Leaves linear to narrowly oblong; stems hairy when young; inland W.A., S.A., Vic.

9 Leaves very tightly revolute; S.A., Vic. **37. D. rosmarinifolia**

9: Leaves not very tightly revolute; inland W.A., S.A. and N.S.W. **34. D. lanceolata**

8: Leaves oblong to elliptic; stems glabrous when young; south-western W.A.

10 Bracteoles shorter than connate part of abaxial corolla lobes; corolla hairs grey **39. D. lavandulacea**

10: Bracteoles equalling connate part of abaxial corolla lobes; corolla hairs silvery white **32. D. fitzgeraldensis**

GROUP 6

(Leaves flat or concave, glabrous)

1 Corolla glabrous outside **24. D. diversifolia**

1: Corolla hairy outside

2 Bracteole below flower at least half as long as flower; W.A.

3 Corolla hairs appressed **22. D. scaevolina**

3: Corolla hairs loose

4 Leaves elliptic to cuneate, to 16 mm wide **42. D. linearis**

4: Leaves linear to narrowly oblong, to 4 mm wide **43. D. pedunculata**

2: Bracteole below flower much less than half as long as flower; widespread

5 Corolla hairs appressed

6 Corolla less than 8 mm long; leaves less than 5 mm wide; W.A. **23. D. tenuicaulis**

6: Corolla more than 8 mm long; leaves mostly more than 5 mm wide; Vic. **49. D. galbraithiana**

5: Corolla hairs not appressed

7 Corolla 8–10 mm long, the hairs type i and ii; widespread

8 Leaves flat; Qld, N.S.W. **33. D. adpressa**

8: Leaves slightly recurved; W.A. **35. D. eriantha**

7: Corolla 13–20 mm long, the hairs mostly type i; W.A.

9 Corolla hairs slate grey, closely tomentose **41. D. juncea**

9: Corolla hairs silvery grey, loosely tomentose **40. D. oligophylla**

GROUP 7

(Leaves flat or concave, hairy below)

1 Leaves with petiole more than 2 mm long **28. D. hederacea**

1: Leaves sessile or with petiole less than 2 mm long

2 Fruit bent or curved; W.A.

3 Corolla hairs whitish, short **17. D. altissima**

3: Corolla hairs grey, long

4 Corolla tomentose, with type i hairs **19. D. tephrea**

4: Corolla silky, with type ii hairs **18. D. salahae**

2: Fruit straight; widespread

5 Inflorescence branches with brownish grey or yellowish hairs

6 Corolla hairs grey or white

7 Corolla hairs white **31. D. stenophylla**

7: Corolla hairs grey

8 Corolla hairs not paler towards tips; eastern Australia **15. D. purpurea**

8: Corolla hairs paler towards tips; W.A. **27. D. haematotricha**

6: Corolla hairs brownish, greyish brown, or yellowish

9 Corolla yellow; W.A. **30. D. luteiflora**

9: Corolla blue

10 Lower leaves to 11 mm long, mostly less than 6 mm wide; S.A., south-eastern N.S.W., Vic. **26. D. marifolia**

10: Lower leaves more than 11 mm long, more than 6 mm wide; Qld **16. D. ferruginea**

5: Inflorescence branches with grey or white hairs

11 Leaves mostly recurved or concave

12 Stems conspicuously ribbed

13 Stems tomentose; corolla hairs silvery; petal wings ±equal; coastal N.S.W. **36. D. scottiana**

13: Stems glabrescent; corolla hairs grey; petal wing above auricle reduced; south-western W.A. **40. D. oligophylla**

12: Stems not conspicuously ribbed; central arid areas of W.A., S.A. and N.S.W. **34. D. lanceolata**

11: Leaves flat

14 Leaves elliptic to obovate; corolla closely tomentose, the hairs not paler towards tip **25. D. roycei**

14: Leaves obovate to orbicular; corolla loosely tomentose, the longer hairs at ±90°, paler towards tip **27. D. haematotricha**

GROUP 8

(Leaves flat or concave, hairy on both surfaces)

1 Bracteole beneath flower at least as long as corolla tube **27. D. haematotricha**

1: Bracteole beneath flower much shorter than corolla tube

2 Corolla hairs appressed **21. D. orchardii**

2: Corolla hairs not appressed

3 Corolla yellow **30. D. luteiflora**

3: Corolla blue

4 Lower leaf surface clearly visible beneath hairs **25. D. roycei**

4: Lower leaf surface completely hidden beneath dense hairs

5 Corolla hairs short, branched along length; fruit and ovary bent or oblique to stem **20. D. incana**

5: Corolla hairs long, silky, branched mostly near base; fruit and ovary straight, not oblique to stem **29. D. tomentosa**

Sect. 1. Linschotenia

Dampiera sect. **Linschotenia** (Vriese) Benth., *Fl. Austral.* 4: 106, 108 (1868); *Linschotenia* Vriese in T.L.Mitchell, *J. Exped. Trop. Australia* 345 (1848).

Type: *D. discolor* (Vriese) K.Krause

Dampiera sect. *Cephalantha* Benth., *Fl. Austral.* 4: 108, 120 (1868). T: *D. eriocephala* Vriese

Undershrubs. Stem terete, sometimes ribbed. Phyllotaxis 2/5. Flowers in a raceme, thyrse or head; inflorescence usually branched at base. Ovary not gibbous, 1-locular. Ovule erect, usually oblong or linear, basally or laterally attached.

A section of 14 species occurring in western and northern Australia.

1. **Dampiera discolor** (Vriese) K.Krause, *Pflanzenr.* 54: 180 (1912)

Linschotenia discolor Vriese in T.L.Mitchell, *J. Exped. Trop. Australia* 346 (1848); *D. linschotenii* F.Muell., *Fragm.* 6: 28 (1867), *nom. illeg.* T: Mt Faraday, Qld, *T.L.Mitchell 314*; holo: BM; iso: MEL.

Illustration: W.H.de Vriese, *Natuurk. Verh. Holl. Maatsch. Wetensch. Haarlem* ser. 2, 10: t. 22 (1854).

Erect, branched subshrub to 70 cm high, closely pale grey-tomentose. Leaves cauline, ovate to elliptic, mostly entire, thick, glabrous above when mature, tomentose below; lamina 24–64 mm long, 9–32 mm wide; petiole 7–16 mm long. Flowers in leafless racemes 13–29 cm long; bracteoles linear, 1–1.5 mm long or absent; pedicel 1–2 mm long. Sepals linear-oblong, to 1 mm long, hidden beneath hairs. Corolla 8–10 mm long; hairs outside ±appressed, mostly type i, pale grey; wings to 1.5 mm wide, very short or obsolete above auricle; calli absent. Ovary 2.5–3 mm long. Fruit not seen. $n = 9$, W.J.Peacock, *Proc. Linn. Soc. New South Wales* 88: 11 (1963). Figs 17A, 29F.

Occurs in central eastern Qld from Blackdown Tableland to Jericho area, mostly on sandstone. Flowers summer. Map 31.

Qld: Blackdown Tableland, *C.H.Gittins S/62* (NSW); c. 25 km on the Jericho–Alice road, *G.W.Althofer* (NSW); Isla Gorge, *S.L.Everist 8042* (BRI); Eidsvold, May 1914, *T.L.Bancroft* (BRI, NSW); Rockland Spring, c. 48 km ENE of Springsure, *M.Lazarides & R.Story 122* (AD, CANB).

Bracts are linear, and 4–8 mm long. Chromosome vouchers are *W.J.Peacock 6111.361*, and *6111.36.1* (SYD).

2. **Dampiera atriplicina** C.Gardner ex Rajput & Carolin, *Telopea* 3: 211 (1988)

T: 4 miles [c. 6 km] N of Well no. 12, Canning Stock Route, W.A., 5 Sept. 1942, *H.M.Wilson 6*; holo: PERTH.

Erect, branched subshrub to 50 cm tall, greyish to brownish tomentose. Leaves cauline, ovate to elliptic, entire or dentate, tomentose on both surfaces; lamina 17–65 mm long, 12–28 mm wide; petiole to 11 mm long. Flowers in leafless racemes to 9 cm long; bracteoles 3 or 4 together, linear, 5–8 mm long; pedicel 1–1.5 mm long. Sepals linear, 5–7 mm long, with a tuft of long silky hairs. Corolla 8–10 mm long; hairs outside type ii, greyish to brownish; wings c. 0.5 mm wide, obsolete above auricle; calli obsolete. Ovary 2–3 mm long. Fruit not seen. Fig. 17B.

Occurs on sand dunes in the Gibson and Great Sandy Deserts, W.A. Flowering not known. Map 32.

Figure 29. A–C, *Anthotium*. **A–B**, *A. rubriflorum*. **A**, habit X0.5; **B**, flower X5 (**A–B**, R.Royce 10250, PERTH). **C**, *A. junciforme*, fruit X5 (G.Keighery 3844, PERTH). **D–F**, *Dampiera*. **D**, *D. candicans*, habit X0.5 (R.Carolin 7922, SYD). **E**, *D. spicigera*, habit X0.5 (B.Benn, Sept. 1963, SYD). **F**, *D. discolor*, habit X0.5 (W.Peacock 6111.36.1, SYD). Drawn by D.Mackay.

W.A.: E of Gregory Ra., *R.D.Royce 1890* (PERTH); c. 30 km NE of Jupiter Well, Gibson Desert, *A.S.George 9066* (PERTH).

Bracts are linear, and 6–8 mm long.

3. Dampiera krauseana Rajput & Carolin, *Telopea* 3: 212 (1988)

T: Ajana, W.A., 27 Oct. 1926, *C.A.Gardner*; holo: PERTH.

Erect, branched subshrub to c. 60 cm tall, with loose brownish hairs when young, becoming grey. Leaves cauline, sessile, obovate to oblanceolate, usually dentate at apex, usually glabrescent above, hairy below; lamina 11–28 mm long, 3–15 mm wide. Flowers in loose leafless racemes to 20 cm long; bracteoles linear, to 2 mm long; pedicel 1–4 mm long. Sepals almost obsolete, covered by tuft of long silky hairs. Corolla 8–10 mm long; hairs outside much-branched, not appressed, type i, grey; wings 2–3 mm wide, shorter above auricle, with small appendage below auricle; calli obsolete. Ovary to 2 mm long. Fruit ovoid, to 4 mm long. $n = 9$ W.J.Peacock, *Proc. Linn. Soc. New South Wales* 88: 11 (1963) as *D. stenophylla*. Fig. 17C.

Occurs from Geraldton to lower Murchison R. and inland to Mullewa, W.A., in sandy heath. Flowers Aug.–Nov. Map 33.

W.A.: East Yuna Reserve, *A.C.Burns* 2 (PERTH); Geraldton to Mullewa road, c. 32 km from Geraldton, *J.Galbraith WA170* (MEL); c. 42 km N of Murchison R. crossing, *E.N.S.Jackson 3130* (AD, SYD); c. 11 km E of Ajana, *W.J.Peacock 60853.2* (SYD); Mullewa, *W.J.Peacock 60846.1* (SYD).

Bracts are oblong-elliptic, and 1–3 mm long. Similar to *D. spicigera* which has smaller blunter teeth on leaves, and pale grey or yellowish hairs on the young growth. Chromosome vouchers are *W.J.Peacock 60846.1* and *60853.2* (SYD).

4. Dampiera spicigera Benth., *Fl. Austral.* 4: 109 (1868)

T: south-western W.A., *J.Drummond 3: 154*; holo: K; iso: BM, MEL, W.

D. spicigera var. *lanata* Benth., *loc. cit.* (1868). T: Murchison River, W.A., *A.Oldfield*; holo: K; iso: MEL.

Illustrations: K.Krause, *Pflanzenr.* 54: 182, fig. 33d–f (1912); R.Erickson *et al.*, *Fl. Pl. W. Australia* 106, t. 317 (1973).

Erect, branched subshrub to 60 cm tall, loosely pale grey or yellowish tomentose. Leaves cauline, sessile, obovate to lanceolate or cuneate, usually bluntly toothed towards apex, ±glabrescent above, usually tomentose below; lamina 9–26 mm long, 2–10 mm wide. Flowers in leafless racemes 13–28 cm long, towards ends of branches; bracteoles 1–2 mm long; pedicel to 3 mm long. Sepals triangular, 0.5 mm long, with long ±erect silky hairs. Corolla 8–12 mm long; hairs outside white to grey, type ii; wings 2–3 mm wide, usually smaller above auricle. Calli almost obsolete. Ovary c. 2 mm long. Fruit ovoid, to 3 mm long. $2n = 36$, W.J.Peacock, *Proc. Linn. Soc. New South Wales* 88: 11 (1963). Figs 17D, 29E.

Occurs between the lower Murchison R. and Perth, and inland to Tammin, W.A., in sandy soil. Flowers Aug.–Dec. Map 34.

W.A.: 436 mile peg [c. 700 km N of Perth, North-West Coastal Hwy] S of Carnarvon, *J.V.Blockley 1015* (PERTH); between Moore and Murchison Rivers, *E.Pritzel 992* (AD); c. 16 km N of Northampton, *R.C.Carolin* 33277 (SYD); Marchagee, *L.A.S.Johnson w52* (NSW); c. 24 km S of Tammin, *R.D.Royce 9367* (PERTH).

The corolla is blue. Similar to *D. stenostachya* which has entire leaves, and more closely appressed hairs on the outside of the corolla. There is some variation in the colour and size of the hairs. Also close to *D. krauseana*. Chromosome voucher is *W.J.Peacock 60847.1* (SYD).

5. Dampiera stenostachya E.Pritzel in F.L.E.Diels & E.Pritzel, *Bot. Jahrb. Syst.* 35: 577 (1905)

T: E of Southern Cross, W.A., Oct. 1901, *E.Pritzel 864*; lecto: K, *fide* M.T.M.Rajput & R.C.Carolin, *Telopea* 3: 212 (1988); isolecto: BM, L, MEL, W.

Illustration: K.Krause, *Pflanzenr.* 54: 182, fig. 33G–H (1912).

Erect, branched subshrub to 40 cm tall, grey-hairy. Leaves cauline, sessile, lanceolate to oblong-elliptic, entire, hairy in patches or ±glabrescent; lamina 8–25 mm long, 2–6 mm wide. Flowers in leafless racemes to 14 cm long; bracteoles oblong to elliptic, 1–2 mm long; pedicel 1–3 mm long. Sepals 1–3 mm long. Corolla 8–10 mm long; hairs outside closely appressed, short, type i and iii, grey; wings 1–1.5 mm wide, smaller above auricle, with prominent appendage below auricle; calli obsolete. Ovary to 3 mm long. Fruit ovoid, to 4 mm long. $n = 9$, W.J.Peacock, *Proc. Linn. Soc. New South Wales* 88: 11 (1963). Fig. 17E.

Occurs from the Murchison R. region to the Eastern Goldfields of W.A., in sandy soil. Flowers chiefly Aug.–Dec. Map 35.

W.A.: Murchison R., 1895, *Mrs Markes* (MEL); Pindar, *W.E.Blackall 655* (PERTH); between Mollerin and Koorda, *W.E.Blackall 3473* (PERTH); Booraan, *C.A.Gardner 9543* (PERTH); 11.6 km from Mt Palmer towards Parker Ra., *B.H.Smith 1244* (NSW).

Similar to *D. spicigera*. Chromosome voucher is *W.J.Peacock 6099.4* (SYD).

6. Dampiera teres Lindley, *Sketch Veg. Swan R.* xxvii (1839)

T: not designated.

Erect perennial to 40 cm tall, with close greyish or brownish tomentum. Leaves cauline, sessile, linear-terete to narrowly lanceolate, entire, thick, glabrescent; lamina 8–25 mm long, 2–6 mm wide. Flowers in leafless racemes 8–15 cm long; bracteoles linear, 2–2.5 mm long; pedicel 1–3 mm long. Sepals 0.5–2 mm long, closely tomentose with short grey hairs. Corolla 10–12 mm long; hairs outside short, type i to iv, grey; wings 2.5–3 mm wide, short and tapering towards apex above auricle; calli 2–4 in each row. Ovary 2–3 mm long. Fruit obovoid, to 4 mm long. Fig. 17F.

Occurs northwards from Perth to the Arrowsmith R. area, W.A., in sand in heathland. Flowers Aug.–Nov. Map 36.

W.A.: c. 1.5 km from Arrowsmith R. towards Three Springs, 22 Sept. 1968, *E.M.Canning* (CBG, SYD); near Jurien Bay, *W.E.Blackall 3628* (PERTH); Dinner Hill, W of Watheroo, *R.D.Royce 5133* (PERTH); between Cockleshell Gully and Mt Peron, *C.A.Gardner 9399* (PERTH).

Distinguished from all other species in this section by its thick, narrow leaves which are often terete.

7. Dampiera cinerea Ewart & O.Davies, *Fl. N. Terr. 269* (1917)

T: NW by N of Meyers Hill, Macdonnell Ra., N.T., 1 June 1911, *G.F.Hill 212*; lecto: MEL, *fide* M.T.M.Rajput & R.C.Carolin, *Telopea* 3: 212 (1988).

Spreading subshrub to 50 cm tall, with close, pale grey tomentum. Leaves cauline, oblong to narrowly elliptic, entire, tapering towards base, tomentose on both surfaces; lamina 11–60 mm long, 4–18 mm wide; petiole to 12 mm long. Flowers in leafless spikes and spike-like thyrses 6–20 cm long; bracteoles oblong-elliptic, to 1.5 mm long; pedicel to 1.5 mm long or obsolete. Sepals 0.5–1.2 mm long. Corolla 4–6 mm long; hairs outside much-branched, type i, pale grey; wings 0.5–0.6 mm wide, ±obsolete above auricle, with an appendage below auricle; calli obsolete. Ovary 2.5–3 mm long. Fruit obovoid, c. 2 mm long. Fig. 17G.

Occurs in the desert areas of northern W.A. and south-western N.T.; usually on sand dunes. Flowers chiefly May–Aug. Map 37.

W.A.: Great Sandy Desert, *J.S.Beard 3222* (PERTH); c. 11 km W of Dovers Hills, *A.S.George 9015* (PERTH). N.T.: Campbell Ra., *P.K.Latz 2083* (CANB, DNA); near Charlotte Waters, Aug. 1887, *R.F.Thornton* (MEL); c. 24 km SW of Mt Currie, *R.C.Carolin 5252* (SYD).

The corolla is pink to red or whitish. Similar to *D. candicans* which has brownish tomentum, leaves glabrescent above, and corolla purplish-blue.

8. Dampiera conospermoides W.Fitzg., *J. & Proc. Roy. Soc. W. Australia* 3: 216 (1918)

T: Dillon Springs, W.A., Oct. 1906, *W.V.Fitzgerald*; lecto: NSW, *fide* M.T.M.Rajput & R.C.Carolin, *Telopea* 3: 212 (1988); isolecto: BM, NSW.

Spreading shrub to 1.2 m tall, with close, greyish or pale brownish tomentum. Leaves cauline, oblong to oblong-lanceolate, usually dentate, glabrescent above, hairy below; lamina 15–75 mm long, 4–13 mm wide; petiole 3–7 mm long. Flowers in spikes or thyrses 12–30 cm long; bracteoles 0.5–1 mm long; pedicel to 1.5 mm long. Sepals c. 0.5 mm long, hidden beneath hairs. Corolla 4–5 mm long; hairs outside much-branched, type i, white to yellowish; wings c. 0.5 mm wide, obsolete above auricle, with an appendage below auricle; calli obsolete. Ovary 1.7–2.2 mm long. Fruit ovoid, c. 3 mm long, ribbed. Fig. 17H.

Grows in the north Kimberley, W.A., and Arnhem Land, N.T.; often on sandstone. Flowers chiefly Mar.–July. Map 38.

W.A.: near junction of Drysdale R. and Mogurnda Ck, *A.S.George 13539* (PERTH); N.T.: c. 8 km E of Mudginberry Stn, *L.G.Adams 2773* (CANB); Deaf Adder Gorge, *C.R.Dunlop 444* (NSW); Jasper Gorge, *N.Byrnes NB765* (AD, DNA).

The corolla is pinkish, while bracts are linear-oblong, and 2–4 mm long. This species is similar to *D. candicans* which has ovate or obovate leaves, and greyish-brown hairs on the outside of the purplish-blue corolla.

9. Dampiera candicans F.Muell., *Fragm.* 10: 86 (1876)

T: between the Alfred and Marie and Rawlinson Ranges, W.A., June 1876, *E.Giles*; lecto: MEL, *fide* M.T.M.Rajput & R.C.Carolin, *Telopea* 3: 212 (1988); isolecto: K, NSW.

Erect, branched subshrub to 70 cm tall, brownish or greyish tomentose. Leaves cauline, sessile or petiolate, ovate or obovate to orbicular, dentate, usually glabrescent above or with few large coarse hairs, always tomentose below; lamina 3–35 mm long, 6–19 mm wide; petiole 3–10 mm long. Flowers in leafless thyrses or spikes 13–28 cm long; bracteoles oblong, 0.5–1 mm long; pedicel to 0.5 mm long. Sepals c. 0.1 mm long, hidden beneath hairs. Corolla 5–7 mm long; hairs outside much-branched, not appressed, type i, greyish brown; wings to 0.5 mm, very small or obsolete above auricle, with an appendage below auricle; calli obsolete. Ovary 2–4 mm long. Fruit ellipsoidal to obovoid, c. 2 mm long. Figs 17I, 29D.

Figure 30. *Dampiera*. **A–B**, *D. salahae*. **A**, habit X0.5; **B**, hair X25 (**A–B**, B.Benn, 10 Sept. 1963, SYD). **C**, *D. altissima,* fruit X5 (B.Benn, 14 Sept 1963, SYD). **D–F**, *D. ramosa*. **D**, habit X0.5; **E**, corolla X4; **F**, hair X25 (**D–F**, A.George 8122, SYD). **G**, *D. dentata*, habit X0.5 (A.George 12176, SYD). Drawn by D.Mackay.

Occurs from north-western W.A. to western N.T., in sandy and lateritic soil often with *Triodia* and *Plectrachne* species. Flowers Apr.–Aug. Map 39.

W.A.: c. 8 km SE of Gordon Downs Stn, *R.A.Perry & M.Lazarides 2464* (AD, CANB); Port Hedland, *N.T.Burbidge 5887* (PERTH); Gorge creek, base of Gorge Ra., *R.C.Carolin 7644* (SYD); Terry Ra., northern Gibson Desert, *A.S.George 9075* (PERTH). N.T.: 39 km from Tanami toward W.A. border, *C.H.Gittins 2306* (NSW).

The corolla is purplish-blue. Differs from *D. cinerea* in dentate leaf margins and longer flower spikes.

10. **Dampiera ramosa** Rajput & Carolin, *Telopea* 3: 213 (1988)

T: 3 miles [c. 5 km] NE of Beegull rock hole, Laverton–Warburton road, W.A., 29 Sept. 1966, *A.S.George 8122*; holo: PERTH; iso: SYD.

Erect, perennial herb 15 to 40 cm tall, with thick, tufted stock. Leaves basal, obovate or elliptic, crenulate, pale yellowish tomentose below, crenulate, glabrescent above, with conspicuous, long, white hairs in leaf axils; lamina 2.5–9.0 cm long, 12–41 mm wide; petiole to 6 cm long. Flowers in leafless, branched thyrses or racemes 2.2–4.5 cm long; scape to 40 cm tall; bracteoles linear-oblong, 2–7 mm long; pedicel 2–3 mm long. Sepals linear, 2–4.5 mm long, tomentose. Corolla to 7 mm long; hairs outside not appressed, type i, pale yellowish; wings 1–1.5 mm wide, smaller above auricle; calli obsolete. Ovary slightly oblique, 2–3 mm long. Fruit ovoid, c. 4 mm long. Figs 17J, 30D–F.

Grows in the Great Victoria Desert, W.A., usually on sand dunes. Flowers chiefly Aug.–Sept. Map 40.

W.A.: c. 48 km W of Neale Junction, Great Victoria Desert, *A.S.George 8401* (PERTH); between Tjidilchurra rock hole and W.A./S.A. border, Connie Sue Hwy, *D.E.Symon 12672* (AD, SYD); 11.9 km S of Neale Junction, *H.R.Toelken 6027* (AD, SYD).

The ovule is oblong, and broader towards the apex; the corolla is purplish.

11. **Dampiera plumosa** S.Moore, *J. Bot.* 41: 99 (1903)

T: Coolgardie district, W.A., 1900, *L.C.Webster*; holo: BM; iso: NSW.

Erect perennial to 20 cm tall, with thick tufted stock. Leaves basal, sessile or shortly petiolate, oblong-lanceolate or spathulate, entire, densely pale grey tomentose below, ±glabrous above; lamina 40–65 mm long, 5–15 mm wide; petiole to 5 mm long. Flowers almost sessile in erect, spreading thyrses or loose heads 6–32 cm long, glabrescent or with scattered hairs; bracteoles oblong to narrowly elliptic, 1.5–2.5 mm long; pedicel obsolete. Sepals linear, 2–4 mm long, concealed by long, erect, silky hairs. Corolla 7–9 mm long; hairs outside long, silky, type ii, white; wings 1–1.5 mm wide, smaller above auricle; calli obsolete. Ovary 1–1.5 mm long. Fruit unknown. Fig. 17K.

Uncommon in the goldfields area of W.A. and northwards in sandy soil. Flowering not known. Map 41.

12. **Dampiera eriocephala** Vriese, *Natuurk. Verh. Holl. Maatsch. Wetensch. Haarlem* ser. 2, 10: 118, t. 21 (1854)

T: south-western W.A., 1843, *J.Drummond 397*; lecto: MEL, *fide* M.T.M.Rajput & R.C.Carolin, *Telopea* 3: 214 (1988).

Illustration: W.H.de Vriese, *loc. cit.* (1854); R.Erickson, *et al.*, *Fl. Pl. W. Australia* 127, t. 396, 397 (1973).

Erect perennial to 40 cm tall, ±tomentose-pubescent; rootstock thick, tufted. Leaves basal, obovate or elliptic, entire, glabrescent above, white-tomentose below, with conspicuous tuft of axillary hairs; lamina 3–7 cm long, 11–31 mm wide; petiole to 6 cm long. Flowers numerous, in loose thyrsoid heads elongating in fruiting stage to 12 cm long; bracteoles oblong, to 5 mm long; pedicel obsolete. Sepals minute, with conspicuous tuft of silky hairs to 3 mm long. Corolla 8–12 mm long; hairs outside silky, type ii, white; wings 2–4 mm wide, smaller above auricle; calli 5–8 in each row. Ovary 1–1.7 mm long. Fruit ellipsoidal, c. 2 mm long. Figs 17L, 23J.

Occurs in south-western W.A. from the upper W coast inland to the Goldfields area and S to the Stirling Ra., in heath and mallee in sandy soil. Flowers Sept.–Dec. Map 42.

W.A.: c. 115 km on Mullewa–Yalgoo road, *E.M.Bennett 1531* (PERTH, SYD); Watheroo, *M.Koch 1304* (NSW); Muntadgin, *E.T.Bailey 961* (PERTH); Red Gum Pass, Stirling Ra., *A.Morrison* (PERTH); c. 20 km NE of Hyden, *K.Newbey 1092* (PERTH).

Bracts are oblong-elliptic, to 10 mm long and 3 mm wide, acute, glabrous outside and villous inside. Similar to *D. wellsiana* which has entirely glabrous leaves.

13. **Dampiera wellsiana** F.Muell., *Fragm.* 10: 12 (1876)

D. eriocephala var. *concolor* F.Muell. ex Benth., *Fl. Austral.* 4: 120 (1868). T: south-western W.A., *J.Drummond 4: 162*; lecto: MEL, *fide* M.T.M.Rajput & R.C.Carolin, *Telopea* 3: 214 (1988).

D. humilis F.Muell. ex E.Pritzel in F.L.E.Diels & E.Pritzel, *Bot. Jahrb. Syst.* 35: 582 (1905). T: Parker Ra., W.A., 1890, *E.Merrall*; neo: MEL, *fide* M.T.M.Rajput & R.C.Carolin, *Telopea* 3: 214 (1988).

Perennial herb to c. 25 cm tall, glabrous or glabrescent except flowers; stock thick, tufted. Leaves basal, obovate to elliptic, entire; lamina 2–7 cm long, 9–40 mm wide; petiole to 6 cm long. Flowers numerous, in loose heads usually branched at base and elongating into spikes 13–30 cm long when fruiting; bracteoles to 5 mm long; pedicel almost obsolete. Sepals reduced to tufts of silky hairs on ovary. Corolla 7–9 mm long; hairs outside silky, type ii, white; wings c. 1.5 mm wide, smaller above auricle; calli 4–11 per row. Ovary 1.5–2 mm long. Fruit ellipsoidal, c. 2 mm long. $n = 9$ (as *D. humilis*), W.J.Peacock, *Proc. Linn. Soc. New South Wales* 88: 11 (1963). Figs 14, 17M.

Occurs in W.A. from the Murchison R. region S to the Darling Ra. and E to Lake Grace, in gravel, sandy loam and lateritic soil. Flowers Aug.–Nov. Map 43.

W.A.: c. 20 km E of Mullewa, *E.A.Shaw 636* (AD); 21 km N of Bencubbin, *W.E.Blackall 3318* (PERTH); Cowcowing district, *M.Koch 1258* (PERTH, NSW); Ardath, *R.C.Carolin 3130* (SYD); c. 1.5 km E of Lake Grace, *K.Newbey 1021* (PERTH)

Bracts are oblong-elliptic, 6–10 mm long, and obtuse. See note under *D. eriocephala*. Chromosome vouchers are *W.J.Peacock 6097.1, 60844.7, 60815.1, 6096.1* (SYD).

14. **Dampiera dentata** Rajput, *Telopea* 2: 57 (1980)

T: Dean Ra., N.T., 26 Aug. 1973, *P.K.Latz 4147*; holo: PERTH; iso: AD, BRI, CANB, DNA, K, MEL, NA, NSW.

Perennial herb to 40 cm tall, glabrous except inflorescence; rootstock thick. Leaves basal, sessile or attenuate at base, narrowly oblong to oblanceolate, conspicuously dentate; lamina 5–16 cm long, 3–15 mm wide. Flowers sessile, loosely arranged in heads elongating into spikes to 15 cm long when fruiting; bracteoles to 5 mm long; pedicel almost obsolete. Sepals reduced to tufts of silky hairs to 3 mm long on the ovary. Corolla 5–6 mm long; hairs outside silky, type ii, white; wings to 1 mm wide, smaller above auricle, with a small appendage opposite; calli obsolete. Ovary 2–2.5 mm long. Fruit ellipsoidal, c. 2 mm diam. Figs 17N, 30G.

Occurs from central W.A. to far south-western N.T., on scree, gravel and sandy soil. Flowers chiefly Sept.–Nov. Map 44.

W.A.: c. 19 km S of Mt Brockman homestead, Hamersley Ra., *J.V.Blockley 312* (PERTH); c. 55 km W of Warburton, *A.S.George 3973* (PERTH); c. 38 km E of Laverton, *G.J.Keighery 525* (PERTH). N.T.: Ewaalinga rockhole, c. 32 km E of Docker Creek settlement, *J.R.Maconochie 768* (AD, BRI, DNA, NSW, PERTH).

Bracts are lanceolate, 5–8 mm long and acute. Close to *D. wellsiana* which has shorter, entire leaves which are obovate to elliptic.

Sect. 2. Dampiera

Dampiera R.Br. sect. **Dampiera**

Many-stemmed perennials; stems ribbed or angled or grooved; phyllotaxis 2/5, 1/3, 1/2. Leaves cauline. Flowers in cyme-like panicles or clustered in axils. Ovary sometimes gibbous, mostly 1-locular. Ovule solitary in each locule, erect or curved, basifixed.

Fifty two species in southern Australia.

Subsect. 1. Dampiera

Dampiera R.Br. sect. **Dampiera** subsect. **Dampiera**

Stems terete, ribbed. Phyllotaxis 2/5. Ovary 1-locular, not gibbous. Ovule erect or curved, basifixed.

Twenty nine species in southern Australia.

15. Dampiera purpurea R.Br., *Prodr.* 588 (1810)

T: banks of the Grose, Forest Land, N.S.W., 1803, *R.Brown*; lecto: BM, *fide* M.T.M.Rajput & R.C.Carolin, *Telopea* 3:199 (1988); isolecto: K, MEL.

D. ovalifolia R.Br., *Prodr.* 588 (1810); *D. brownii* var. *ovalifolia* (R.Br.) Domin, *Biblioth. Bot.* 86: 648 (1930). T: Lane Cove, N.S.W., Nov. 1803, *R.Brown*; lecto: BM, *fide* M.T.M.Rajput & R.C.Carolin, *Telopea* 3: 200 (1988); isolecto: K.

D. rotundifolia R.Br., *Prodr.* 587 (1810). T: Grose and Portland Head, N.S.W., *R.Brown*; lecto: BM, *fide* M.T.M.Rajput & R.C.Carolin, *Telopea* 3: 200 (1988); isolecto: K.

D. undulata R.Br., *Prodr.* 587 (1810); *D. brownii* var. *rotundifolia* (R.Br.) Domin, *Biblioth. Bot.* 89: 648 (1930). T: George's River, N.S.W., Oct. 1803, *R.Brown*; lecto: BM, *fide* M.T.M.Rajput & R.C.Carolin, *Telopea* 3: 200 (1988); isolecto: K.

D. bicolor Vriese, *Ned. Kruidk. Arch.* 2: 11 (1850). T: N.S.W., *R.Cunningham 44*; holo: W.

D. omissa Vriese, *Ned. Kruidk. Arch.* 2:10 (1850); *D. melanopogon* R.Br. ex Vriese, *Natuurk. Verh. Holl. Maatsch. Wetensch. Haarlem* ser. 2, 10: 87 (1854), *nom. illeg.* T: Argyle–Paramatta, N.S.W., *C.A.A. von Hügel*; lecto: W, *fide* M.T.M.Rajput & R.C.Carolin, *Telopea* 3: 200 (1988).

D. nervosa Vriese, *Ned. Kruidk. Arch.* 2: 12 (1850). T: Glenbrook, N.S.W., *R.Cunningham;* lecto: W, *fide* M.T.M.Rajput & R.C.Carolin, *Telopea* 3: 200 (1988).

D. brownii F.Muell., *Fragm.* 6: 29 (1867), *nom. illeg.*

Erect perennial to 1 m tall, tomentose to scabrid. Leaves sessile or shortly petiolate, obovate to elliptic, entire or dentate, glabrescent or scabrid above when mature, grey or golden-brown tomentose below; lamina 9–60 mm long, 5–42 mm wide. Flowers in panicles; inflorescence branches 5–23 mm long, usually brownish hairy, 2–9 together, each 3–5-flowered; bracteoles linear-oblong, 2–4 mm long; pedicel 1–3 mm long. Sepals obscured beneath hairs, to 0.5 mm long. Corolla 12–15 mm long; hairs outside loose, type i, usually dark grey; wings 1.5–2.5 mm wide, slightly smaller above auricle; calli 4–6 per row. Ovary 1.5–2.2 mm long; ovule oblong. Fruit obovoid, 4–5 mm long, ribbed beneath hairs. Figs 15, 18A.

Occurs in eastern Australia from Wide Bay, Qld, southwards through N.S.W. to Vic., on the coast and the Great Dividing Ra., mostly in sclerophyll forest and heath. Flowers chiefly Aug.–Jan. Map 45.

Qld: Stanthorpe, 15 Sept. 1962, *G.Word* (MEL). N.S.W.: Mt Wilson, 23 Sept. 1949, *L.A.S.Johnson* (NSW); on banks of Nepean R., 1 Dec. 1963, *H.Salasoo* (NSW); Adams lookout, Bungonia Gorge, *M.Evans 2523* (NSW). Vic.: East Gippsland, Snowy R., *A.C.Beauglehole ACB 37666* (MEL).

Similar to *D. ferruginea* which, however, has brown hairs on both the corolla and the inflorescence branches. There is a considerable range of variation, particularly in the shape and size of the leaves, and the colour and size of the hairs on the peduncle and on the outside of the corolla. In general the specimens collected from Wallangra and Gibraltar Ra. have unusually long inflorescence branches with dirty, white hairs, and ovate-lanceolate or oblong to narrowly elliptic leaves.

16. **Dampiera ferruginea** R.Br., *Prodr.* 588 (1810)

T: Shoalwater Bay passage, [Qld], 26 Aug. 1802, *R.Brown*; lecto: BM, *fide* M.T.M.Rajput & R.C.Carolin, *Telopea* 3: 200 (1988); isolecto: MEL.

Erect perennial to 1 m tall, tomentose with loose, brownish hairs. Leaves sessile or shortly petiolate, entire or crenulate, ovate to elliptic, glabrescent above; lamina 11–46 mm long, 6–25 mm wide. Flowers in cymo-panicles; inflorescence branches solitary, 5–11 mm long, each with 2–4 flowers; bracteoles leaf-like, elliptic to ovate, 7–10 mm long; pedicel 1–2 mm long. Sepals obscured by indumentum, 1–1.2 mm long. Corolla 10–12 mm long; hairs outside dense, not appressed, type i, brown; wings 1.5–2 mm wide, smaller above auricle; calli to 4 per row. Ovary 2.5–3 mm long; ovule erect. Fruit not seen. Fig. 18B.

Occurs from southern Cape York Peninsula to central Qld; on coast and the Great Dividing Ra., in open sclerophyll forest and paperbark swamps. Flowers most of the year. Map 46.

Qld: Hinchinbrook Is., *P.Sharp 1673* (BRI); 24 km S of Cardwell, *M.Thorsborne 212* (BRI); Wathanda, 16 km WNW of Bowen, 10 Aug. 1962, *W.G.Trapnell & K.Williams* (BRI); Byfield road, *N.Michael 3098* (BRI); Mainfold Stn, *R.C.Carolin 4513* (NSW, SYD).

See note under *D. purpurea*.

17. **Dampiera altissima** F.Muell. ex Benth., *Fl. Austral.* 4: 113 (1868)

T: White Peak, Murchison River, W.A., *A.Oldfield*: holo: K; iso: MEL.

Illustration: K.Krause, *Pflanzenr.* 54: 195, fig. 34A–D (1912).

Decumbent to ascending perennial to 50 cm tall, greyish tomentose. Leaves sessile or petiolate, flat, narrowly oblong to lanceolate, dentate, glabrescent above, mostly tomentose below; lamina 8–65 mm long, 3–18 mm wide. Flowes in leafy cymo-panicles; inflorescence branches usually solitary, 3–9-flowered, 9–21 mm long; bracteoles oblong, 4–6 mm long; pedicel 2–4 mm long. Sepals concealed beneath hairs, 0.5–1 mm long or obsolete. Corolla 10–17 mm long; hairs outside type ii, pale grey; wings 3–4 mm wide, smaller above auricle; calli 6–11 per row. Ovary 2–2.5 mm long; ovule erect. Fruit obloid, slightly curved, to 4 mm long, glabrescent, ±ribbed. n = 18, W.J.Peacock, *Proc. Linn. Soc. New South Wales* 88: 10 (1963). Figs 18C, 30C.

Occurs in W.A., from Murchison R. area southwards to Gingin, in sandy soil and laterite. Flowers chiefly June–Oct. Map 47.

W.A.: Murchison House Stn, *C.H.Gittins 1634* (AD, PERTH); c. 14 km from Kalbarri on North-West Coastal Hwy, *C.H.Gittins 1604* (BRI); Galena, *R.C.Carolin 3232* (SYD); c. 11 km N of Geraldton, *J.W.Green 1371* (PERTH); Gingin, *C.A.Gardner 65* (PERTH).

See note under *D. salahae*. Cytological voucher is *W.J.Peacock 60848.1* (SYD).

18. Dampiera salahae Rajput & Carolin, *Telopea* 3: 198 (1988)

T: between Yuna–Mullewa road, W.A., 14 July 1965, *A.M.Ashby 1591*; holo: MEL.

Ascending perennial to 50 cm tall, greyish tomentose. Leaves sessile, ovate to elliptic, entire with thick, papillate, slightly recurved margin, glabrescent above, white tomentose below; lamina 7–39 mm long, 5–21 mm wide. Flowers in cymo-panicles; inflorescence branches usually 1 or rarely 3 together, 1–3-flowered, to 5.5 cm long; pedicel 1–3 mm long; bracteole oblong, 3–5 mm long. Sepals obsolete. Corolla 7–10 mm long; hairs outside dense, silky, type ii, grey; wings 1.5–3 mm wide, usually shorter above auricle; calli 5–9 per row. Ovary 2–2.5 mm long; ovule curved. Fruit obovoid, slightly oblique, c. 4 mm long, glabrescent. n = 18 (as *D. altissima*), W.J.Peacock, *Proc. Linn. Soc. New South Wales* 88: 10 (1963). Figs 18D, 30A–B.

Occurs in Geraldton area of W.A., on sand plain and in pebbly lateritic soil. Flowers chiefly Aug.–Nov. Map 48.

W.A.: East Yuna, NE of Geraldton, *A.C.Burns 15* (PERTH); Pindar, *F.A.Sharr 2716* (PERTH); Wilroy c. 96 km E of Geraldton, *A.M.Ashby 328* (AD); Canna, 28 Aug. 1959, *J.L.McMullan* (PERTH); c. 27 km E of Mingenew, *K.Newbey 2124* (PERTH).

Close to *D. tephrea*, which has grey or yellowish hairs on abaxial surface of leaves. Close to *D. altissima*, which has greyish tomentose leaves becoming glabrescent above, and silky hairs on the corolla. Cytological voucher is *W.J.Peacock 60841.1* (SYD).

19. Dampiera tephrea Rajput & Carolin, *Telopea* 3: 199 (1988)

T: 7 miles [c. 11 km] from Dongara towards Eneabba, W.A., *M.E.Phillips*; holo: CBG; iso: NSW (incorrectly given in the protologue as CANB and SYD respectively).

Ascending to erect perennial to 60 cm tall, whitish or yellowish tomentose. Leaves sessile, elliptic to obovate, often dentate, glabrescent above, closely grey or yellowish tomentose below; lamina 15–47 mm long, 5–15 mm wide. Flowers in cymo-panicles; inflorescence branches 9–20 mm long, 2 or 3 together; pedicel 1–1.5 mm long; bracteoles linear-oblong, 1.5–6 mm long. Sepals obscured by the hairs, 0.7–1.4 mm long. Corolla 10–12 mm long; hairs outside loose, type i and ii, grey; wings 1.7–2 mm wide, smaller above auricle; calli 2–7 per row. Ovary oblong, 2.2–2.5 mm long; ovule curved. Fruit obovoid, slightly oblique, 3.5–4 mm long. Fig. 18E.

Occurs between Dongara and Eneabba, W.A., among heath in sandy soil. Flowers chiefly in spring. Map 49.

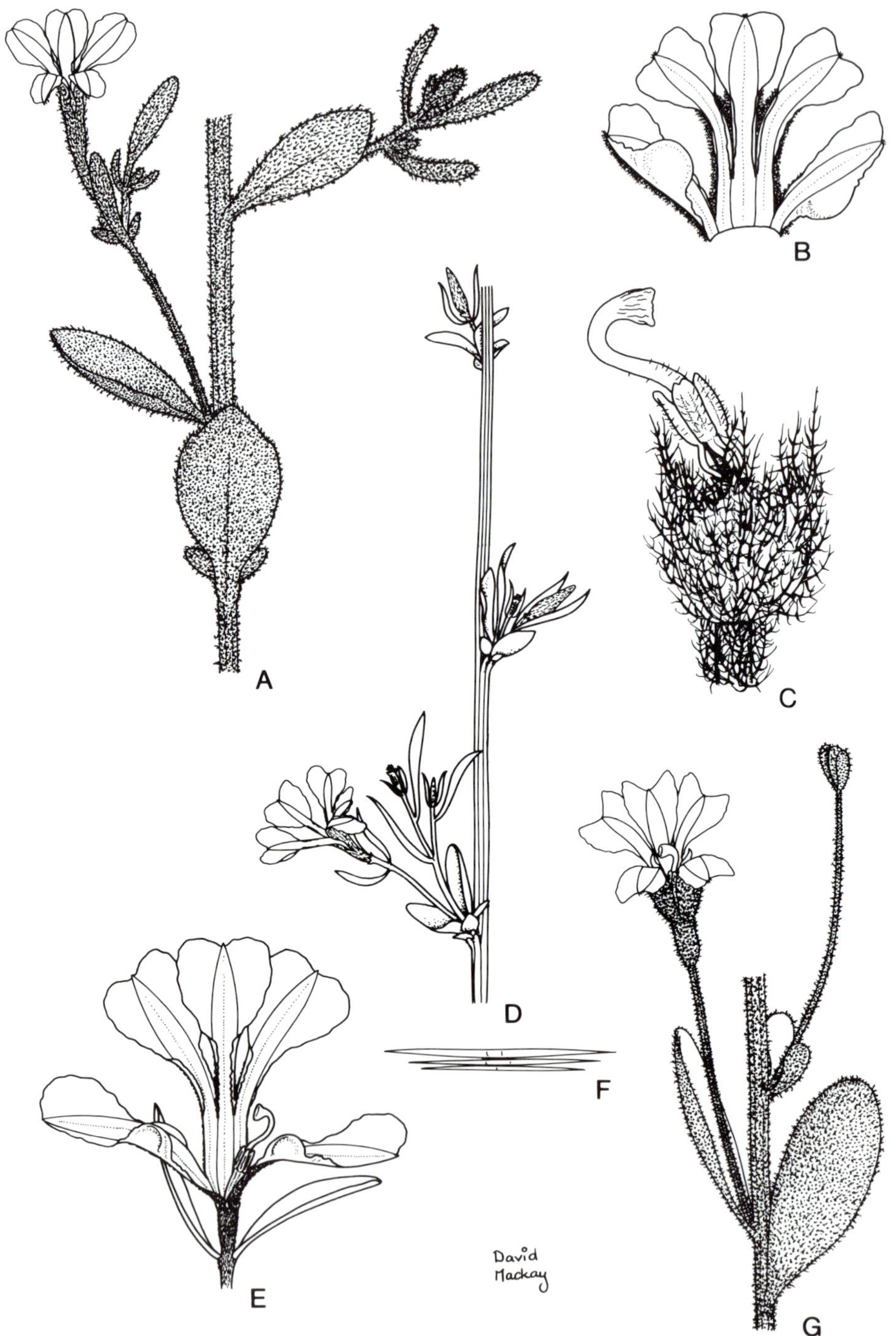

Figure 31. *Dampiera*. **A–C**, *D. incana var. fuscescens*. **A**, habit X2; **B**, corolla X4; **C**, fruit X10 (**A–C**, M.Trudgen 1334, SYD). **D–F**, *D. scaevolina*. **D**, habit X2; **E**, flower X4; **F**, hair X50 (**D–F**, W.Blackall 3422, SYD). **G**, *D. roycei*, habit X2 (A.Beauglehole 60826 & E.Errey, SYD). Drawn by D.Mackay.

W.A.: Dongara, *R.C.Carolin 10634* (SYD); c. 10 km N of Eneabba, *R.C.Carolin 10628* (SYD).

See note under *D. salahae*. Note the correction given above to the location of the types.

20. **Dampiera incana** R.Br., *Prodr*. 588 (1810)

T: Shark Bay area, [W.A.], Aug. 1699, *W.Dampier*, Herb. *W.Sherard 24*; lecto: BM, *fide* M.T.M.Rajput & R.C.Carolin, *Telopea* 3: 195 (1988).

Illustration: K.Krause, *Pflanzenr*. 54: 176, fig. 32D–F (1912).

Erect, much-branched perennial 30–45 cm tall, with ivory-white to grey hairs. Leaves sessile, obovate to oblong-elliptic, entire or dentate, tomentose on both surfaces; lamina 7–63 mm long, 4–15 mm wide. Flowers in cymo-panicles; inflorescence branches 4.5 mm long, usually solitary; pedicel 0.5–1.5 mm long; bracteoles oblong-elliptic, 2.5–4 mm long. Sepals obscured beneath indumentum, 0.7–1 mm long. Corolla 6–9 mm long; hairs outside dense, type i, ivory-white to grey; wings 2.7–3.5 mm wide, smaller above auricle; calli 1–3 per row or obsolete. Ovary to 2.5 mm long; ovule very curved. Fruit ±oblique, obovoid, 2–3 mm long, glabrescent in patches.

Two varieties are recognised.

Vegetative parts tightly tomentose **20a**. var. **incana**

Vegetative parts loosely tomentose **20b**. var. **fuscescens**

20a. **Dampiera incana** var. **incana**

Indumentum usually dense, ivory-white to pale grey; hairs outside corolla not much-branched.

Occurs from the Geraldton area to the Pilbara region, W.A., in sandy soil. Flowers chiefly June–Sept. Map 50.

W.A.: Bullara Stn, 320 km N of Carnarvon, 2 July 1970, *R.O.Farrell* (PERTH); c. 8 km N of Yardie Ck, *A.S.George 6669* (PERTH); Mt Bruce, 17 Aug. 1974, *J.H.Willis* (PERTH); c. 29 km along Dorre Is., Shark Bay, 15 July 1959, *R.D.Royce* (PERTH); mouth of Murchison R., *R.C.Carolin 3372* (SYD).

20b. **Dampiera incana** var. **fuscescens** Benth., *Fl. Austral*. 4: 111 (1869)

T: Murchison R., W.A., *A.Oldfield*, ex Herb. *F.Mueller*; holo: K; iso: MEL.

Indumentum loose, light to dark grey, sometimes yellowish on young stems and leaves; hairs outside corolla usually much-branched. $n = 18$, W.J.Peacock, *Proc. Linn. Soc. New South Wales* 88: 10 (1963) as *D. altissima* aff. Figs 18G, 31A–C.

Occurs in W.A. in Murchison R. and Geraldton areas, and northwards into the Pilbara, in sandy soil. Flowers chiefly June–Sept. Map 51.

W.A.: N side of Passage Paddock, Dirk Hartog Is., *A.S.George 11382* (MEL, PERTH); N of Northampton, *J.Galbraith WA 390* (MEL); Geraldton, *J.Galbraith WA 181* (MEL).

Cytological voucher is *W.J.Peacock 60844.8* (SYD).

21. **Dampiera orchardii** Rajput & Carolin, *Telopea* 3: 197 (1988)

T: 58 km N of mouth of Oldfield R., W.A., 21 Oct. 1968, *A.E.Orchard, 1709*; holo: PERTH.

Erect perennial to 40 cm tall, yellowish tomentose when young; stem whitish hairy at nodes. Leaves sessile, oblong to elliptic, entire; lamina 0.5–1.5 mm long, 0.5–0.7 mm wide. Flowers solitary or in 1–3-flowered cymes; inflorescence branches to 2 mm long, solitary or in clusters; pedicel 0.5–1 mm long; bracteoles oblong, 1–1.2 mm long. Sepals to 0.5 mm long. Corolla to 10 mm long; hairs outside appressed, type iv, yellowish; wings 1.5–2 mm wide, smaller above auricle; calli 5–9 per row. Ovary c. 4 mm long; ovule erect. Fruit not seen. Fig. 18H.

Recorded near the Tone R., NW of Ravensthorpe and near the upper Oldfield R., W.A. Flowers chiefly in spring. Map 52.

W.A.: towards Tone R., 1880, *Muir* (MEL); Lake King to Ravensthorpe, 7 Sept. 1963, *B.Benn* (SYD).

22. **Dampiera scaevolina** C.Gardner ex Rajput & Carolin, *Telopea* 3: 196 (1988)

T: Canna Siding, W.A., Sept. 1934, *C.A.Gardner*; holo: PERTH; iso: PERTH.

Erect or ascending perennial to 25 cm tall, ±glabrous except flowers. Leaves sessile, ±concave, fasciculate, linear-oblong, entire; lamina 5–10 mm long, 1–1.5 mm wide. Flowers solitary or 2 or 3 together in panicles; inflorescence branches 2–4 mm long, single-flowered; pedicel 0.5–1 mm long; bracteoles 2, leaf-like, narrowly oblong to elliptic, 4–6 mm long. Sepals 0.2–0.5 mm long. Corolla 8–10 mm long; hairs outside appressed, type iv, silvery-white; wings veined, 1–1.5 mm wide, shorter above auricle; calli 2–4 per row or obsolete. Ovary c. 2.5 mm long; ovule erect. Fruit not seen. Figs 18J, 31D–F.

Occurs in sandy and gravelly soil, in inland south-western W.A. Flowers chiefly in spring. Map 53.

W.A.: Waddouring, *W.B.Alexander 1241* (PERTH); Beacon, *C.F.H.Jenkins* (PERTH); c. 5 km E of Bencubbin, *W.E.Blackall 3422* (PERTH); approaching Wyalkatchem, *M.E.Phillips* (NSW); NE of Newdegate and SE of Hyden, *W.E.Blackall 1380* (PERTH).

The corolla lobes are more spreading than in any other species of *Dampiera,* making the corolla resemble that of a *Scaevola* (fan-flower). See also *D. orchardii.*

23. **Dampiera tenuicaulis** E.Pritzel in F.L.E.Diels & E.Pritzel, *Bot. Jahrb. Syst.* 35: 580 (1905)

T: Coolgardie district, W.A., 1898, *C.L.Webster*; holo: B (destroyed); neo: Coolgardie, W.A., Sept. 1934, *C.A.Gardner* (PERTH), *fide* M.T.M.Rajput & R.C.Carolin, *Telopea* 3: 199 (1988); isoneo: K.

Wiry perennial to 40 cm tall, ±papillate. Leaves sessile, sometimes fasciculate in axils, oblong to oblong-elliptic, entire or slightly dentate, glabrescent on both sides; lamina 2–19 mm long, 0.5–5 mm wide. Flowers in almost leafless racemes or panicles; inflorescence branches 1–3 together, 1–4-flowered, mostly to 15 mm long; pedicel 0.5–1.5 mm long; bracteoles oblong, 3–5 mm long. Sepals hidden by hairs, to 0.7 mm long. Corolla 6–8 mm long; hairs outside appressed, type ii, black to pale grey; wings 1.5–2 mm wide, slightly smaller above auricle; calli to 8 per row. Ovary 1–2 mm long; ovule erect. Fruit ovoid, 3–4 mm long.

Two varieties are recognised.

Hairs on corolla dark grey to black **23a.** var. **tenuicaulis**

Hairs on corolla pale grey **23b.** var. **curvula**

23a. **Dampiera tenuicaulis** var. **tenuicaulis**

Stem usually glabrous. Outside of corolla with very short, dark grey to black hairs. $n = 18$, W.J.Peacock, *Proc. Linn. Soc. New South Wales* 88: 11 (1963). Fig. 18K.

Occurs in eastern Goldfields region of W.A., in sandy and lateritic soil. Flowers chiefly Aug.–Nov. Map 54.

W.A.: c. 11 km N of Beacon, *W.E.Blackall 3346* (PERTH); c. 48 km N of Bullfinch, *W.J.Peacock 60874* (SYD); c. 10 km ENE of Coolgardie, *A.E.Orchard 1237* (AD); Coolgardie, *T.E.H.Aplin 1878* (PERTH); c. 5 km NW of Norseman, 4 Sept. 1968 *M.E.Phillips* (AD, CANB).

Cytological voucher is *W.J.Peacock 60874.1* (SYD).

23b. **Dampiera tenuicaulis** var. **curvula** (K.Krause) Rajput & Carolin, *Telopea* 3: 210 (1988)

D. curvula K.Krause, *Pflanzenr.* 54: 197 (1912). T: near Gnarlbine, W.A., Nov. 1891, *R.Helms*; holo: K; iso: AD.

D. helmsii K.Krause, *loc. cit.* T: Warangering, W.A., 14 Nov. 1891, *R.Helms*; holo: ?B (destroyed) *n.v.*; iso: AD, K, NSW.

Stem usually irregularly pubescent. Outside of corolla with pale grey, dendritic hairs. Fig. 18I.

Occurs from Mullewa to the Eastern Goldfields, W.A. Flowers chiefly Aug.–Nov. Map 55.

W.A.: c. 20 km E of Mullewa, *E.A.Shaw 637* (AD); c. 115 km [sic.] on Mullewa–Yalgoo road, *E.M.Bennett 1530* (PERTH); Coolgardie, 21 Oct. 1901, *L.C.Webster* (BRI, PERTH); Yerbillon, *M.Koch 2903* (MEL).

Close to *D. lavandulacea* which usually has wider stems, revolute leaves, and loose hairs on the outside of the corolla. Also see *D. oligophylla.*

24. **Dampiera diversifolia** Vriese in J.G.C.Lehmann, *Pl. Preiss.* 1: 403 (1845)

T: south-western W.A., no date, *L.Preiss 1469*; lecto: LD, *fide* M.T.M.Rajput & R.C.Carolin, *Telopea* 3: 203 (1988); isolecto: L, MEL.

Prostrate perennial to 50 cm long, glabrous except axillary bristles and sepals. Leaves sessile, fasciculate in axils, linear-oblong to oblong-lanceolate, entire or coarsely dentate; lamina 8–48 mm long, 1.5–8 mm wide. Flowers solitary or in small cymes terminating leafy branches 4–6 mm long; pedicel mostly 0.5–1 mm long; bracteoles 2–4, elliptic, 5–6 mm long, concave. Sepals unequal, with a few white bristles on margins and in axils, 0.7–1 mm long. Corolla 9–12 mm long, glabrous outside; wings 2–2.2 mm wide, smaller above auricle; calli 3–6 per row. Ovary c. 2 mm long; ovule erect. Fruit obloid, c. 4 mm long. Figs 13, 18L.

Occurs in south-western W.A., in sandy heath. Flowers chiefly Sept.–Nov. Map 56.

W.A.: Mt Barker, Nov. 1907, *W.V.Fitzgerald* (NSW); Hamersley R., near Ravensthorpe, Mar. 1931, *C.A.Gardner* (PERTH); Fitzgerald R. crossing, Ongerup–Ravensthorpe *R.C.Carolin 3565* (SYD); c. 1.5 km S of Tambellup, *R.D.Royce 7681* (PERTH); c. 30 km N of Denmark on Denmark road, *S.Carlquist 3787* (NSW).

The only species of *Dampiera* which is glabrous on the outside of the corolla.

25. **Dampiera roycei** Rajput, *Telopea* 2: 57 (1980)

T: 40 miles [c. 65 km] west of Sandstone towards Mt Magnet, W.A., 17 Oct. 1972, *R.D.Royce 10482*; holo: PERTH.

Decumbent to ascending perennial to 40 cm high, tomentose. Leaves sessile, oblong-obovate or oblong-elliptic, the lower ones oblanceolate, entire, sometimes glabrescent above; lamina 7–35 mm long, 3–13 mm wide. Flowers in axillary cymes or racemes or solitary; inflorescence branches solitary, 1–3-flowered, 6–23 mm long; pedicel 1–3 mm long; bracteoles oblong to narrowly elliptic, 2–5 mm long. Sepals with long not appressed hairs, 0.2–0.7 mm long. Corolla 9–12 mm long; hairs outside not appressed, type i and ii, dark grey; wings 1.2–1.5 mm wide, usually shorter above auricle; calli 4 or 5 per row. Ovary 2.5–3.5 mm long; ovule erect. Fruit ellipsoidal, 4–5 mm long, ribbed beneath hairs. Figs 18M, 31G.

Occurs from central W.A. to south-western N.T., in sandy soil. Flowers chiefly Sept.–Oct. Map 57.

W.A.: White Cliffs Stn, between 35–100 km NE of Laverton, *N.Harper 8* (PERTH); between Kunnunoppin and Mt Marshall, winter-spring 1919, *W.V.Fitzgerald* (NSW); Parker Ra., *E.Merrall 1892* (MEL). N.T.: Bloods Ra. area, *P.K.Latz 2389* (AD); c. 80 km W of Mt Olga on road to Docker R. Mission, *M.Lazarides 8302* (NSW).

26. **Dampiera marifolia** Benth., *Fl. Austral.* 4: 114 (1868)

T: Wimmera, Vic., *J.Dallachy*; lecto: K, *fide* M.T.M.Rajput & R.C.Carolin, *Telopea* 3: 203 (1988); isolecto: MEL.

Spreading perennial to 60 cm long, brownish grey to rusty tomentose. Leaves sessile, ovate-oblong to oblong-elliptic, entire with margin recurved to almost revolute, thick, sometimes glabrescent; lamina 7–18 mm long, 3–6 mm wide. Flowers in cymes; inflorescence branches not longer than leaves, mostly 1 per axil, 1–3-flowered, 5–8 mm long; pedicel 1–2 mm long; bracteoles oblong to narrowly elliptic, 3–4.5 mm long. Sepals with rusty-brown hairs, 0.5–1.5 mm long. Corolla 9–12 mm long; hairs outside loose, type ii, yellowish or brown-grey; wings 1.5–2 mm wide, almost equal; calli 4 or 5 per row. Ovary 1.2–1.5 mm long; ovule usually broader towards apex. Fruit ovoid, to 4 mm long. Fig. 18F.

Occurs from the Mt Lofty Ra., S.A., to north-western Vic.; often in mallee communities. Flowers Sept.–Dec. Map 58.

S.A.: Murray Mallee, Billiatt Natl Pk, *R.V.Southcott* (AD); Braendlers Scrub, *A.G.Spooner 4110* (AD); SE of Shipley Hill, c. 26 km NE of Keith, *N.N.Donner* 176 (AD). Vic.: Tempy, c. 23 km S of Ouyen, 7 Oct. 1966, *T.Henshall* (NSW); Hattah Lakes Natl Park, 11 Sept. 1960, *A.C.Beauglehole* (MEL).

27. **Dampiera haematotricha** Vriese, *Natuurk. Verh. Holl. Maatsch. Wetensch. Haarlem* ser. 2, 10: 94 (1854)

T: south-western W.A., 1843, *J.Drummond 'suppl. 56, no. 105'*; neo: MEL, *fide* M.T.M.Rajput & R.C.Carolin, *Telopea* 3: 203 (1988)

[*D. ferruginea* auct. non R.Br.: W.H.de Vriese, *Ned. Kruidk. Arch.* 2: 12 (1851)]

Ascending to decumbent perennial to 40 cm long, red-brown to grey-hairy. Leaves sessile or nearly so, obovate to orbicular, entire or dentate, glabrescent above, tomentose in patches below; lamina 9–42 mm long, 5–20 mm wide. Flowers in panicles; inflorescence branches 1–3 together, 3–9-flowered, 9–45 mm long; pedicel 1–5 mm long; bracteoles oblong-elliptic, 2–3.5 mm long. Sepals concealed beneath hairs, to 0.4 mm long. Corolla 8–10 mm long; hairs outside loose, type ii, grey or red-brown with paler tips; wings 2.5–3.2 mm wide, shorter above auricle; calli 4–7 per row. Ovary 2–3 mm long; ovule erect. Fruit cylindrical, 4–5 mm long.

Similar to *D. hederacea* which however has petiolate leaves with cordate bases. G.Bentham and J.W.Krause included *D. haematotricha* with *D. triloba*. Two subspecies are recognised.

Hairs on inflorescence branches rusty with greyish tips **27a.** subsp. **haematotricha**

Hairs on inflorescence branches grey **27b.** subsp. **dura**

27a. Dampiera haematotricha Vriese subsp. haematotricha

Leaves ovate-orbicular or obovate, entire. Inflorescence branch hairs rusty. Corolla hairs greyish. Fig. 19A.

Occurs in inland south-western W.A. Flowers chiefly Aug.–Nov. Map 59.

W.A.: c. 32 km N of Merredin, *C.F.Davies 221* (PERTH); Wickepin area, *E.Wittwer W880* (PERTH); 32 km W of Lake Grace, *W.E.Blackall 1328* (PERTH); c. 8 km W of Tarin Rock, *K.Newbey 1497* (PERTH); near Wagin, *C.A.Gardner 2021* (PERTH).

27b. Dampiera haematotricha subsp. dura (Benth.) Rajput & Carolin, *Fl. Australia* 35: 330 (1992)

D. altissima var. *dura* Benth., *Fl. Austral.* 4: 113 (1868). T: south-western W.A., *J.Drummond 5: 71*; holo: K; iso: BM, MEL, NSW.

D. dura Pritzel, *Bot. Jahrb. Syst.* 35: 579 (1905). T: near White Peak, near Geraldton, W.A., July 1901, *E.Pritzel 428*; *n.v.*

Illustration: K.Krause, *Pflanzenr.* 54: 195, fig. 34E–F (1912) as *D. dura.*

Leaves oblong-elliptic, elliptic or cuneate, dentate. Inflorescence branch hairs and corolla hairs grey to dark grey. $n = 9$, W.J.Peacock, *Proc. Linn. Soc. New South Wales* 88: 11 (1963) as *D. triloba.* Fig. 19B.

Occurs in south-western W.A. mainly between Bencubbin and Katanning. Flowers chiefly Aug.–Nov. Map 60.

W.A.: Wilgoyne Hill, *C.A.Gardner 6464* (PERTH); c. 16 km E of Mukinbudin on Bencubbin road, *W.J.Peacock 60813.1* (SYD); c. 5 km E of Baandee, *T.E.H.Aplin 2248* (PERTH); c. 3 km S of Tammin, *K.Newbey 1952* (PERTH); Katanning, *W.E.Blackall 2986* (PERTH).

Cytological voucher is *W.J.Peacock 60813.1* (SYD).

28. Dampiera hederacea R.Br., *Prodr.* 588 (1810)

T: King George Sound, [W.A.], Dec. 1801, *R.Brown*; holo: BM.

Illustration: R.Erickson *et al.*, *Fl. Pl. W. Australia* 58, fig. 149 (1973).

Spreading to decumbent perennial to 40 cm long, tomentose with brownish type i hairs, glabrescent. Leaves lanceolate to broadly ovate, ±cordate at base, lobed or toothed, tomentose on both surfaces; lamina 9–40 mm long, 3–32 mm wide; petiole 2–7 mm long. Flowers in cymo-panicles; inflorescence branches 1–3 together, 3–7-flowered, 5–46 mm long; pedicel 2–4 mm long; bracteoles oblong-elliptic, 2–3 mm long. Sepals 0.5–0.7 mm long. Corolla 7–10 mm long; hairs outside not appressed, type i and ii, grey; wings 1.5–2.2 mm wide, much smaller above auricle; calli obsolete. Ovary 1.2–1.7 mm; ovule erect. Fruit cylindrical, to 3 mm long. Figs 19C, 23I, 32E.

Occurs in extreme south-western W.A. S of Waroona, in Karri (*Eucalyptus diversicolor*) forest and similar sites. Flowers chiefly Aug.–Jan. Map 61.

Figure 32. *Dampiera*. **A–D**, *D. tomentosa*. **A**, habit X2; **B**, corolla X2.5; **C**, hair X25; **D**, hair X25 (**A–D**, NSW 83455, NSW). **E**, *D. hederacea*, habit X2 (B.Benn, 27 Sept. 1963, SYD). **F–G**, *D. scottiana*. F, habit X2; **G**, hair X25 (**F–G**, F.Rodway, Oct. 1927, NSW). Drawn by D.Mackay.

W.A.: near Waroona Dam, Darling Ra., *P.G.Wilson 3724* (PERTH); Big Brook, between Margaret River and Cowaramup, *T.A.Halliday 235* (AD, PERTH); c. 9 km S of Northcliffe, *A.S.George 2643* (PERTH); Denmark, *C.A.Gardner 5001* (PERTH); near road from Denmark to Walpole, *H.Eichler 16116* (AD).

Similar to *D. haematotricha* which has sessile leaves without cordate bases.

29. **Dampiera tomentosa** K.Krause, *Pflanzenr.* 54: 181 (1912)

T: Camp 39, Great Victoria Desert, W.A., Sept. 1891, *R.Helms*; holo: K; iso: AD.

Erect perennial to 1 m tall, densely tomentose with whitish grey type i hairs. Leaves sessile, sometimes fasciculate in axils, elliptic-obovate or oblong-lanceolate, entire, densely hairy on both surfaces; lamina 8–60 mm long, 4–31 mm wide. Flowers in cymo-panicles or clusters; inflorescence branches 1–3 together, 3–5-flowered, 9–31 mm long or sometimes shorter; pedicel 2–4 mm long; bracteoles oblong-elliptic, 1.5–2.5 mm long. Sepals obsolete. Corolla 9–12 mm long; hairs outside long, not appressed, type i and ii, grey; wings 1.5–2.5 mm wide; calli 7–11 per row or rarely obsolete. Ovary c. 3 mm long; ovule erect. Fruit not seen. $n = 9$, W.J.Peacock, *Proc. Linn. Soc. New South Wales* 88: 11 (1963). Figs 19D, 23H, 32A–D.

Occurs in inland south-western W.A. and from 2 locations in the western Great Victoria Desert, in sandy soils. Flowers chiefly Aug.–Nov. Map 62.

W.A.: between Perenjori and Dalwallinu, *W.E.Blackall 2850* (PERTH); Bonnie Rock, *W.J.Peacock 608811.1* (SYD); Merredin, *M.Koch 2734* (MEL, NSW); Muntadgin, *E.T.Bailey 952* (PERTH); Narembeen, *E.Ashby 14950* (AD).

Cytological vouchers are *W.J.Peacock 60814.3*, *6087.3*, and *6089.1* (SYD). This species is similar to *D. luteiflora* but the yellow corolla distinguishes the latter.

30. **Dampiera luteiflora** F.Muell., *Fragm.* 10: 11 (1876)

T: near Ularing, W.A., 13–17 Oct. 1875, *J.Young*; holo: MEL.

Illustrations: K.Krause, *Pflanzenr.* 54: 182, fig. 33A–C (1912); R.Erickson *et al.*, *Fl. Pl. W. Australia* 113, fig. 339 (1973).

Erect to decumbent perennial to 1 m tall, loosely yellowish tomentose. Leaves sessile, fasciculate, elliptic-obovate to oblong, entire, tomentose on both surfaces; lamina 9–38 mm long, 4–14 mm wide. Flowers in panicles; inflorescence branches stout, 1–3 together, 4–7-flowered, 10–75 mm long; pedicel 2–4 mm long; bracteoles 1–2 mm long. Sepals obsolete. Corolla 10–12 mm long; hairs outside dense, silky-hairy, type i and ii, yellowish; wings 1.2–1.7 mm wide, smaller above auricle; calli indistinct. Ovary c. 3 mm long; ovule erect. Fruit not seen. $n = 9$, W.J.Peacock, *Proc. Linn. Soc. New South Wales* 88: 10 (1963). Figs 12, 19E.

Occurs in inland areas of south-western W.A., in sandy soil. Flowers chiefly Sept.–Nov. Map 63.

W.A.: c. 29 km W of old Gidgee homestead N of Sandstone, *R.D.Royce 10444* (PERTH); near Warangering, Nov. 1891, *R.Helms* (AD); c. 42 km N of Beacon, NW of Merredin, *W.E.Blackall 3344* (PERTH); Carrabin, *C.A.Gardner 603* (PERTH); Boorabbin, *J.S.Beard 2439* (PERTH).

This is the only species of *Dampiera* known to have a yellow corolla. The cytological voucher is *W.J.Peacock 6099.1* (SYD).

31. **Dampiera stenophylla** K.Krause, *Pflanzenr.* 54: 187 (1912)

T: Camp 58, Great Victoria Desert, W.A., Sept. 1891, *R.Helms*; lecto: K, *fide* M.T.M.Rajput & R.C.Carolin, *Telopea* 3: 195 (1988); isolecto: AD, MEL.

Ascending perennial to 60 cm long, pale grey tomentose. Leaves sessile, sometimes fasciculate in axils, oblong to narrowly elliptic, recurved, dentate, glabrescent above; lamina 6–23 mm long, 2.7 mm wide. Flowers in cymo-panicles or solitary; inflorescence branches 2–4 mm long, usually solitary with 1 or 2 flowers; pedicel 0.5–1.5 mm long; bracteoles oblong-elliptic, 1–1.7 mm long. Sepals hidden beneath hairs, linear-oblong, 0.5–0.7 mm long. Corolla 6–7 mm long; hairs outside short, silky, type i and few long type ii, white; wings 1–2 mm wide, short or obsolete above auricle; calli ±obsolete. Ovary 1.5–2 mm long; ovule erect. Fruit not seen. Fig. 19F.

Occurs in the Eastern Goldfields and western Great Victoria Desert, W.A., in sandy soil. Flowers chiefly Aug.–Nov. Map 64.

W.A.: c. 120 km N of Sandstone, *A.S.George 5651* (PERTH); Wanarra, *C.A.Gardner 12072* (PERTH); c. 16 km S of Queen Victoria Spring, *R.D.Royce 5533* (PERTH); Queen Victoria Spring, *W.H.Butler* (PERTH); 20 km E of Zanthus, *P.G.Wilson 7626* (PERTH).

Similar to *D. lavandulacea*, but that species has glabrous stems and leaves linear to oblong-elliptic.

32. **Dampiera fitzgeraldensis** Rajput & Carolin, *Telopea* 3: 196 (1988)

T: Fitzgerald R. crossing, Ongerup–Ravensthorpe road, W.A., 11 Sept. 1961, *R.C.Carolin 3564*; holo: NSW.

Erect perennial probably to 50 cm tall; stem tomentose in grooves. Leaves sessile, fasciculate, spathulate or oblong-elliptic, slightly recurved, dentate, pale grey or pale brown hairy below, glabrescent above; lamina 4–16 mm long, 1.5–6 mm wide. Flowers in cymo-panicles; inflorescence branches 1–3 together, 2–5-flowered, very short; bracteoles linear-oblong, 1–3 mm long; pedicel 1–3 mm long. Sepals unequal, tomentose, 0.5–1 mm long. Corolla 10–12 mm long; hairs outside ±appressed, short, type iv, silvery or pale grey; wings 2–2.2 mm wide, slightly shorter above auricle; calli 6–13 per row. Ovary 1.2–1.5 mm long; ovule erect. Fruit not seen. Fig. 19G.

Known only from the type locality, W.A., in sand plain heath. Flowers chiefly in spring. Map 65.

Similar to *D. scaevolina* but that species has leaves that are neither spathulate nor oblong-elliptic, and are glabrous on both surfaces. Also similar to *D. lavandulacea* which has narrower leaves.

33. **Dampiera adpressa** A.Cunn. ex DC., *Prodr.* 7(2): 503 (1839)

T: Crokers Ra., N.S.W., *A.Cunningham*; holo: G-DC; iso: BM, K.

[*D. lanceolata auct. non* A.Cunn.: H.W. de Vriese, *Natuurk. Verh. Holl. Maatsch. Wetensch. Haarlem* ser. 2, 10: 101 (1845)].

Illustration: W.H.de Vriese, *op. cit.* t. 17 (1845) as *D. lanceolata*.

Erect perennial to 1 m tall, glabrescent. Leaves sessile, sometimes fasciculate in axils, usually ovate-elliptic to lanceolate or sometimes linear-oblong, usually glabrescent; lamina 11–55 mm long, 2–23 mm wide. Flowers in panicles; inflorescence branches to 3 together, 3–5-flowered, 3–12 mm long; pedicel 1–3 mm long; bracteoles linear-oblong, 3–5 mm long. Sepals unequal, hidden by hairs, 1–1.5 mm long. Corolla 8–10 mm long; hairs outside dense, type ii, silvery to grey; wings 1.5–2.2 mm wide, smaller above auricle; calli obsolete. Ovary 2–3 mm long; ovule erect. Fruit 4–5 mm long, glabrescent, ±ribbed, rugose. *n* = 9, W.J.Peacock, *Proc. Linn. Soc. New South Wales* 88: 10 (1963). Frontispiece, Fig. 19H.

Occurs in southern Qld and N.S.W. on the Great Dividing Ra. and western slopes and plains, in sclerophyll woodlands and forests. Flowers chiefly June–Jan. Map 66.

Qld: top of Main Ra. near Gurulmundi, *C.E.Hubbard 5084* (BRI); c. 100 km S of Charleville, Boatman road, *K.Williams 132* (BRI). N.S.W.: between Gerara and Lia Springs Bourke, *D.J.McGillivray 2981* (NSW); Tullamallee SE of Boradine, *R.C.Carolin 9550* (SYD); c. 8 km S of Mendooran *E.F.Biddiscombe 45* (CANB).

Similar to *D. lanceolata* which has linear to oblong leaves and woolly or silky hairs on the outside of corolla. The colour of the hairs on the corolla varies from silvery to grey. Cytological vouchers are *W.J.Peacock 6011.2.4* and *61111.39.1* (SYD).

34. **Dampiera lanceolata** A.Cunn. ex DC., *Prodr.* 7(2): 503 (1839)

T: towards the Wellington Valley, Bathurst, N.S.W., *A.Cunningham*; lecto: BM, *fide* M.T.M.Rajput & R.C.Carolin *Telopea* 3: 202 (1988); isolecto: K.

D. maideniana K.Krause, *Pflanzenr.* 54: 189 (1912). T: west from Wellington Valley, *A.Cunningham*, Dec. 1825; lecto: W.

Erect perennial to 1 m tall, whitish or grey-tomentose; stem often papillate. Leaves sessile, sometimes fasciculate in axils, linear to oblong, sometimes slightly revolute, entire or dentate, thick, glabrescent above, usually closely tomentose below, papillate; lamina 5–47 mm long, 2–26 mm wide. Flowers in cymo-panicles; inflorescence branches 1–3 together, 3–9-flowered, 3–16 mm long; pedicel 1–4 mm long; bracteoles mostly linear-oblong, 1–3 mm long. Sepals almost concealed beneath hairs, 0.3–0.5 mm long. Corolla 8–12 mm long; hairs outside woolly or silky, types i and ii, grey; wings 2–3.5 mm wide, ±equal; calli to 6 per row. Ovary c. 2 mm long; ovule erect. Fruit cylindrical, 4–5 mm long.

Three varieties are recognised.

1 Hairs on corolla pale grey

2 Hairs on corolla not silky, types i and ii — **34a.** var. **lanceolata**

2: Hairs on corolla silky, mostly type ii — **34b.** var. **intermedia**

1: Hairs on corolla dark grey, mostly type i — **34c.** var. **insularis**

34a. **Dampiera lanceolata** A.Cunn. ex DC. var. **lanceolata**

[*D. adpressa auct. non* A.Cunn.: W.H.de Vriese, *Natuurk. Verh. Holl. Maatsch. Wetensch. Haarlem* ser. 2, 10: 100, t. 16 (1854)].

Illustrations: W.H.de Vriese, *Natuurk. Verh. Holl. Maatsch. Wetensch. Haarlem* ser. 2, 10: t. 17 (1854) as *D. lanceolata*; G.R.Cochrane *et al.*, *Fl. Pl. Victoria & Tasmania* 47, fig. 184 (1980).

Stem usually glabrescent, papillate. Leaves linear to oblong-elliptic, usually not papillate. Hairs outside corolla type i or ii, mostly type i, pale grey to grey. *n* = 18, W.J.Peacock, *Proc. Linn. Soc. New South Wales* 88: 11 (1963) as *Dampiera* sp.

Occurs from the margin of the Great Victoria Desert, S.A., through Vic. to the Great Dividing Ra., N.S.W., and just into Qld; grows mostly in sclerophyllous woodland and forest. Flowers chiefly Aug.–Jan. Map 67.

S.A.: c. 20 km NE of Blanchetown on road to Waikerie, *R.Hill 1104* (AD). Qld: Leichhardt, *A.D.Cribb* (BRI). N.S.W.: Timor Rock, W of Coonabarabran, *H.Salasoo 2234* (NSW); Shepherds Hill, *G.Cunningham* (NSW). Vic.: c. 2.4 km S of Hattah station in Kulkyne Natl Pk, *J.H.Willis* (MEL, NSW).

Cytological voucher is *W.J.Peacock 6110.5.1* (SYD).

34b. **Dampiera lanceolata** var. **intermedia** Rajput & Carolin in J.P.Jessop & H.R.Toelken, *Fl. South Australia* 4th edn, 3: 1387 (1986)

T: Aldinga Bay, Southern Mt Lofty Ra., S.A., 2 Dec. 1923, *J.B.Cleland*; holo: AD.

Stem tomentose, not papillate. Leaves oblong to ovate-elliptic, slightly papillate. Hairs on outside of corolla loose, silky, mostly type ii, pale grey.

Occurs in the southern part of Mt Lofty Ra., S.A., often on sandstone cliffs and in sand. Flowers chiefly Aug.–Jan. Map 68.

S.A.: eastern side of Gulf St Vincent, *T.Smith 841* (AD); Aldinga Bay, 2 Dec. 1923, *J.B.Cleland* (AD).

34c. **Dampiera lanceolata** var. **insularis** Rajput & Carolin in J.P.Jessop & H.R.Toelken, *Fl. South Australia* 4th edn, 3: 1387 (1986)

T: Mt Taylor, Kangaroo Is., S.A., 7 Nov. 1958, *P.G.Wilson 782*; holo: AD.

Stem glabrescent, papillate. Leaves oblong to ovate-elliptic, papillate. Hairs outside corolla not appressed, type i, grey to very dark grey. Fig. 19I.

Occurs on Kangaroo Is., S.A. Flowers chiefly Aug.–Jan. Map 69.

S.A.: Hundred of Newland, S of Mt Stockdale, *H.Eichler 18565* (AD); Muston, 11 Dec. 1964, *H.M.Cooper* (AD); Eumalla, Sept. 1933, *A.B.Cashmore* (AD).

35. **Dampiera eriantha** K.Krause, *Pflanzenr.* 54: 185 (1912)

T: Camp 54, Great Victoria Desert, W.A., 17 Sept. 1891, *R.Helms*; holo: ?B (destroyed) *n.v.*; iso: AD, MEL, NSW.

Erect perennial to 60 cm tall, glabrous or glabrescent except inflorescence. Leaves sessile, linear-oblong, revolute, entire, thick, papillate; lamina 11–37 mm long, 2–4 mm wide. Flowers in axillary clusters; branches 1–3 together, 3–7-flowered, 9–21 mm long; pedicel 2–5 mm long; bracteoles linear-oblong, 4–6 mm long. Sepals obsolete. Corolla 8–10 mm long; hairs outside fine, not appressed, types i and ii, pale grey; wings 1.7–2.2 mm wide, smaller above auricle; calli 2–5 per row. Ovary c. 2 mm long; ovule erect. Fruit not seen. Fig. 19J.

Occurs in the Great Victoria Desert, W.A. Flowering unknown. Map 70.

W.A.: Camp 43, Great Victoria Desert, *R.Helms* (MEL, NSW).

Many S.A. specimens referred to *D. lanceolata* appear closer to this species, although in *D. lanceolata* the leaves are ±tomentose especially on the lower surface. There are few specimens agreeing well with the type.

36. **Dampiera scottiana** F.Muell., *Fragm* 11: 120 (1881)

T: Port Jackson, N.S.W., *W.Woolls*; lecto: MEL, *fide* M.T.M.Rajput & R.C.Carolin, *Telopea* 3: 206 (1988); isolecto: K, NSW.

D. rodwayana Rajput & Carolin, *Telopea* 3: 201 (1988). T: c. 15 miles [c. 25 km] SW of Nowra, Nowra–Nerriga road, Turpentine Ra., N.S.W., 27 Oct. 1957, *E.F.Constable*; holo: NSW; iso: MEL.

Erect perennial to 60 cm tall; stem conspicuously ribbed, pale grey tomentose. Leaves sessile, sometimes fasciculate in axils, linear to elliptic, recurved, entire or with few teeth, glabrous above, pale grey tomentose below; lamina 9–30 mm long, 1–6 mm wide. Flowers in cymo-panicles; branches 1 or 2 together, 1–3-flowered, 1–5 mm long; pedicel 1–2 mm long; bracteoles linear-oblong, 3–6 mm long. Sepals hidden beneath hairs, 0.5–1 mm long. Corolla 9–12 mm long; hairs outside fine, mostly type ii, silvery grey; wings 1.5–2.5 mm wide, almost equal; calli obsolete. Ovary 2–2.5 mm long; ovule erect. Fruit obovoid, to 4 mm long. $n = 18$ (as *Dampiera* sp.), W.J.Peacock, *Proc. Linn. Soc. New South Wales* 88: 11 (1963). Figs 19K, 32F–G.

Occurs near Nowra and in Maramara Natl Park, in the Hawkesbury region, N.S.W.; in sclerophyll woodland and heath on sandstone. Flowers chiefly Aug.–Dec. Map 71.

N.S.W.: 8 km W of Nowra on Yalwal road, *D.F.Blaxell 447* (NSW); c. 8 km W of Nowra, *Herbarium F.A.Rodway* (NSW); Flat Rock Dam, c. 3 km SW of Nowra, *Herbarium F.A.Rodway 14612* (NSW); Jervis Bay, *F.A.Rodway 5488* (NSW); Box Point to Barbers Creek, Oct. 1898, *J.H.Maiden* (NSW).

Similar to *D. rosmarinifolia* and its relatives but these have glabrous stems and usually narrower leaves. Cytological voucher *W.J.Peacock, 6110.17.1* (SYD).

37. **Dampiera rosmarinifolia** Schldl., *Linnaea* 20: 603 (1847)

T: S.A., Nov. c. 1845, *Dr. Behr 86*; holo: HAL.

Erect to prostrate perennial to 60 cm long; stem conspicuously ribbed, tomentose, glabrescent. Leaves sessile, usually fasciculate in axils, linear to linear-oblong, usually so revolute as to hide lower surface, entire, thick, glabrous and glossy above, tomentose below; lamina 9–26 mm long, 2–5 mm wide. Flowers in cymo-panicles or clusters; inflorescence branches usually single, 3–11 mm long, mostly 1-flowered, usually tomentose; pedicel 3–5 mm long; bracteoles linear-elliptic, 1–2 mm long. Sepals tomentose 1–1.5 mm long. Corolla 10–14 mm long; hairs outside type i, short, grey to black, and type ii, longer, paler; wings 1.7–3.2 mm wide, almost equal; calli 2–9 per row. Ovary 2–3 mm long; ovule erect. Fruit obovoid, to 4 mm long. Figs 10, 19L.

Occurs from the Eyre Peninsula, S.A., to north-western Vic. Flowers chiefly Aug.–Jan. Map 72.

S.A.: Mt St Mungo [sic.], *J.Carrick 2425* (AD); Hinks Natl Park, *A.C.Alcock 2284* (AD); 2 km W of Curramulka, York Peninsula, *K.Czornij 856* (MEL); between Stansbury and Minlaton, *H.Eichler 14211* (AD); c. 20 km SE of Murray Bridge, *J.B.Cleland* (AD).

Very variable in colour, length and nature of the hairs on the outside of the corolla. It is close to *D. dysantha* and *D. lavandulacea* and there appear to be intermediates between it and the latter species.

38. **Dampiera dysantha** (Benth.) Rajput & Carolin in J.P.Jessop & H.R.Toelken, *Fl. South Australia* 4th edn, 3: 1386 (1986)

D. rosmarinifolia var. *dysantha* Benth., *Fl. Austral.* 4: 114 (1868). T: The Grampians, Vic., *J.F.C.Wilhelmi*; lecto: K, *fide* M.T.M.Rajput & R.C.Carolin, *Telopea* 3: 201 (1988).

[*D. lavandulacea auct. non* Lindley: E.M.Robertson in J.M.Black, *Fl. South Australia* 2nd edn, 4: 833 (1965)].

Erect or decumbent perennial to 70 cm long; stem conspicuously ribbed, glabrous or glabrescent, papillate. Leaves sessile, oblong-elliptic, usually revolute but not so as to hide lower surface, entire or dentate, glabrous above, tomentose below; lamina 5–16 mm long, 2–5 mm wide. Flowers in cymo-panicles; inflorescence branches 1–3 together, 5–7-flowered, 11–17 mm long; pedicel 1–2 mm long; bracteoles linear-oblong, 3–4.5 mm long. Sepals covered by hairs, 0.2–0.5 mm long. Corolla 9–11 mm long; hairs outside type ia, ±short and dark grey, and longer and yellow-grey; wings 1.7–2.2 mm wide, smaller above auricle; calli 4–7 per row. Ovary c. 2.5 mm long; ovule erect. Fruit obovoid, to 4 mm long. Fig. 19M.

Occurs from the Flinders Ranges, S.A., to north-western Vic. Flowers chiefly Sept.–Jan. Map 73.

S.A.: Wilpena Pound, Northern Flinders Ra., *J.B.Cleland* (AD); Boston Point, Eyre Peninsula, *P.G.Wilson 388* (AD); SE Bunns Bore, c. 52 km N of Bordertown, *N.N.Donner 210* (AD). Vic.: NW Little Desert, S of Kaniva, *A.C.Beauglehole ACB 43268* (MEL); Longwood, *R.Filson 1544* (MEL).

Close to *D. rosmarinifolia*, but can be distinguished by its glabrescent papillate stem, entire or dentate leaves which are less revolute and the yellowish hairs on the corolla. Has been identified in S.A. as *D. lavandulacea*, but *D. dysantha* can be separated by its glabrescent papillate stem, its longer inflorescence branches, and the yellowish hairs on the outside of the corolla. Also close to *D. marifolia* which has pubescent stems, shorter inflorescence branches, and loose yellowish to brownish corolla hairs.

39. **Dampiera lavandulacea** Lindley, *Sketch Veg. Swan R.* xxvii (1839)

T: Swan River, W.A., 1839, *J.Drummond s.n.*; lecto: CGE, *fide* M.T.M.Rajput & R.C.Carolin, *Telopea* 3: 201 (1988).

D. preissii Vriese in J.G.C.Lehmann, *Pl. Preiss.* 1: 403 (1845). T: near York, W.A., 12 Sept. 1839, *L.Preiss 1481*; lecto: LD, *fide* M.T.M.Rajput & R.C.Carolin, *Telopea* 3: 201 (1988); isolecto: MEL.

D. rupicola S.Moore, *J. Linn. Soc., Bot.* 45: 186 (1920). T: Bruce Rock, W.A., *F.Stoward 720*; holo: BM.

D. repens DC., *Prodr.* 7(2): 503 (1839). T: Swan River, W.A., *J.Drummond s. n.*; holo: K.

[*D. rosmarinifolia auct. non* Schldl.: C.A.Gardner, *Enum. Pl. Austral. Occid.* 127 (1930)].

Erect or ascending perennial to 70 cm tall; stem conspicuously ribbed, glabrous. Leaves sessile, linear to oblong-elliptic, revolute, entire or sometimes with few teeth, glabrescent above, white-tomentose below; lamina 9–21 mm long, 2–11 mm wide. Flowers in panicles; inflorescence branches solitary in lower axils, up to 3 clustered in upper axils, to 10 mm long, mostly 1-flowered; pedicel 0.5–1 mm long; bracteoles linear to oblong, 4–7 mm long. Sepals hidden by hairs, 0.2–0.5 mm long. Corolla 10–15 mm long; hairs outside long, type v, grey; wings 2.5–3 mm wide, slightly smaller above auricle; calli 6–10 per row. Ovary 1.5–1.7 mm long; ovule erect. Fruit obovoid, to 4 mm long. Figs 19N, 23M.

Widespread throughout south-western W.A. from near-coastal areas to the Eastern Goldfields; grows in sand in woodland and heath. Flowers chiefly July–Oct. Map 74.

W.A.: between Mullewa and Morawa, *W.E.Blackall 2815* (PERTH); Dinner Hill c. 48 km W of Watheroo, *R.C.Carolin 3401* (SYD); Bencubbin, *W.E.Blackall 3388* (PERTH); Wongan Hills, *N.T.Burbidge 0348* (CANB); c. 37 km E of Lake King, *J.S.Beard 3696* (PERTH).

Similar to *D. rosmarinifolia* which has more revolute leaves which always lack teeth.

However, in some specimens the character combination used to separate *D. rosmarinifolia* and *D. lavandulacea* breaks down. Nearly all specimens from south-eastern districts and some from further W have more tightly revolute leaves so that it is sometimes difficult to see the lower surface. One specimen in particular (c. 3 km NE of Howick Hill, *H.Eichler 19871*, AD) has tightly revolute leaves and tomentose stems and appears quite intermediate between *D. lavandulacea* and *D. rosmarinifolia*.

40. Dampiera oligophylla Benth., *Fl. Austral.* 4: 115 (1868)

T: south-western W.A., *J.Drummond 4: 193*; lecto: K, *fide* M.T.M.Rajput & R.C.Carolin, *Telopea* 3: 195 (1988); isolecto: BM, MEL.

Erect perennial to 60 cm tall, glabrescent; stem conspicuously ribbed. Leaves sessile, linear-terete to elliptic or spathulate, entire or with few teeth, thickened or slightly revolute at margin, thick, glabrous or whitish pubescent in patches, plants often seemingly leafless; lamina 3–56 mm long, 1–19 mm wide. Flowers in cymo-panicles; inflorescence branches mostly solitary, 2–3-flowered, 5–27 mm long; pedicel 1–6 mm long; bracteoles oblong to narrowly elliptic, 1.5–3 mm long. Sepals hidden by hairs, 0.3–0.7 mm long. Corolla 13–18 mm long; hairs outside loose, mostly type i, silvery-grey; wings 2–5 mm wide, smaller above auricle; calli usually 5–10 per row. Ovary 2–4.5 mm long; ovule erect. Fruit cylindrical to ovoid, to 7 mm long. $n = 9$ (as *Dampiera juncea*), W.J.Peacock, *Proc. Linn. Soc. New South Wales* 88: 10 (1963). Fig. 20A.

Occurs in W.A., from the Murchison R. southwards to Moora, in sandy heath. Flowers chiefly Sept.–Jan. Map 75.

W.A.: Murchison House, *R.C.Carolin 3329* (SYD); Hutt R., *A.C.Burns 11* (PERTH); c. 16 km N of Northampton, *C.A.Gardner 2042* (PERTH); Eradu, 24 Oct. 1965, *A.C.Burns 1* (PERTH); c. 33 km S of Three Springs, *A.S.George 3066* (PERTH).

See note under *D. juncea*. Cytological voucher is *W.J.Peacock 60847.2* (SYD).

41. Dampiera juncea Benth., *Fl. Austral.* 4: 115 (1869)

T: south-western W.A., *J.Drummond 168*; lecto K, *fide* M.T.M.Rajput & R.C.Carolin, *Telopea* 3: 195 (1988); isolecto: MEL.

D. stowardii S.Moore, *J. Linn. Soc., Bot.* 45: 186 (1920). T: locality uncertain, probably Belka, W.A. *F.Stoward 306*; holo: BM.

Illustration: K.Krause, *Pflanzenr.* 54: 195, fig. 34G–K (1912).

Erect perennial to 60 cm high, glabrescent except flowers; stems obscurely ribbed. Leaves sessile, linear-terete to lanceolate, entire, thick, glabrous or pubescent in patches; lamina 3–50 mm long, 1–20 mm wide. Flowers in cymo-panicles; inflorescence branches usually solitary, to 3-flowered, to 12 mm long; pedicel to 2 mm long; bracteoles oblong, to 3 mm long. Sepals almost obsolete. Corolla 15–20 mm long; hairs outside much-branched, type i, slate grey; wings 2–5 mm wide, smaller above auricle. Calli to 10. Ovary c. 3 mm long; ovule erect. Fruit cylindrical, to 7 mm long. $n = 9$ (as *Dampiera sp.*), and $n = 18$ W.J.Peacock, *Proc. Linn. Soc. New South Wales* 88: 10 (1963) as *D. loranthifolia*. Fig. 20B.

Occurs in inland areas of south-western W.A., on sand plains. Flowers chiefly Aug.–Nov. Map 76.

W.A.: c. 16 km SW of Three Springs on road between Latham and Maya, SE of Morawa, *W.E.Blackall 3752* (PERTH); W of Coorow, *J.S.Beard 1940* (PERTH); near Koorda, *W.E.Blackall 3478* (PERTH); c. 8 km W of Southern Cross, *H.Doing* (CANB); c. 35 km S of Narembeen, *R.H.Kuchel 2046* (AD).

Similar to *D. oligophylla* which has looser hairs on the corolla, and much more prominent ribs on the stem. It occurs to the south of the distribution of *D. oligophylla*. Cytological vouchers are *W.J.Peacock 6096.2*, *6093.1*, *6087.4* and *60910.1* (SYD).

42. Dampiera linearis R.Br., *Prodr.* 588 (1810)

T: King George Sound, [W.A.], Dec. 1801, *R.Brown*; lecto: BM, *fide* M.T.M.Rajput & R.C.Carolin, *Telopea* 3: 195 (1988).

D. cuneata R.Br., *Prodr.* 588 (1810). T: King George Sound, [W.A.], Dec. 1801, *R.Brown*; lecto: BM, *fide*

M.T.M.Rajput & R.C.Carolin, *Fl. Australia* 35: 330 (1992).

D. azurea Vriese in J.G.C.Lehmann, *Pl. Preiss.* 1: 400 (1845). T: near Swan R., W.A., 25 June 1839, *L.Preiss 1475*; lecto: LD, *fide* M.T.M.Rajput & R.C.Carolin, *Telopea* 3: 195 (1988); isolecto: L, MEL, W.

D. erecta Vriese in J.G.C.Lehmann, *Pl. Preiss.* 1: 401 (1845). T: near Albany, W.A., 22 Sept. 1840, *L.Preiss 1487*; holo: BM.

D. eriophora Vriese in J.G.C.Lehmann, *Pl. Preiss.* 1: 400 (1845). T: Stirling Terrace, Albany, W.A., Sept. 1842, *L.Priess 1500*; lecto LD, *fide* M.T.M.Rajput & R.C.Carolin, *Telopea* 3: 195 (1988); isolecto: MEL, W.

D. lanuginosa Vriese, *Natuurk. Verh. Holl. Maatsch. Wetensch. Haarlem* ser. 2, 10: 81 (1854). T: Swan River, W.A., *J.Drummond 127*; lecto: K, *fide* M.T.M.Rajput & R.C.Carolin, *Telopea* 3: 195 (1988).

D. linearis f. *latifolia* K.Krause, *Pflanzenr.* 54: 192 (1912). T: near Parkerville, W.A., Nov. 1900, *F.L.E.Diels 1632*; holo: B.

D. linearis f. *humilis* K.Krause, *loc. cit.* T: King George Sound, W.A., *Wawra 915*; holo: W.

D. linearis f. *elongata* K.Krause, *loc. cit.* T: Albany, W.A., *B.T.Goadby;* holo: K.

Illustration: R.Erickson *et al.*, *Fl. Pl. W. Australia* 53, fig. 133 (1973).

Erect perennial to 60 cm tall, glabrescent except inflorescence; stems obscurely ribbed. Leaves sessile, sometimes in pseudowhorls, obovate to elliptic, entire to lobed; lamina 6–39 mm long, 2–17 mm wide. Flowers in cymo-panicles; inflorescence branches 1–6 together, 1–7-flowered, 13–47 mm long; pedicel 2–5 mm long; bracteoles linear to elliptic, 4–10 mm long, hairy or glabrescent. Sepals hidden by hairs, to 0.5 mm long. Corolla 10–15 mm long; hairs outside loose, silky, type ii, silvery-grey to almost black; wings 2–3.5 mm wide, usually smaller above auricle; calli 4–11 per row. Ovary c. 3 mm long; ovule erect. Fruit cylindrical, to 5 mm long. $n = 9$; $n = 18$; $n = 27$ W.J.Peacock, *Proc. Linn. Soc. New South Wales* 88: 10 (1963) as *D. cuneata*. Figs 9, 20C.

Common in south-western W.A. from the Geraldton area southwards; in sandy soil. Flowers chiefly Aug.–Nov. Map 77.

W.A.: Koorda, *S.B.Rosier 306* (PERTH); Solomons Well, Conical Hill, Stirling Ra., 27 Sept. 1902, *A.Morrison* (CANB); SE of Mundaring, *J.J.Havel H474* (PERTH); Wooroloo, *R.C.Carolin 3425* (SYD); Ravensthorpe, Sept. 1925, *C.A.Gardner & W.E.Blackall* (PERTH); Mt Barker, *A.Oldfield 232* (MEL).

There is a remarkable variation in the leaf shape, leaf margin, size of peduncle, colour and nature of the indumentum on the outside of the corolla, much of which has been the basis for taxa described in the past. However, it is impossible in our present state of knowledge to circumscribe these subdivisions. This species and *D. pedunculata* have bracteoles well exceeding the ovary.

Cytological vouchers are *W.J.Peacock 60883.2*, *60877.1*, *60830.1*, *60879.1*, *60881.2*, and *60820.1* (SYD).

L.R.Brousfield & S.H.James (The behaviour and possible cytoevolutionary significance of ß chromosomes in *Dampiera linearis* (Angiospermae: Goodeniaceae), *Chromosoma* 55: 309–323, 1976) have suggested that diploids occur on old lateritic land surfaces whilst tetraploids occur on adjacent coastal sandplains. They suggested that the tetraploid is 'a distinct biological species' but gave no morphological differences between the cytodemes and cited no voucher specimens.

43. **Dampiera pedunculata** Rajput & Carolin, *Telopea* 3: 200 (1988)

T: Ruabon, W.A., 29 Sept. 1953, *R.D.Royce 4516*; holo PERTH.

Ascending to decumbent perennial to 70 cm long, glabrous except flowers; stem not much-branched, terete, ribbed below, ±triangular above. Leaves sessile, fasciculate in axils, linear to linear-oblong, dentate, glabrous; lamina 10–35 mm long, 1.5–4 mm wide. Flowers in numerous lateral cymo-panicles; inflorescence branches 1–4 together, 3–4-flowered, to 4 cm long; pedicel to 10 mm long; bracteoles oblong, 5–7 mm long. Sepals minute. Corolla 9–12 mm long; hairs outside type i, short and dark grey to black, and type ii, long silky and paler grey; wings c. 2.5 mm wide, smaller above auricle; calli 7–12 per row. Ovary c. 3 mm long; ovule erect. Fruit not seen. Fig. 20D.

Occurs in south-western W.A. between Perth and Albany, in swampy and sandy soil. Flowers chiefly Aug.–Nov. Map 78.

W.A.: Welshpool to Kalamunda, Sept. 1909, *J.H.Maiden* (NSW); NE of Albany at Oyster Harbour, *A.M.Ashby 1998* (AD).

Similar to *D. linearis* which has obovate to elliptic leaves.

Subsect. 2. Angulares

Dampiera sect. **Dampiera** subsect. **Angulares** Rajput & Carolin, *Telopea* 3: 194 (1988)

Type: *D. stricta* (Smith) R.Br.

Stems triangular, angular or flattened; phyllotaxis 1/3 or 1/2. Ovary usually 1-locular, sometimes gibbous; ovule erect, slightly curved or horseshoe-shaped, basifixed.

A subsection with 23 species, in southern Australia.

Ser. 1. Angulares

Dampiera sect. **Dampiera** subsect. **Angulares** ser. **Angulares** Rajput & Carolin, *Fl. Australia* 35: 329 (1992).

Type: *D. stricta* (Smith) R.Br.

Dampiera sect. *Dicoelia* Benth., *Hooker's Icon. Pl.* 11: 19, t. 1026 (1866). T: *D. trigona* Vriese; lecto, *fide* M.T.M.Rajput & R.C.Carolin, *Telopea* 3: 194 (1988).

Ovary cylindrical or slightly oblique; ovule oblong-linear, erect.

This series contains 15 species, in southern Australia.

44. Dampiera latealata (E.Pritzel) Rajput & Carolin, *Telopea* 3: 204 (1988)

D. trigona var *latealata* E.Pritzel in F.L.E.Diels & E.Pritzel, *Bot. Jahrb. Syst.* 35: 578 (1905). T: Dundas, W.A., *F.L.E.Diels 5257*; holo: ?B (destroyed) *n.v.*; 27 miles [c. 43 km] north on Eyre Highway, W.A., 3 Oct. 1961, *J.H.Willis*; neo: MEL, *fide* M.T.M.Rajput & R.C.Carolin, *Telopea* 3: 204 (1988).

Illustration: K.Krause, *Pflanzenr.* 54: 172, fig. 31D–G (1912) as *D. trigona* var. *latealata*.

Erect perennial to 1 m tall, glabrous except flowers; stems 3-winged, each wing 2.5–5 mm wide, distinctly thickened on angle of each wing. Leaves sessile, linear-oblong to lanceolate, entire or with 2–4 small lobes; lamina 11–29 mm long, 2.5–8 mm wide. Flowers in much-branched cymo-panicles; inflorescence branches 1–3 together, each bearing to 12 flowers, to 14 cm long, much exceeding leaves; pedicel 3–4 mm long; bracteoles oblong, 0.7–1.2 mm long. Sepals 0.4–0.5 mm long. Corolla 10–15 mm long; hairs outside fine, appressed, type iii and iv, dark grey; wings 2.7–3 mm wide, smaller above auricle; calli 3 or 4 per row. Ovary bilocular, 2–2.5 mm long. Fruit to 4 mm long. Fig. 20E.

Occurs in the Eastern Goldfields region of W.A. in lateritised and red sandy soil. Flowers chiefly Sept.–Dec. Map 79.

W.A.: c. 55 km N of Kalgoorlie, *R.C.Carolin 5760* (SYD); Widgiemooltha, *C.A.Gardner 9521* (PERTH); Beacon Hill, *M.E.Phillips 681673* (PERTH); c. 8 km N of Norseman, *R.D.Royce 3475* (PERTH); at 450–451 mile peg, [i.e. c. 720 km from Perth] N of Gibson, *S.Carlquist 3425* (NSW).

See note under *D. trigona.*

45. **Dampiera decurrens** Rajput & Carolin, *Telopea* 3: 203 (1988)

T: Lucky Bay, E of Esperance, W.A., 10 Sept. 1966, *E.M.Bennett [Scrymgeour] 882*; holo: PERTH.

D. prostrata Vriese, *Natuurk. Verh. Holl. Maatsch. Wetensch. Harlem* ser. 2, 10: 83 (1854), *nom. illeg.* non Vriese (1845). T: south-western W.A., *J.Drummond 364.*

Illustration: K.Krause, *Pflanzenr.* 54: 172, fig. 31A–C (1912) as *D. prostrata.*

Stiff, robust perennial to 1 m tall, glabrous except flowers; stem with narrow wings. Leaves sessile, ovate to elliptic with broad base, dentate; lamina 12–41 mm long, 5–23 mm wide. Flowers in cymes; inflorescence branches to 3 together, each bearing to 12 flowers, to 6 cm long; pedicel 3.5–5.2 mm long; bracteoles linear-oblong, 3.2–4.7 mm long. Sepals almost obsolete. Corolla to 12 mm long; hairs outside fine, appressed, type iv, grey; wings 3.5–4 mm wide, smaller or obsolete above auricle; calli 2–5 per row. Ovary bilocular, glabrous, 3.5–4 mm long. Fruit not seen. Figs 20F, 33A–B.

Occurs on the south coast of W.A. from Cheyne Beach eastwards; on granite. Flowers chiefly Sept.–Jan. Map 80.

W.A.: Cheyne Beach fishery, *G.Maxwell* (MEL); Sandy Hook Is., Archipelago of the Recherche, 10 Nov. 1950, *J.H.Willis* (MEL); Mt Gardner, *G.Maxwell* (MEL); Lucky Bay, *A.S.George 7461* (PERTH, SYD); Cape le Grand, 30 Oct. 68, *J.W.Wrigley* (CBG, SYD).

See note under *D. trigona.* G.Bentham, *Fl. Austral.* 4: 110 (1868) applied the name *D. prostrata* Vriese (1854, non 1845) to this species.

46. **Dampiera trigona** Vriese in J.G.C.Lehmann, *Pl. Preiss.* 1: 401 (1845)

T: near Maddington, W.A., 2 Nov. 1839, *L.Preiss 1471*; lecto: LD, *fide* M.T.M.Rajput & R.C.Carolin, *Telopea* 3: 204 (1988); isolecto: L, MEL, W.

D. biloculata F.Muell., *Fragm.* 2: 17 (1860). T: Vasse, W.A., *A.Oldfield*; holo: MEL.

D. trigona var. *tenuis* Benth., *Fl. Austral.* 4: 110 (1868). T: south-western W.A., *J.Drummond 4: 192*; holo: K; iso: MEL.

Ascending, often delicate herb to 40 cm tall, glabrous except flowers; stem with acute angles. Leaves sessile, mostly linear to lanceolate, entire to slightly lobed; lamina 11–28 mm long, 2–6 mm wide. Flowers in loose, terminal cymo-panicles; inflorescence branches usually solitary, to 2 cm long, usually bearing only 2 flowers; pedicel 1–2 mm long; bracteoles linear to linear-elliptic, c. 4 mm long. Sepals obsolete, or reduced to few stiff hairs. Corolla 7–8 mm long; hairs outside fine, appressed, type iv, grey to blackish; wings 3–3.5 mm wide, ±smaller above auricle; calli 3–7 per row. Ovary bilocular, glabrous, 4.2–5 mm long. Fruit to 5 mm long. Fig. 20G.

Occurs in near-coastal areas between the Moore R. and Albany, south-western W.A. Flowers chiefly Sept.–Dec. Map 81.

W.A.: Mogumber, Aug. 1929, *W.E.Blackall* (PERTH); S of Lake Clifton, Yalgorup Natl Park, *S.Paust 1379* (PERTH); Kudardup, *R.D.Royce 4636* (PERTH); c. 1.5 km NW of Shannon R., *A.C.Beauglehole 12660* (SYD); Cape Leschenault, *A.Oldfield* (MEL).

Dampiera latealata, *D. decurrens* and *D. trigona* differ from all other members of this series in having 2-locular ovaries. *Dampiera latealata* has conspicuous wings on the stem; *D. decurrens* has narrow wings; *D. trigona* has an angled but not winged stem.

47. **Dampiera loranthifolia** F.Muell. ex Benth., *Fl. Austral.* 4: 115 (1868)

T: Phillips R., W.A., 1861, *G.Maxwell 140*; lecto: K, *fide* M.T.M.Rajput & R.C.Carolin, *Telopea* 3: 207 (1988); isolecto: BM, photo SYD.

Stiff, erect perennial to 50 cm high, glabrous except inflorescence; stem with acute angles. Leaves sessile, lanceolate to elliptic, entire or dentate, thick; lamina 11–57 mm long, 5–32 mm wide. Flowers in cymes in upper axils; branches clustered, each 2–many-flowered, 3–15 mm long; pedicel 1–3 mm long; bracteoles oblong-elliptic, 1.5–2.5 mm long. Sepals obsolete. Corolla to 12 mm long; hairs outside loose, type iv, grey, brown or black; wings 2–2.5 mm wide, smaller above auricle; calli 2–6 per row. Ovary c. 2.5 mm long. Fruit not seen. Fig. 20H.

Occurs from the Stirling Ra. to Cape le Grand, south-western W.A. Flowers chiefly Sept.–Dec. Map 82.

W.A.: Mt Drummond, SW of Ravensthorpe, *K.Newbey 2699* (PERTH); Annie Peak, Eyre Ra., *A.S.George 7256* (PERTH); above creek on NE side of Whoogarup Ra., *A.S.George 1892* (PERTH); Cape le Grand, c. 1.5 km from beach, *J.W.Wrigley 029962* (SYD); Thumb Peak range, SW of Ravensthorpe, *A.S.George 7142* (PERTH).

Close to *D. leptoclada* which has appressed hairs on the corolla. See also *D. fasciculata*. Most specimens collected at Thumb Peak south-west of Ravensthorpe have darker brown leaves and longer hairs on the outside of the corolla compared to other specimens of this species. One specimen collected near Bluff Knoll, Stirling Ra. (*R.D.Royce 6049*), possibly to be referred here, differs from the rest as follows: broader leaves, upper leaves whorled and black hairs on the outside of the corolla. There are some specimens from the Mt Barren area which have red-brown hairs on the outside of the corolla.

48. **Dampiera leptoclada** Benth., *Fl. Austral.* 4: 116 (1868)

T: King George Sound, W.A., *F.Mueller*; lecto: K, *fide* M.T.M.Rajput & R.C.Carolin, *Telopea* 3: 207 (1988).

D. subspicata Benth., *op. cit.* 117. T: near the base of Mt Bland, W.A., *G.Maxwell*; holo: K; iso: MEL.

D. leptoclada var. *parviflora* Benth., *op. cit.* 116. T: Cape Arid, W.A., *G.Maxwell*; holo: K.

Erect perennial to 60 cm tall, glabrous except flowers; stems with acute angles. Leaves sessile, sometimes in pseudowhorls, oblong to lanceolate, entire or dentate; lamina 16–40 mm long, 3–10 mm wide. Flowers in cymo-panicles; inflorescence branches to 5 together, 1–3-flowered, to 3 cm long, usually not exceeding the leaves; pedicel 3–4 mm long; bracteoles oblong, c. 3 mm long. Sepals c. 0.5 mm long. Corolla 10–14 mm long; hairs outside appressed, type iv, dark grey to black; wings 2.5–3 mm wide, smaller above auricle; calli to 3 per row. Ovary 1.7–2.5 mm long, glabrous or glabrescent. Fruit to 5 mm long. $n = 9$, W.J.Peacock. *Proc. Linn. Soc. New South Wales* 88: 10 (1963). Figs 20I, 26H.

Occurs from Perth to Augusta and E to Esperance, south-western W.A., in sandy and often damp soil. Flowers chiefly Sept.–Jan. Map 83.

W.A.: Pingelly, *T.E.H.Aplin 760* (PERTH); E of Esperance, *R.D.Royce 8718* (PERTH); West Mt Barren, *C.A.Gardner 2208* (PERTH); c. 16 km W of Albany, on Denmark road, *W.J.Peacock 608842* (SYD); King George Sound, *B.T.Goadby 101* (PERTH).

Similar to *D. fasciculata* which has to 7 flowers per branch, and the ovary not glabrous. The leaves of *D. fasciculata* are almost always in pseudowhorls, while those of *D. leptoclada* are not usually so arranged. Cytological voucher *W.J.Peacock 60885.2* (SYD).

49. **Dampiera galbraithiana** Rajput & Carolin, *Telopea* 3: 204 (1988)

T: 2.5 miles [c. 4 km] E of Cheynes Bridge on Macalister R., Vic., 20 Oct. 1973, *J.H.Willis*; holo: MEL; iso: AD, CANB.

[*D. scottiana auct. non* F.Muell.: J.Galbraith, *Victorian Naturalist* 93: 161 (1976)].

Erect perennial to 60 cm tall, glabrous except flowers or with scattered hairs; stems triangular but also with ribs alternating with the more prominent angles. Leaves oblong-elliptic to lanceolate, dentate; lamina 12–45 mm long, 3–17 mm wide. Flowers in panicles; inflorescence branches 1–3 together, usually 1-flowered, to 10 mm long, usually not exceeding leaves; pedicel 3–9 mm long; bracteoles linear-oblong, to 2 mm long. Sepals often unequal, 0.5–1 mm long. Corolla 9–11 mm long; hairs outside, appressed, type iv, dark grey; wings 1–1.5 mm wide, ±equal; calli 5 or 6 per row. Ovary 2.5–3 mm long. Fruit to 7 mm long. Figs 20J, 33E.

Occurs in the east central highlands of Vic. Flowers chiefly Sept.–Jan. Map 84.

Vic.: c. 15 km SSE of Licola, *A.C.Beauglehole 43382, J.H.Willis & E.Chester* (SYD).

Bracteoles are inserted well below ovary articulation.

50. **Dampiera triloba** Lindley, *Sketch Veg. Swan R.* xxvii (1839)

T: Swan River, W.A., 1837, *J.Mangles*; holo: CGE.

D. repandra Vriese in J.G.C.Lehmann, *Pl. Preiss.* 1: 400 (1845). T: between Perth and Guildford, W.A., 22 Nov. 1839, *L.Preiss 1518*; lecto: LD, *fide* M.T.M.Rajput & R.C.Carolin, *Telopea* 3: 204 (1988); isolecto: K, L, W.

D. drummondii Vriese, *Ned. Kruidk. Arch.* 2: 8 (1850). T: Swan River, W.A., 1845, *J.Drummond 37 suppl. coll. 1845*; holo: K.

Erect perennial to 50 cm tall, brown tomentose when young, glabrescent; stem with acute angles. Leaves sessile, sometimes in pseudowhorls, obovate-oblong to cuneate, dentate or lobed; lamina 16–55 mm long, 3–35 mm wide. Flowers in panicles; inflorescence branches up to 7 together, to 7-flowered, to 6 cm long; pedicel 3–7 mm long; bracteoles oblong-elliptic, 1.7–2.5 mm long. Sepals hidden beneath hairs, 0.2–0.3 mm long. Corolla c. 8 mm long; hairs outside ±appressed, type v, golden or greyish brown; wings 1–2.5 mm wide, smaller above auricle; calli 4–9 per row. Ovary 1.5–2.5 mm long. Fruit to 5 mm long. Fig. 20K.

Occurs in south-western W.A. between Perth and Cunderdin and S almost to Albany. Flowers chiefly Aug.–Dec. Map 85.

W.A.: Bayswater, Lower Swan R., 4 Nov. 1911, *A.Morrison* (NSW); no precise locality, *J.Drummond 57* (MEL); Keenan, *C.E.Carter 235* (CANB); Cunderdin, Aug. 1903, *W.V.Fitzgerald* (PERTH); Gnangara, Oct. 1945, *C.A.Gardner* (PERTH).

The bracteoles are alternate. There is considerable variation in the colour of the hairs and indumentum of the leaves.

Figure 33. *Dampiera*. **A–B**, *D. decurrens*. **A**, habit X2; **B**, hair X50 (**A–B**, A.Strid 21175, SYD). **C–D**, *D. parvifolia*. **C**, habit X2; **D**, hair X50 (**C–D**, R.Kuchel 1678, SYD). **E**, *D. galbraithiana*, habit X2 (A.Beauglehole 43382, SYD). Drawn by D.Mackay.

51. **Dampiera parvifolia** R.Br., *Prodr.* 589 (1810)

T: Bay 1 [Lucky Bay], [W.A.], 13 Jan. 1802, *R.Brown*; lecto: BM, *fide* M.T.M.Rajput & R.C.Carolin, *Telopea* 3: 204 (1988); isolecto: MEL.

Erect perennial to 60 cm tall, glabrous except inflorescence; stem bluntly triangular, grooved. Leaves sessile, oblong to elliptic, entire or dentate, thick; lamina 4–40 mm long, 2–12 mm wide. Flowers in cymo-panicles; inflorescence branches solitary or few together in axillary clusters, 1–3-flowered, very short, the flowers almost sessile in the axils; pedicel obsolete; bracteoles numerous, scarious, ovate-deltoid, brownish, concave, 2.5–3.5 mm long. Sepals obscured by hairs, to 0.5 mm long. Corolla 8–10 mm long; hairs outside loose, type iv or v, silvery; wings 2–2.5 mm wide, usually smaller above auricle; calli 1–3 per row. Ovary c. 2.5 mm long. Fruit to 5 mm long. Figs 20L, 33C–D.

Occurs in southern W.A. from near Esperance E to Point Culver. Flowers chiefly Sept.–Dec. Map 86.

W.A.: c. 16 km W of Point Culver, *M.G.Brooker 3699* (PERTH); c. 11 km out of Esperance on Norseman road, 13 Oct. 1963, *B.Benn* (SYD); 15 km N of Esperance, *P.G.Wilson 3053* (AD); near Howick Hill, *H.Eichler 19846* (AD, PERTH); Lucky Bay, *A.S.George 7504* (PERTH).

Distinguished from *D. sericantha* by the numerous bracteoles beneath the flowers.

52. **Dampiera sericantha** F.Muell. ex Benth., *Fl. Austral.* 4: 118 (1868)

T: Lucky Bay, W.A., *G.Maxwell*; holo: K.

Erect, perennial herb to 40 cm tall, glabrous except flowers; stem with blunt angles. Leaves sessile, oblong-lanceolate to ovate-elliptic, entire or toothed; lamina 2.5–9 mm long, 1.5–3 mm wide. Flowers in panicles; inflorescence branches mostly solitary, usually 1- or 2-flowered, to 2.5 cm long; pedicel 1.7–2.5 mm long; bracteoles c. 3 mm long. Sepals to 0.3 mm long. Corolla 9–11 mm long; hairs outside fine, appressed, type iii and iv, silvery-grey; wings 1.2–2 mm wide, smaller above auricle; calli to 3 per row. Ovary 2–2.5 mm long. Fruit not seen. $n = 9$, W.J.Peacock, *Proc. Linn. Soc. New South Wales* 88: 11 (1963). Fig. 20M.

Occurs near Esperance, W.A. Flowers chiefly Aug.–Nov. Map 87.

W.A.: c. 10 km NW of Gibson, *P.J.Cole 433* (AD); between Oldfield and Young R., Esperance road, *B.Benn* (SYD); Lucky Bay, *F.Mueller* (MEL); south-western W.A., *G.Maxwell* (MEL).

See note under *D. parvifolia*. Cytological voucher *W.J.Peacock 60884.3* (SYD).

53. **Dampiera fasciculata** R.Br., *Prodr.* 588 (1810)

T: King George Sound, [W.A.], Dec. 1801, *R.Brown*; lecto BM, *fide* M.T.M.Rajput & R.C.Carolin, *Telopea* 3: 207 (1988); isolecto: K.

D. subverticillata Vriese in J.G.C.Lehmann, *Pl. Preiss.* 1: 403 (1845). T: Konkoberup hills [Mt Melville, near Cape Riche], W.A., 19 Nov. 1840, *L.Preiss 1510*; lecto: LD, *fide* M.T.M.Rajput & R.C.Carolin, *Telopea* 3: 207 (1988); isolecto: L, MEL.

D. fasciculata var. *angustifolia* Benth., *Fl. Austral.* 4: 117 (1868). T: Cape Arid, W.A., *G.Maxwell*; holo: K.

[*D. stricta auct. non* R.Br.: F.W.Sieber in A.P.de Candolle, *Prodr.* 7(2): 504 (1839)].

Erect or ascending perennial to 60 cm tall, glabrous except flowers; stem with rounded angles and 3 distinct grooves. Leaves sessile, often fasciculate in axils and in pseudowhorls, oblong to oblanceolate, entire or slightly dentate; lamina 14–39 mm long,

4–13 mm wide. Flowers in panicles; inflorescence branches clustered in upper axils, to 7-flowered, to 2 cm long, mostly scarcely exceeding leaves; pedicel to 5 mm long; bracteoles linear to oblong-elliptic, 2–3 mm long. Sepals obsolete. Corolla 10–12 mm long; hairs outside appressed, type iv, light brown to grey; wings 2.5–3 mm wide, smaller above auricle; calli 2–5 per row. Ovary 2–2.5 mm long. Fruit to 5 mm long. Fig. 21B.

Occurs along the south coast and a little inland from Cape Leeuwin to Mt Ragged, south-western W.A., in sandy soil. Flowers chiefly Aug.–Nov. Map 88.

W.A.: c. 1.5 km W of Newdegate, *F.Lullfitz 13694* (PERTH); rabbit proof fence, Esperance, 10 Oct. 1963, *B.Benn* (SYD); Lort R. crossing, *E.N.S.Jackson 1358* (AD, PERTH); near West Mt Barren, *C.A.Gardner 2215* (PERTH); Albany, 1898, *R.Helms* (PERTH).

See notes under *D. loranthifolia* and *D. leptoclada.*

54. Dampiera stricta (Smith) R.Br., *Prodr.* 589 (1810)

Goodenia stricta Smith, *Trans. Linn. Soc. London, Bot.* 2: 349 (1794). T: New South Wales, *J.White*; holo: LINN; iso: BM.

D. oblongata R.Br., *Prodr.* 588 (1810); *D. stricta* var. *oblongata* (R.Br.) Benth., *Fl. Austral.* 4: 116 (1868). T: near Kingston, Newcastle, N.S.W., Oct. 1804, *R.Brown*; holo: BM.

D. stricta var. *laxa* Benth., *loc. cit.* T: Bunyip Creek, Mt Macedon, Plenty Range, Vic., *F.Mueller*; lecto: K, *fide* M.T.M.Rajput & R.C.Carolin, *Telopea* 3: 206 (1988).

[*D. fasciculata auct. non* R.Br.: A.P.de Candolle, *Prodr.* 7(2): 504 (1839), *p.p.*].

Illustration: E.R.Rotherham *et al.*, *Fl. Pl. New South Wales & S. Queensland* 28, fig. 38 (1975).

Erect perennial 20–60 cm high, glabrous or glabrescent except flowers; stem with acute angles. Leaves sessile, often in pseudowhorls, linear to elliptic or lanceolate, entire or with few teeth; lamina 16–45 mm long, 2–19 mm wide. Flowers in panicles; inflorescence branches 1 or 2 together, 1- or 2-flowered, to 3 cm long; pedicel 1–2 mm long; bracteoles linear, 2.5–3 mm long. Sepals 0.7–1.2 mm long. Corolla 10–12 mm long; hairs outside type v, rusty; wings 2–2.7 mm wide, smaller above auricle; calli 3–9 per row. Ovary 2–2.5 mm long. Fruit to 5 mm long. Figs 11, 21A, 23L.

Occurs from Port Curtis, Qld, through N.S.W. and Vic. to Tas.; grows on the Great Dividing Ra. and in coastal regions, usually among heath in sandy soil. Flowers chiefly Aug.–Jan. Map 89.

Qld: near Wallangarra, *L.Pedley 1582* (BRI). N.S.W.: Bundarra, *C.W.E.Moore 3591* (CANB, NSW). Vic.: Blackwood Ra., *A.C.Beauglehole 43263* (MEL). Tas.: Freycinet Peninsula, *E.R.Rodway 2651* (NSW).

See notes under *D. fusca*, *D. ferruginea* and *D. sylvestris.*

55. Dampiera fusca Rajput & Carolin, *Telopea* 3: 205 (1988)

T: Kydra Peaks, N.S.W., 11 Jan. 1970, *K.C.Rogers & J.H.Willis*; holo: MEL.

Erect, branched perennial to 30 cm tall, brownish pubescent or glabrescent; stem with acute angles, papillate. Leaves sessile, upper ones usually in pseudowhorls, slightly recurved, oblong to oblanceolate, dentate, papillate; lamina 8–22 mm long, 2–8 mm wide. Flowers in condensed panicles; inflorescence branches clustered, 1–3-flowered, to 2 cm long; pedicel 1–2.5 mm long; bracteoles oblong, 6–7.5 mm long. Sepals obscured by hairs, 0.4–0.7 mm long. Corolla c. 8 mm long; hairs outside loose, type v, brownish grey; wings 1–2 mm wide, slightly smaller above auricle; calli to 2 per row. Ovary 1.5–2 mm long. Fruit not seen.

Figure 34. *Dampiera*. **A–B**, *D. sylvestris*. **A**, habit X1; **B**, hair X25 (**A–B**, P.Weston 30 & J.Ford, SYD). **C–E**, *D. obliqua*. **C**, habit X1; **D**, hair X25; **E**, fruit X10 (**C–E**, K.Allan 136, SYD). **F–H**, *D. heteroptera*. **F**, fruit X10; **G**, flower X3; **H**, habit X1 (**F–H**, A.Ashby 3696, SYD). Drawn by D.Mackay.

Occurs in southern N.S.W. and eastern Vic. along the Great Dividing Ra. Flowers chiefly Dec.–Jan. Map 90.

Vic.: c. 2.5 km from Diggers Hole track, *A.C.Beauglehole 41403* (MEL); Kybeyan, *R.H.Cambage 1999* (NSW); E of Kybeyan, 15 May 1949, *A.B.Costin* (NSW).

56. **Dampiera sylvestris** Rajput & Carolin, *Telopea* 3: 206 (1988)

T: Peach Mountain, Whian Whian State Forest, 15 miles [c. 25 km] north of Lismore, N.S.W., 3 Oct. 1967, *K.Grieves*; holo: NSW.

Erect perennial to 70 cm tall, glabrous except inflorescence; stem with acute angles. Leaves sessile, rarely in pseudowhorls, narrowly oblong to lanceolate, usually entire; lamina 50–80 mm long, 8–30 mm wide. Flowers in panicles; inflorescence branches 1–4 together, 4- or 5-flowered, 21–31 mm long; pedicel 1–3 mm long; bracteoles oblong-elliptic, 3–3.5 mm long. Sepals linear or linear-oblong, 1.5–2 mm long. Corolla 14–17 mm long; hairs outside long, not appressed, type v, grey; wings 3.8–4.2 mm wide, smaller above auricle; calli 3–7 per row. Ovary c. 3 mm long. Fruit ellipsoidal, c. 4 mm long, glabrescent, ribbed. Figs 21C, 34A–B.

Occurs from Port Curtis, Qld, southwards to N.S.W.; grows in coastal regions, in forest and woodland. Flowers chiefly Aug.–Feb. Map 91.

Qld: c. 3 km S of Tewantin, *P.Baxter & B.Lebler 1104* (CANB). N.S.W.: Chatsworth to Woodburn, *J.H.Maiden & J.L.Boorman* (NSW); Elimbah, *H.S.Mckee 9724* (CANB, NSW); Green Hills, c. 16 km directly N of Woolgoolga, *D.J.McGillivray 16* (NSW); Booti Booti, 13 Oct. 1953, *H.Johnson* (NSW).

Similar to *D. stricta* which has rusty ±appressed hairs on the outside of the corolla.

57. **Dampiera glabrescens** Benth., *Fl. Austral.* 4: 119 (1868)

T: south-western W.A., *J.Drummond 4: 194*; lecto: K; isolecto: BM, MEL, fide R.C.Carolin, *Fl. Australia* 35: 330 (1992).

Erect perennial to 20 cm tall, hairy in patches to glabrescent, hairs whitish; stem triangular. Leaves sessile, oblong to lanceolate, entire, pubescent in patches or glabrescent; lamina 9–28 mm long, 2–7 mm wide. Flowers in panicles; inflorescence branches 1–3 together, 1–3-flowered, 8–32 mm long; pedicel 3–11 mm long; bracteoles 3–4 mm long. Sepals hidden beneath hairs, linear-oblong, 0.5–0.7 mm long. Corolla 7–8 mm long; hairs outside ±appressed, type iii, brown to grey; wings 2.5–3.5 mm wide, usually smaller above auricle; calli obsolete. Ovary 2.5–3 mm long. Fruit to 5 mm long. Fig. 21D.

Occurs near Wongan Hills, W.A., in gravelly and sandy soil. Flowering not known. Map 92.

W.A.: c. 1.5 km S of Ballidu, *K.Newbey 2002* (PERTH); Wongan Hills, Oct. 1932, *E.H.Ising* (AD).

58. **Dampiera obliqua** Rajput & Carolin, *Telopea* 3: 208 (1988)

T: 17 miles [c. 27 km] east of Pingelly, W.A., 19 Sept. 1962, *R.D.Royce 7601*; holo: PERTH.

Erect perennial to 70 cm tall, glabrous except flowers; stem with acute angles. Leaves sessile, linear to oblong-elliptic, entire; lamina 11–65 mm long, 1.5–21 mm wide. Flowers in panicles; inflorescence branches clustered, 2–5 together, to 5 cm long, to 3-flowered; pedicel to 3 mm long; bracteoles oblong to narrowly elliptic, 1–1.5 mm long. Sepals unequal, 0.4–0.6 mm long. Corolla 8–10 mm; hairs outside appressed, type iv, silvery to grey; wings 2.4–2.6 mm wide, almost obsolete above auricle; calli 1–3 per row. Ovary oblique, 1.6–2 mm long. Fruit oblique, to 7 mm long. Figs 21E, 34C–E.

Occurs in inland south-western W.A. between Pingelly, Narrogin and Kukerin, on sandplains. Flowers chiefly Sept.–Dec. Map 93.

W.A.: Congelin, W of Narrogin, *G.Heinsohn 129* (PERTH); West Popanyinning, *F.Lullfitz L1725* (PERTH); Dryandra State Forest, Narrogin, *E.C.Nelson ANU 16899* (CANB); Kukerin to Kalgan road, *B.Benn* (SYD).

Ser. 2. Camptospora

Dampiera sect. **Dampiera** subsect. **Angulares** ser. **Camptospora** (Benth.) Rajput & Carolin, *Telopea* 3: 194 (1988).

Dampiera sect. *Camptospora* Benth. in Hooker, *Icon. Pl.* 11: 19, t. 1027 (1867). T: *D. alata* Lindley, lecto *fide* M.T.M.Rajput & R.C.Carolin, *Telopea* 3: 194 (1988).

Ovary and fruit gibbous; ovule curved.

This series contains 8 species, in south-western Australia.

59. Dampiera angulata Rajput & Carolin, *Telopea* 3: 209 (1988)

T: NE of Ravensthorpe, W.A., 22 Sept. 1925, *W.E.Blackall & C.A.Gardner 1854*; holo: PERTH.

Erect perennial to 50 cm tall, glabrous except flowers; stem triangular with acute angles, 2–3.5 mm wide. Leaves sessile, linear to linear-oblong, entire; lamina 15–30 mm long, 2–4 mm wide. Flowers in panicles; inflorescence branches to 3 together, to 5-flowered, 8–13 mm long; pedicel 3–5.5 mm long, bracteoles 1–1.5 mm long. Sepals hidden by hairs, c. 0.5 mm long. Corolla to 10 mm long; hairs outside appressed, type iv, pale grey; wings 2.5–2.7 mm wide, slightly smaller above auricle; calli 1–3 per row or obsolete. Ovary gibbous, 1.5–1.7 mm long. Fruit orbicular, 2–3 mm diam., gibbous, compressed, wrinkled, glabrous. n = 18 W.J.Peacock, *Proc. Linn. Soc. New South Wales* 88: 10 (1963) as *D. sacculata*. Fig. 21F.

Occurs in south-western W.A. Flowers chiefly Aug.–Dec. Map 94.

W.A.: W of Elphin, near Wongan Hills, *R.D.Royce 6637* (PERTH); c. 2 km N of Lake King, *R.H.Kuchel 1862* (AD, SYD); Stokes Inlet, c. 75 km W of Esperance, *A.E.Orchard 1638* (AD, PERTH).

The cytological voucher is *W.J.Peacock 60880.1* (SYD).

60. Dampiera heteroptera Rajput & Carolin, *Telopea* 3: 210 (1988)

T: near Augusta, W.A., 21 Oct. 1970, *A.M.Ashby 3696*; holo: PERTH; iso: AD, SYD.

Erect perennial to 60 cm tall, glabrous except flowers; stem compressed, scarcely winged but with distinct ribs. Leaves sessile, linear, entire; lamina 10–55 mm long, to 3 mm wide. Flowers in panicles; inflorescence branches usually 1–3 per axil, 2–5-flowered, to 8 cm long; pedicel 8–10 mm long; bracteoles linear, 1.5–1.7 mm long. Sepals glabrous, c. 0.5 mm long. Corolla 10–12 mm long; hairs outside appressed, type iv, grey; wings c. 2.5 mm wide, obsolete above auricle; calli to 3 per row. Ovary gibbous, tomentose except the glabrescent gibbosity, 1.2–1.5 mm long. Fruit orbicular, c. 3 mm diam., gibbous, compressed, wrinkled, glabrous. Figs 21G, 34F–H.

Occurs in far south-western W.A., in damp places. Flowers chiefly Sept.–Jan. Map 95.

W.A.: c. 50 km W of Nannup, *A.R.Fairall 826* (PERTH); Cowaramup, *R.D.Royce* 2832 (PERTH); S of Blackwood R., *R.D.Royce 2948* (PERTH).

61. **Dampiera coronata** Lindley, *Sketch Veg. Swan R.* xxvii (1839)

T: Swan River, W.A., 1839, *J.Drummond*; holo: CGE; iso: K.

D. cauloptera DC., *Prodr.* 7(2): 504 (1839). T: Swan River, W.A., *J.Drummond s.n.*; holo: G-DC.

D. trialata Vriese in J.G.C.Lehmann, *Pl. Preiss.* 1: 401 (1845). T: near Perth, W.A., 6 Aug. 1839, *L.Preiss 1444*; lecto: L, *fide* M.T.M.Rajput & R.C.Carolin, *Telopea* 3: 208 (1988); isolecto: MEL, W.

Illustration: K.Krause, *Pflanzenr.* 54: 176, fig. 32A–C (1912).

Erect perennial to 40 cm tall, glabrous except flowers; stem 3-winged, 2–3 mm wide, ribbed on margins. Leaves sessile, narrowly oblong to spathulate, dentate; lamina 17–70 mm long, 6–19 mm wide. Flowers in panicles; inflorescence branches to 3 together, to 5-flowered, to 8 cm long; pedicel 4–11 mm long; bracteoles linear, to 4 mm long. Sepals glabrous, 0.8–1.2 mm long. Corolla 12–15 mm long; hairs outside appressed, type iv, dark grey or almost black; wings 3–4 mm wide, ±obsolete above auricle; calli obsolete. Ovary gibbous, tomentose but gibbous expansion glabrescent. Fruit orbicular, 3–4 mm diam, gibbous, compressed, glabrous, wrinkled. Figs 21H, 23K.

Occurs in south-western W.A. from Jurien to Busselton, with a record near Ravensthorpe; grows in sandy soil. Flowers chiefly Aug.–Dec. Map 96.

W.A.: Jurien Bay, *R.D.Royce 7721* (PERTH); Gillingarra, *A.Morrison 16054* (NSW); between Moora and Gingin via Mogumber, *W.E.Blackall 2945* (PERTH); Capel, *R.D.Royce 2691* (PERTH); c. 16 km S of Ravensthorpe, *W.J.Peacock 6095.9* (SYD).

62. **Dampiera carinata** Benth., *Fl. Austral.* 4: 111 (1868)

T: south-western W.A., *J.Drummond 2: 397*; lecto: MEL, *fide* R.C.Carolin, *Fl. Australia* 35: 329 (1992); isolecto: BM.

D. mooreana E.Pritzel in F.L.E.Diels & E.Pritzel, *Bot. Jahrb. Syst.* 35: 579 (1905). T: near Watheroo, W.A., Nov. 1901, *E.Pritzel 993*; lecto: AD, *fide* M.T.M.Rajput & R.C.Carolin, *Telopea* 3: 210 (1988); isolecto: ?B (destroyed) *n.v.*, BM, K, L.

Illustration: R.Erickson *et al.*, *Fl. Pl. W. Australia* 109, fig. 329 (1973).

Erect perennial 12–23 cm tall, glabrous except inflorescence; stem bluntly triangular, much-branched, woody below. Leaves sessile; lower leaves oblong-lanceolate, entire, with lamina 37–45 mm long, to 8 mm wide; upper leaves reduced to minute scales. Flowers in panicles; inflorescence branches usually solitary, 1-flowered, to 4 cm long; pedicel 1–1.5 mm long; bracteoles oblong, 1–2 mm long. Sepals unequal, ovate, 2.5–3 mm long. Corolla 10–12 mm long; hairs outside closely appressed, type iv, silvery; wings 1.2–1.5 mm wide, almost equal; calli obsolete. Ovary gibbous, keeled, 1–2 mm long. Fruit not seen. Fig. 21I.

Occurs between Three Springs and York, south-western W.A. Flowers chiefly Dec.–Mar. Map 97.

W.A.: c. 5 km S of Three Springs, *S.Carlquist 3941* (AD, NSW); near Badgerabbie Hill, SW of Moora, *A.S.George 4376* (PERTH); c. 24 km S of Tammin, *R.D.Royce 9423* (PERTH); c. 13 km E of Mt Brown, York, *O.H.Sargent 628* (MEL, PERTH).

63. **Dampiera alata** Lindley, *Sketch Veg. Swan R.* xxvii (1839)

T: Swan River, W.A., 1839, *J.Drummond*; holo: CGE; iso: K

D. epiphylloidea Vriese in Lehm., *Pl. Preiss.* 1: 402 (1845). T: south-western W.A., *J.A.L.Preiss 1494*; lecto: LD, *fide* M.T.M.Rajput and R.C.Carolin, *Telopea* 3: 208 (1988); isolecto: L, MEL, W.

Erect, perennial herb to 40 cm tall, glabrous except inflorescence; stem flat, winged, 3–13 mm wide with a distinct rib on margin of each wing, sometimes constricted at nodes.

Leaves sessile, linear-lanceolate to oblanceolate, entire or dentate; lamina 18–49 mm long, 3.5–16 mm wide. Flowers in panicles; inflorescence branches usually solitary, to 3-flowered, to 5 cm long; pedicel 3–3.5 mm long; bracteoles tomentose, 2.5–4 mm long, Sepals usually hidden by hairs, ovate-elliptic, 1–1.5 mm long. Corolla 12–16 mm long; hairs outside not appressed, type v, grey; wings 2.7–3.2 mm wide, smaller above auricle; calli to 4 per row. Ovary gibbous, ±tomentose, 2–2.5 mm long. Fruit orbicular, 3–4 mm diam, gibbous, compressed, wrinkled, hairy. *n* = 9 (as *D. epiphylloidea*), *n* = 18, *n* = 27 (as *D. epiphylloidea* and *D. coronata*), W.J.Peacock, *Proc. Linn. Soc. New South Wales* 88: 10 (1963). Fig. 21J.

Occurs in south-western W.A. between Eneabba and Bremer Bay, growing on sand plains. Flowers chiefly Aug.–Nov. Map 98.

W.A.: Badgingarra, *A.M.Ashby 3041* (SYD); Moora, *R.C.Carolin 3243* (SYD); Wooroloo, *M.Koch 1392* (PERTH); Mundaring, *C.A.Gardner 8642* (PERTH); c. 3.5 km SE of Armadale, near the road to Gleneagle, *H.Eichler 15774* (AD).

Cytological vouchers are *W.J.Peacock 60822.1*, *60883.1*, *60885.3*, *60878.1* and *60875.2* (SYD).

64. **Dampiera deltoidea** Rajput & Carolin, *Telopea* 3: 209 (1988)

T: Mt Drummond, SW of Ravensthorpe, W.A., 13 Aug. 1967, *K.Newbey 2697*; holo: PERTH.

Perennial herb 25–35 cm tall, glabrous except flowers; stem flat, winged, slightly constricted at nodes, 6–8 mm wide, with a distinct midrib. Leaves sessile, triangular-ovate, scarcely narrowing at base, entire; lamina 10–16 mm long, 6–9 mm wide. Flowers usually in clusters of 3 in upper axils; pedicel 2.2–2.5 mm long; bracteoles oblong, 0.7–1.1 mm long. Sepals ovate-elliptic, glabrous, 0.5–0.7 mm long. Corolla 10–11 mm long; hairs outside appressed, type iv, brown-grey; wings 2.7–3.2 mm wide, smaller above auricle; calli to 3 per row. Ovary gibbous, 1.7–2 mm long; hairs similar to petals, but fewer. Fruit not seen. Fig. 21K.

Occurs between the Fitzgerald R. and Ravensthorpe, W.A. Flowers chiefly Sept.–Nov. Map 99.

W.A.: Fitzgerald R., c. 110 km ESE of Ongerup, *T.E.H.Alpin, I.Lethbridge & R.Coveny 3258* (NSW); c. 0.8 km E of Elverton Mine, *K.Newbey 945* (PERTH).

Similar to *D. alata* which has spreading hairs on the outside of the corolla.

65. **Dampiera lindleyi** Vriese in J.G.C.Lehmann, *Pl. Preiss.* 1: 402 (1845)

T: York, W.A., 13 Sept. 1839, *L.Preiss 1514*; lecto: L, *fide* M.T.M.Rajput & R.C.Carolin, *Telopea* 3: 208 (1988); isolecto: L, MEL, W.

D. lindleyi var. *angusta* E.Pritzel in F.L.E.Diels & E.Pritzel, *Bot. Jahrb. Syst.* 35: 578 (1905). T: near Mongerup [Mondurup, Stirling Ra.], W.A., *F.L.E.Diels 5097*; holo: B destroyed.

Erect, perennial herb to 40 cm high, glabrous except flowers; stem flattened, with 2 wings and marginal ribs, 3–7 mm wide, usually constricted at nodes. Leaves sessile, linear or linear-oblong, contracted at base, entire; lamina 5–12 mm long, 1–1.6 mm wide. Flowers in panicles; inflorescence branches 1–3 together, 1–3-flowered, to 8 cm long; pedicel 1–1.5 mm long; bracteoles linear-oblong, 1.5–2 mm long. Sepals c. 2.5 mm long, obscured by hairs. Corolla 12–15 mm long; hairs outside appressed, type iv, dark grey; wings 2.5–2.7 mm wide, smaller above auricle; calli obsolete. Ovary gibbous, hairy as corolla, 1.7–2.2 mm long. Fruit orbicular, 2–3 mm diam., gibbous, compressed. *n* = 27, W.J.Peacock, *Proc. Linn. Soc. New South Wales* 88: 10 (1963). Fig. 21L.

Occurs from the Murchison R. to Busselton and inland to Pingelly, south-western W.A., usually in heath. Flowers chiefly June–Oct. Map 100.

W.A.: Murchison House, *R.C.Carolin 3376* (SYD); Ogilvie, *C.A.Gardner 8590* (PERTH); Coorow, *E.Wittwer W817* (PERTH); c. 27 km E of Pingelly, *R.D.Royce 7551* (PERTH); c. 18 km E of Calingiri, along the Wongan Hills road, *T.E.H.Aplin 164* (PERTH).

Similar to *D. alata* and *D. deltoidea* which have shorter inflorescence branches; *D. alata* also has loose hairs on the corolla. Also similar to *D. sacculata* which has longer leaves. Cytological vouchers are *W.J.Peacock 6088.1* and *60836.1* (SYD).

66. **Dampiera sacculata** F.Muell. ex Benth., *Fl. Austral.* 4: 111 (1868)

T: upper Kalgan River, W.A., *A.Oldfield*; lecto: K, *fide* M.T.M.Rajput & R.C.Carolin, *Telopea* 3: 210 (1988).

Erect perennial to 35 cm tall, glabrous except flowers; stem compressed, 2–3 mm wide, with ribs either side. Leaves sessile, linear to oblong, entire; lamina 13–37 mm long, 1–2.5 mm wide. Flowers in panicles; inflorescence branches usually 3 or more together, 1–3-flowered, curved, to 6 cm long; pedicel 1.2–1.7 mm long; bracteoles oblong-elliptic, 1.5–2 mm long. Sepals c. 0.5 mm long, obscured by hairs. Corolla 10–12 mm long; hairs outside appressed, type iii and iv, dark grey; wings 3.5–4 mm wide, smaller above auricle; calli mostly obsolete. Ovary gibbous, tomentose, 1.7–2 mm long. Fruit orbicular, c. 2 mm diam., gibbous, compressed, hairy. $n = 9$, W.J.Peacock, *Proc. Linn. Soc. New South Wales* 88: 10 (1963). Fig. 21M.

Occurs in south-western W.A. in heath. Flowers chiefly Aug.– Nov. Map 101.

W.A.: N of Narembeen, Hyden, *M.Barrow 30* (PERTH); Wittenoom Hill, c. 3 km W of Mt Burdett, *A.E.Orchard 1327* (AD, PERTH); Pingerup, *W.E.Blackall 3064* (PERTH); 39 km SW of Fitzgerald, c. 106 km WSW of Ravensthorpe, *A.C.Beauglehole ACB49195* (SYD).

Similar to *D. lindleyi* which has flattened stems and shorter acute leaves. Cytological voucher is *W.J.Peacock 6092.2* (SYD).

Unplaced and excluded names

Dampiera cunninghamii Vriese, *Ned. Kruidk. Arch.* 2: 11 (1850)

T: Port Jackson, N.S.W., *A.Cunningham*; *n.v.*

Vriese's description of this species is poor and no specimen of the type collection has been found.

Dampiera dielsii E.Pritzel in F.L.E.Diels & E.Pritzel, *Bot. Jahrb. Syst.* 35: 581 (1905)

T: near Greenough River, beside Mullewa bridge, W.A., *F.L.E.Diels 3288*; holo: ?B (destroyed) *n.v.*

No specimen of the type collection has been found, nor any other specimen that agrees with the description.

Dampiera glabriflora F.Muell., *Fragm.* 1: 120 (1859)

T: Blackwood [R.] valley and beside Gardiner R., W.A., *coll. unknown*; *n.v.*

The description is poor and no specimen of the type collection has been found.

Dampiera inundata Vriese in J.G.C.Lehmann, *Pl. Preiss.* 1: 404 (1845.)

T: near Albany, W.A., 29 Jan. 1840, *L.Preiss 1523*; syn: L (2 sheets).

This is *Stylidium crassifolium* R.Br., see W.H.de Vriese, *Pl. Preiss.* 2: 242 (1848).

Dampiera restiacea E.Pritzel in F.L.E.Diels & E.Pritzel, *Bot. Jahrb. Syst.* 35: 580 (1905.)

T: near Watheroo, W.A., *F.L.E.Diels 2089*; holo: ?B (destroyed) *n.v.*

No specimen of the type collection, nor any other specimen that agrees with the description, has been found.

4. COOPERNOOKIA

R.C.Carolin

Coopernookia Carolin, *Proc. Linn. Soc. New South Wales* 92: 209 (1968); named for the Coopernook State Forest in N.S.W. where *C. chisholmii* is common.

Type: *C. barbata* (R.Br.) Carolin

Perennial undershrubs with stellate and usually glandular hairs, often viscid becoming varnished. Leaves sessile or nearly so. Inflorescences terminal leafy thyrses or racemes, bracteolate; pedicels articulate. Sepals adnate to ovary. Corolla scarcely bilabiate, with long, stiff, retrorse bristles inside, white to mauve or pinkish; corolla pouch obscure; auricle obsolete; lobes broadly winged. Stamens free, epigynous. Ovary incompletely 2-locular; style simple; indusium ±horizontal, with long bristles on lips; ovules 2–8. Fruit a capsule; valves 2, entire or bifid. Seeds ovoid, scarcely compressed, glossy, strophiolate, not winged. $x = 7$, (3 species), W.J.Peacock, *Proc. Linn. Soc. New South Wales* 88: 2–27 (1963).

A genus of 6 species, all endemic in Australia.

In this treatment, the length of the character *pedicel* is the distance from bract to flower, thus equivalent to *peduncle + pedicel* of other genera.

R.C.Carolin, *Coopernookia*: a new genus of Goodeniaceae. *Proc. Linn. Soc. New South Wales* 92: 209–216 (1968).

1 Adaxial leaf surface tomentose with stellate or substellate hairs

2 Leaf margin revolute; abaxial leaf surface glabrescent; inflorescence a compact head **1. C. polygalacea**

2: Leaf margins flat or slightly recurved; abaxial leaf surface tomentose; inflorescence loose **6. C. chisholmii**

1: Adaxial leaf surface glabrous or glandular-hairy with stellate hairs very few or none

3 Leaves elliptic to ovate with flat or slightly recurved margins

4 Corolla white, 7–12 mm long **2. C. strophiolata**

4: Corolla mauve to pink, 15–20 mm long

5 Leaves almost glabrous even when young **3. C. georgei**

5: Leaves glandular-hairy when young, becoming scabrid **5. C. scabridiuscula**

3: Leaves linear to narrowly elliptic, with revolute margins **4. C. barbata**

1. **Coopernookia polygalacea** (Vriese) Carolin, *Proc. Linn. Soc. New South Wales* 92: 210 (1968)

Dampiera polygalacea Vriese, *Natuurk. Verh. Holl. Maatsch Wetensch. Haarlem* ser. 2, 10: 115 (1854). T: south-western Australia, *J.Drummond 356;* holo: K; iso: MEL.

Goodenia phylicoides F.Muell., *Fragm.* 1: 206 (1859). T: towards the Gardiner R., W.A., *G.Maxwell*; holo: MEL.

Spreading undershrub to 70 cm tall. Leaves linear to elliptic, revolute, entire, stellate-tomentose but upper surface glabrescent with scurfy scars, and sometimes varnished; lamina 12–30 mm long, 3–6 mm wide. Inflorescences with pedicels very short; bracteoles close under flower. Sepals linear to narrowly elliptic, 6–8 mm long, adnate to lower half of ovary. Corolla to 15 mm long, stellate-tomentose outside, adaxial lobes with bristles near both margins, lilac-pink to white; lobes ±equal; wings 2–3 mm wide. Ovules 6–8, axile. Capsule ovoid, c. 5 mm long; valves deeply 2-fid. $2n$ = 14, W.J.Peacock, *Proc Linn. Soc. New South Wales* 88: 7 (1963). Figs 23F, 25E, 35G–H.

Occurs in near-coastal districts from the Stirling Ra. to Esperance, south-western W.A., in heathland. Flowers chiefly Aug.–Nov. Map 102.

W.A.: between Salmon Gums and Grass Patch, *W.E.Blackall 1011* (PERTH); Jerdacuttup R., *T.E.H.Aplin 26806* (PERTH); Young R., *R.C.Carolin 3375* (SYD); c. 20 km SSE of Ravensthorpe on Hopetown road, *L.Haegi 1032* (AD); Coolgardie–Esperance road, *E.M.Bennett 799* (PERTH).

The revolute leaf margins, and dense stellate indumentum on the undersurface of the leaves, distinguish this species from the others. Chromosome voucher is *W.J.Peacock 6095.4* (SYD).

2. **Coopernookia strophiolata** (F.Muell.) Carolin, *Proc. Linn. Soc. New South Wales* 92: 211 (1968)

Goodenia strophiolata F.Muell., *Fragm.* 1: 119 (1859). T: Fitzgerald Ranges, W.A., *coll. unknown* (probably *G.Maxwell 286*); lecto: MEL, *fide* R.C.Carolin, *op. cit.* 212.

Spreading shrub to 1 m tall, viscid. Leaves obovate to spathulate or elliptic, attenuate towards base, dentate, with few stellate hairs; lamina 10–35 mm long, 2–15 mm wide. Inflorescence with pedicels to 12 mm long; bracteoles just above middle. Sepals lanceolate to elliptic, 3–7 mm long, adnate to ovary only towards base. Corolla to 12 mm long, stellate hairy outside, white with dark veins; lobes unequal; wings to 1.5 mm wide; adaxial lobes with bristles near posterior margin only. Ovules 4–6, almost basal. Capsule globular, 5–7 mm diam.; valves bifid or entire. Seeds ellipsoidal, 3 mm long. Figs 35A–B, 50.

Occurs in W.A. from Merredin to Queen Victoria Spring and southwards, and in S.A. near Maralinga and on Eyre Peninsula. Flowers chiefly Sept.–Dec. Map 103.

W.A.. Merredin, *M.Koch 2866* (NSW); c. 32 km S of Queen Victoria Spring, *R.D.Royce 5333* (PERTH); 90 mile tank, c. 80 km W of Daniell, *P.Wilson 3203* (AD); c. 2.5 km towards Barkers Inlet, S of Esperance, 2 Nov. 1968, *E.M.Canning* (CBG). S.A.: Mt Beadell near Maralinga, c. 30 km N of Watson, 4 Sept. 1960, *H.Turner* (AD).

Distinguished from all other species by the viscid varnish which is exuded over all parts, concealing the short glandular hairs. The plant thus appears almost glabrous.

Figure 35. *Coopernookia*. **A–B**, *C. strophiolata*. **A**, habit X1; **B**, hair X40 (**A–B**, A.Strid 21262, SYD). **C–D**, *C. scabridiuscula*. **C**, habit X1; **D**, hairs (stellate and glandular) X40 (**C–D**, M.Olsen 374, SYD). **E–F**, *C. chisholmii*. **E**, habit X1; **F**, hair X40 (**E–F**, J.Peacock 6012.4.1, SYD). **G–H**, *C. polygalacea*. **G**, habit X1; **H**, hair X40 (**G–H**, L.Haegi 1032, SYD). Drawn by D.Mackay.

3. Coopernookia georgei Carolin, *Proc. Linn. Soc. New South Wales* 92: 213 (1968)

T: north eastern side of Whoogarup Ra., 30 miles [48 km] SSW of Ravensthorpe, 1 Nov. 1965, *A.S.George 7201*; holo: PERTH; iso: B, K, SYD.

Slender shrub to 1.5 m tall, almost glabrous. Leaves elliptic to narrowly elliptic, serrate or dentate; lamina 2–5 cm long, 8–25 mm wide. Inflorescence with pedicels to 15 mm long; bracteoles at middle. Sepals linear to lanceolate, 7–8 mm long, adnate to ovary for more than 1/2 its length. Corolla 15–20 mm long, stellate-pubescent outside, mauve; lobes unequal; wings to 2 mm wide; adaxial lobes with bristles near posterior margin only. Ovules 2, almost basal. Capsule cylindrical, c. 10 mm long, valves entire or bifid. Seeds obloid, 5 mm long.

Endemic in the ranges SW of Ravensthorpe, W.A. Flowers in spring and summer. Map 104.

W.A.: western side of Thumb Peak range, c. 80 km SW of Ravensthorpe, *A.S.George 7168* (PERTH); Phillips Ra., *coll. unknown* (possibly *G.Maxwell*) *139* (MEL).

Distinguished from *C. strophiolata* by the larger flowers and leaves, the lack of varnish and the mauve corolla.

4. Coopernookia barbata (R.Br.) Carolin, *Proc. Linn. Soc. New South Wales* 92: 213 (1968)

Goodenia barbata R.Br., *Prodr.* 576 (1810). T: Port Dalrymple, Tas., *R.Brown*; lecto: BM, *fide* R.C.Carolin, *op. cit.* 215 (1967).

Goodenia cistiflora A.Cunn. ex DC., *Prodr.* 7(2): 516 (1839). T: near Hunter R., N.S.W., *A.Cunningham*; holo: G; iso: K, P.

Scaevola scaberula Summerhayes, *Bull. Misc. Inform. 1927*: 356 (1927). T: Ettrema R., N.S.W., Feb. 1927, *F.A.Rodway*; holo: K; iso: NSW.

Illustration: E.R.Rotherham *et al.*, *Fl. Pl. New South Wales & S. Queensland* 70, fig. 197 (1975).

Erect undershrub to 1 m tall, glandular-hairy, ±viscid. Leaves linear, revolute, entire or nearly so, with scattered stellate hairs, becoming scabrid; lamina 1–3 cm long, mostly 1–2 mm wide. Inflorescence with pedicels to 20 mm long; bracteoles above the middle. Sepals lanceolate to narrowly elliptic, 3–4 mm long, adnate to lower 1/2 of ovary. Corolla to 15 mm long, glandular- and stellate-pubescent outside, blue to mauve; lobes unequal; wings c. 1 mm wide; adaxial lobes with bristles near posterior margin only. Ovules 2, almost basal. Capsule cylindrical to ovoid, 5–7 mm long, valves entire. Seeds ellipsoidal, 4–5 mm long. $2n = 14$, W.J.Peacock, *Proc. Linn. Soc. New South Wales* 88: 7 (1963). Fig. 16.

Occurs on the eastern highlands and along the E coast from New England, N.S.W., through Vic., to northern Tas. Flowers most of the year. Map 105.

N.S.W.: Putty, *W.J.Peacock 6011* (SYD); Moruya, Nov. 1911, *J.L.Boorman* (NSW); Eden, 17 Dec. 1903, *E.Cheel* (NSW); Belougery Split Rock, 25 Sept. 1973, *Lassak & Southwell* (NSW). Vic.: 4.6 km along unnamed track from Mallacoota–Genoa road, *S.J.Forbes 2926* (MEL); WB Line track, 5.5 km from Cann Valley Hwy, *P.Weston 87* (SYD).

See notes under *C. chisholmii* and *C. scabridiuscula*. Chromosome vouchers are *W.J.Peacock 6012.5.1, 6012.26.2, 6110.16.1* (SYD).

5. Coopernookia scabridiuscula Carolin, *Telopea* 2: 74 (1980)

T: Mt Maroon, Qld, 23 Sept. 1976, *M.Olsen 374*; holo: BRI; iso: SYD.

Ascending shrub to 1 m tall. Leaves flat, narrowly elliptic to oblong, attenuate towards base, dentate, glandular-hairy becoming scabrid; lamina 4–8 cm long, 5–10 mm wide. Inflorescence with pedicels to 15 mm long; bracteoles above the middle. Sepals ovate to lanceolate, 5–6 mm long, adnate to lower half of ovary. Corolla to 16 mm long, stellate- and glandular-hairy outside, 'purple-pink'; lobes unequal; wings c. 2 mm wide; adaxial lobes with bristles near posterior margin only. Ovules 2–4, almost basal. Capsule and seeds not seen. Fig. 35C–D.

Only known from the Mt Maroon area in southern Qld. Flowers in summer. Map 106.

Qld: Mt Maroon, *S.L.Everist 7053* (BRI).

This species is similar to some forms of *C. barbata* but the leaves are broader and larger, and the margins are not revolute; see also *C. chisholmii*.

6. Coopernookia chisholmii (Blakely) Carolin, *Proc. Linn. Soc. New South Wales* 92: 215 (1968)

Goodenia chisholmii Blakely, *Proc. Linn. Soc. New South Wales* 53: 684 (1929). T: Kendall, N.S.W., Sept. 1929, *F.M.Bailey*; lecto: NSW, *fide* R.C.Carolin, *op. cit.* 216 (1968).

Erect undershrub to 1.5 m tall. Leaves elliptic, dentate or entire with margin almost flat, stellate-pubescent with a few glandular hairs; lamina 4–9 cm long, 10–25 mm wide. Inflorescence with pedicels to 20 mm long; bracteoles close under flower. Sepals linear-elliptic, 4–6 mm long, adnate to ovary towards base. Corolla to 15 mm long, stellate and glandular pubescent outside, mauve to almost pink; lobes unequal; wings c. 1 mm wide; adaxial lobes with bristles near posterior margin only. Ovules 2, basal. Capsule obovoid, 5–6 mm long, valves entire. Seeds obloid, 4–5 mm long. $2n = 14$, W.J.Peacock, *Proc. Linn. Soc. New South Wales* 88: 7 (1963). Figs 26A, 35E–F.

Occurs in the coastal ranges of N.S.W. from Hunter R. valley to Port Macquarie, in damp *Eucalyptus* forest. Flowers chiefly Sept.–Mar. Map 107.

N.S.W.: Wollombi Ck ford, Broke Rd, Dec. 1961, *C.Burgess* (NSW); Bird Tree Reserve, Middle Brother State Forest, *R.Coveny 12240 & W.Bishop* (NSW); the Comboyne plateau, Jan. 1926, *E.C.Chisholm* (NSW); Manning R. Natl Forest no 1, 28 Mar. 1962, *C. & I.P.Burgess* (CBG); Middle Brother State Forest, *R.C.Carolin 7962* (AD, SYD).

Distinguished from the 2 previous species, which have very few stellate hairs on the vegetative parts, by the dense stellate indumentum. Chromosome voucher is *W.J.Peacock 6012.4.1* (SYD).

5. SCAEVOLA

R.C.Carolin

Scaevola L., *Mant. Pl.* 2: 145 (1771); from *scaevola*, the Latin for little hand. The (dried) flowers are supposed to resemble a withered hand.

Type: *S. plumieri* (L.) M.Vahl

Cebera Lour., *Fl. Cochinch.* 136 (1790). T: *C. salutaris* Lour.

Roemeria Dennst., *Schluss. Hort. Malab.* 24 (1818), *nom. illeg.* non Medikus (1792) T: *R. lobelia* Dennst. = *Scaevola taccada* (Gaertner) Roxb., (see D.H.Nicolson, C.R.Suresh, and K.S.Manital, *Interpr. van Rheede's Hort. Malabari*. Koeltz, Konigstein, B.D.R., 1988).

Crossotoma (G.Don) Spach, *Hist. Nat. Vég.* 9: 583 (1838); *Scaevola* sect. *Crossotoma* G.Don, *Gen. Hist.* 3: 730 (1834). T: *S. spinescens* R.Br.

Xerocarpa (G.Don) Spach, *Hist. Nat. Vég.* 9: 583 (1838). T: *S. crassifolia* Labill.

Pogonanthera (G.Don) Spach, *Hist. Nat. Vég.* 9: 583 (1838); *Scaevola* sect. *Pogonanthera* G.Don, *Gen. Hist.* 3: 729 (1834). T: *S. striata* R.Br.

Temminckia Vriese, *Ned. Kruidk. Arch.* 2: 140 (1851). T: *T. mollis* Vriese

Camphusia Vriese, *Ned. Kruidk. Arch.* 2: 148 (1851). T: *C. glabra* (Hook. & Arn.) Vriese

Merkusia Vriese, *Ned. Kruidk. Arch.* 2: 150 (1851). T: *M. crassifolia* (Labill.) Vriese

Molkenboeria Vriese, *Natuurk. Verh. Holl. Maatsch. Wetensch. Haarlem* ser. 2, 10: 38 (1854). T: *M. pilosa* (Benth.) Vriese

Nigromnia Carolin, *Nuytsia* 1: 292 (1974). T: *N. globosa* Carolin

Perennial herbs, scramblers, shrubs or small trees to 3 m tall, variously hairy. Leaves alternate (rarely opposite), sessile or petiolate, usually with axillary hair tufts. Inflorescence with flowers in terminal and axillary thyrses, racemes, spikes, or apparent cymes, bracteolate; pedicels articulate. Sepals adnate to full length of ovary. Corolla tubular, completely slit adaxially, opening along slit, white, blue, mauve, rarely yellow; throat often with hairs inside, the hairs simple and/or long multicellular (barbulae), the barbulae simple or branched; lobes usually equal, winged, not auriculate; corolla pouch absent. Stamens free, epigynous. Ovary 1–4-locular; style unbranched; indusium ±horizontal, usually pubescent; ovules 2–4. Fruit indehiscent; mesocarp fleshy, corky or ±dry; endocarp hard. Seeds ovoid, not strophiolate, wingless; embryo terete. $x = 8$, W.J.Peacock, *Proc. Linn. Soc. New South Wales* 88: 22–27 (1962). *Fan-flowers.*

The genus contains c. 96 species occurring in Australia and tropical areas of the Indo-Pacific region; 71 in Australia, 70 of which are endemic.

Two types of tactile pollinator guides occur in this genus. In sect. *Scaevola*, flat hairs are attached to the tops of laciniae on the wings (Fig. 40C). In other sections, the laciniae are narrow and referred to as barbulae; their apices may be hairy, papillate (Fig. 26C) or simple (Fig. 26D).

G.Bentham, *Scaevola*, *Fl. Austral.* 4: 83–104 (1868); K.Krause, *Pflanzenr.* 54: 117–168 (1912); R.C.Carolin, Nomenclatural notes, new taxa and the systematic arrangement in the genus *Scaevola* and its synonyms, *Telopea* 3: 477–515 (1990).

KEY TO SECTIONS, SUBSECTIONS AND SERIES

1 Fruit with fleshy or corky mesocarp; growth of inflorescence axis continuing after flowering

2 Leaves alternate — sect. 1. **Scaevola**

2: Leaves opposite — sect. 2. **Enantiophyllum**

1: Fruit usually with dry mesocarp; growth of inflorescence axis terminating with flowering — sect. 3. **Xerocarpa**

3 Flowers usually sessile — subsect. 1. **Biloculatae**

4 Indusium glabrous or with scattered weak hairs — ser. 1. **Macrostachyae**

4: Indusium with a stiff basal beard on adaxial surface which may be reduced to short bristles — ser. 2. **Pogogynae**

3: Flowers usually pedunculate, or if sessile in the bract, then cauline leaves reduced

5 Anthers with hairs at tip (except *S. hookeri*); cauline leaves well developed — subsect. 2. **Pogonanthera**

5: Anthers glabrous at tip (except sometimes *S. depauperata*); cauline leaves usually reduced — subsect. 3. **Parvifoliae**

KEY TO SPECIES

1 At least the lower flowers with stalk below bracteoles more than 5 mm long — **GROUP 1**

1: All flowers with stalk below bracteoles obsolete or to 5 mm long

2 Adaxial surface of indusium with basal tuft of prominent bristles exceeding lips — **GROUP 2**

2: Adaxial surface of indusium glabrous, or with short basal bristles which do not exceed lips

3 Plant glabrous or glabrescent (except flowers), or with very few scattered hairs, or viscid and varnished — **GROUP 3**

3: Plant variously hairy or scabrous — **GROUP 4**

GROUP 1

1 Flowers in axillary cymes

2 Leaves opposite; scrambler or vine; corolla yellow — **7. S. enantophylla**

2: Leaves alternate; shrub, subshrub or small tree; corolla white to blue

3 Leaves usually 2–9 cm wide — **1. S. taccada**

3: Leaves less than 6 mm wide

4 Fruit dry, green, grooved — **15. S. cunninghamii**

4: Fruit fleshy, black, smooth — **3. S. acacioides**

1: Flowers solitary in bract axils or sometimes paired on stalk

5 Bracteoles connate along one margin — **6. S. tomentosa**

5: Bracteoles free from each other

6 Branchlets on main stem bearing flowers and often spinescent — **2. S. spinescens**

6: Flowers borne on main stem; stems not spinescent

7 Leaves auriculate, at least the larger ones ±stem-clasping

8 Hairs on stem appressed — **52. S. microphylla**

8: Hairs on stem loose or at 90 °

9 Peduncles more than 10 mm long — **58. S. pilosa**

9: Peduncles less than 10 mm long

10 Racemes loose; bracts not overlapping at flowering; plant ascending or decumbent; leaves 15–35 mm wide — **53. S. auriculata**

10: Racemes compact; bracts overlapping at flowering; plant erect; leaves to 20 mm but mostly to 15 mm wide — **54. S. macrophylla**

7: Leaves not stem-clasping

11 Sepals free or connate only at base

12 Stems with conspicuous ridges (except *S. parvifolia*); at least upper, cauline leaves reduced

13 Stems glabrous or nearly so **66. S. depauperata**

13: Stems variously hairy

14 Stems tortuous **71. S. tortuosa**

14: Stems straight **65. S. parvifolia**

12: Stems without conspicuous ridges; cauline leaves well developed

15 Bracteoles dentate **61. S. calliptera**

15: Bracteoles entire

16 Leaves with revolute margins; sepals to 3 mm long **59. S. tenuifolia**

16: Leaves with recurved margins or flat; sepals usually more than 3 mm long

17 Petal wings not striate (eastern Australia) **56. S. ramosissima**

17: Petal wings striate (W.A.)

18 Bracteoles linear; petals deep purple **62. S. phlebopetala**

18: Bracteoles elliptic to obovate; petals blue to purplish **60. S. striata**

11: Sepals connate at least in lower third, or reduced to rim, or obsolete

19 Plant prostrate **57. S. hookeri**

19: Plant erect or ascending

20 Cauline leaves 25–90 mm long, to 6 mm wide; shrub or subshrub (Pilbara region, W.A.)

21 Plant scurfy with substellate hairs; corolla with substellate hairs outside **3. S. acacioides**

21: Plant ±viscid, glabrous or with simple, antrorse hairs; corolla with simple hairs outside **15. S. cunninghamii**

20: At least upper cauline leaves reduced to c. 4 mm long; many-stemmed perennial (central arid regions)

22 Stems mostly glabrous; corolla densely puberulous outside **66. S. depauperata**

22: Stems mostly puberulous and with scattered, long, simple hairs; corolla with short and long, stiff, arcuate to appressed hairs outside **64. S. basedowii**

GROUP 2

1 Leaves ±stem-clasping

2 Outside of corolla with dense, silvery, closely appressed hairs hiding surface **55. S. platyphylla**

2: Outside of corolla, at least at base, with hairs loose, surface clearly visible between hairs; or glabrous

3 Hairs on stem ±appressed **52. S. microphylla**

3: Hairs on stem not appressed

4 Inflorescence loose, the bracts not overlapping at flowering; plant ascending or decumbent; leaves 15–35 mm wide **53. S. auriculata**

4: Inflorescence compact, the bracts overlapping at flowering; plant erect; leaves to 20 mm but mostly to 15 mm wide **54. S. macrophylla**

1: Leaves not stem-clasping

5 Corolla glabrous outside

6 Leaves linear to narrowly oblong, less than 2 mm wide, usually entire **50. S. graminea**

6: Leaves elliptic to cuneate, 3–23 mm wide, mostly dentate **51. S. laciniata**

5: Corolla hairy outside

7 Flowers mostly in dense thyrses, i.e. several flowers present in axils of at least lower bracts

8 Leaves with scattered fine white hairs, the surface clearly visible **43. S. cuneiformis**

8: Leaves with surface hidden beneath dense, silvery hairs **44. S. argentea**

7: Flowers in spikes or spike-like thyrses

9 Stems ridged or lateral stems often spinescent; cauline leaves usually reduced

10 Indusium with basal tuft of brown bristles on adaxial surface **68. S. chrysopogon**

10: Indusium with basal tuft of white or purple bristles on adaxial surface

11 Hairs on ovary and fruit golden **67. S. restiacea**

11: Hairs on ovary and fruit white **69. S. oxyclona**

9: Stems not ridged; not spinescent; leaves well developed

12 Hairs on stem coarse, mostly yellowish **45. S. aemula**

12: Hairs on stem fine and white, or stems glabrous

13 Stems with mostly simple, antrorse, often appressed hairs

14 Prostrate plant with ±recurved leaves (south-western W.A.) **44. S. argentea**

14: Plant ascending or erect; leaves flat (not in W.A.)

15 Hairs on stem mostly arcuate, sometimes present only on young stems **49. S. humilis**

15: Hairs on stem straight, loosely appressed **39. S. ovalifolia**

13: Stems glabrous or with loose, simple hairs and/or glandular hairs

16 Leaves glabrous or almost so **47. S. collina**

16: Leaves densely hairy

17 Leaves obovate, narrowing abruptly toward base **48. S. obovata**

17: Leaves narrowly elliptic to obovate, tapering gradually toward base **46. S. amblyanthera**

GROUP 3

1 Fruit beaked, sometimes shortly so; endocarp like a woody sponge or with woody plates or spines surrounded by a membranous skin **8. S. collaris**

1: Fruit not beaked; endocarp solid except for fertile locules, and (in some species) 2 sterile cavities

2 Leaves glaucous; upper leaves stem-clasping **26. S. brookeana**

2: Leaves not both glaucous and stem-clasping

3 Plant prostrate

4 Leaves to 2 mm wide; corolla with white hairs outside **33. S. pulvinaris**

4: Leaves 2–15 mm wide; corolla with golden hairs outside **31. S. repens**

3: Plant erect or ascending

5 Fruit with coarse, arcuate, simple, golden hairs **67. S. restiacea**

5: Fruit with soft, white, simple hairs, or glabrous

6 Leaves basal or towards base of stems, linear to oblanceolate, 6–20 cm long, tapering gradually to a broadened base **23. S. lanceolata**

6: Leaves scattered along stem

7 Sepals free, acute

8 Divaricate shrub; indusium with scattered hairs (northern Australia) **70. S. angulata**

8: Herb, not divaricate; indusium with stiff basal hairs or glabrous (south-eastern Australia) **37. S. albida**

7: Sepals an undulate rim or obtuse and connate

9 Fruit smooth or grooved

10 Fruit with swollen ribs, often gibbous **12. S. thesioides**

10: Fruit without swollen ribs, not gibbous

11 Plant not viscid **14. S. porocarya**

11: Plant viscid on young parts

12 Fruit dry; leaves sessile **11. S. nitida**

12: Fruit fleshy; leaves shortly petiolate

13 Leaves elliptic to oblong, 17–40 mm long **5. S. myrtifolia**

13: Leaves obovate-elliptic to spathulate, 10–14 mm long **4. S. bursariifolia**

9: Fruit rugose or tuberculate

14 Flowers in lateral spikes mostly shorter than leaves **32. S. oldfieldii**

14: Flowers in terminal spikes mostly longer than leaves

15 Indusium with short, stiff hairs towards base on upper surface **40. S. glabrata**

15: Indusium glabrous or with a few long hairs on upper surface

16 Leaves distinctly petiolate, ovate to orbicular **9. S. crassifolia**

16: Leaves sessile

17 Bracts distinctly different from foliage leaves **10. S. angustata**

17: Bracts intergrading with leaves

18 Leaves elliptic to obovate or narrowly oblong, 9–34 mm long **18. S. macrostachya**

18: Leaves linear to narrowly oblanceolate, 15–120 mm long **13. S. globulifera**

GROUP 4

1 Plant with conspicuous, glandular hairs

2 Plant brownish from glandular hairs with brown heads, and long, simple, yellowish hairs **19. S. glandulifera**

2: Plant not brownish; glandular hairs yellowish and simple hairs white

3 Leaves linear to narrowly elliptic, recurved **35. S. linearis**

3: Leaves oblanceolate to obovate, flat

4 Bracts similar to foliage leaves but smaller

5 Petal wings 1–2 mm wide; hairs on back of indusium few, long, not appressed (south-western W.A.) **32. S. oldfieldii**

5: Petal wings to 10 mm wide; hairs on back of indusium many, short, appressed (Qld) **41. S. glutinosa**

4: Bracts very different from foliage leaves, much shorter **17. S. revoluta**

1: Plant without conspicuous glandular hairs, sometimes with minute ones scarcely visible to naked eye

6 Ovary and fruit glabrous or nearly so (sometimes ovary and fruit hidden by long hairs from below base)

7 Bracteoles connate along one margin **6. S. tomentosa**

7: Bracteoles free

8 Leaves basal or toward base of stems, 6–20 cm long, tapering gradually to a broadened base **23. S. lanceolata**

8: Leaves scattered along stems

9 Lateral inflorescences shorter than or equal to leaves

10 Flowers in woolly heads, to 3 mm long **29. S. globosa**

10: Flowers in spikes, more than 6 mm long

11 Plant prostrate or decumbent

12 Fruit dry (W.A.) **31. S. repens**

12: Fruit fleshy (S.A., Qld, N.S.W., Vic.) **25. S. calendulacea**

11: Plant erect or ascending

13 Leaves glabrescent or with scattered hairs **32. S. oldfieldii**

13: Leaves with dense silky hairs **30. S. sericophylla**

9: Inflorescences much longer than leaves

14 Prostrate or decumbent plant of coastal dunes

15 Fruit fleshy, purple **25. S. calendulacea**

15: Fruit dry, greenish **37. S. albida**

14: Erect or ascending shrubs, rarely on coastal dunes

16 Bracteoles mostly to 4 mm long

17 Upper flowers at least twice as long as their subtending bracts **17. S. revoluta**

17: Upper flowers equalling or slightly longer than their subtending bracts **24. S. virgata**

16: Bracteoles mostly more than 4 mm long

18 Upper bracts ovate, very different in size and shape from foliage leaves **24. S. virgata**

18: Upper bracts either linear to narrowly elliptic, different in size and shape from foliage leaves, or broader and similar to foliage leaves

19 Leaves linear to oblanceolate, mostly more than 6 cm long

20 Fruit rugose, to 4 mm long, with 2 fertile locules **21. S. anchusifolia**

20: Fruit grooved, otherwise smooth, to 11 mm diam., with 4 fertile locules **14. S. porocarya**

19: Leaves obovate to oblong, to 5.5 cm long

21 Plant softly, densely hairy **16. S. browniana**

21: Plant loosely hairy with stiff, coarse hairs **18. S. macrostachya**

6: Ovary and fruit hairy

22 Flowers in lateral spikes mostly shorter than leaves

23 Sepals free; leaves ±hispid, hairs both simple and minute glandular **36. S. paludosa**

23: Sepals reduced to rim; leaves tomentose with dense, simple or felted hairs

24 Axillary hairs silky, not felted, scarcely distinguishable from other hairs

25 Inflorescence elongate to 20 cm long (W.A.) **27. S. spicigera**

25: Inflorescence condensed to 6 cm long (S.A.) **35. S. linearis**

24: Axils woolly, hairs felted, tangled, clearly different from other hairs **28. S. canescens**

22: Flowers in terminal spikes or thyrses usually longer than leaves

26 Fruit and ovary with two lateral protuberances and vertical rows of tubercles **42. S. densifolia**

26: Fruit and ovary irregularly rugose or tuberculate or smooth

27 Plant prostrate forming mats

28 Axils conspicuously woolly **34. S. humifusa**

28: Axillary hairs ±inconspicuous

29 Leaves linear, entire (south-western Australia) **33. S. pulvinaris**

29: Leaves broader than linear (south-eastern Australia)

30 Flowers sessile **37. S. albida**

30: Flowers with stalks below bracteoles **57. S. hookeri**

27: Plant erect, ascending, or decumbent, not forming mats

31 Barbulae clavate, with swollen papillate head on short stalk **20. S. pulchella**

31: Barbulae simple or with papillate head on long stalk

32 Hairs on outside of corolla stiff and brownish yellow towards top

33 Leaves mostly more than 2 mm wide; barbulae simple **63. S. hamiltonii**

33: Leaves less than 2 mm wide; barbulae papillate apically **22. S. eneabba**

32: Hairs on outside of corolla white

34 Fruit covered with arcuate, simple, golden hairs; leaves mostly reduced **67. S. restiacea**

34: Fruit with whitish hairs; leaves well developed

35 Leaves with recurved to revolute margins **35. S. linearis**

35: Leaves flat

36 Sepals acute (northern Australia) **70. S. angulata**

36: Sepals obtuse or forming a rim (mostly western, southern and central arid regions)

37 Hairs on back of indusium sparse, usually neither stiff nor straight, sometimes absent

38 Leaves basal or towards base of scapose stems, more than 60 mm long **23. S. lanceolata**

38: Leaves scattered along stems, to 60 mm long **24. S. virgata**

37: Hairs on back of indusium numerous, straight, stiff

39 Beard in corolla throat sparse; ovary 1-locular except at base (mostly E of Great Dividing Ra. and along S coast) **37. S. albida**

39: Beard in corolla throat dense; ovary 2-locular except near top (mostly drier areas W of Great Dividing Ra.)

40 Hairs soft, short, usually very dense; leaves with small teeth or entire **39. S. ovalifolia**

40: Hairs coarse, long, not appressed; leaves usually prominently triangular-toothed **38. S. parvibarbata**

Sect. 1. Scaevola

Scaevola L. sect. **Scaevola**

Scaevola sect. *Sarcocarpa* G.Don, *Gen. Hist.* 3: 727 (1834); *Scaevola* sect. *Sarcocarpeae* DC., *Prodr.* 508 (1839), *orth. var.* T: *S. plumieri* (L.) M.Vahl; lecto, *fide* R.C.Carolin, *Telopea* 3: 489 (1990).

Scaevola sect. *Crossotoma* G.Don, *Gen. Hist.* 3: 730 (1834); *Crossotoma* (G.Don) Spach, *Hist. Nat. Vég.* 9: 583 (1840). T: *S. spinescens* R.Br.; lecto, *fide* R.C.Carolin, *Telopea* 3: 489 (1990).

Scaevola sect. *Phacelophyllum* K.Krause, *Pflanzenr.* 54: 118 (1912). T: *S. haianensis* Hance

Mostly shrubs or small trees; inflorescence axis continuing vegetative growth after flowering. Leaves alternate, often large. Flowers pedunculate in axillary cymes or racemes, or solitary in leaf axils; bracts leaf-like; bracteoles usually small; pedicel obsolete or to 12 mm long. Corolla white to pale lilac or brownish (red and yellow in some non-Australian species); barbulae various. Anthers glabrous at tip. Ovary 2-locular; indusium with scattered hairs or glabrous on adaxial surface. Fruit with mesocarp ±fleshy or corky; endocarp with 2 fertile locules and often 2 sterile cavities.

This section contains c. 24 species, mostly occurring in tropical and subtropical areas of the Pacific; 6 species are found in Australia, 5 endemic and 1 common in strand communities throughout the tropical areas of the Indo-Pacific region including Christmas Is. and the Cocos (Keeling) Is.

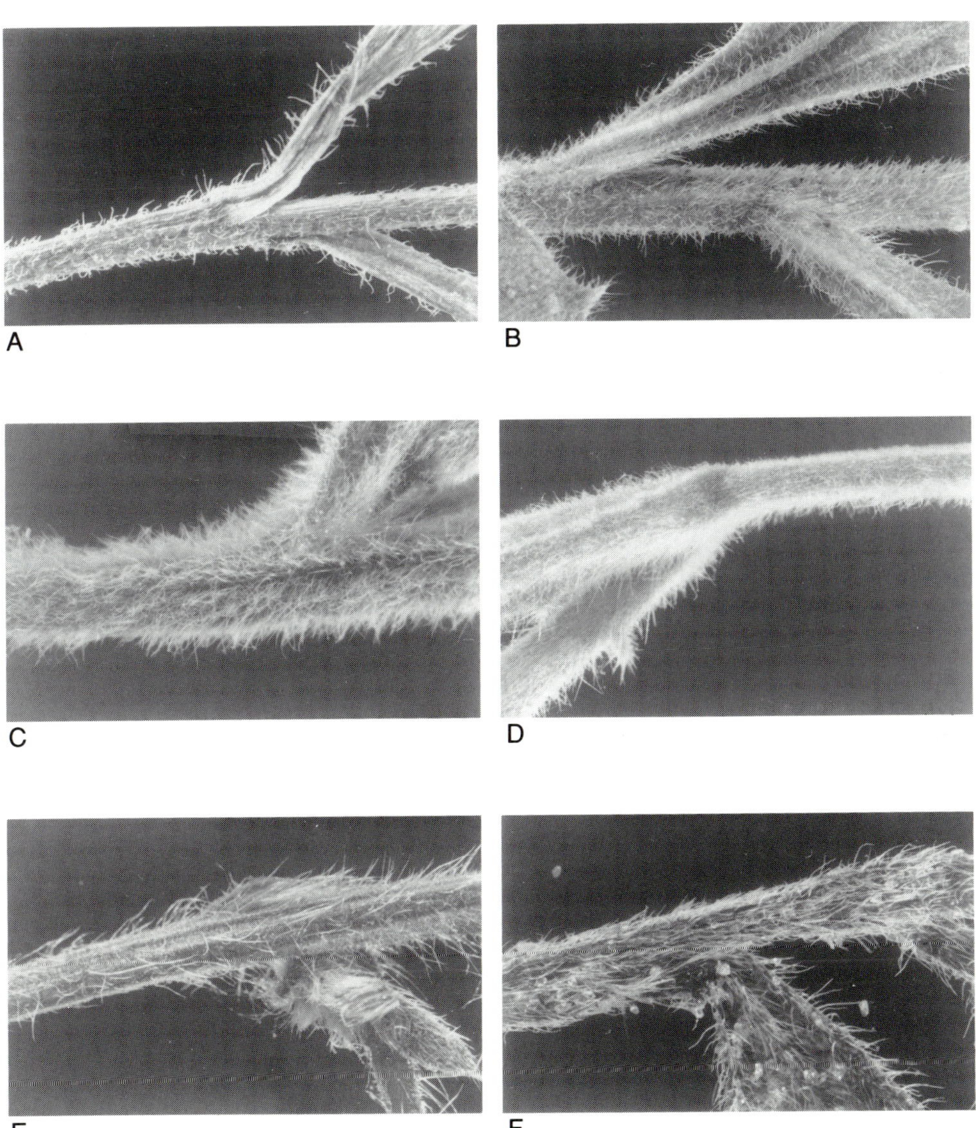

Figure 36. Indumentum on the stems of *Scaevola* ser. *Pogogynae*, all X5. **A**, *S. albida* (R.Cambage, 12 Jan. 1907, SYD). **B**, *S. parvibarbata* (R.Carolin 13034, SYD). **C–D**, *S. ovalifolia* (**C**, R.Carolin 9406, SYD; **D**, P.Latz 1124, SYD). **E**, *S. densifolia* (-.Allen 180, SYD). **F**, *S. cuneiformis* (A.Beauglehole 12997, SYD). Photographs by M.Ricketts.

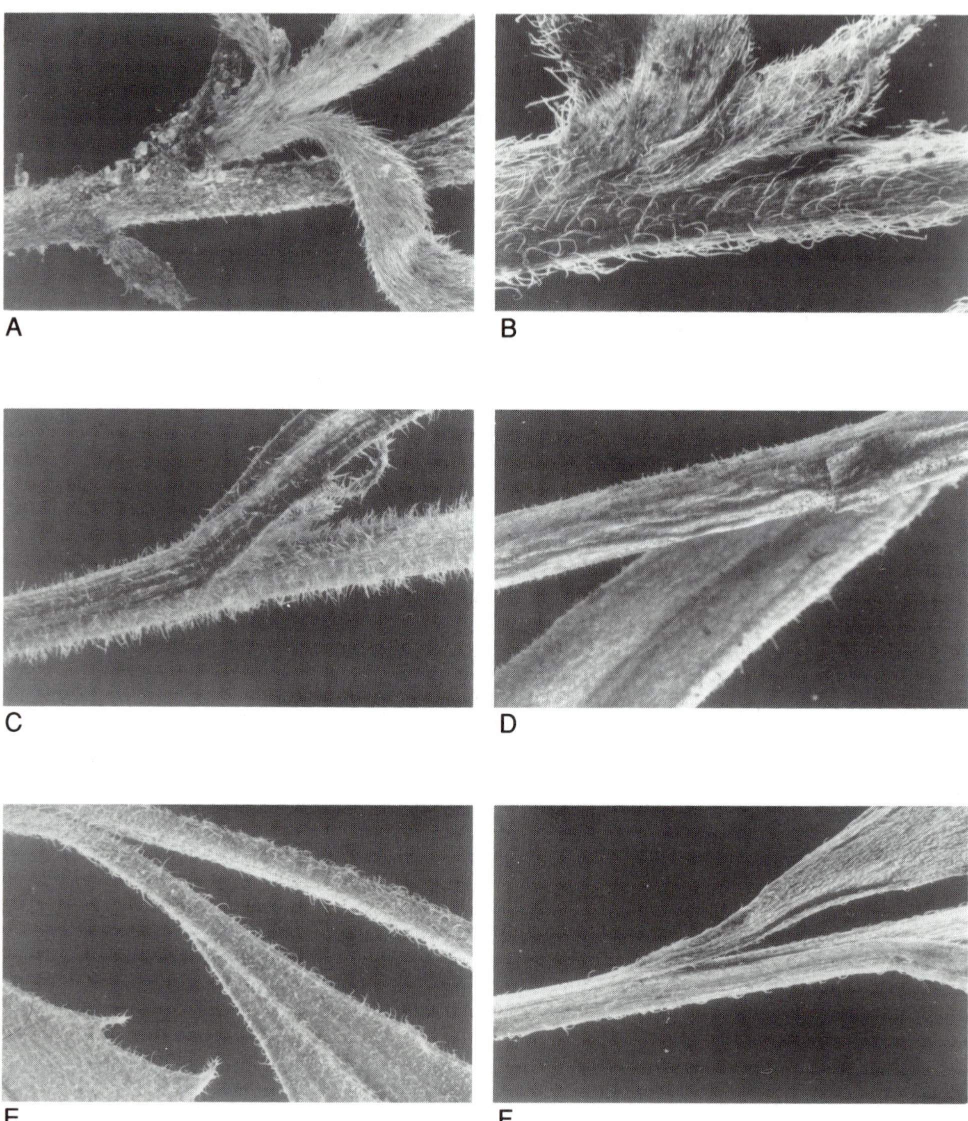

Figure 37. Indumentum on the stems of *Scaevola* ser. *Pogogynae*, all X5. **A**, *S. argentea* (A.George 7018, SYD). **B**, *S. aemula* (P.Myerscough, 9 June 1984, SYD). **C**, *S. amblyanthera* var. *amblyanthera* (-.Henry 441, SYD). **D**, *Scaevola amblyanthera var. centralis* (S.Jacobs, 27 May 1967, SYD). **E–F**, *S. humilis* (**E**, M.Crisp 062910, SYD; **F**, J.West 4092, SYD). Photographs by M.Ricketts.

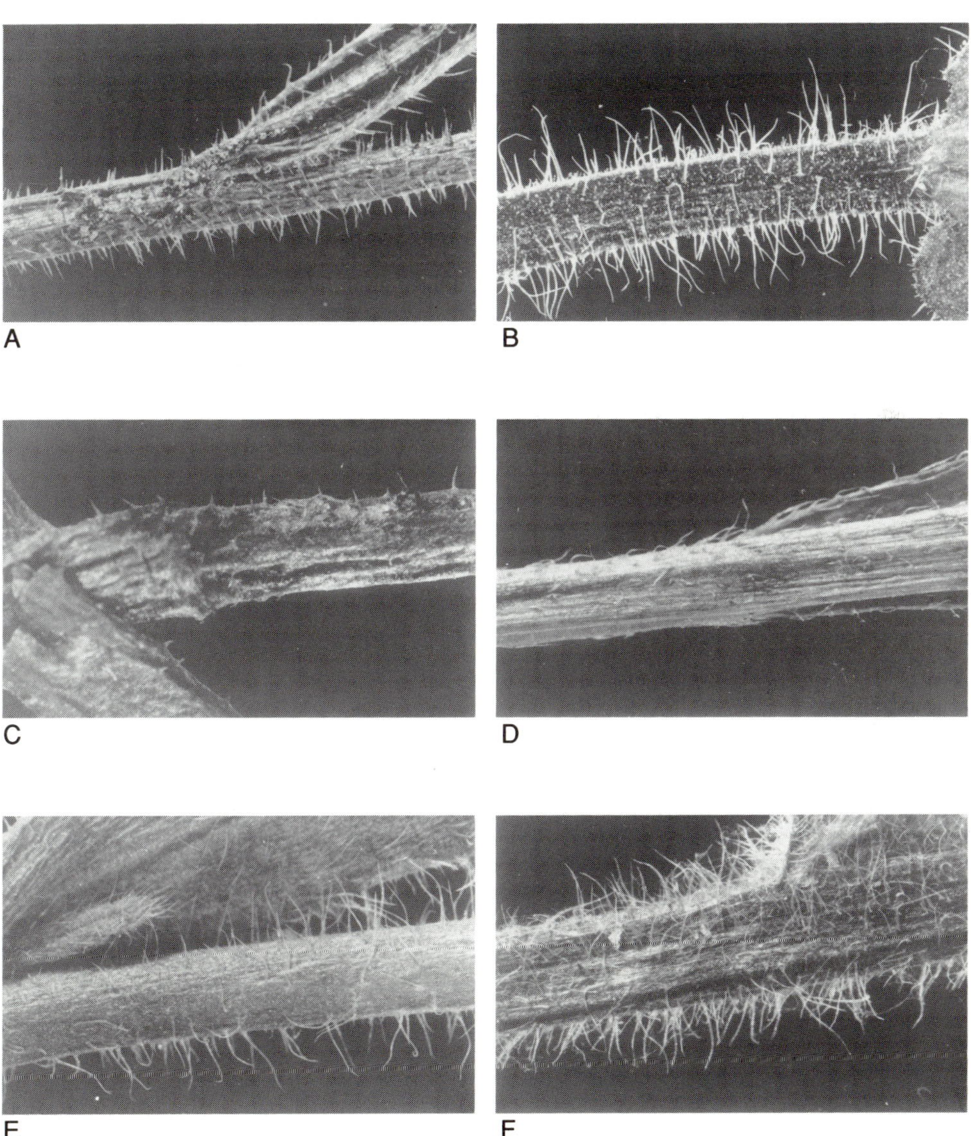

Figure 38. Indumentum on the stems of *Scaevola* subsect. *Pogonanthera*, all X5. **A**, *S. ramosissima* (R.Carolin 1525, SYD). **B**, *S. pilosa* (M.Phillips CBG 030160, SYD). **C**, *S. tenuifolia* (A.Beauglehole 12993, SYD). **D**, *S. striata* var. *striata* (W.Peacock 60884.1, SYD). **E**, *S. calliptera* (N.Donner 1452, SYD). **F**, *S. phlebopetala* (B.Benn, 14 Sept. 1963, SYD). Photographs by M.Ricketts.

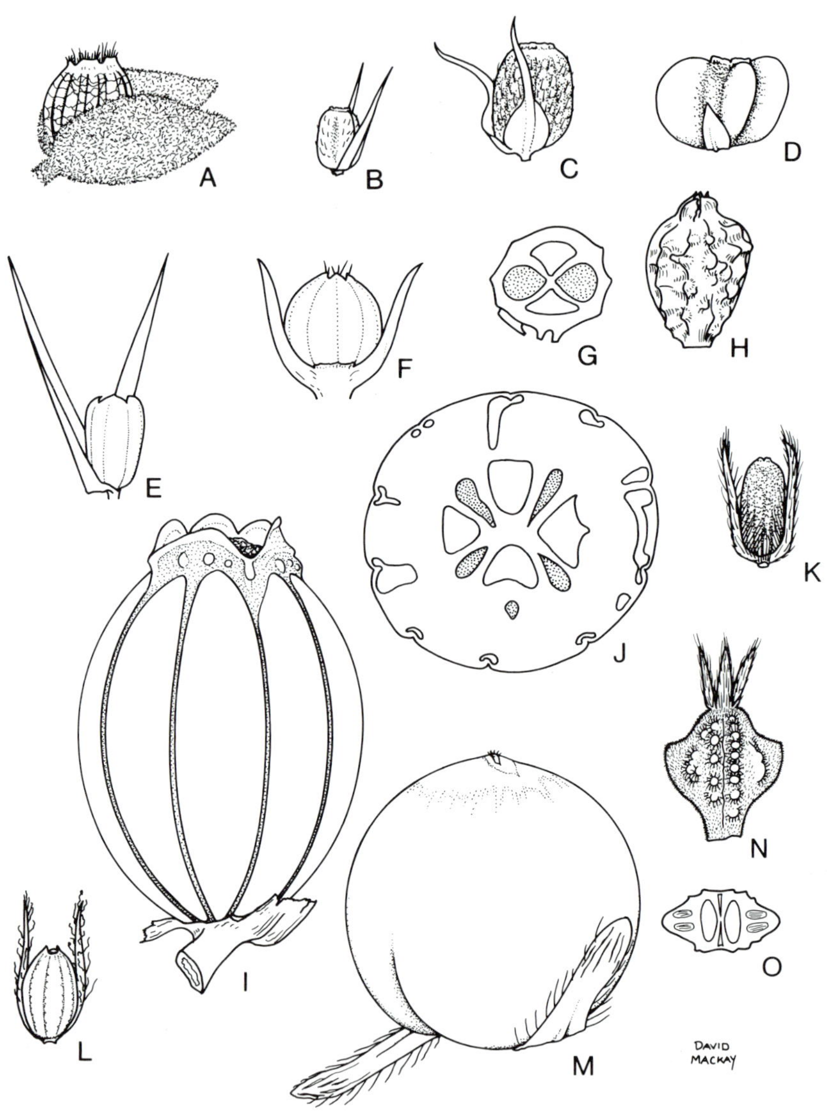

Figure 39. Fruits of *Scaevola,* all X4. **A**, *S. tomentosa* (R.Carolin 10652, SYD). **B**, *S. thesioides* (E.M.Canning, 20 Aug. 1968, SYD). **C**, *S. angustata* (T.Reichstein 856, AD). **D**, *S. crassifolia* (M.E.Phillips, 18 Sept. 1962, SYD). **E**, *S. nitida* (S.Jackson, 22 Dec. 1912, SYD). **F**, *S. cunninghamii* (A.George 10316, SYD). **G–H**, *S. globulifera*. **G**, T.S. (**G–H**, M.E.Phillips, 22 Sep. 1968, SYD). **I–J**, *S. porocarya*. **J**, T.S. (**I–J**, R.Carolin 3358, SYD). **K**, *S. browniana* subsp. *browniana* (A.Beauglehole 47147, SYD). **L**, *S. virgata* (R.Carolin 3240, SYD). **M**, *S. calendulacea* (fresh material, Sydney, Aug. 1984). **N–O**, *S. densifolia*. **O**, T.S. (**N–O**, C.A.Gardner, Nov. 1944, PERTH). Drawn by D.Mackay.

1. **Scaevola taccada** (Gaertner) Roxb., *Hort. Bengal.* 15 (1814)

Lobelia taccada Gaertner, *Fruct. Sem. Pl.* 1: 119, t. 25, fig. 5 (1788). T: 'e collect. sem. hort. lugdb.'; holo: L *n.v.*, *fide* C.Jeffrey, *Kew Bull.* 34: 543 (1980).

S. sericea M.Vahl, *Symb. Bot.* 2: 37 (1791). T: 'Habitat in Savage Island, Dn. *Prof. Fabricius*'; *n.v.*

S. koenigii M.Vahl, *Symb. Bot.* 3: 36 (1794); *Lobelia sericea* var. *koenigii* (M.Vahl) Kuntze, *Revis. Gen. Pl.* 2: 377 (1891); *S. frutescens* var. *koenigii* (M.Vahl) Domin, *Biblioth. Bot.* 89: 646 (1929); *S. lobelia* L. ex Vriese, *Ned. Kruidk. Arch.* 2: 20 (1850), *nom. illeg.* T: India, *J.G.Koenig*; holo: C *n.v.*, *fide* C.Jeffrey, *Kew Bull.* 34: 543 (1980).

S. frutescens K.Krause, *Pflanzenr.* 54: 125 (1912), *nom. illeg.* T: Niue Is., *G.Forster*; holo: C *n.v.*, *fide* C.Jeffrey, *Kew Bull.* 34: 543 (1980).

Illustration: A.B.Cribb & J.W.Cribb, *Pl. Life Great Barrier Reef & Adjacent Shores* 136 (1985); T.Low, *Bush Medicine* 168 (1990).

Erect, spreading shrub or small tree to 4 m tall. Leaves obtuse, oblong to obovate, tapering basally, usually entire, thick, glabrescent; lamina 4–23 cm long, 2–9 cm wide; axillary hairs long, silky or woolly. Flowers in axillary cymes; peduncle 5–20 mm long; bracteoles lanceolate-ovate, 2–8 mm long; pedicel 3–12 mm long. Sepals oblong-linear, 3–6 mm long, free. Corolla 14–28 mm long, glabrous or pubescent outside, densely bearded inside, white to pale lilac; barbulae flat, with long apical hairs; wings to 1.5 mm wide, laciniate. Indusium glabrous above or sparsely hairy. Fruit ovoid or globular, 7–13 mm long, grooved, ribbed, glabrous or silky hairy, whitish. $n = 8$, C.J.F.Skottsberg, *Arch. für Botanik* 3: 63 (1955). Figs 26B, 52.

Occurs in tropical W.A., N.T. and Qld, along beaches; also through the Indo-Pacific region including Christmas Is. and the Cocos (Keeling) Is. Flowers most of the year. Map 108.

W.A.: vicinity of Cone Hill, Cape Domett, *T.G.Hartley 14749* (PERTH). N.T.: Wessel Islands, *P.K.Latz 3271* (DNA); Nightcliff, Darwin, *R.L.Specht 15* (AD). Qld: Morris Is., *D.R.Stoddart 4975* (BRI); Slade Point near Mackay, *R.C.Carolin 4545* (SYD).

The argument around the legitimate name of this species is pursued particularly in the following publications: C.Jeffrey, *Kew Bull.* 34: 543 (1980); P.S.Green, *Taxon* 40: 118–122 (1991). The latter argument is accepted here. The combination of broad fleshy leaves, shrubby habit and axillary cymes distinguishes this species from all other species likely to be found on beaches in Australia. Various parts of the plant used by Aborigines for medicinal purposes (T.Low, *Bush Medicine* 1990; E.V.Lassak & T.McCarthy, *Austral. Medicinal Pl.* 138, 1983; Aboriginal Communities of the N. Territory of Australia, *Traditional Bush Medicines* 194–195, 1988).

2. **Scaevola spinescens** R.Br., *Prodr.* 586 (1810)

Crossotoma spinescens (R.Br.) Vriese, *Natuurk. Verh. Holl. Maatsch. Wetensch. Haarlem* ser. 2, 10: 36 (1854); *Lobelia spinescens* (R.Br.) Kuntze, Revis. *Gen. Pl.* 2: 378 (1891). T: near the shores of Fowler Bay, [S.A.], *R.Brown*; lecto: BM, *fide* R.C.Carolin, *Telopea* 3: 491 (1990); isolecto: K.

S. oleoides DC., *Prodr.* 7(2): 512 (1839); *Crossotoma oleoides* (DC.) Vriese, *Natuurk. Verh. Holl. Maatsch. Wetensch. Haarlem* ser. 2, 10: 37 (1854). T: interior of N.S.W., *A.Cunningham*; holo: G-DC; iso: BM?, K.

S. lycoides DC., *Prodr.* 7(2): 512 (1839); *Crossotoma lycoides* (DC.) Vriese, *Natuurk. Verh. Holl. Maatsch. Wetensch. Haarlem* ser. 2, 10: 38 (1854). T: Dampier Archipelago, [W.A.], *A.Cunningham*; holo: G-DC; iso: K.

S. spinescens var. *rufa* E.Pritzel in F.L.E.Diels and E.Pritzel, *Bot. Jahrb. Syst.* 35: 568 (1905). T: near Kalgoorlie, W.A., Nov. 1900, *L.Diels 1685*; holo: ?B (destroyed).

Illustrations: K.Krause, *Pflanzenr.* 54: 137, fig. 26H–L (1912); E.R.Rotherham *et al.*, *Fl. Pl. New South Wales & S. Queensland* 150, fig. 488 (1975); G.M.Cunningham *et al.*, *Pl. W. New South Wales* 638 (1981).

Figure 40. *Scaevola*. **A–D**, *S. myrtifolia*. **A**, indusium X7.5; **B**, anther X10; **C**, corolla X3; **D**, habit X1.33 (**A–D**, D.Whibley 1983, SYD). **E**, *S. spinescens,* habit X0.5 (J.West 4008, SYD). **F–H**, *S. pulchella*. **F**, anther X10; **G**, indusium X7.5; **H**, corolla X2.5 (**F–H**, J.Beard 2980, PERTH). **I**, *S. browniana* subsp. *browniana,* habit X0.5 (G.Carr 47511, SYD). **J**, *S. revoluta* subsp. *revoluta*, habit X0.5 (R.Carolin 9244, SYD). Drawn by D.Mackay.

Rigid, divaricate shrub to 2 m tall, glabrous or with substellate scurfy hairs, greyish when young; dwarf branchlets often spinescent. Leaves often clustered on branchlets, sessile, obovate to linear, entire, thick; lamina 9–36 mm long, 1–6 mm wide. Flowers solitary in axils; peduncle slender, 5–20 mm long; bracteoles linear, 2–5 mm long, ±equal to ovary. Sepals rim-like, to 1 mm long. Corolla 9–16 mm long, scurfy or glabrous outside, densely bearded inside, white or yellowish, occasionally with purple veins; barbulae broad, prominent, simple; wings to 1 mm wide, laciniate. Indusium to 2 mm wide, with scattered hairs above. Fruit ovoid, 5–8 mm long, glabrous, black or purplish. n = 8, W.J.Peacock, *Proc. Linn. Soc. New South Wales* 88: 8 (1963). *Currant Bush, Maroon Bush.* Figs 40E, 53.

Occurs throughout the drier parts of all mainland States; grows on hillsides and on stony plains. Flowers most of the year. Map 109.

W.A.: c. 14.5 km N of Learmonth, *A.S.George 1278* (PERTH). N.T.: 3 km S of Yuendumu, *T.S.Henshall 2896* (DNA). S.A.: Dalhousie Springs, *D.E.Symon 9324* (AD). Qld: Whynot Stn, c. 38 km WSW of Quilpie, 15 Nov. 1954, *L.S.Smith* (BRI). N.S.W.: Mt Oxley near Bourke, 13 July 1958, *C.K.Ingram* (NSW). Vic.: Barneys Track, Hattah Lakes Natl Park, *G.W.Anderson 16* (MEL).

A very variable species. The dwarf branchlets distinguish it from all other shrubby species except *S. tomentosa* but that species has markedly stellate hairs. Used by Aborigines for various medicinal purposes including, in W.A., an infusion from the form on the Eastern Goldfields for alleviating the pain of cancer (E.Reid, *Rec. W. Austral. Pl. Used by Aboriginals as Medicinal Agents*, Western Australian Institute of Technology, 1977; E.V.Lassak & T.McCarthy, *Austral. Medicinal Pl.* 138, 1983).

3. **Scaevola acacioides** Carolin, *Telopea* 3: 491 (1990)

T: Bee Gorge, 22°16'S, 118°15'E, Wittenoom area, W.A., *J.V.Blockley 21-9*; holo: PERTH.

Shrub to 1 m high; hairs on young parts substellate, scurfy. Leaves sessile, narrowly oblong to elliptic, tapering towards base, entire, thick, greyish green; lamina 25–50 mm long, 2–5 mm wide. Flowers in dichasia or monochasia, or flowers solitary in axils; peduncle 8–15 mm long; bracteoles deltoid, 1–1.5 mm long. Sepals rim-like, sinuate, c. 0.5 mm long. Corolla 10–13 mm long, scurfy stellate-glandular-hairy outside, densely bearded inside, white to cream; barbulae prominent, papillate apically; wings nearly obsolete. Indusium 0.5 mm long, hairs above few, white, not exceeding lips. Fruit ovoid, to 5 mm long, glabrous, black.

Rare in the Pilbara region of W.A. Flowers about May. Map 110.

W.A.: 38 km W of Wittenoom, *K.Newbey 10051* (PERTH).

The branched inflorescences and the long narrow leaves together distinguish this species from others in sect. *Scaevola.*

4. **Scaevola bursariifolia** J.M.Black, *Trans. & Proc. Roy. Soc. South Australia* 51: 385 (1927)

T: Bunda Plateau (N of Fowler's Bay towards Eucla), S.A., Feb. 1879, *R.Tate*; lecto: AD, *fide* R.C.Carolin, *Telopea* 3: 492 (1990); isolecto: K.

Shrub to 70 cm tall, viscid becoming varnished; hairs minute, substellate, scurfy. Leaves shortly petiolate, obtuse, obovate-elliptic or spathulate, entire, thick; lamina 10–14 mm long, 2–5 mm wide. Flowers solitary in leaf axils; peduncle to 1 mm long; bracteoles linear, c. 1 mm long. Sepals broadly ovate, to 0.5 mm long, connate. Corolla 8–10 mm long, with few varnished hairs outside, densely bearded inside, cream; barbulae distinct, mostly simple; wings nearly obsolete. Indusium 1 mm long, with scattered, white hairs above. Fruit ovoid, c. 4 mm long, glabrous, black or purple.

Occurs from the Mt Ragged area, W.A., to the Fowlers Bay area, S.A. Flowers Oct.–Feb. Map 111.

W.A.: c. 60 km E of Salmon Gums, *K.Newbey 6692* (PERTH); N of Pine Hill, c. 22 km NW of Mt Ragged, *A.S.George 7413* (PERTH). S.A.: Koonalda, *D.E.Symon 4580* (AD); S of Coorabie post office, 135 km WNW of Ceduna, 18 Oct. 1953, *J.B.Cleland* (AD).

Ovary incompletely 2-locular. Resembles *S. spinescens*, which has leaves clustered on dwarf branches that are usually spinescent, and thus a very different habit, and which has bracteoles nearly as long as the ovary.

5. **Scaevola myrtifolia** (Vriese) K.Krause, *Pflanzenr.* 54: 136 (1912)

Merkusia myrtifolia Vriese, *Natuurk. Verh. Holl. Maatsch. Wetensch. Haarlem* ser. 2, 10: 72 (1854); *Lobelia myrtifolia* (Vriese) Kuntze, *Revis. Gen. Pl.* 2: 378 (1891). T: south-western W.A., *J.Drummond 363*; holo: K; iso: BM, MEL, P, W.

S. groeneri F.Muell., *Fragm.* 6: 15 (1866). T: south-western W.A., *J.Drummond 363*; lecto: MEL, *fide* R.C.Carolin, *Telopea* 3: 492 (1990); isolecto: BM, K, P, W.

Spreading shrub to 1.5 m tall, slightly viscid when young; hairs scattered, substellate, scurfy. Leaves shortly petiolate, acute, elliptic to oblong, entire or few-toothed, thick; lamina 17–40 mm long, 1–12 mm wide. Flowers in cymes or solitary in leaf axils; peduncle to 5 mm long; bracteoles linear, 5–8 mm long. Sepals united in a tube, c. 1 mm long. Corolla 13–16 mm long, glabrous or sparsely scurfy-glandular-hairy outside, bearded inside, white; barbulae prominent, numerous, papillate apically; wings sinuate, c. 1 mm wide. Indusium 1 mm long, nearly glabrous except lips. Fruit cylindroid, c. 3 mm long, rugose, glabrous, green. Fig. 40A–D.

Occurs from near Albany, W.A., to the Eyre Peninsula, S.A.; grows in stony soil, often over limestone, in mallee. Flowers chiefly May–Oct. Map 112.

W.A.: Hamersley R., *E.Wittwer 414* (PERTH); Twilight Cove, S of Cocklebiddy, *B.C.Crisp* 65 (AD). S.A.: c. 10 km N of Port Neill, *D.J.E.Whibley 1983* (AD); N of Arno Bay, *C.R.Alcock 546* (AD); c. 5 km E of Butler, *B.Copley 2980* (AD).

Ovary c. 1.5 mm long. Closely related to *S. bursariifolia* which, however, has smaller obtuse leaves and shorter bracteoles which barely exceed the sepals.

6. **Scaevola tomentosa** Gaudich., *Voy. Uranie* 460 (1826)

Temminckia tomentosa (Gaudich.) Vriese, *Ned. Kruidk. Arch.* 2: 148 (1851). T: Shark Bay, [W.A.], *C.Gaudichaud*; holo: P.

S. atriplicina F.Muell., *Fragm.* 2: 18 (1860); *Lobelia atriplicina* (F.Muell.) Kuntze, *Revis. Gen. Pl.* 2: 378 (1891); *S. tomentosa* var. *atriplicina* (F.Muell.) K.Krause, *Pflanzenr.* 54: 138 (1912). T: towards Port Gregory, W.A., *A.Oldfield*; holo: MEL; iso: K.

Illustrations: C.Gaudichaud, *op. cit.* t. 81; K.Krause, *Pflanzenr.* 54: 137, fig. 26A–G (1912).

Divaricate shrub to 1.5 m tall, often with dwarf branchlets, stellate-pubescent also with minute glandular hairs. Leaves often crowded, obtuse, oblong-elliptic to narrowly oblong, tapering towards base, slightly recurved, entire or dentate; lamina 15–50 mm long, 3–15 mm wide. Flowers solitary in axils; peduncle 2–14 mm long; bracteoles ovate, 8–12 mm long. Sepals rim-like, sinuate, with long hairs. Corolla 15–23 mm long, gibbous at base and curved backwards above the swelling, stellate-pubescent outside, bearded inside, yellowish or orange-brown; barbulae simple, 0.5 mm wide; wings to 1 mm wide or obsolete. Indusium glabrous except lips. Fruit c. 5 mm diam., glabrous, black. Fig. 39A.

Occurs in W.A. from Dongara northwards to near Carnarvon and inland; in calcareous soils of various types, sometimes in sands which probably have calcium in subsoil. Flowers May–Oct. Map 113.

W.A.: c. 1.5 km S of Geraldton on Dongara road, *S.Carlquist 2954* (AD); Denham, Shark Bay, *R.C.Carolin 10652* (SYD); Passage Paddock, Dirk Hartog Is., *A.S.George 11379* (PERTH); c. 772 km N of Perth on Carnarvon road, *H.Demarz 3883* (PERTH); Kalbarri, *R.C.Wemm 1449* (PERTH).

A very characteristic species showing affinities with *S. spinescens* but easily distinguished by its dense, stellate indumentum and broader bracteoles that are fused along one margin. There is some variation in *S. tomentosa*, particularly in the indumentum. In most specimens it is very dense but in some it is rather sparse; often in these less hairy forms the minute glandular hairs are more numerous.

Sect. 2. Enantiophyllum

Scaevola sect. **Enantiophyllum** Miq., *Ann. Mus. Bot. Lugduno-Batavum* 1: 210 (1864)

Type: *S. amboinensis* Miq.

Scramblers or vines; inflorescence axis continuing vegetative growth after flowering. Leaves opposite. Flowers in pedunculate, axillary cymes; bracts leaf-like; bracteoles minute, much shorter than pedicel; pedicel distinct. Corolla yellow; barbulae simple. Anthers glabrous at tip. Ovary 2-locular. Indusium with basal tuft of long, stiff hairs on adaxial surface. Fruit with fleshy mesocarp; endocarp hard, with 2 fertile locules.

A section of c. 10 species that are not well distinguished and may be only 1 very variable species in northern Australia and northwards to the Philippines; the Australian representative of this complex is, however, treated here as a separate species. See P.W.van Leenhouts, *Fl. Males.* ser. I, 5: 342 (1957).

7. Scaevola enantophylla F.Muell., *Fragm.* 8: 58 (1873)

Based on *S. oppositifolia sensu* F.Mueller, *Fragm.* 6: 225 (1868), non Roxb.; *Lobelia enantophylla* (F.Muell.) Kuntze, *Revis. Gen. Pl.* 2: 378 (1891). T: Rockingham Bay, Qld, *J.Dallachy*; lecto: MEL *fide* R.C.Carolin, *Fl. Australia* 35: 333 (1992); isolecto: K.

S. scandens Bailey, *Rep. New Pl.* 2 (1889); *Rep. Exped. Bellenden-Ker* 47 (1889); *S. enantophylla* var. *scandens* (Bailey) Ewart, *Proc. Roy. Soc. Victoria* ser. 2, 19(2): 45 (1907). T: Bellenden Ker at about 3000 ft, Qld, *F.M.Bailey;* holo: BRI.

Scrambler or vine to 4 m long, glabrous or pubescent. Leaves shortly petiolate, acuminate, ovate to lanceolate, dentate; lamina 4–15 cm long, 20–60 mm wide. Flowers in few-flowered, axillary cymes; peduncle to 16 mm long; bracteoles deltoid, mostly to 1 mm long; pedicel to 5 mm long, articulate below ovary. Sepals linear, to 5 mm long, free. Corolla 14–20 mm long, glabrous outside, densely bearded inside, yellow; wings to 1.5 mm wide. Fruit narrowly obovoid, 10–12 mm long, grooved, glabrous or with scattered hairs, black. Fig. 51.

Occurs on forest margins near the east coast of northern Qld. Flowers chiefly July–Nov. Map 114.

Qld: Thornton Peak, Atherton, *B.Hyland 7092* (BRI); E of Ravenshoe, *S.T.Blake 9900* (BRI); Topaz near Malanda, *L.S.Smith 3307* (BRI, NSW); Mt Fox, Nov. 1949, *M.S.Clemens* (BRI); Mt Spec near Paluma, *C.H.Gittins 492* (BRI).

The endocarp is tuberculate. The only species in the genus with opposite leaves.

Sect. 3. Xerocarpa

Scaevola sect. **Xerocarpa** G.Don, *Gen. Hist.* 3: 728 (1834).

Xerocarpa (G.Don) Spach, *Hist. Nat. Vég.* 9: 583 (1838). T: *S. crassifolia* Labill.; lecto, *fide* R.C.Carolin, *Telopea* 3: 489 (1990).

Scaevola sect. *Gymnostegia* Benth. in S.L.Endlicher *et al.*, *Enum. Pl.* 68 (1837). T: *S. thesioides* Benth.

Shrubs to perennial herbs with basal stock; inflorescence axis ceasing growth after flowering. Leaves alternate. Flowers in terminal or axillary thyrses, racemes or spikes; bracts and bracteoles leaf-like or reduced; peduncle and pedicel obsolete or distinct. Corolla white, bluish or pink; barbulae simple, papillate or branched. Anthers glabrous or hairy apically. Ovary 1–4-locular; indusium glabrous or hairy, with tuft of stiff or flexible bristles towards base on adaxial surface. Fruit with mesocarp drying rapidly and green, or rarely fleshy and white to purplish; endocarp hard, with 1–4 fertile locules and sometimes 2–4 sterile cavities.

This section contains 66 species; 64 endemic in Australia; *S. gracilis* occurs in New Zealand and *S. porrecta* occurs in Tonga.

Subsect. 1. Biloculatae

Scaevola sect. **Xerocarpa** subsect. **Biloculatae** K.Krause, *Pflanzenr.* 54: 160 (1912).

Type: *S. crassifolia* Labill.; lecto, *fide* R.C.Carolin, *Telopea* 3: 490 (1990).

Scaevola ser. *Monospermae* Benth., *Fl. Austral.* 4: 86, 100 (1868); *Scaevola* sect. *Xerocarpa* subsect. *Uniloculatae* K.Krause, *loc. cit.* T: *S. canescens* Benth.; lecto, *fide* R.C.Carolin, *Telopea* 3: 490 (1990).

Scaevola sect. *Xerocarpa* subsect. *Xerocarpa* R.C.Carolin, *Telopea* 3: 490 (1990), *nom. illeg.*

Leaves well developed. Flowers usually in a spike; peduncle and pedicel obsolete. Anthers glabrous at tip. Ovary 1–4-locular; indusium glabrous or with stiff or soft basal hairs.

This subsection contains 50 species; 48 in Australia, all endemic.

Ser. 1. Globuliferae

Scaevola sect. **Xerocarpa** ser. **Globuliferae** Benth., *Fl. Austral.* 4: 85, 92 (1868).

Type: *S. globulifera* Labill.; lecto, *fide* R.C.Carolin, *Telopea* 3: 490 (1990).

Scaevola sect. *Xerocarpa* subsect. *Biloculatae* ser. *Macrostachyae* Benth., *Fl. Austral.* 4: 85, 96 (1868). T: *S. macrostachya* (Vriese) Benth.; lecto, *fide* R.C.Carolin, *Telopea* 3: 490 (1990).

Scaevola sect. *Xerocarpa* subsect. *Xerocarpa* ser. *Xerocarpa* R.C.Carolin, *Telopea* 3: 490 (1990), *nom. illeg.* T: *S. crassifolia* Labill.; lecto, *fide* R.C.Carolin, *loc. cit.*

Flowers mostly in spikes. Indusium glabrous or often with scattered weak hairs on adaxial surface. Ovary 1–4-locular.

This series contains 31 species; 29 in Australia, all endemic.

8. Scaevola collaris F.Muell., *Rep. Pl. Babbage's Exped.* 15 (1858)

Lobelia collaris (F.Muell.) Kuntze, *Revis. Gen. Pl.* 2: 378 (1898). T: near Wonnomulla, S.A., *Babbage Expedition*; holo: MEL.

S. decurrens W.Fitzg., *J. W. Austral. Nat. Hist. Soc.* 1: 26 (1904). T: Nannine, W.A., Sept. 1903, *W.V.Fitzgerald*; holo: PERTH.

Erect, perennial subshrub to 50 cm tall, glabrous. Leaves succulent, sessile, linear to oblanceolate, entire or dentate; lamina 1–8.5 cm long, 1–9 mm wide. Flowers in terminal, leafy spikes, or solitary, or clustered in axils; bracteoles triangular, 0.5–2 mm long. Sepals ovate to triangular, 1.5–2 mm long, free. Corolla 6–17 mm long, glabrous outside, with short vesicular hairs inside, yellow to cream or mauve; barbulae vesicular or absent; wings 0.5–1 mm wide. Ovary 2-locular, with beak 1–6 mm long, contracted below; indusium 1–2 mm wide, folded above, with short vesicular hairs in channel and on lips. Fruit ellipsoidal, 8–26 mm long, with beak to 6 mm long, ribbed, glabrous. Fig. 55.

Occurs in arid areas of southern W.A., southern N.T., S.A., western Qld and north-western N.S.W., in saline soil. Flowers chiefly May–Nov. Map 115.

W.A.: Well 40, Canning Stock Route, *M.Howe & P.Smith 82* (PERTH). N.T.: Lake Mackay, *G.Chippendale 3396* (AD, DNA, NSW). S.A.: Margaret Ck crossing, *C.R.Alcock 6492* (AD). Qld: c. 16 km S of Gregory R., *B.Copley 2223* (AD). N.S.W.: Yandama Ck track on Callabonna–Frome outflow, *R.C.Carolin 13055* (SYD).

Even though the beak may be very short, its presence in the fruit usually distinguishes this species from all others in the genus.

This is a very variable species especially in the teeth on the leaf margins, the size of the flowers and the length of the beak on the fruit (it may be almost absent). The fruit has a thin outer skin which usually disintegrates revealing a sponge-like but woody endocarp. The endocarp is lacunate like a woody sponge or with woody plates or spines surrounded by an evanescent skin and very hard. Its spongy appearance is unique in the genus. In some W.A. specimens this sponge-like structure is so loose as to appear as woody plates and is rarely even more reduced to conical tubercles. The last condition gives the fruit a very different appearance from the common state.

9. Scaevola crassifolia Labill., *Nov. Holl. Pl.* 1: 56 (1804)

Merkusia crassifolia (Labill.) Vriese, *Ned. Kruidk. Arch.* 2: 151 (1851); *Lobelia crassifolia* (Labill.) Kuntze, *Revis. Gen. Pl.* 2: 378 (1891). T: south-western [W.A.], *J.J.H.de Labillardière*; syn: BM, P.

Illustrations: J.J.H.de Labillardière, *op. cit.* t. 79; K.Krause, *Pflanzenr.* 54: 151, fig. 28A–E (1912).

Spreading shrub to 1.5 m high and 3 m wide, apparently glabrous, viscid when young. Leaves ovate to orbicular, mostly dentate, thick; lamina 1.5–8 cm long, 3–43 mm wide; petiole ±decurrent. Flowers in terminal spikes to 5.5 cm long; bracts leafy below, reduced above, mostly not exceeding flowers; bracteoles linear to triangular, to 4 mm long. Sepals united in a tube, sinuate, 0.3–1 mm long. Corolla 8–11 mm long, glabrous or shortly hairy basally outside, bearded inside, white, blue or pale purple; barbulae papillate apically; wings 0.5–1 mm wide. Ovary 2-locular, 1–2 mm long; indusium 1.5–2 mm wide, mostly glabrous except lips. Fruit globular, to 5 mm diam., rugose, with 3–5 swollen ribs, usually glabrous. Figs 39D, 54.

Occurs along the south-west and south coast from Carnarvon, W.A., to Robe, S.A.; on coastal sand dunes and limestone. Flowers Aug.–Jan. Map 116.

W.A.: Cape Cuvier, *R.Storey 8224* (MEL); Claremont, *A.Morrison 17083* (NSW). S.A.: Fowler Bay, *R.J.Chinnock 2735* (AD); Proper Bay, *C.R.Alcock 803* (AD); Wallaroo Sandhills, *B.Copley 1601* (AD).

The corolla is obtuse in bud. A form with leaves and flowers smaller than normal occurs on Dorre Island N of Shark Bay, W.A. The species resembles *S. nitida* which has thinner, narrower and sessile leaves, longer bracteoles that exceed the sepals, and cylindrical grooved fruit. Also similar to *S. angustata* which has sessile leaves.

10. **Scaevola angustata** Carolin in J.P.Jessop & H.R.Toelken, *Fl. S. Australia* 4th edn, 3: 1410 (1986)

T: Aldinga, c. 40 km SSW of Adelaide, S.A., 31 Aug. 1966, *J.R.Wheeler 21*; holo: AD.

[*Scaevola nitida auct. non* R.Br.: E.L.Robertson in J.M.Black, *Fl. S. Australia* 2nd edn, 4: 830 (1965)]

Decumbent or ascending shrub to 1 m tall, glabrous, viscid when young. Leaves sessile, oblanceolate to elliptic, dentate; lamina 2–10 cm long, 2–40 mm wide. Flowers in loose, terminal spikes to 9.5 cm long; bracts linear, 5–24 mm long; bracteoles linear to narrowly triangular, 2–9 mm long, exceeding sepals, keeled. Sepals rim-like, c. 0.3 mm high. Corolla 11–20 mm long, mostly viscid-glabrous outside, bearded inside, pale blue or violet; barbulae short, papillate apically; wings 1 mm wide. Ovary 2-locular; indusium c. 2 mm wide, shortly hairy on both surfaces. Fruit cylindroid, to 4 mm long, tuberculate-rugose, ±ribbed, glabrous or with scattered hairs. Fig. 39C.

Occurs along the coast of south-eastern S.A. including Eyre Peninsula, in heathland and on sand dunes. Flowers chiefly Aug.–Nov. Map 117.

S.A.: Proper Bay Rd, c. 10 km from Port Lincoln, *D.Freckelton 12* (AD); Brighton, 26 Apr. 1937, *E.H.Ising* (AD); Sturt Bay, *B.Copley 2921* (AD); Rocky R., c. 25 km SSE of Cape Borda, Kangaroo Is., 24 Nov. 1945, *J.B.Cleland* (AD); between S end of Lake Robe and beach, *D.E.Symon 4464* (AD, NSW).

The corolla is acuminate in bud. Distinguished from *S. nitida* and *S. crassifolia* by the more attenuate narrower leaves and the rugose fruit.

11. **Scaevola nitida** R.Br., *Prodr.* 584 (1810)

Merkusia nitida (R.Br.) Vriese, *Natuurk. Verh. Holl. Maatsch. Wetensch. Haarlem* ser. 2, 10: 73 (1854); *M. aemula* Vriese, *Natuurk. Verh. Holl. Maatsch. Wetensch. Haarlem* ser. 2, 10: 74 (1854), ?*nom. illeg.*; *Lobelia nitida* (R.Br.) Kuntze, *Revis. Gen. Pl.* 2: 378 (1891). T: Princess Royal Harbour, [W.A.], Dec. 1801, *R.Brown*; holo: BM.

S. attenuata R.Br., *Prodr.* 583 (1810); *Merkusia attenuata* (R.Br.) Vriese, *Ned. Kruidk. Arch.* 2: 162 (1851); *Lobelia attenuata* (R.Br.) Kuntze, *Revis. Gen. Pl.* 2: 378 (1891). T: unlabelled [probably from King George Sound, W.A., Dec. 1801, *R.Brown*]; lecto: BM, *fide* R.C.Carolin, *Fl. Australia* 35: 333 (1992); isolecto: K.

S. drummondii DC., *Prodr.* 7(2): 508 (1839). T: Swan River, W.A., *J.Drummond s.n.*; holo: G-DC.

S. multiflora Lindley, *Sketch Veg. Swan R.* xxvi (1840); *Merkusia multifora* (Lindley) Vriese, *Ned. Kruidk. Arch.* 2: 152 (1851). T: not designated; decision based on the illustration of W.H.de Vriese, *op. cit.* t. 10.

S. multiflora var. *microstachya* Vriese in J.G.C.Lehmann, *Pl. Preiss.* 1: 407 (1845). T: Oyster Harbour, [Albany], W.A., 23 Sept. 1840, *L.Preiss 1488*; lecto: LD, *fide* R.C.Carolin, *Telopea* 3: 496 (1990).

S. fastigiata Vriese in J.G.C.Lehmann, *Pl. Preiss.* 1: 406 (1845); *Merkusia fastigiata* (Vriese) Vriese, *Ned. Kruidk. Arch.* 2: 152 (1851). T: near Baldhead, W.A., 16 Oct. 1840, *L.Preiss 1491*; lecto: LD, *fide* R.C.Carolin, *Telopea* 3: 496 (1990); isolecto: L.

Illustration: W.H.de Vriese, *Natuurk. Verh. Holl. Maatsch. Wetensch. Haarlem* ser. 2, 10: t. 10 (1854) as *Merkusia multiflora*.

Spreading shrub to 3 m tall, glabrous, viscid when young. Leaves sessile, obovate to narrowly elliptic, dentate; lamina 2–8.7 cm long, 7–40 mm wide. Flowers in terminal spikes to 6.5 cm long; bracts lanceolate to ovate, 8–23 mm long; bracteoles linear, 5–9 mm long. Sepals rim-like, 0.3 mm high. Corolla 13–20 mm long, pilose or glabrous outside, bearded inside, blue to lilac; barbulae long, papillate apically; wings 1–2 mm wide. Ovary 2-locular; indusium c. 2 mm wide, sparsely shortly hairy on both surfaces basally. Fruit cylindrical, to 4 mm long, grooved, otherwise smooth, glabrous. *n* = 8, W.J.Peacock, *Proc. Linn. Soc. New South Wales* 88: 8 (1963). Fig. 39E.

Occurs in coastal communities in south-western W.A. from near Albany to north of Perth. Flowers Aug.–Dec. Map 118.

W.A.: Point Peron, *N.T.Burbidge 1969* (AD, PERTH); Yallingup Natl Park, *A.E.Orchard 4316* (AD); Hamelin Bay, *R.D.Royce 3189* (PERTH); Channel Point SE of Cheyne Beach, *N.G.Marchant 71/715* (PERTH); Nanarup beach, Albany, *S.Carlquist 3764* (NSW).

The corolla is acutely acuminate in bud. Similar to *S. crassifolia* which has petiolate leaves, globular fruit, and shorter bracteoles to 4 mm long. See also *S. angustata*. Cytological voucher is *W.J.Peacock 60884.4* (SYD).

12. **Scaevola thesioides** Benth. in S.L.Endlicher *et al.*, *Enum. Pl.* 68 (1837)

Merkusia thesioides (Benth.) Vriese, *Ned. Kruidk. Arch.* 2: 155 (1851); *Lobelia thesioides* (Benth.) Kuntze, *Revis. Gen. Pl.* 2: 378 (1891). T: Swan River, W.A., *C.A.Hügel*; holo: W; iso: K.

S. polystachya DC., *Prodr.* 7(2): 508 (1839). T: Swan River, W.A., *J.Drummond s.n.*; holo: G-DC.

S. squarrosa Lindley, *Sketch Veg. Swan R.* xxvi (1840). T: none cited; *n.v.*

S. flaccida Vriese in J.G.C.Lehmann, *Pl. Preiss.* 1: 407 (1845). T: near Fremantle, W.A., 13 Dec. 1838, *L.Preiss 1521*; lecto: LD, *fide* R.C.Carolin, *Telopea* 3: 497 (1990); isolecto: K, L, MEL, P, W.

S. paniculata Vriese in J.G.C.Lehmann, *Pl. Preiss.* 1: 407 (1845). T: near Limekiln, Perth, W.A., 5 Nov. 1839, *L.Preiss 1516*; lecto: LD, *fide* R.C.Carolin, *Telopea* 3: 497 (1990); isolecto: K, L, P, W.

S. dielsii E.Pritzel in F.L.E.Diels & E.Pritzel, *Bot. Jahrb. Syst.* 35: 571 (1905). T: near mouth of Hutt River, W.A., *L.Diels 7516*; holo: ?B (destroyed) *n.v.*

Spreading subshrub to perennial herb to 1 m tall, mostly glabrous. Leaves sessile, filiform to oblanceolate, entire or denticulate; lamina 1.5–7 cm long, 1–5 mm wide. Flowers in terminal spikes to 4 cm long; bracts narrowly triangular and keeled or filiform, 4–14 mm long; bracteoles lanceolate to filiform, 2–6 mm long. Sepals rim-like, 0.3 mm high. Corolla 5–9 mm long, hairy or glabrous outside, bearded inside, pale blue to white; barbulae flat, papillate apically; wings 0.6 mm wide. Ovary 2-locular; indusium c. 1 mm wide, hairy basally, with bristles on lips curved into orifice. Fruit cylindrical to ellipsoidal, obovoid, or depressed ovoid, to 3 mm diam., with swollen ribs, smooth, rarely pubescent.

Occurs in coastal areas of south-western W.A. The conspicuously swollen ribs of the fruit distinguish this species from others in the series. *Scaevola squarrosa* is placed in synonymy here following K.Krause, *Pflanzenr.* 54: 148 (1912). Sometimes some ribs are larger than others and the fruit thus appears gibbous. Two subspecies are recognised.

Leaves linear to oblanceolate **12a.** subsp. **thesioides**

Leaves filiform **12b.** subsp. **filiformis**

12a. **Scaevola thesioides** Benth subsp. **thesioides**

Illustration: K.Krause, *Pflanzenr.* 54: 151, fig. 28F–H (1912) as *S. dielsii*; W.H.de Vriese, *Natuurk. Verh. Holl. Maatsch. Wetensch. Haarlem* ser. 2, 10: t. 11 (1854) as *Merkusia thesioides*.

Plant not usually viscid when young. Leaves linear to oblanceolate, attenuate. Flowers mostly clustered at top of spike; bracts narrowly triangular; bracteoles lanceolate. Corolla 6–9 mm long, with few, short hairs outside. Ovary glabrous or hairy. Fruit mostly depressed-ovoid, to 3 mm diam., ±glabrous. *n* = 8, W.J.Peacock, *Proc. Linn. Soc. New South Wales* 88: 8 (1963) as *S. dielsii*. Fig. 39B.

Occurs in W.A. from the Murchison R. to Yalgorup Natl Park; in heath or scrub, in sand or limestone soil. Flowers Sept.–Mar. Map 119.

W.A.: Murchison River House, *J.S.Beard 6879* (PERTH); N of Arrowsmith Lake, *A.S.George 9773* (PERTH); c. 9 km SE of Jurien Bay, 3 Oct. 1960, *B.Briggs* (NSW, SYD); Yanchep near pine plantation, *F.Lullfitz l3017* (PERTH); Cottesloe, Dec. 1921, *C.A.Gardner* (PERTH).

Cytological voucher is *W.J.Peacock 60862.1* (SYD).

12b. **Scaevola thesioides** subsp. **filifolia** (E.Pritzel) Carolin, *Fl. Australia* 35: 333 (1992)

S. thesioides var. *filifolia* E.Pritzel in F.L.E.Diels & E.Pritzel, *Bot. Jahrb.* 35: 571 (1905). T: near Esperance, W.A., Nov. 1901, *F.L.E.Diels 5937*; holo: ?B (destroyed); neo: 4 miles [6.4 km] S of Trulove [Truslove], W.A., 15th Oct. 1931, *W.E.Blackall 1041*; PERTH, *fide* R.C.Carolin, *Telopea* 3: 497 (1990).

Plant usually viscid when young. Cauline leaves filiform, mostly entire. Flowers loosely arranged towards top of spike; bracts filiform; bracteoles narrowly triangular to filiform. Corolla mostly 5–7 mm long, glabrous. Ovary with short bristles at 90° and glandular-peltate hairs. Fruit oblong-ellipsoidal, to 1 mm diam., densely pilose.

Occurs along the south coast of south-western W.A. between the Hamersley R. and Point Malcolm. Flowers Sept.–Dec. Map 120.

W.A.: c. 650 km on Esperance–Ravensthorpe road, *E.M.Bennett 917* (PERTH); SW corner of location 38, c. 30 km NNE of Stokes Inlet, *H.Eichler 20328* (AD); Point Malcolm, *W.E.Blackall 1192* (PERTH); Hamersley R., 12 Nov. 1935, *C.A.Gardner* (PERTH).

13. **Scaevola globulifera** Labill., *Nov. Holl. Pl.* 1: 55 (1804)

Merkusia globulifera (Labill.) Vriese, *Ned. Kruidk. Arch.* 2: 156 (1851); *Lobelia globulifera* (Labill.) Kuntze, *Revis. Gen. Pl.* 2: 387 (1891). T: south-western [W.A.], *J.J.H.de Labillardière*; syn: BM, P.

S. caespitosa R.Br., *Prodr.* 585 (1810); *Merkusia caespitosa* (R.Br.) Vriese, *Ned. Kruidk. Arch.* 2: 164 (1851). T: near King George Sound, [W.A.], 6 Jan. 1802, *R.Brown*; lecto: BM, *fide* R.C.Carolin, *Telopea* 3: 497 (1990); isolecto: K.

S. revoluta var. *strigosa* Vriese in J.G.C.Lehmann, *Pl. Preiss.* 1: 409 (1845). T: towards Point Possession, [Swan R., near Perth], W.A., 16 Oct. 1840, *L.Preiss 1506*; lecto: LD, *fide* R.C.Carolin, *Telopea* 3: 497 (1990); isolecto: MEL.

S. globulifera var. *humilis* Benth., *Fl. Austral.* 4: 94 (1868). T: W.A., *J.Drummond s.n.*; lecto: MEL, *fide* R.C.Carolin, *Telopea* 3: 497 (1990); isolecto: K.

[*S. revoluta auct. non* R.Br.: W.H.de Vriese in J.G.C.Lehmann, *Pl. Preiss.* 1: 408 (1845)]

Illustration: J.J.H.de Labillardière, *op. cit.*, t. 78.

Ascending shrub to 1.2 m tall, usually glabrous, sometimes viscid when young. Leaves sessile, linear to narrowly oblanceolate, entire or dentate; lamina 1.5–12 cm long, 1–25 mm wide. Flowers in terminal spikes to 20 cm long; bracts linear 6–45 mm long; bracteoles narrowly lanceolate, 3–7 mm long. Sepals ovate, 1 mm long, connate. Corolla 8–25 mm long, glabrous outside or sparsely shortly hairy, bearded inside, blue to white; barbulae papillate apically; wings c. 1 mm wide. Ovary 2-locular; indusium 2 mm wide, shortly hairy basally. Fruit ellipsoidal, to 5 mm diam., with 2 sterile cavities, rugose, glabrous. Fig. 39G–H.

Occurs in coastal W.A. from Northampton to Israelite Bay, in heath in sand or on limestone. Flowers chiefly Sept.–Nov. Map 121.

W.A.: c. 17 km N of Northampton, *K.Newbey 2194* (PERTH); Dongara, *C.A.Gardner 12850* (PERTH); c. 16 km S of Knobby Head, *J.S.Beard 7196* (PERTH); Wanneroo to Yanchep, *H.Salasoo 4210* (NSW); Cape le Grand Natl Park, *R.D.Royce 8781* (PERTH).

The corolla is acuminate in bud. This species has been confused in the past with *S. cunninghamii* and *S. porocarya*. It differs from both in its rugose fruit.

14. **Scaevola porocarya** F.Muell., *Fragm.* 2: 19 (1860)

Lobelia porocarya (F.Muell.) Kuntze, *Revis. Gen. Pl.* 2: 378 (1891). T: Murchison R., W.A., *A.Oldfield*; holo: MEL; iso: K.

Weak, ascending shrub to 1.8 m tall, glabrous or hairy. Leaves sessile, linear to oblanceolate, broad basally, usually entire; lamina 2–12 cm long, 3–17 mm wide. Flowers in loose, terminal spikes to 20 cm long; bracts linear-elliptic, 1–4 cm long; bracteoles lanceolate, mostly more than 5 mm long. Sepals united in a tube, undulate, to 1 mm long, fringed by glandular hairs. Corolla 18–25 mm long, glabrous outside, densely bearded inside, pale blue to white; barbulae numerous, papillate apically. Ovary 4-locular; indusium 2–3 mm wide, hairs above short, white. Fruit globular to ovoid, to 11 mm diam., grooved, otherwise smooth, glabrous; locules 4 fertile and 4 sterile. n = 8, W.J.Peacock, *Proc. Linn. Soc. New South Wales* 88: 8 (1963). Fig. 39I–J.

Occurs in W.A., in the Murchison R.–Geraldton region; usually in rocky or clayey soil. Flowers July–Oct. Map 122.

W.A.: Kalbarri, *B.Bellairs 1450* (PERTH); 19 km W of Northampton, *C.A.Gardner* (PERTH); Northampton, *R.C.Carolin 10707* (SYD); 8 km W of Lake Indoon, *G.J.Keighery 2471* (PERTH); Port Gregory road, *A.M.Ashby 1019* (AD, MEL).

The corolla is acuminate in bud. This species is distinguished from all others in the genus by its ovary, which has 4 locules each containing an ovule. See *S. globulifera*. Cytological vouchers are *W.J.Peacock 60850.1 & 60852.1* (SYD).

15. **Scaevola cunninghamii** DC., *Prodr.* 7(2): 508 (1839)

Lobelia cunninghamii (DC.) Kuntze, *Revis. Gen. Pl.* 2: 378 (1891). T: Dampier Archipelago, [W.A.], *A.Cunningham*; holo: G-DC; iso: BM, K.

S. maitlandii F.Muell., *Trans. Edinburgh Bot. Soc.* 7: 497 (1863); *Edinburgh New Philos. J.* ser. 2, 17(2): 231 (1863). T: Nickol Bay, W.A., (either *M.Brown* or *P.Wallcott*, Gregory's expedition); holo: MEL.

S. cunninghamii var. *hispida* Benth., *Fl. Austral.* 4: 92 (1869). T: Depuech Is., W.A., *B.Bynoe*; lecto: K, *fide* R.C.Carolin, *Telopea* 3: 497 (1990).

Spreading, perennial subshrub to 80 cm tall, sometimes viscid when young, with antrorse hairs or glabrous. Leaves sessile, linear to narrowly oblanceolate, entire or dentate; lamina 3.5–9 cm long, 1–6 mm wide. Flowers in loose terminal spikes, thyrses or pedunculate axillary cymes to 10 cm long; bracts linear, 5–15 mm long; bracteoles lanceolate, 2.5–4 mm long, ±equalling sepals. Sepals united in a tube, undulate, c. 1 mm long. Corolla c. 12 mm long, glabrous-viscid or hairy outside, bearded inside, blue or white; barbulae papillate apically; wings c. 1 mm wide. Ovary 2-locular; indusium 1.5 mm wide, shortly hairy on both surfaces. Fruit globular, to 4 mm wide, grooved, otherwise smooth, glabrous or hairy. Fig. 39F.

Occurs in the Pilbara region, north-western W.A., usually near the coast, in sand. Flowers May–Sept. Map 123.

W.A.: W of Withnell Bay, Burrup Peninsula, *K.F.Kenneally 7289* (PERTH); Legendre Is., Dampier Archipelago, *R.D.Royce 7336* (PERTH, SYD); behind beach NE of North West Cape lighthouse, *A.S.George*

1409 (PERTH); Rough Ra., c. 7 km by road S of Exmouth homestead on Exmouth–Carnarvon road, *W.R.Barker 2136* (AD); Mia Mia, Minilya R., *C.A.Gardner 3205* (PERTH).

It can also be confused with *S. crassifolia*, but its narrower sessile leaves distinguish it from that species. See also *S. globulifera*.

Some unusually hairy forms occur, particularly on the islands. The length and density of the axillary hairs, and the type of hairs, (particularly the proportions of simple and glandular), are very variable and not apparently correlated. Therefore Bentham's variety (*loc. cit.*) is not clearly recognisable.

16. Scaevola browniana Carolin, *Telopea* 3: 498 (1990)

T: 6 miles [c. 9.6 km] W of Louisa Downs homestead, W.A., 27 May 1971, *J.R.Maconochie 1150*; holo: DNA; iso: CANB, K, PERTH.

[*S. revoluta auct. non* R.Br. *p.p.*: G.Bentham, *Fl. Austral.* 4: 96 (1869)].

Shrub to 1 m tall, tomentose. Leaves sessile, broadly obovate to oblong, usually entire, with prominent, silky, axillary hairs; lamina 6–55 mm long, 2–17 mm wide. Flowers in terminal spikes to 20 cm long; bracts narrowly lanceolate to narrowly elliptic, 4–25 mm long; bracteoles linear to ovate, 4–9 mm long. Sepals rim-like, to 0.3 mm long. Corolla 5–18 mm long, straight silky hairs outside, bearded inside, blue to white; barbulae simple or papillate apically; wings 0.2–1 mm wide. Ovary incompletely 2-locular; indusium 1–2 mm wide, hairs scattered on both surfaces. Fruit cylindrical, 3–5 mm long, rugose, glabrous, usually 1-seeded.

Occurs from the Pilbara, W.A., to far north-western Qld. This species is distinguished from *S. revoluta* particularly in the narrower bracteoles, in the lack of the long ciliate bristles on the bracts and bracteoles and in the bracts which frequently intergrade with the leaves. Two subspecies are recognised.

Corolla 5–11 mm long **16a.** subsp. **browniana**

Corolla 15–18 mm long **16b.** subsp. **grandior**

16a. Scaevola browniana Carolin subsp. browniana

Leaves 6–36 mm long, 2–12 mm wide. Inflorescence bracts narrowly lanceolate to ovate, 3.5–25 mm long; bracteoles ovate, 2–7 mm long. Corolla 5–11 mm long, usually white; barbulae simple or papillate apically. Indusium c. 1 mm wide, sparsely shortly hairy on both surfaces, shortly bristled on lips. Fruit 3–4 mm long. Figs 39K, 40I.

Occurs from the Hamersley Ra. through the Kimberley, W.A., to the Barkly Tableland, N.T., and Qld; in open woodland savannah, usually on sandstone. Flowers Jan.–Sept. Map 124.

W.A.: SE of Cape Londonderry, *A.S.George 13367* (PERTH). N.T.: 17.5 km NNE of Jabiru, *L.A.Craven 6052* (CANB, MEL); c. 120 km SE of Mt Gilruth, *M.Lazarides 7921* (CANB, DNA, MEL); Jasper Gorge hills on north side, *R.C.Carolin 6716* (SYD). Qld: c. 5 km W of Westmoreland Stn, *R.A.Perry 1348* (AD, MEL, NSW).

This subspecies circumscribes a considerable variation but there appear to be no distinct discontinuities within it. Specimens from the Hamersley Ra. and the southern Kimberley, W.A., are very densely tomentose, have attenuate leaves, broadly ovate bracts only 3.5–5 mm long, lanceolate to ovate bracteoles up to 3 mm long and flowers only 5 mm long. Their corollas have a very dense indumentum outside. Specimens from the north-eastern Kimberley, W.A., have narrowly lanceolate bracts 5.5–7 mm long, narrowly lanceolate to linear bracteoles 3–4.3 mm long and flowers about 10 mm long.

Specimens from the upper Nicholson R. and Westmoreland Stn, Qld, and southern Durack Ra., W.A., have the bracts distinctly intergrading with the leaves, which are linear to narrowly oblong and up to 25 mm long, while specimens from the headwaters of the East and South Alligator Rivers, N.T., are more silky, have more lanceolate bracts and otherwise have intermediate features. The specimens from Jasper Gorge, N.T., are distinguished by their oblong to obovate, obtuse leaves and their 4–6.5 mm long bracteoles.

Other specimens have features intermediate between those described above.

16b. Scaevola browniana subsp. grandior Carolin, *Telopea* 3: 499 (1990)

T: Hidden Valley, 3.2 km E of Kununurra, East Kimberley, W.A., 3 Aug. 1974, *K.F.Kenneally 1909*; holo: PERTH.

Leaves 12–55 mm long, 3–17 mm wide. Inflorescence bracts obovate, 11–22 mm long; bracteoles lanceolate, 5–9 mm long. Corolla 15–18 mm long, blue; barbulae papillate. Indusium c. 2 mm wide, with long, slender hairs on adaxial surface basally. Fruit 5 mm long.

Occurs in the Kimberley, W.A., and Arnhem Land, N.T. Flowers Jan.–Sept. Map 125.

W.A.: Emu Ck, E of Kununurra, *D.E.Symon 12122* (AD, SYD); junction of Hann and Fitzroy Rivers, *W.V.Fitzgerald 1173* (PERTH); Hidden Valley, Kununurra, *P.Ollerenshaw 1674* (CBG, SYD). N.T.: Keep R., *A.S.Mitchell 316* (DNA, NSW).

This differs from the typical subspecies in having more leaf-like bracts, large corollas with longer and more distinct barbulae, generally larger floral parts and longer, more distinct hairs at the base of the indusium.

17. Scaevola revoluta R.Br., *Prodr.* 586 (1810)

Merkusia revoluta (R.Br.) Vriese, *Ned. Kruidk. Arch.* 2: 164 (1851); *Lobelia revoluta* (R.Br.) Kuntze, *Revis. Gen. Pl.* 2: 378 (1891). T: Carpentaria Islands, [N.T.], 20–21 Dec. 1802, *R.Brown*; lecto: BM, *fide* R.C.Carolin, *Telopea* 3: 499 (1990); isolecto: K, NSW.

S. paniculata Ewart & O.Davies, *Fl. N. Territory* 268 (1917), *nom. illeg.*, *non* Vriese (1845). T: sandstone ranges near Western Creek, N.T., *G.F.Hill 774*; holo: NSW.

Spreading subshrub to 70 cm tall, scabrous to tomentose or glandular-hairy. Leaves sessile, oblanceolate to narrowly elliptic, entire or few-toothed, with prominent axillary hairs; lamina 3–60 mm long, 1–12 mm wide. Flowers in terminal spikes to 15 cm long; bracts ovate, 2–9.5 mm long, acute to acuminate, fringed with stiff hairs to 1 mm long; bracteoles ovate, 2–5 mm long, equalling bracts, with silky hairs basally. Sepals obsolete. Corolla 6–10 mm long, ±appressed hairy outside, bearded inside, white to pale blue; barbulae simple or papillate apically; wings 0.1–0.5 mm wide. Ovary incompletely 2 locular; indusium c. 1 mm wide, shortly hairy basally on both surfaces. Fruit cylindrical, 2.5–4 mm long, ±rugose, glabrous, usually 1-seeded.

Occurs from the Kimberley, W.A., to north Qld. Distinguished from *S. browniana* particularly by the ovate glume-like bracts and bracteoles which are fringed by stiff hairs. Two subspecies are recognised.

Leaves and stems softly hairy or glandular; leaves oblanceolate to narrowly elliptic ..... **17a. subsp. revoluta**

Leaves and stems hispid to scabrous; leaves mostly narrowly oblanceolate ..... **17b. subsp. stenostachya**

17a. **Scaevola revoluta** R.Br. subsp. **revoluta**

Leaves oblanceolate to narrowly elliptic, tomentose, with small glandular and long simple hairs or mostly glandular hairs; lamina 1–6 cm long, 1–12 mm wide. Inflorescence bracts acuminate, ovate, 3–9.5 mm long; bracteoles ovate, 2.5–5 mm long. Corolla 7–10 mm long, pale blue to white; barbulae mostly papillate apically; wings 0.3–0.5 mm wide. Fruit 3.5–4 mm long. Fig. 40J.

Two varieties are recognised.

Leaves softly hairy, glandular hairs small and hidden beneath simple hairs **17a1** . var. **revoluta**

Leaves glandular-hairy **17a2** . var. **viscida**

17a1 . **Scaevola revoluta** R.Br. subsp. **revoluta** var. **revoluta**

Plant covered with long, slender, simple hairs and ±dense, small, glandular hairs.

Occurs on the Barkly Tableland, in northern N.T. and adjacent Qld; grows in sandy soil in open forest. Flowers Jan.–July. Map 126.

N.T.: Maria Is., *C.R.Dunlop 2876* (DNA); Cox River Stn, *T.S.Henshall 1588* (DNA); Sir Edward Pellew Group, *L.A.Craven 3750* (CANB, DNA). Qld: Settlement Creek, *L.J.Brass 281* (BRI); Adels Grove via Camooweal, *A.de Lestang 274* (BRI).

17a2 . **Scaevola revoluta** subsp. **revoluta** var. **viscida** Carolin, *Telopea* 3: 499 (1990)

T: near Dry Creek Gorge in China Wall, Nicholson R. area, Barkly Tableland, N.T., 11 June 1974, *A.Kanis 1815*; holo: DNA; iso: DNA, K, L, US.

Plant viscid, covered with very dense, small, glandular hairs, with long, simple hairs few or absent.

Occurs on the Barkly Tableland, N.T.; grows in sandy soil in open forest. Flowers May–June. Map 127.

N.T.: c. 13 km SW of Calvert Hills on Creswell Downs road, *R.C.Carolin 9260* (SYD); c. 32 km SW of Calvert Hills on Creswell Downs road, *R.C.Carolin 9270* (SYD).

17b. **Scaevola revoluta** subsp. **stenostachya** (W.Fitzg.) Carolin, *Telopea* 3: 500 (1990)

S. stenostachya W.Fitzg., *J. & Proc. Roy. Soc. W. Australia* 3: 215 (1918). T: near Isdell River, between Isdell Ra. and Graces Knob, W.A., *W.V.Fitzgerald 880*; holo: NSW.

Leaves mostly narrowly oblanceolate, hispid to scabrous; lamina 3–45 mm long, 1–7 mm wide. Inflorescence bracts ovate, 2–5 mm long, mostly keeled; bracteoles lanceolate, 2–4.5 mm long. Corolla c. 6 mm long, white; barbulae few, mostly simple; wings c. 0.1 mm wide. Fruit 1.5–3.5 mm long.

Occurs in the Kimberley, W.A., mostly in open woodland on sandstone. Flowers Mar.–Oct. Map 128.

W.A.: Nymphaea Ck, Drysdale R. Natl Park, *K.F.Kenneally 4294* (PERTH); headwaters of Nyia, *T.G.Hartley 14732* (PERTH); Cockburn Ra., 46 km SSW of Wyndham, *M.Lazarides 8596* (BRI, CANB, DNA); Pentecost Ra., *J.R.Maconochie 151* (DNA); N of Inglis Gap, 18 km beyond Beverley Springs, *D.E.Symon 12074* (AD, SYD).

Differs from the typical subspecies in its indumentum, its acuminate bracts and usually having fewer, smaller leaves.

18. Scaevola macrostachya (Vriese) Benth., *Fl. Austral.* 4: 97 (1868)

Merkusia macrostachya Vriese, *Ned. Kruidk. Arch.* 2: 154 (1851). T: no precise locality, 'various parts of N.W.C.', [north-west coast, W.A.], *A.Cunningham*; lecto: K, *fide* R.C.Carolin, *Telopea* 3: 497 (1990); isolecto: [Prince Regent River, W.A., *A.Cunningham 523*] BM.

S. scabrida W.Fitzg., *J. & Proc. Roy. Soc. W. Australia* 3: 215 (1918). T: Mt Herbert, King Leopold Ra., W.A., *W.V.Fitzgerald 788*; lecto: NSW, *fide* R.C.Carolin, *Telopea* 3: 497 (1990).

Spreading shrub to 60 cm tall, with long, stiff, not appressed hairs or almost glabrous. Leaves sessile, narrowly oblong to obovate, entire or few-toothed; lamina 10–30 mm long, 1–5 mm wide. Flowers in terminal spikes to 15 cm long; bracts mostly lanceolate, 6–11 mm long, fringed by long, stiff hairs; bracteoles narrowly lanceolate to linear, 5–6 mm long, longer than fruit, silky basally. Sepals rim-like, shortly undulate, c. 1 mm long. Corolla 8–10 mm long, with sparse, stiff, very long hairs outside, bearded inside, pale blue; barbulae simple or papillate apically; wings 0.3 mm wide. Ovary incompletely 2-locular; indusium 1 mm long, glabrous or hairs not reaching lips. Fruit cylindrical, c. 3 mm long, tuberculate, glabrous.

Occurs in the northern and western Kimberley, W.A. Flowers Mar.–Oct. Map 129.

W.A.: near Longini Landing, Kalumburu, *D.E.Symon 10200* (AD, PERTH, SYD); headwaters of the Helby R., *T.G.Hartley 14831* (DNA); Heywood Is., *P.G.Wilson 10900* (PERTH); 85 km N of Mount House, *D.E.Symon 10260* (AD, PERTH, SYD); King Leopold Ra., *C.A.Gardner 981* (PERTH).

W.H.de Vriese referred the type collection to New Caledonia. He was mistaken in his interpretation of 'N.W.C.' (= north-west coast of Australia) as western New Caledonia.

Similar to *S. anchusifolia*. *Scaevola eneabba* is also similar but that species has brownish hairs on the outside of the corolla and the ovary and fruit are pubescent.

19. Scaevola glandulifera DC., *Prodr.* 7(2): 510 (1839)

Merkusia glandulifera (DC.) Vriese, *Ned. Kruidk. Arch.* 2: 166 (1851); *Lobelia glandulifera* (DC.) Kuntze, *Revis. Gen. Pl.* 2: 378 (1891). T: Swan River, W.A., *J.Drummond 394*; holo: G-DC; iso: BM, K, MEL, W.

S. rufa Vriese in J.G.C.Lehmann, *Pl. Preiss.* 1: 405 (1844); *Merkusia glandulifera* var. *eglandulosa* Vriese, *Ned. Kruidk. Arch.* 2: 166 (1851). T: towards foot of Darling Ra., near Perth, W.A., 23 Sept. 1839, *L.Preiss 1513*; lecto: LD, *fide* R.C.Carolin, *Telopea* 3: 500 (1990); isolecto: L, P, W.

S. glandulifera var. *tenuis* E.Pritzel in F.L.E.Diels & E.Pritzel, *Bot. Jahrb. Syst.* 35: 570 (1905). T: Moore R., W.A., *E.Pritzel 739*; lecto: AD, *fide* R.C.Carolin, *Telopea* 3: 500 (1990); isolecto: AD, K, W.

Erect subshrub to 50 cm tall, with large, brown-headed, glandular hairs and simple, yellowish hairs at 90°. Leaves sessile, oblanceolate to linear, entire or dentate; lamina 2–9 cm long, 1–10 mm wide. Flowers in terminal spikes; bracts lanceolate to linear, 5–30 mm long; bracteoles narrowly lanceolate, 4–12 mm long. Sepals rim-like, to 0.5 mm long. Corolla 10–30 mm long, with short simple and glandular hairs and occasionally long, stiff, yellow hairs outside, bearded inside, deep blue-purple; barbulae branched apically; wings 1–2.5 mm wide. Ovary 2-locular; indusium 1.5–3 mm wide, hairy on both surfaces. Fruit obovoid, to 4 mm long, striate-tuberculate, glabrous or pubescent. n = 8, W.J.Peacock, *Proc. Linn. Soc. New South Wales* 88: 8 (1963).

Occurs in south-western districts of W.A. from E of Geraldton to the Stirling Ra., in woodland or heath in heavy soil. Flowers chiefly Sept.–Dec. Map 130.

Figure 41. *Scaevola*. **A–D**, *S. anchusifolia*. **A**. anther X10; **B**, indusium X7.5; **C**, corolla X2.5; **D**, habit X0.5 (**A–D**, T.Aplin 3348, PERTH). **E**. *S. virgata*, habit X1 (R.Carolin 3320, SYD). **F**, *S. canescens*, habit X1 (F.Smith 1820, SYD). **G**, *S. lanceolata*, habit X1 (T.Aplin 1285, SYD). Drawn by D.Mackay.

W.A.: c. 1 km W of Casuarinas Rd, on Mullewa road between Indarra and Ambania, *A.M.Ashby 5163* (AD); Watheroo Natl Park, *R.D.Royce 9760* (PERTH); Swan View, *C.A.Gardner 1569* (PERTH); Bald Hill, Avon Valley Natl Park, 28 Oct. 1971, *B.Maslin* (NSW, PERTH); c. 19 km E of Twin Peaks, *K.Newbey 1702* (PERTH).

A very distinct species which is easily recognised by the deep blue corolla and the large, dark-headed, glandular hairs. Cytological voucher is *W.J.Peacock 60876.1* (SYD).

20. **Scaevola pulchella** Carolin, *Telopea* 3: 496 (1990)

T: 80–85 miles [c. 130 km] southwards from Onslow, W.A., 24 Aug. 1963, *J.S.Beard 2980*; holo: PERTH.

Perennial herb, decumbent to ascending, many-stemmed, with silvery hairs; stems to 90 cm long. Leaves sessile, linear to elliptic-oblanceolate, entire to dentate; lamina 17–41 mm long, 4–10 mm wide. Flowers in terminal spikes to 10 cm long; bracteoles narrowly elliptic to linear, 4–8 mm long, with conspicuous, silky, axillary hairs. Sepals ovate-triangular, 1.5 mm long, free, ±obscured by hairs. Corolla 1.3–2.2 cm long, with simple hairs outside, ±bearded inside, bluish-mauve; barbulae clavate, heads papillate. Ovary 2-locular; indusium 1–2 mm wide, glabrous except lips. Fruit ellipsoidal to globular, to 4 mm diam., with simple hairs at 90°. Fig. 40F–H.

Occurs in W.A. from near Carnarvon to east of Exmouth Gulf, on spinifex plains. Flowers June–Oct. Map 131.

W.A.: Onslow, 24 July 1964, *F.W.Humphreys* (PERTH); E of Pitgramunne Well on Yardie Creek Stn, *Y.Chadwick 1436* (PERTH); c. 5 km W of Giralia homestead, Aug. 1963, *J.Tonkinson* (PERTH); Mia Mia, North West Coastal Hwy, *A.M.Ashby 5168* (AD); c. 3 km S of Lyndon R., on North West Coastal Hwy, *A.M.Ashby 3209* (AD).

The presence of free sepals, ridged stems and the hairs on the ovary are features which may indicate an affinity with subsect. *Parvifoliae*, but this species has well-developed cauline leaves which are densely hairy.

21. **Scaevola anchusifolia** Benth. in S.L.Endlicher *et al.*, *Enum. Pl.* 68 (1837)

Merkusia anchusifolia (Benth) Vriese, *Ned. Kruidk. Arch.* 2: 165 (1851) as *anchusaefolia*; *Lobelia anchusifolia* (Benth.) Kuntze, *Revis. Gen. Pl.* 2: 378 (1891). T: Swan River, W.A., *C.A.Hügel*; holo: W; iso: K.

S. holosericea Vriese in J.G.C.Lehmann, *Pl. Preiss.* 1: 408 (1845); *Merkusia anchusifolia* var. *holosericea* (Vriese) Vriese, *Ned. Kruidk. Arch.* 2: 165 (1851); *Lobelia holosericea* (Vriese) Kuntze, *Revis. Gen. Pl.* 2: 378 (1891). T: Mt Eliza, Perth, W.A., 23 Sept. ?1838, *L.Preiss 1478*; lecto: LD, *fide* R.C.Carolin, *Telopea* 3: 500 (1990); isolecto: L, MEL, W.

S. sphaerocarpa Vriese in J.G.C.Lehmann, *Pl. Preiss.* 1: 409 (1845); *Merkusia anchusifolia* var. *sphaerocarpa* (Vriese) Vriese, *Ned. Kruidk. Arch.* 2: 165 (1851). T: near Fremantle, W.A., 14 Dec. 1838, *L.Preiss 1512*; lecto: LD, *fide* R.C.Carolin, *Telopea* 3: 500 (1990); isolecto: L, MEL, P, W.

Decumbent, shrubby plant to 1.4 m tall, scabrous to hirsute. Leaves oblanceolate, tapering towards base, entire or dentate; lamina to 9 cm long, to 1.8 cm wide. Flowers in loose, terminal spikes to 15 cm long; bracts linear to narrowly elliptic, 8–76 mm long, acuminate; bracteoles triangular 4–16 mm long. Sepals united in a 5-lobed tube, to 1 mm long. Corolla 10–22 mm long, hairy outside, bearded inside, pale blue; barbulae with compact papillate heads; wings c. 1 mm wide. Ovary 2-locular; indusium 2–3 mm wide, sparsely hairy basally, shortly hairy on lips. Fruit ellipsoidal, to 4 mm long, rugose, glabrous, with 2 sterile cavities. $n = 8$, W.J.Peacock, *Proc. Linn. Soc. New South Wales* 88: 8 (1963). Fig. 41A–D.

Occurs in coastal areas of south-western W.A. from the Murchison R. to Yalgorup Natl Park. Flowers July–Nov. Map 132.

W.A.: 537 mile peg [c. 860 km N of Perth] on road to Denham, *T.E.H.Aplin 3348* (PERTH); near Claremont, Oct. 1900, *W.V.Fitzgerald* (NSW); N of Mt Benia, NE of Jurien, *E.A.Griffin 2350* (PERTH); Murchison R., c. 40 km above mouth, *J.S.Beard 2061* (PERTH); W of Quoin Bluff South, Dirk Hartog Is., *A.S.George 11463* (PERTH).

Similar to *S. macrostachya*, differing in the longer leaves which taper towards the base. Also similar to *S. eneabba* and *S. pulchella* which have a pubescent ovary and fruit. Cytological voucher is *W.J.Peacock 60838.1* (SYD).

22. **Scaevola eneabba** Carolin, *Telopea* 3: 500 (1990)

T: 40 miles [c. 65 km] from Eneabba, W.A., 15 Oct. 1964, *F.W.Humphreys*; holo: PERTH.

Shrubby plant, probably to 60 cm tall, hispid, glabrescent at base. Leaves sessile, linear-oblanceolate, entire, thick; lamina to 30 mm long, to 2 mm wide. Flowers in terminal spikes to 25 mm long; bracts lanceolate, to 8 mm long, with long bristles marginally; bracteoles resembling bracts but smaller. Sepals deltoid, c. 1 mm long, connate basally. Corolla c. 9 mm long, with long, stiff, brownish hairs outside, pubescent inside on lobes and in throat, colour unknown; barbulae few, papillate apically; wings c. 1.5 mm wide. Ovary densely covered with long, white hairs, 2-locular; indusium c. 1 mm long, with a few scattered hairs on surface. Fruit not seen.

Occurs in W.A. in heath near Eneabba. Flowers about Dec. Map 133.

Known only from the type collection.

The densely hairy ovary is also found in *S. pulchella* but that species has a denser silky indumentum on the rest of the plant, and paler hairs outside the corolla. See also *S. anchusifolia* and *S. macrostachya*.

23. **Scaevola lanceolata** Benth. in S.L.Endlicher *et al.*, *Enum. Pl.* 69 (1837)

Merkusia anchusifolia var. *lanceolata* (Benth.) Vriese, *Ned. Kruidk. Arch.* 2: 165 (1851); *Lobelia lanceolata* (Benth.) Kuntze, *Revis. Gen. Pl.* 2: 378 (1891). T: Swan River, W.A., *C.A.Hügel*; holo: W.

S. longifolia Vriese in J.G.C.Lehmann, *Pl. Preiss.* 1: 410 (1845); *Merkusia longifolia* (Vriese) Vriese, *Natuurk. Verh. Holl. Maatsch. Wetensch. Haarlem* ser. 2, 10: 69 (1854); *Lobelia longifolia* (Vriese) Kuntze, *Revis. Gen. Pl.* 2: 378 (1891). T: Vasse R., W.A., 7 Dec. 1839, *L.Preiss 1472*; lecto: LD, *fide* R.C.Carolin, *Telopea* 3: 501 (1990); isolecto: L, MEL, P, W.

S. lasiantha F.Muell., *Fragm.* 1: 207 (1859). T: Phillips R., W.A., *G.Maxwell*; holo: MEL.

Illustration: K.Krause, *Pflanzenr.* 54: 151, fig. 28J–N (1912).

Erect, many-stemmed perennial to 50 cm tall, pubescent to glabrous; stock thick, retaining dead leaf bases. Leaves mostly towards base of stem, linear to oblanceolate, broad basally, entire or few-toothed, thick; lamina 6–20 cm long, 1–10 mm wide. Flowers in terminal, scapose spikes to 10 cm long; bracts lanceolate, to 4 cm long; bracteoles linear-lanceolate, to 6 mm long. Sepals united in a tube, toothed, to 2 mm long. Corolla 7–15 mm long, with straight hairs outside, bearded inside, white to pale blue with brownish markings; barbulae prominent, papillate apically. Ovary 2-locular; indusium to 2 mm wide, sparsely hairy on both surfaces. Fruit subglobular, to 4 mm diam., rugose, pubescent or glabrous, with 2 sterile cavities. Fig. 41G.

Occurs in south-western W.A. from the Murchison R. to the Fitzgerald R., in heath. Flowers chiefly Aug.–Dec. Map 134.

W.A.: Murchison R. mouth, *J.S.Beard 2052* (PERTH); Hill R., c. 13 km WNW of Badgingarra, *P.G.Wilson 3790* (NSW, PERTH); Piawaning, 3 Oct. 1962, *M.E.Phillips* (CBG, SYD); Cannington, *C.A.Gardner 8262* (PERTH); Drakesbrook, *C.A.Gardner 917* (PERTH).

This species is similar to *S. virgata* which also has scapose inflorescences but in *S. lanceolata* the flowers are larger, the corolla is more densely hairy, the bracts are broader and less acuminate, the plants less woody with the leaves more clustered towards the base of the plant, and the hairs throughout the plant are coarser. Specimens from between the Moore and Murchison Rivers have somewhat larger corollas than those from further south.

24. **Scaevola virgata** Carolin, *Telopea* 3: 501 (1990)

T: between Northampton and Geraldton, W.A., 23 Aug. 1964, *A.M.Ashby 1026*; holo: AD; iso: SYD.

Subshrub, usually much-branched, to 45 cm high, with fine hairs. Leaves sessile, linear to oblanceolate, broad basally, entire or dentate; lamina 2–6 cm long, 3–8 mm wide. Flowers in terminal, scapose spikes to 10 cm long; bracts ovate, 3–27 mm long; bracteoles narrowly lanceolate, 4–6 mm long. Sepals rim-like, sinuate, 0.3 mm long. Corolla 5–8 mm long, with fine hairs outside, few simple hairs inside, white to pale blue; barbulae simple or papillate apically; wings 0.5–1 mm wide. Ovary 2-locular; indusium c. 1 mm wide, base with sparse, short hairs. Fruit globular to cylindrical, to 3 mm diam., rugose, glabrous, with 2 sterile cavities. Figs 39L, 41E.

Occurs between the Ogilvie Plains and Watheroo, W.A., in scrub or heath in rocky soil. Flowers July–Nov. Map 135.

W.A.: Ogilvie Plains, 75 km N of Geraldton, *W.E.Blackall 4502* (PERTH); Northampton, *R.C.Carolin 3240* (SYD); 27 km SE of Walkaway on Burma Road, *A.S.George 7853* (PERTH); 15 km E of Eneabba, *E.A.Griffin 1520* (PERTH); near Coorow, *W.E.Blackall 2593* (PERTH).

This species has been confused with *S. lanceolata* in the past. The distinguishing features are given under that species.

25. **Scaevola calendulacea** (Kenn.) Druce, *Rep. Bot. Exch. Club Brit. Isles 1916 Suppl.* 2: 644 (1917)

Goodenia calendulacea Kenn. in H.C.Andrews, *Bot. Repos.* 1: t. 22 (1798); *Lobelia calendulacea* (Kenn.) Kuntze, *Revis. Gen. Pl.* 2: 378 (1891). T: plant cultivated at Hammersmith Nursery, 1797, from Port Jackson, N.S.W., *W.Paterson*; *n.v.*

S. suaveolens R.Br., *Prodr.* 585 (1810) *nom. superfl.*; *Merkusia suaveolens* (R.Br.) Vriese, *Ned. Kruidk. Arch.* 2: 163 (1851). T: Botany Bay, N.S.W., *R.Brown*; lecto: BM, *fide* R.C.Carolin, *Telopea* 3: 502 (1990).

Illustration: P.B.Kennedy, *loc. cit.*; E.R.Rotherham *et al.*, *Fl. Pl. New South Wales & S. Queensland* fig. 12 (1975).

Prostrate shrub with flowering stems to 40 cm high, with appressed hairs. Leaves oblanceolate to obovate, tapering basally, entire; lamina to 80 mm long, to 27 mm wide; axillary hairs white. Flowers in terminal spikes to 8 cm long; lowest bracts mostly oblong-oblanceolate, 5–40 mm long, half length of flower; bracteoles linear to narrowly elliptic, 3–6 mm long. Sepals united in a tube, sinuate, ciliate, 0.5–1 mm long. Corolla 12–18 mm long, pubescent outside, bearded inside, bright blue; barbulae numerous, broad, often papillate apically; wings 1–2 mm wide. Ovary 2-locular; indusium 2 mm wide, with few hairs basally. Fruit globular, to 12 mm diam., glabrous, white and purplish. Figs 39M, 57.

Occurs along the coast of S.A., Qld, N.S.W. and Vic., from Encounter Bay, S.A., to Mackay, Qld; on sand dunes near the coast, where it acts as a stabiliser. Flowers most of the year. Map 136.

Qld: Fraser Is., *S.T.Blake 14383* (BRI); 10 km S of Tangalooma, Moreton Is., *P.Sharpe & L.Durrington 1159* (BRI). N.S.W.: Corindi Beach, 30 Apr. 1956, *E.F.Constable* (NSW); One Mile Beach, Anna Bay, *K.McDonald 5128* (NSW). Vic.: Portland to Bridgewater Lakes, *A.C.Beauglehole 43281* (MEL).

This species differs from all other Australian members of the series in the thick, fleshy mesocarp with purplish white epicarp, and tuberculate endocarp. *Scaevola sericea,* which occurs in similar habitats, has flowers with white to cream corollas in axillary cymes, and whitish, fleshy fruit.

Variable in indumentum and in the barbulae which may or may not possess papillate heads. Closely related to *S. gracilis* from N.Z. and *S. porrecta* from Tonga, both of which occur in similar coastal situations.

26. Scaevola brookeana F.Muell., *Victorian Naturalist* 1: 122 (1884)

Lobelia brookeana (F.Muell.) Kuntze, *Revis. Gen. Pl.* 2: 378 (1891) as '*brooksiana*'. T: Israelite Bay, W.A., *S.J.Brooke*; holo: MEL; iso: K.

[*S. brooksiana* J.H.Willis, *Muelleria* 1: 91 (1959), *orth. var.*]

Subshrub to 80 cm tall, glabrous, glaucous; stems glossy when mature. Leaves sessile, orbicular to elliptic, dentate; lamina to 20 mm long, to 15 mm wide; upper leaves stem-clasping. Flowers in terminal spikes; bracts leafy; bracteoles deltoid, to 1 mm long. Sepals ovate, to 1 mm long, free almost to base. Corolla c. 10 mm long, glabrous outside, bearded inside, blue; barbulae numerous, papillate apically; wings c. 1.5 mm wide. Ovary 2-locular; indusium c. 1.5 mm long, ±glabrous, hairs on lips short. Fruit ellipsoidal, 4 mm long, wrinkled.

Occurs only on Mt Ragged, W.A. Flowers about Dec. Map 137.

W.A.: Mt Ragged, *A.S.George 2089* (PERTH); saddle between peaks of Mt Ragged, *J.Powell 3463, J.Everett & D.Bedford* (NSW).

Distinguished from all other species in the genus by its shrubby habit and glaucous, stem-clasping leaves.

27. Scaevola spicigera Carolin, *Telopea* 3: 502 (1990)

T: 79 miles [c. 125 km] S of Learmonth, W.A., 2 June 1961, *A.S.George 2399*; holo: PERTH.

Subshrub to 50 cm tall, tomentose, with simple and glandular hairs. Leaves sessile, narrowly elliptic to oblanceolate, entire; lamina 33–65 mm long, 3–8 mm wide; axillary hairs conspicuous, silky. Flowers in spike-like, axillary thyrses to 20 cm long; bracts leaf-like but smaller; bracteoles lanceolate, 4–6 mm long. Sepals rim-like, minute, sinuate, hairy. Corolla 5–6 mm long, with appressed hairs outside, pubescent inside, white; barbulae simple, hair-like; wings to 0.5 mm wide. Ovary 1-locular except basally; indusium 0.5–1 mm diam., base with sparse, short hairs. Fruit cylindrical, 3 mm long, pubescent, rugose, ribbed, usually 1-seeded.

Occurs between Learmonth and Lake Macleod, W.A. Flowers June–Feb. Map 138.

W.A.: c. 16 km E of Ningaloo, *A.S.George 10235* (PERTH); Learmonth road, c. 70 km S of Bullara turnoff, *A.S.George 3287* (PERTH); c. 88 km N of Minilya R. on road to Learmonth, *A.S.George 1434* (PERTH); Cardabia Stn turnoff on Learmonth road, *J.S.Beard 3536* (PERTH).

Similar to *S. canescens* but in *S. spicigera* the indumentum is much less dense and incorporates small glandular hairs, the axillary hairs are silky and not felted, and the inflorescences are more spike-like and with longer internodes.

28. **Scaevola canescens** Benth. in S.L.Endlicher *et al.*, *Enum. Pl.* 69 (1837)

Dampiera canescens (Benth.) Vriese, *Natuurk. Verh. Holl. Maatsch. Wetensch. Haarlem* ser. 2, 10: 114 (1854); *Lobelia canescens* (Benth.) Kuntze, *Revis. Gen. Pl.* 2: 378 (1891). T: King George Sound, W.A., 1833, *C.A.Hügel*; holo: W.

S. glaucescens Vriese in J.G.C.Lehmann, *Pl. Preiss.* 1: 410 (1845). T: near Perth, W.A., 15 Apr. 1839, *L.Preiss 1477*; lecto: LD, *fide* R.C.Carolin, *Telopea* 3: 503 (1990); isolecto: K, L, MEL, P, W.

S. trinervis Vriese in J.G.C.Lehmann, *Pl. Preiss.* 1: 407 (1845). T: near Lake Keiermulu [probably Lake Monger], W.A., 16 July 1839, *L.Preiss 1479*; lecto: LD, *fide* R.C.Carolin, *Telopea* 3: 503 (1990); isolecto: K, ?MEL.

Illustration: W.H.de Vriese, *Natuurk. Verh. Holl. Maatsch. Wetensch. Haarlem* ser. 2, 10: t. 19 (1854) as *Dampiera canescens*.

Shrub to 60 cm high, tomentose, with some long, stiff hairs; stems brittle. Leaves sessile, narrowly oblong to oblanceolate, entire; lamina 12–85 mm long, 4–15 mm wide; axillary hairs softly felted. Flowers in condensed, axillary spikes to 10 cm long; bracts ovate, 4–6.5 mm long, concave; bracteoles ovate, 2.5–4 mm long. Sepals rim-like, undulate, ciliate, to 0.5 mm long. Corolla 9–11 mm long, pubescent outside, bearded inside, white with brownish veins; barbulae simple, hair-like; wings to 0.5 mm wide. Ovary 1-locular except basally; indusium 1–1.5 mm long, sparsely weakly hairy above. Fruit obovoid, 2–4 mm long, rugose, pubescent, usually 1-seeded. $n = 8$, W.J.Peacock, *Proc. Linn. Soc. New South Wales* 88: 8 (1963). Fig. 41F.

Occurs in south-western W.A. from S of Shark Bay to near Perth, in open forest and heath in sandy soil. Flowers Mar.–Nov. Map 139.

W.A.: Ogilvie, *C.A.Gardner 8582* (PERTH); 48 km N of Ajana, *C.A.Gardner & W.E.Blackall 608* (PERTH); c. 40 km N of Murchison R., *J.S.Beard 6746* (PERTH); c. 3 km N of Watheroo, *J.S.Beard 1661* (PERTH); Jarrah Rd, South Perth, *R.J.Cranfield R120* (PERTH).

This species is related to *S. sericophylla* and *S. spicigera*. Cytological voucher is *W.J.Peacock 60891.1* (SYD).

29. **Scaevola globosa** (Carolin) Carolin, *Telopea* 3: 504 (1990)

Nigromnia globosa Carolin, *Nuytsia* 1: 292 (1974). T: between Yuna and Dartmoor, W.A., 20 Sept. 1940, *W.E.Blackall 4833*; holo: PERTH.

Sprawling shrub to 70 cm high and 1 m across, pubescence felt-like, hairs yellowish white when young, becoming grey. Leaves tapering into short petiole, obovate to elliptic, entire; lamina 3–6 cm long, 10–30 mm wide; axillary hairs conspicuous, silky. Inflorescence a sessile, globular, axillary head, to 15 mm diam.; flowers buried in soft hairs; bracts and bracteoles c. 2 mm long, villous inside. Sepals a minute lobed rim. Corolla to 3 mm long, mostly with simple hairs outside, almost glabrous inside, yellow; barbulae indistinguishable from hairs or absent; wings obsolete. Ovary 1-locular; indusium broadly deltoid, to 1 mm wide, glabrous. Fruit ellipsoidal, to 2 mm long, ribbed, otherwise smooth, glabrous, usually 1-seeded. Fig. 42A–D.

Known from 3 localities, near Yuna, Mingenew and Carnamah, W.A. Flowers about Oct. Map 140.

W.A.: SW of Carnamah on road to Eneabba, *A.S.George 16350* (PERTH, SYD).

A very distinctive species with the minute flowers buried in a globose mass of soft hairs. Corolla lobes are deltoid-ovate, to 0.8 mm long.

30. **Scaevola sericophylla** F.Muell. ex Benth., *Fl. Austral.* 4: 102 (1868)

Lobelia sericophylla (Benth.) Kuntze, *Revis. Gen. Pl.* 2: 378 (1891). T: Murchison R., W.A., *A.Oldfield*; lecto: K, *fide* R.C.Carolin, *Telopea* 3: 503 (1990).

S. sericophylla var. *decumbens* K.Krause, *Pflanzenr.* 54: 162 (1912). T: near Wageri Lake [Wagin Lake],

W.A., Jan. 1901, *L.Diels 2414*; holo: ?B (destroyed) *n.v.*

Spreading shrub to 150 cm tall, silky hairy; branches hard, woody. Leaves sessile, elliptic to obovate, broad basally, entire; lamina 12–45 mm long, 3–10 mm wide; axillary hairs silky. Flowers in compact, axillary spikes to 15 mm long; bracts oblanceolate, to 10 mm long; bracteoles ovate to orbicular, 4–6 mm long. Sepals rim-like, undulate, to 0.5 mm long. Corolla 10–19 mm long, silky hairy outside, bearded inside, white with purplish lines; barbulae simple, hair-like; wings to 0.8 mm wide. Ovary 1-locular except basally; indusium 1–1.5 mm long, shortly hairy basally. Fruit ovoid, 3–5 mm long, glabrous or nearly so, 1-seeded. $n = 8$, W.J.Peacock, *Proc. Linn. Soc. New South Wales* 88: 9 (1963).

Occurs in W.A. in near coastal areas of the Pilbara region southwards to Geraldton with one inland collection from the Rudall R. area; grows in sand heath and on sand dunes. Flowers about May–June. Map 141.

W.A.: c. 9 km from N end of Burma road, *A.C.Burns 6* (PERTH); c. 32 km E of Onslow, *J.S.Beard 2967* (PERTH); Minderoo, Ashburton R., 8 Oct. 1903, *A.Morrison* (NSW, PERTH); 25 km S of Rudall R., *P.G.Wilson 10466* (PERTH); Point Sampson, *C.A.Gawler 6307* (PERTH).

Differs from *S. canescens* in the silky, not tomentose, indumentum, the silky, not felted, axillary hairs, the glabrous ovary and the branches which have a harder wood. Cytological voucher is *W.J.Peacock 60833.2* (SYD).

31. **Scaevola repens** Vriese in J.G.C.Lehmann, *Pl. Preiss.* 1: 406 (1845)

Dampiera repens (Vriese) Vriese, *Natuurk. Verh. Holl. Maatsch. Wetensch. Haarlem* ser. 2, 10: 114 (1854); *S. paludosa* var. *prostrata* Benth., *Fl. Austral.* 4: 102 (1868), *nom. illeg.*; *S. paludosa* var. *repens* (Vriese) C.Chr. & Ostenf., *Biol. Meddel. Kongel. Danske Vidensk. Selsk.* 3(2): 125 (1921). T: near Perth, W.A., 20 Oct. 1839, *L.Preiss 1519*; lecto: LD, *fide* R.C.Carolin, *Telopea* 3: 503 (1990); isolecto: L, MEL, P, W.

?*S. paludosa* var. *pilosa* E.Pritzel in F.L.E.Diels & E.Pritzel, *Bot. Jahrb. Syst.* 35: 572 (1905). T: near Cottesloe, W.A., *Nov. 1900, L.Diels 1511*; holo: ?B (destroyed) *n.v.*

[*S. paludosa auct. non* R.Br.: G.Bentham, *Fl. Austral.* 4: 102 (1868); W.E.Blackall & B.J.Grieve, *How to Know W. Austral. Wildfl.* 443 (1965)]

Prostrate shrub, hairy or glabrous; branches to 50 cm long. Leaves sessile, linear to elliptic, broad basally, sometimes recurved, entire, with or without prominent axillary hairs; lamina to 90 mm long and 15 mm wide. Flowers in compact, axillary spikes to 30 mm long; bracts leaf-like but smaller; bracteoles lanceolate, to 14 mm long. Sepals rim-like, undulate, to 1 mm long. Corolla 8–15 mm long, with dense, appressed, golden or yellowish hairs outside, hairy inside on lobes and throat, white to cream or rarely mauve; barbulae fimbriate; wings c. 1 mm wide. Ovary 1-locular except basally; indusium to 2 mm wide, base with long, loose, white hairs. Fruit ovoid, c. 4 mm long, tuberculate, glabrous.

Occurs in south-western W.A. Distinguished from other species with flowers in axillary spikes by the hairs on the outside of the corolla, which are yellowish to almost golden. Similar to *S. oldfieldii*. Two varieties are recognised.

Leaves 4–15 mm wide **31a.** var. **repens**

Leaves 1–4 mm wide **31b.** var. **angustifolia**

31a. **Scaevola repens** Vriese var. **repens**

Prostrate shrub, hairy. Leaves elliptic to oblanceolate; lamina to 90 mm long, 4–15 mm wide. Corolla 12–15 mm long, white to cream, with golden hairs outside.

Occurs between Kojonup and Badgingarra, south-western W.A., in sand plain heath. Flowers chiefly Aug.–Jan. Map 142.

W.A.: c. 25 km NW of Badgingarra, *P.G.Wilson 3884* (PERTH); W of Kojonup, *K.Newbey 2718* (PERTH); c. 50 km S of Mandurah, *V.Mann & A.S.George 52* (PERTH); Forrestfield, *R.J.Cranfield 860* (PERTH); Moore River Natl Park, *R.D.Royce 9466* (PERTH).

31b. **Scaevola repens** var. **angustifolia** Vriese in J.G.C.Lehmann, *Pl. Preiss.* 1: 406 (1845)

Dampiera repens var. *angustifolia* (Vriese) Vriese, *Natuurk. Verh. Holl. Maatsch. Wetensch. Haarlem* ser. 2, 10: 114 (1854); *S. paludosa* var. *angustifolia* (Vriese) K.Krause, *Pflanzenr.* 54: 164 (1912). T: Swan R., Jan. 1839, *L.Preiss 1493*; lecto: LD, *fide* R.C.Carolin, *Telopea* 3: 503 (1990).

Prostrate shrub, glabrous or hairy. Leaves linear to linear-lanceolate; lamina to 50 mm long, 1–4 mm wide. Corolla 8–11 mm long, white to mauve, with yellow or whitish hairs outside.

Occurs between Perth and Jurien Bay, W.A.; in sand plain heath, often over limestone. Flowers chiefly Aug.–Jan. Map 143.

W.A.: Wanneroo, *A.S.George 2627* (PERTH); c. 13 km N of Gingin turnoff on Lancelin road, *A.S.George 7762* (PERTH); c. 22 km from Trigg Is., 23 Sep. 1968, *E.M.Canning* (CBG, SYD); Jurien Bay road, 5.8 km W of Cockleshell Gully road, *P.Weston 280* (SYD).

No duplicates of the lectotype have been located in other herbaria.

32. **Scaevola oldfieldii** F.Muell., *Fragm.* 2: 19 (1860)

Lobelia oldfieldii (F.Muell.) Kuntze, *Revis. Gen. Pl.* 2: 378 (1891). T: towards Murchison R., W.A., *A.Oldfield*; lecto: MEL, *fide* R.C.Carolin, *Telopea* 3: 503 (1990); isolecto: K, MEL.

S. oldfieldii var. *tomentosa* E.Pritzel in F.L.E.Diels & E.Pritzel, *Bot. Jahrb. Syst.* 35: 571 (1905). T: near Northampton, W.A., *L.Diels 5727*; holo: ?B (destroyed) *n.v.*

Erect shrub, to 60 cm tall, glabrous or glandular. Leaves sessile, entire, elliptic; lamina 3–12 cm long, 3–25 mm wide; axillary hairs ±inconspicuous. Flowers in axillary spikes to 30 mm long; bracts leaf-like but smaller; bracteoles lanceolate, to 8 mm long. Sepals rim-like, undulate, to 1 mm long. Corolla 10–20 mm long, with ±appressed, whitish hairs outside, hairy inside on lobes and throat, white; barbulae fimbriate apically; wings 1–2 mm wide. Ovary glabrous, 1-locular except basally; indusium depressed-ovate, 2–3 mm wide, basal hairs long, loose. Fruit ellipsoidal, 3–4 mm long, tuberculate, glabrous.

Occurs between Geraldton and the Murchison R., W.A., in loam and clay and occasionally in sandy soil. Flowers Aug.–Dec. Map 144.

W.A.: Greenough R., Nov. 1877, *coll. unknown* (MEL); Champion Bay, 1877, *F.Mueller* (MEL); Hutt R., 11 Nov. 1972, *B.M.J.Hussey* (PERTH); c. 19 km N of Northampton, *A.S.George 10131* (PERTH).

Similar to *S. repens* but *S. oldfieldii* has a more erect habit, leaves which are more acute and broader in comparison to their length, and axillary hairs which are usually hidden by the leaf base.

33. **Scaevola pulvinaris** (E.Pritzel) K.Krause, *Pflanzenr.* 54: 164 (1912)

S. humifusa var. *pulvinaris* E.Pritzel in F.L.E.Diels & E.Pritzel, *Bot. Jahrb. Syst.* 35: 572 (1905). T: near Cranbrook, W.A., *L.Diels 4403*; holo: ?B (destroyed) *n.v.*; neo: 28.5 miles [45.6 km] N of Ravensthorpe, 13 Sept. 1959, *A.S.George 314* (PERTH), *fide* R.C.Carolin, *Telopea* 3: 504 (1990).

Prostrate shrub with much-branched, woody stems to 15 cm high, glabrous or with scattered hairs. Leaves sessile, linear, ±terete, thick; lamina 8–40 mm long, 0.7–2.0 mm wide; axillary hairs ±inconspicuous. Flowers in axillary or terminal spikes to 25 mm long; bracts leafy; bracteoles ovate, 6–9 mm long, ciliate on margin and midrib. Sepals rim-like, lobed, to 3 mm long. Corolla 6–11 mm long, stiffly white hairy outside, stiffly retrorse hairy inside, white or pale blue; barbulae prominent, papillate apically; wings to 1 mm wide. Ovary 1-locular except basally; indusium 0.7–1.2 mm long, ±glabrous. Fruit ellipsoidal, 2.5–3.5 mm long, ridged, otherwise smooth, pubescent, usually 1-seeded. *n* = 8, W.J.Peacock, *Proc. Linn. Soc. New South Wales* 88: 9 (1963).

Occurs in south-western W.A. between Ravensthorpe, the Stirling Range and Frankland, in sand heath. Flowers Aug.–Nov. Map 145.

W.A.: c. 65 km W of Ravensthorpe, *R.D.Royce 3688* (PERTH); c. 20 km from Gnowangerup towards Borden, 22 Oct. 1968, *J.W.Wrigley* (CBG, SYD); 31 km N of Frankland on Kojonup road, *A.S.George 15251* (PERTH); c. 43 km from Cranbrook towards Albany, 11 Oct. 1968, *J.W.Wrigley* (CBG, SYD).

Similar to *S. humifusa*. Cytological voucher is *W.J.Peacock 60881.1* (SYD).

34. Scaevola humifusa Vriese in J.G.C.Lehmann, *Pl. Preiss.* 1: 410 (1845)

Merkusia humifusa (Vriese) Vriese, *Ned. Kruidk. Arch.* 2: 162 (1851); *Lobelia humifusa* (Vriese) Kuntze, *Revis. Gen. Pl.* 2: 378 (1891). T: towards Avon R., W.A., 10 Sept. 1839, *L.Preiss 1480*; lecto: LD, *fide* R.C.Carolin, *Telopea* 3: 503 (1990); isolecto: L, MEL, P, W.

S. depressa Vriese in J.G.C.Lehmann, *Pl. Preiss.* 1: 410 (1845); *Merkusia depressa* (Vriese) Vriese, *Natuurk. Verh. Holl. Maatsch. Wetensch. Haarlem* ser. 2, 10: 70 (1854). T: interior, south-western W.A., Nov. 1840, *L.Preiss 1502*; lecto: LD, *fide* R.C.Carolin, *Telopea* 3: 503 (1990).

Merkusia molluginea Vriese, *Natuurk. Verh. Holl. Maatsch. Wetensch. Haarlem* ser. 2, 10: 71 (1854). T: Swan R., W.A., 1843, *J.Drummond*; holo: K.

S. arenaria E.Pritzel in F.L.E.Diels & E.Pritzel, *Bot. Jahrb. Syst.* 35: 572 (1905). T: near Tammin, W.A., *E.Pritzel 754*; lecto: K, *fide* R.C.Carolin, *Telopea* 3: 504 (1990); isolecto: AD, ?B (destroyed) *n.v.*, BM, W.

Decumbent to prostrate shrub, pubescent; stems to 40 cm long. Leaves sessile, narrowly oblanceolate, entire, thick; lamina 5–20 mm long, 1–4 mm wide; axillary hairs conspicuous, woolly. Flowers in axillary racemes to 10 mm long; bracts leafy; bracteoles lanceolate, 4–6 mm long, sometimes ±obscured by axillary wool. Sepals c. 0.5 mm long, almost free. Corolla 7–11 mm long, with hairs not appressed outside, pubescent in throat, white or cream with greenish or brownish lines; barbulae prominent, papillate apically; wings to 0.5 mm wide. Ovary 1-locular except basally; indusium 1–1.5 mm long, glabrous or sparsely hairy basally. Fruit obovoid, 2–4 mm long, ±tuberculate, ±ribbed, almost glabrous, 1-seeded. Figs 26C, 42J.

Occurs in south-western W.A. from Coorow to Quairading, in open heath and mallee, in sandy soil. Flowers Aug.–Nov. Map 146.

W.A.: c. 9 km S of Coorow, *A.S.George 9518* (PERTH); between Pithara and Miling, *W.E.Blackall 2884* (PERTH); Perth–Geraldton Hwy, 168 mile post [c. 270 km N of Perth], *W.J.Peacock 60840.1* (SYD); W of Cunderdin, *C.A.Gardner 7655* (PERTH); Quairading, SW of Merredin, *W.E.Blackall 3275* (PERTH).

The dense axillary wool readily distinguishes this species from *S. pulvinaris*. The distributions of the two species are very different. The type of *S. arenaria* is a particularly hairy specimen but otherwise it seems to agree with *S. humifusa*.

35. Scaevola linearis R.Br., *Prodr.* 586 (1810)

Merkusia linearis (R.Br.) Vriese, *Ned. Kruidk. Arch.* 2: 166 (1851); *Lobelia linearis* (R.Br.) Kuntze, *Revis. Gen. Pl.* 2: 378 (1891). T: Bay X, South Coast, [Port Lincoln, S.A.], Mar. 1802, *R.Brown*; lecto: BM, *fide* R.C.Carolin, *Telopea* 3: 504 (1990); isolecto: K.

Ascending to prostrate shrub, with simple and often glandular hairs; stems to 60 cm long. Leaves sometimes fasciculate, sessile, linear to narrowly elliptic, recurved to revolute, sometimes with prominent, axillary, silvery hairs; lamina 2–85 mm long, 1–7 mm wide. Inflorescence a terminal or axillary spike, some short and dense, others longer and less dense, to 6 cm long; bracts leafy; bracteoles elliptic, 4–9 mm long. Sepals rim-like, undulate, to 1 mm long. Corolla to 18 mm long, pubescent outside, shortly bearded inside, blue; barbulae prominent, papillate apically. Ovary 1-locular except basally; indusium glabrous or ±pubescent on both surfaces. Fruit ellipsoidal, 4–6 mm long, tuberculate, variously pubescent, usually 1-seeded.

Occurs in S.A. Related to *S. paludosa*. Two subspecies are recognised. The subspecies are not absolutely clearly defined. For instance on Yorke Peninsula the specimens often have a number of intermediate characteristics but in general these fall within the variation assigned here to subsp. *linearis*.

Leaves not fasciculate **35a.** subsp. **linearis**

Leaves fasciculate **35b.** subsp. **confertifolia**

35a. Scaevola linearis R.Br. subsp. linearis

Stems with prominent, short, glandular hairs and antrorse or not appressed, usually straight, simple hairs. Leaves not fasciculate, linear to narrowly elliptic, recurved; lamina 6–85 mm long, 1–7 mm wide. Flowers in loose, leafy spikes; bracts leaf-like, ovate and flattened towards base, ciliate; bracteoles elliptic, 4–9 mm long, less than 1/2 as long as flower. Corolla 12–18 mm long.

Occurs on Eyre Peninsula and in the Gawler Ranges, S.A. Flowers chiefly Aug.–Dec. Map 147.

S.A.: Marble Ra., *J.Z.Weber 5960* (AD); Mt Wedge, 60 km W of Lock, *N.N.Donner 2379* (AD); Hincks Natl Park, *J.R.Wheeler 999* (AD); 30 km NNW of Port Lincoln, *K.B.Warnes 112* (AD); c. 1.5 km N of Port Lincoln, *M.Tindale 485* (AD, SYD).

35b. Scaevola linearis subsp. confertifolia (J.Black) Carolin, *Telopea* 3: 504 (1990)

S. linearis var. *confertifolia* J.Black., *Fl. S. Australia* 565 (1929). T: Kangaroo Is., 16 Nov. 1924, *J.B.Cleland s.n.*; lecto: AD, *fide* R.C.Carolin, *Telopea* 3: 504 (1990); isolecto: ?K.

Stems tomentose with simple hairs which conceal the minute, glandular ones. Leaves mostly fasciculate on very short, lateral branches, linear, revolute but sometimes flattened towards base; lamina 2–15 mm long, 1–3 mm wide. Inflorescence a spike condensed to appear axillary amongst a tuft of leaves; bracts leafy; bracteoles ovate-elliptic, 5–7 mm long, c. 1/2 as long as flower. Corolla 7–10 mm long.

Occurs in S.A. on Yorke Peninsula, Mt Lofty Ranges and on Kangaroo Is. Flowers chiefly Aug.–Jan. Map 148.

S.A.: on road between Milang and Finniss, *D.Hunt 2646* (AD); N of Victor Harbor, 21 Aug. 1947, *J.B.Cleland* (AD); Waitpinga, *R.Bates 1016* (AD); 13 km S of Kingscote, *P.G.Wilson 916* (AD); Flinders Chase, Kangaroo Is., *J.R.Wheeler 1301* (AD).

36. Scaevola paludosa R.Br., *Prodr.* 586 (1810)

Merkusia paludosa (R.Br.) Vriese, *Ned. Kruidk. Arch.* 2: 167 (1851); *Lobelia paludosa* (R.Br.) Kuntze, *Revis. Gen. Pl.* 2: 378 (1891). T: Bay 1 [Lucky Bay], South Coast, [W.A.], 12 Jan. 1802, *R.Brown*; lecto: BM, *fide* R.C.Carolin, *Telopea* 3: 504 (1990).

Erect shrub to 50 cm tall, ±hispid, with simple hairs and minute, glandular hairs. Leaves narrowly elliptic to oblanceolate, slightly recurved, entire or with small teeth; lamina mostly 3–4.5 cm long, to 10 mm wide. Flowers in condensed or elongate, axillary spikes to 5 cm long; bracts leaf-like but smaller; bracteoles lanceolate, c. 5 mm long. Sepals broadly ovate, c. 0.3 mm long, free. Corolla 7–10 mm long, with coarse, simple hairs outside and scattered hairs inside, white(?); barbulae simple; wings c. 0.3 mm wide. Ovary 1-locular, hairy; indusium with tuft of irregular hairs basally. Fruit not seen.

Occurs near Mt le Grand, W.A., in black, peaty soil. Flowers probably about Dec. Map 149.

W.A.: N of Mt le Grand, *A.S.George 2237* (PERTH).

Differs from *S. linearis* in its very short scarcely visible glandular hairs and much coarser simple hairs. The name has been misapplied to *S. repens* Vriese in the past.

Ser. 2. Pogogynae

Scaevola sect. **Xerocarpa** subsect. **Biloculatae** ser. **Pogogynae** Benth., *Fl. Austral.* 4: 85, 98 (1868).

Type: *S. aemula* R.Br.; lecto: *fide* R.C.Carolin, *Telopea* 3: 490 (1990).

Scaevola sect. *Xerocarpa* subsect. *Xerocarpa* ser. *Pogogynae* Benth. *sensu* R.C.Carolin, *Telopea* 3: 490 (1990), *nom. illeg.*

Inflorescence a raceme, or rarely a spike-like thyrse or panicle. Indusium with basal beard of stiff, straight bristles on back. Ovary 1 or 2-locular.

Nineteen species confined to Australia.

37. Scaevola albida (Smith) Druce, *Rep. Bot. Exch. Club Brit. Isles* 1916, Suppl. 2: 644 (1917)

Goodenia albida Smith, *Trans. Linn. Soc. London, Bot.* 2: 348 (1794); *S. laevigata* var. *albida* (Smith) Pers., *Syn. Pl.* 1: 195 (1805). T: Port Jackson, N.S.W., *J.White*; holo: LINN-SM.

Goodenia laevigata Curtis, *Bot. Mag.* 8: t. 287 (1795); *Scaevola laevigata* (Curtis) Pers., *Syn. Pl.* 1: 195 (1805). T: Samual Tolfrey's collection ex Botany Bay, N.S.W., 1792, *coll. unknown*; *n.v.*

S. microcarpa Cav., *Anales Hist. Nat.* 1(2): 97, t. 9 (1799), *nom. illeg.*, *nom. superfl.*; *Merkusia microcarpa* (Cav.) Vriese, *Ned. Kruidk. Arch.* 2: 157 (1851); *Lobelia microcarpa* (Cav.) Kuntze, *Revis. Gen. Pl.* 2: 378 (1891). T: none cited.

S. pallida R.Br., *Prodr.* 585 (1810); *Merkusia pallida* (R.Br.) Vriese, *Ned. Kruidk. Arch.* 2: 159 (1851); *S. microcarpa* var. *pallida* (R.Br.) Benth., *Fl. Austral.* 4: 101 (1868); *S. albida* var. *pallida* (R.Br.) Carolin in J.P.Jessop & H.R.Toelken, *Fl. S. Australia* 4th edn, 1410 (1986). T: Port Phillip, [Vic.], 25–28 Jan. 1804, *R.Brown*; lecto: BM, *fide* R.C.Carolin, *Telopea* 3: 506 (1990).

Goodenia pubescens Sieber ex Sprengel, *Syst. Veg.* 4: 76 (1827). T: New Holland, [N.S.W.], *F.W.Sieber 514*; *n.v.*

Illustration: G.R.Cochrane *et al.*, *Fl. Pl. Victoria* 10, fig. 292 (1968) as *pallida*.

Prostrate to ascending herb to 50 cm tall, often woody basally, with curved, antrorse hairs or glabrous. Leaves sessile, obovate to elliptic, dentate to entire, surface usually visible beneath hairs; lamina 6–50 mm long, 1–25 mm wide. Flowers in spikes to 25 cm long; bracts leafy; bracteoles narrowly elliptic, 3–9 mm long, ciliate. Sepals ovate-deltoid, to 0.7 mm long, ±free. Corolla 5–10 mm long, white ±appressed-hairy outside, thinly bearded inside, blue or white; barbulae usually numerous, ±papillate apically; wings to 1 mm wide. Ovary 1-locular except basally; indusium 1–2 mm long, white or purple beard not exceeding bristles on lips. Fruit ellipsoidal, c. 3 mm long, rugose, pubescent or glabrous, usually 1-seeded. $n = 8$, W.J.Peacock, *Proc. Linn. Soc. New South Wales* 88: 8 (1963). Figs 36A, 58.

Widespread in coastal parts and along the Great Dividing Ra. of eastern Australia from Kangaroo Is., S.A., through Vic. and N.S.W. to southern Qld; grows in a variety of habitats. Flowers most of the year. Map 150.

S.A.: Gorge Rd, 13 Oct. 1960, *D.E.Symon* (AD); Upper Sturt, Waverley Ridge, 10 Nov. 1957, *M.Kenny* (AD). N.S.W.: St Peters cemetery, Campbelltown, *E.J.McBarron 7902* (NSW). Vic.: Cape Everard, 10 Jan. 1970, *T.S.Henshall* (MEL); Wilsons Promontory, *J.Stirling 156* (MEL).

See *S. parvibarbata* for distinguishing characters. Cytological voucher is *W.J.Peacock 6012.6.2* (SYD). A very variable species which requires some experimental work to elucidate but apparently includes numerous ecotypes.

38. Scaevola parvibarbata Carolin in J.P.Jessop & H.R.Toelken, *Fl. S. Australia* 4th edn, 3: 1414 (1986)

T: 9 miles [c. 14.5 km] SE Bottom Bore, Hale River, N.T., 23 Sept. 1958, *G.A.M.Chippendale*; holo: DNA.

Illustration: G.M.Cunningham *et al.*, *Pl. W. New South Wales* 637 (1981) as *S. humilis.*

Erect herb to 50 cm tall, woody basally, with curled or not appressed hairs. Leaves sessile, oblanceolate to orbicular, usually prominently triangular-toothed, surface usually visible beneath hairs; lamina 8–40 mm long, 4–20 mm wide. Flowers in spikes to 25 cm long; bracts leaf-like but smaller; bracteoles linear, to 5 mm long. Sepals rim-like, to 1 mm long. Corolla 11–25 mm long, with short, appressed, white hairs outside, densely bearded inside, lilac or greenish; barbulae simple; wings to 1 mm wide. Ovary 2-locular; indusium 1–2.5 mm wide, sparse white beard scacely reaching lips. Fruit obovoid, 3–5 mm long, rugose, hairy. Figs 36B, 42E–G.

Occurs in central Australia in N.T., S.A., Qld and N.S.W. between 130°E and 149°E and 21°S and 32°S; grows in dry open communities, mostly in sandy soil. Flowers chiefly May–Oct. Map 151.

N.T.: c. 26 km E of New Crown homestead, *J.Must 138* (DNA). S.A.: Goyder Lagoon, *J.Z.Weber 4626* (AD); Mt Gason Bore, *E.N.S.Jackson 2791* (AD). Qld: Burenda, Warrego district, *S.L.Everist 1903* (BRI). N.S.W.: c. 27 km SE of Innesowen on Tipa road, *S.Jacobs 191* (SYD).

Similar to *S. albida* which is only thinly bearded inside the corolla, has a smaller corolla (5–10 mm long), has a very short septum so that the ovary appears 1-locular, and often sparse indumentum. *Scaevola parvibarbata* is extremely variable with regard to the degree of hairiness, and size and colour of the corolla; some of the variants appear to show distinct distributions but all intergrade very gradually.

39. Scaevola ovalifolia R.Br., *Prodr.* 584 (1810)

S. ovalifolia var. *cinerascens* R.Br., *Prodr.* 584 (1810); *Merkusia ovalifolia* (R.Br.) Vriese, *Ned. Kruidk. Arch.* 2: 153 (1851); *Merkusia ovalifolia* var. *cinerascens* (R.Br.) Vriese, *Ned. Kruidk. Arch.* 2: 154 (1851); *Lobelia ovalifolia* (R.Br.) Kuntze, *Revis. Gen. Pl.* 2: 378 (1891). T: Carpentaria, [N.T.], *R.Brown*; lecto: BM, *fide* R.C.Carolin, *Telopea* 3: 504 (1990); isolecto: MEL.

S. densevestita Domin, *Biblioth. Bot.* 89: 646 (1929). T: towards Cloncurry, Qld, Feb. 1910, *K.Domin 8799*; holo: PR.

Subshrub to 50 cm tall, with short, soft hairs. Leaves sessile, obovate to narrowly elliptic, dentate or entire, surface usually hidden beneath hairs; lamina 8–55 mm long, 4–21 mm wide. Flowers in terminal spikes to 20 cm long; bracts leaf-like but smaller; bracteoles linear-deltoid, 4–15 mm long. Sepals semiorbicular, c. 1 mm long, connate. Corolla

14–24 mm long, with short, simple hairs outside, bearded inside, blue or mauve; barbulae numerous, simple; wings c. 1 mm wide. Ovary 2-locular; indusium 2–2.5 mm long, short white hairs above often equalling lips. Fruit obovoid, 4–5 mm long, rugose towards base, obscurely grooved, pubescent, with 2 sterile cavities. Figs 36C–D, 42I.

Occurs in N.T. and north-western Qld mostly in sand but occasionally on rocky hill slopes. Flowers most of the year. Map 152.

N.T.: c. 6 km E of Bullman turnoff on Mainoru road, *R.C.Carolin 9406* (SYD); c. 3 km E of Quartz Hill Stn, *P.K.Latz 1124* (DNA); Anningini Stn, *A.S.Mitchell 016* (CANB, DNA, NSW); c. 250 km E of Stuart Hwy on Borroloola road, *C.R.Dunlop 2185* (DNA, NSW). Qld: c. 65 km W of Windorah, *R.C.Carolin 6397* (SYD).

Similar to *S. parvibarbata* which has less dense and looser hairs so that leaf surface is visible below hairs. The names *S. ovalifolia* and *S. ovalifolia* var. *cinerascens* were lectotypified on the same collection.

40. **Scaevola glabrata** Carolin in J.P.Jessop & H.R.Toelken, *Fl. S. Australia* 4th edn, 3: 1412 (1986)

S. ovalifolia var. *glabra* R.Br., *Prodr.* 584 (1810); *Merkusia ovalifolia* var. *glabra* (R.Br.) Vriese, *Ned. Kruidk. Arch.* 2: 154 (1851). T: Carpentaria, [N.T.], *R.Brown*; lecto: BM; isolecto: MEL *fide* R.C.Carolin, *Fl. Australia* 35: 333 (1992)

Erect subshrub to 50 cm tall, glabrous or with few, scattered hairs on stem. Leaves ±sessile, elliptic to obovate, dentate or entire; lamina 9–34 mm long, 3–14 mm wide. Flowers in terminal spikes to 18 cm long; bracts leaf-like but smaller; bracteoles lanceolate to narrowly elliptic, 5–7 mm long. Sepals semiorbicular, c. 1 mm long, connate. Corolla 16–21 mm long, glabrous outside or with some hairs towards top, ±bearded inside, usually blue; barbulae simple; wings to 1 mm wide. Ovary 2-locular; indusium 1–2.5 mm long, scanty beard above scarcely reaching lips. Fruit obovoid, 3–6 mm long, rugose, often pubescent near base. Fig. 42H.

Extends from the N.T., excluding Arnhem Land, south just into S.A., and just over the border into Qld; usually in rocky soil often over limestone, but occasionally in sand. Flowers chiefly May–Oct. Map 153.

N.T.: c. 50 km E of Frewena, Barkly Hwy, *J.R.Maconochie 407* (DNA); c. 62 km SW of Tobermorey homestead, 20 Sept. 1956, *G.A.M.Chippendale* (DNA); Georgina Downs homestead, 1 Oct. 1957, *G.A.M.Chippendale* (DNA). S.A.: Ernabella, 24 July 1966, *F.J.Tuvey* (SYD). Qld: between Dajarra and Boulia, *P. & H.Althofer 8540* (BRI).

All other members of the series which have a short beard on the indusium have some degree of hairiness. *Scaevola laciniata*, which shows some resemblance to *S. glabrata* and is also virtually glabrous, has a long purple beard on the indusium.

41. **Scaevola glutinosa** Carolin, *Telopea* 3: 504 (1990)

T: Granada, about 50 miles [80 km] N of Cloncurry, 11 Apr. 1954, *S.L.Everist 5225*; holo: BRI.

Spreading subshrub to 70 cm tall, viscid, with soft, simple hairs at 90° and pale-headed, glandular hairs ±as long as simple ones. Leaves sessile, often almost stem-clasping, obovate, dentate; lamina 21–68 mm long, 6–26 mm wide. Flowers in spikes to 12 cm long; bracts ovate-elliptic, to 15 mm long; bracteoles linear-lanceolate, 7–11 mm long. Sepals rim-like, sinuate, 0.5–1 mm long. Corolla 14–24 mm long, with hairs at 90° outside, bearded inside, blue; barbulae simple; wings to 10 mm wide. Ovary 2-locular; indusium to 2 mm long, sparse beard above scarcely equalling bristles on lips. Fruit cylindrical, 4–6 mm long, rugose, pubescent.

Figure 42. *Scaevola*. **A–D**, *S. globosa*, **A**, habit X0.5 (A.George 16350, SYD). **B**, fruit X10 **C**, fruit with bracteole X10 **D**, flower with bracteole X10 (**B–D**, A.George 9214, NSW). **E–G**, *S. parvibarbata*. **E**, indusium X7.5; **F**, corolla X2.5; **G**, anther X10; (**E–G**, voucher not recorded). **H**, *S. glabrata*, habit X1 (A.Beauglehole 27367, SYD). **I**, *S. ovalifolia*, habit X1 (D.Sillar, Aug. 1959, BRI). **J**, *S. humifusa*, habit X1 (T.Aplin 2000, PERTH). Drawn by D.Mackay.

Occurs in Qld between the Mt Isa highlands and the Great Dividing Ra., frequently on limestone. Flowers Feb.–Sept. Map 154.

Qld: Digby Peaks Ra., *R.W.Purdie 1047* (BRI); c. 3 km S of Duchess, *P.Ollerenshaw & D.Kratzing 1249* (BRI, CBG); c. 50 km E of Cloncurry, *M.Lazarides 4071* (BRI, CANB, DNA); Granada, c. 80 km N of Cloncurry, *S.L.Everist 5225* (BRI); Flinders R., Aug. 1913, *F.Sulman* (NSW).

The long, viscid, glandular hairs with simple hairs at 90°, and the ±stem-clasping leaves distinguish this species from other members of ser. *Pogogynae* which have sparse, short beards on the back of the indusia.

42. **Scaevola densifolia** Carolin, *Telopea* 3: 505 (1990)

T: Hamersley R., W.A., 8 Nov. 1969, *K.M.Allan 180*; holo: PERTH; iso: SYD.

Prostrate subshrub with stems to 40 cm long, hairy. Leaves sessile, often fasciculate, oblanceolate, entire or with a tooth on either side; lamina 8–30 mm long, 2–9 mm wide. Flowers in terminal racemes; bracts leafy; bracteoles linear, 5–7 mm long. Sepals linear-deltoid, to 2.5 mm long, connate. Corolla 12–15 mm long, with dense ±appressed hairs outside, bearded inside, white or cream; barbulae few, simple, hair-like; wings to 0.8 mm wide. Ovary 2-locular; indusium 1–2 mm long, few basal hairs not exceeding bristles on lips. Fruit cylindrical, 3–4 mm long, with two lateral protuberances and 2 vertical bands of tubercles, pubescent, with 2 sterile cavities. Figs 36E, 39N–O.

Occurs between the Oldfield and Fitzgerald Rivers, W.A., in heath. Flowers chiefly Oct.–Dec. Map 155.

W.A.: c. 58 km N of Oldfield R., *Hj.Eichler 20391* (AD, CANB, PERTH); West R., *K.Newbey 1726* (PERTH); Ravensthorpe, Nov. 1944, *C.A.Gardner* (PERTH); Fitzgerald R., *A.S.George 10561* (PERTH).

This species has peduncles to 2 mm long. The lateral protuberances and rows of tubercles on the fruit separate this species from all others in the genus.

43. **Scaevola cuneiformis** Labill., *Nov. Holl. Pl.* 1: 56 (1804)

Merkusia cuneiformis (Labill.) Vriese, *Ned. Kruidk. Arch.* 2: 157 (1851); *Lobelia cuneiformis* (Labill.) Kuntze, *Revis. Gen. Pl.* 2: 378 (1891). T: south coast [Esperance, W.A.], *J.J.H.de Labillardière*; holo: FI; iso: BM.

Illustration: J.J.H.de Labillardière, *op. cit.* t. 80.

Erect shrub to 60 cm tall, with fine, appressed, white hairs. Leaves sessile, ovate, usually with an acute tooth apically, dentate; lamina 6–35 mm long, 4–16 mm wide. Flowers in dense, leafy thyrses or panicles to 15 cm long; bracts leaf-like but smaller, entire, imbricate; bracteoles linear-elliptic, 4–6 mm long. Sepals semiorbicular, to 6.5 mm long, connate basally. Corolla 8–12 mm long, with fine, appressed, simple, white hairs outside, bearded inside, mauve to blue; barbulae few, simple; wings 1 mm wide. Ovary 2-locular; indusium 1–1.75 mm long, stiff purplish beard ±exceeding bristles on lips. Fruit ovoid, to 3.5 mm long, rugose, pubescent, usually 1-seeded. Figs 36F, 43A.

Occurs between Esperance and the Fitzgerald R., W.A., in sandy heath. Flowers Sept.–Nov. Map 156.

W.A.: near West R., c. 32 km SSW of Ravensthorpe, *A.S.George 9833* (PERTH); Pink Lake, Esperance, 30 Oct. 1968, *J.W.Wrigley* (PERTH); Culham Inlet, *K.Newbey 1731* (PERTH); East Mt Barren, *A.S.George 542* (PERTH); Point Ann, Fitzgerald River Natl Park, *A.S.George 11281* (PERTH).

The type appears to have been mislabelled as having been collected in Tasmania.

The bracteoles are c. 1/4 as long as the flower and the indusium is square or broadly oblong. The inflorescence consists of a series of monochasia combined into thyrses, and thus appears dense. Resembles *S. aemula* which has coarser hairs and the flowers in spikes. See also *S. argentea.*

This species is placed by G.Bentham in series *Monospermae*, and by K.Krause in subsect. *Uniloculatae*, both of which are characterised by an ovary with a single locule. The septum, however, extends almost to the top of the ovary which is thus 2-locular. Previous authors may have obtained their information from a section of the fruit which usually has only one seed, and a sterile locule which is so compressed that it is often difficult to see.

44. **Scaevola argentea** Carolin, *Fl. Australia* 35: 332 (1992)

T: 3 miles [c. 5 km] S of Lake Cobham, W.A., 6 Jan. 1972, *K.Newbey 3467*; holo: PERTH.

Prostrate, much-branched subshrub to 15 cm tall, silvery hairy. Leaves sessile, elliptic to obovate, ±recurved, dentate or entire; lamina usually 8–20 mm long, 3–13 mm wide. Flowers in dense spikes or spike-like thyrses; bracts elliptic, 5–10 mm long, entire, imbricate; bracteoles narrowly elliptic, 3–4 mm long. Sepals semiorbicular, c. 0.5 mm long, free. Corolla 8–13 mm long, with silvery, ±appressed hairs outside, bearded inside, mauve to blue; barbulae simple; wings c. 0.5 mm wide. Ovary 2-locular; indusium 1 mm long, stiff purplish beard ±equalling bristles on lips. Fruit ellipsoidal, c. 2 mm long, ±ribbed, otherwise smooth, ±pubescent. Fig. 37A.

Occurs in W.A. between Ravensthorpe and Ongerup, in sandy heath. Map 157.

W.A.: c. 25 km N of Jerramungup–Ravensthorpe road, c. 45 km W of Ravensthorpe, *A.S.George 7065* (PERTH); N of Needilup, *A.S.George 7018* (PERTH, SYD); c. 15 km N of Ongerup, *K.Newbey 2408* (PERTH); 1 km NE of Lake Cairlocup, *K.Newbey 4836* (PERTH).

Indusium semiorbicular. Resembles *S. cuneiformis* which has an erect habit, fine, appressed, white hairs which are less dense, rugose fruit, and bracteoles only 1/4 the length of the flower (1/2 to 2/3 as long in *S. argentea*).

45. **Scaevola aemula** R.Br., *Prodr.* 584 (1810)

Lobelia aemula (R.Br.) Kuntze, *Revis. Gen. Pl.* 2: 378 (1891). T: Bay X, South Coast, [Port Lincoln, S.A.], Mar. 1802, *R.Brown*; lecto: BM, *fide* R.C.Carolin, *Telopea* 3: 506 (1990); isolecto: K, MEL.

S. sinuata R.Br., *Prodr.* 584 (1810); *Merkusia sinuata* (R.Br.) Vriese, *Ned. Kruidk. Arch.* 2: 160 (1851). T: Goose Island Bay, [W.A.], May 1803, *R.Brown*; lecto: BM, *fide* R.C.Carolin, *Telopea* 3: 506 (1990); isolecto: K, MEL.

Illustration: E.R.Rotherham *et al.*, *Fl. Pl. New South Wales & S. Queensland* 79, fig. 228 (1975).

Ascending to decumbent herb to 50 cm tall; stems coarsely yellowish-brownish hirsute. Leaves obovate, tapering towards base, dentate; lamina 10–88 mm long, 4–31 mm wide. Flowers in spikes to 24 cm long; bracts leaf-like but smaller; bracteoles lanceolate, 4.5–7 mm long. Sepals broadly deltoid, to 0.5 mm long, ciliate, basally connate. Corolla 17–25 mm long, with appressed hairs outside, bearded inside, blue or white; barbulae few, simple; wings 1–1.5 mm wide. Ovary 2-locular; indusium to 1.5 mm long, stiff purplish beard ±equalling bristles on lips. Fruit ovoid, to 4.5 mm long, rugose, pubescent with short hairs, with 2 sterile cavities. Fig. 37B.

Occurs in south-eastern Australia, from Eyre Peninsula, S.A., through Vic. to Mt Warning in northern N.S.W. Flowers chiefly Aug.–Mar. Map 158.

S.A.: Pinkawillinie, *K.D.Rohrlach 313* (AD); near The Gap, Naracoorte, *D.Hunt 2037* (AD). N.S.W.: Bodalla, Nov. 1911, *J.L.Boorman* (NSW); Yadboro, Clyde R., *R.C.Carolin 3927* (SYD). Vic.: Kongal

Rocks, The Grampians, Mt Zero–Stapylton area, *A.C.Beauglehole 17862* (MEL).

The coarse usually yellowish or brownish hairs distinguish this species from others in this series. It has been confused with *S. ramosissima* but that species has a short, white beard on the upper side of the indusium. See also *S. cuneiformis* and *S. humilis*. It is possible that this species occurs on the Archipelago of the Recherche in W.A. but the few specimens from there are not complete enough to be identified with certainty.

46. **Scaevola amblyanthera** F.Muell., *Fragm.* 1: 121 (1859)

Lobelia amblyanthera (F.Muell.) Kuntze, *Revis. Gen. Pl.* 2: 378 (1891). T: near mouth of Nicholson R., Qld, *F.Mueller*; holo: MEL.

S. decipiens W.Fitzg., *J. & Proc. Roy. Soc. W. Australia* 3: 216 (1918). T: Port Hedland, W.A., *W.V.Fitzgerald*; holo: NSW.

Spreading subshrub to 70 cm tall; hairs not appressed, of variable density, sometimes glandular. Leaves sessile, narrowly elliptic to obovate, usually dentate; lamina 8–30 mm long, 2–9 mm wide. Flowers in spikes to 12 cm long; bracts leaf-like, smaller; bracteoles ovate, elliptic or linear, 4–6 mm long. Sepals deltoid, c. 1 mm long, connate basally. Corolla 8–16 mm long, pubescent with ±appressed hairs outside, densely bearded inside, mauve to pale pink or white; barbulae few, simple; wings c. 0.7 mm wide. Ovary 2-locular; indusium oblong to square, 1–2 mm long, erect white or purplish hairs at base exceeding lips. Fruit obovoid, to 4 mm long, tuberculate, pubescent.

Occurs in tropical and central Australia. The relationships of this species with others in this series are not clear. See notes under *S. laciniata* and *S. collina* suggesting features distinguishing these species from *S. amblyanthera*. However when further collections are examined, these distinctions may prove to be unreliable. See also *S. humilis*.

This species is extremely variable. Two varieties are recognised here but this does not deal with the variation entirely satisfactorily. In particular in var. *centralis* the size of the leaves varies, and there may be further taxa to be distinguished here. The hairs on the leaves are also very variable, particularly the proportions of glandular and simple hairs in var. *centralis*.

Glandular hairs numerous, usually visible beneath the simple hairs which may be almost absent **46b.** var. **centralis**

Glandular hairs absent or very sparse **46a.** var. **amblyanthera**

46a. **Scaevola amblyanthera** F.Muell. var. **amblyanthera**

Leaves ±dentate, usually obtuse. Glandular hairs absent or very sparse; simple hairs often curled and tangled, often dense so as almost to hide the leaf surface. Fig. 37C.

Occurs in northern Australia from the Kimberley, W.A., to the Barkly Tablelands, N.T. and Qld; in a variety of habitats. Flowers Mar.–Oct. Map 159.

W.A.: Wolf Ck Meteorite Crater, *G.W.Carr & A.C.Beauglehole 47382* (SYD). N.T.: c. 32 km N of Newcastle Waters, *R.Perry 366* (CANB); c. 35 km N of Daly Waters, 7 Nov. 1957, *G.Chippendale* (DNA); Attack Ck, Brunchilly Stn, Mar. 1966, *G.Chippendale* (DNA). Qld: Calvert R. crossing on Wollogorang–Calvert Hills road, *R.C.Carolin 9256* (SYD).

46b. **Scaevola amblyanthera** var. **centralis** Carolin, *Fl. Australia* 35: 332 (1992)

T: 3 km NW of Kunoth Well, Hamilton Downs Stn, N.T., 11 Nov. 1974, *D.J.Nelson 2383*; holo: DNA.

Leaves usually dentate, acute to obtuse. Glandular hairs numerous (visible at least under hand lens); simple hairs not usually dense or tangled, the leaf surface visible beneath hairs. Fig. 37D.

Occurs in a variety of habitats in the drier parts of Australia in W.A., N.T. and S.A., between 117°E and 139°E, and south of 20°S. Flowers Mar.–Oct. Map 160.

W.A.: Southesk Tableland, *A.S.George 15463* (PERTH). N.T.: Mt Olga, *M.Lazarides 6168* (CANB); Stuart Bluff Ra., *P.E.Conrick 1416* (AD). S.A.: Maralinga, *D.J.E.Whibley 695* (AD).

Similar to *S. collina* but that species is almost glabrous and the corolla is blue, not mauve as in the present variety.

47. Scaevola collina J.M.Black ex E.L.Robertson in E.L.Robertson & J.M.Black, *Fl. S. Australia* 2nd edn, 4: 946, fig. 1144 (1957)

T: near Ernabella, Musgrave Ra., S.A., 24 Sept. 1945, *J.B.Cleland*; holo: AD.

Erect subshrub to 50 cm tall, almost glabrous, with minute glandular hairs on young parts. Leaves sessile, narrowly elliptic to narrowly oblong, acute, entire or dentate; lamina 12–43 mm long, 3–15 mm wide. Flowers in spikes to 14 cm long; bracts leafy; bracteoles narrowly ovate, 8–11 mm long. Sepals broadly ovate, to 0.5 mm long, connate. Corolla 16–24 mm long, with ±appressed hairs outside, bearded inside, blue; barbulae few, simple; wings c. 1 mm wide. Ovary 2-locular; indusium 1–1.5 mm long, stiff purplish beard ±equalling bristles on lips. Fruit cylindrical, 2–4 mm long, rugose, slightly pubescent.

Occurs in the Musgrave and Tomkinson Ranges, north-western S.A., in open communities in rocky soil. Flowers chiefly May–Aug. Map 161.

S.A.: Ernabella Mission, 4 July 1963, *F.Turney* (NSW).

Indusium depressed-obovate. Similar to *S. laciniata* but that species has less dentate leaves and the corolla glabrous outside. Also similar to *S. amblyanthera* which has leaves with at least some hairs of various types.

48. Scaevola obovata Carolin, *Fl. Australia* 35: 333 (1992)

T: Dulgunia Hill, Tomkinson Ra., S.A., 4 Sept. 1978, *J.Z.Weber 5398*; holo: AD.

Spreading shrub to 50 cm tall, often with tough, woody lower branches, with soft, simple hairs at 90° and numerous short, glandular ones. Leaves sessile, obovate to oblong-obovate, dentate; lamina 3–4 cm long, 1–2 cm wide. Flowers in spikes to 15 cm long; bracts leafy; bracteoles linear-lanceolate, 6–8 mm long. Sepals rim-like, sinuate. Corolla 16–18 mm long, pubescent outside with simple hairs at 90° and short, glandular ones, densely bearded inside, blue; barbulae simple. Ovary 2-locular; indusium 1.5 mm long, erect purplish hairs at base exceeding lips. Fruit not seen.

Occurs in the Tomkinson Ranges, S.A., on rocky hillsides. Flowers ?May–Aug. Map 162.

S.A.: Mt Davies, W.S.Reid 103 (AD).

Indusium semiorbicular. Probably related to *S. collina* which has narrower almost glabrous leaves, and bracteoles more than half as long as the flower. It is possible that there are intermediates between these two species. Also resembling the Qld species *S. glutinosa* which has a sparse beard on the back of the indusium.

49. Scaevola humilis R.Br., *Prodr.* 585 (1810)

Merkusia humilis (R.Br.) Vriese, *Ned. Kruidk. Arch.* 2: 161 (1851); *Lobelia humilis* (R.Br.) Kuntze, *Revis. Gen. Pl.* 2: 378 (1891). T: Inlet XII South Coast, [Spencer Gulf, S.A.], Mar. 1802, *R.Brown*; lecto: BM, *fide* R.C.Carolin, *Telopea* 3: 506 (1990); isolecto: K.

Decumbent or ascending herb to 35 cm tall, ±woody at base, almost glabrous or with short, ±arcuate, white hairs. Leaves very acute, elliptic to narrowly obovate, tapering towards base, coarsely dentate; lamina 10–50 mm long, 4–17 mm wide. Flowers in spikes to 15 cm long; bracts leaf-like but smaller; bracteoles narrowly elliptic to narrowly ovate, 6–9 mm long. Sepals broadly ovate, to 0.6 mm long, basally connate. Corolla 12–23 mm long, hairy outside, densely bearded inside, blue or white; barbulae simple; wings to 1.5 mm wide. Ovary 2-locular; indusium to 2 mm long, purplish beard ±equalling bristles on lips. Fruit ovoid, 3–5 mm long, rugose, pubescent. Figs 26D, 37E–F, 43C–F.

Occurs in N.T. and from Eyre Peninsula and Flinders Ranges, S.A., to the drier parts of Qld and N.S.W. Flowers chiefly May–Oct. Map 163.

N.T.: Talipata east, W of Haast Bluff, 13 Dec. 1977, *P.K.Latz 7526* (NT). S.A.: c. 2 km W of Bethany, *Hj.Eichler 16393* (AD); Burra Gorge, c. 4 km W of Robertstown–Burra road, *N.N.Donner 8292* (AD). Qld: c. 25 km N of St George, *K.A.Williams 81231* (BRI, NSW). N.S.W.: c. 22 km E of Rankins Spring, *E.N.S.Jackson 2152* (NSW).

S. humilis is similar to the following

1. *S. cuneiformis*, which is erect, has broader leaves and bracts, appressed indumentum, and the lower flowers in each inflorescence usually in tight axillary cymes; 2. *S. aemula*, with which it has been often confused in the past, but that species has coarse, straight, yellowish brownish hairs on the stem and corolla, wider and more coarsely toothed leaves, and the flowers are single in spikes; 3. *S. laciniata*, which is glabrous; 4. *S. amblyanthera*, which has stem hairs loose, ±tangled, not arcuate, smaller leaves, a mauve to pink corolla, and bracteoles 1/3 to 1/2 as long as the flower.

The specimens from the northern Flinders Ranges, S.A., are significantly more pubescent than others.

50. Scaevola graminea Ewart & A.Petrie, *Proc. Roy. Soc. Victoria* n.s., 38: 181 (1926)

T: Taylor Well, N.T., June 1924, *A.J.Ewart*; holo: MEL.

Illustration: A.J.Ewart & A.H.K.Petrie, *op. cit.* fig. 4.

Erect subshrub to 50 cm tall, glabrous. Leaves sessile, acute, linear to narrowly oblong, usually entire; lamina 7–20 mm long, less than 2 mm wide. Flowers in interrupted spikes to 15 cm; bracts leaf-like; bracteoles linear, 5–7 mm long, c. 1/3–1/2 length of flower. Sepals to 1 mm long, connate. Corolla 14–17 mm long, glabrous outside, densely bearded inside, yellowish to bluish pink sometimes with a yellow throat; barbulae few, simple; wings to 0.4 mm wide. Ovary 2-locular; indusium to 2 mm long, dense purple beard exceeding lips. Fruit ovoid, 4–5 mm long, rugose, pubescent. Fig. 43B.

Occurs in the Kimberley, W.A., and parts of N.T., in rocky situations. Flowering unknown. Map 164.

N.T.: Mt Liebig area, 23°17'S, 131°18'E, *P.K.Latz 2277* (CANB).

This species is most similar to *S. laciniata* which, however, has leaves 3–23 mm wide.

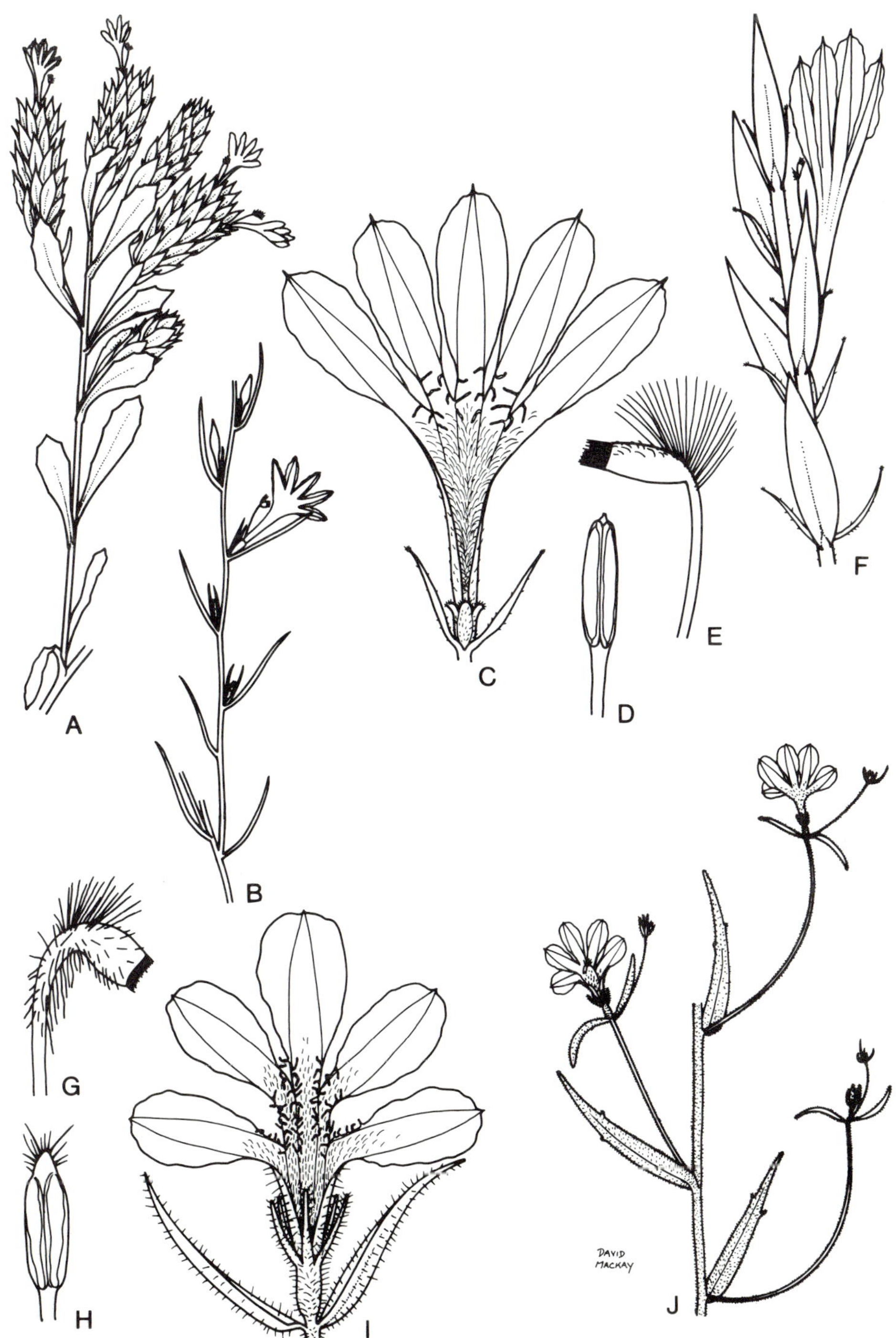

Figure 43. *Scaevola*. **A**, *S. cuneiformis*, habit X1 (A.George 11281, PERTH). **B**, *S. graminea*, habit X1 (P.Latz 2277, NT). **C–F**, *S. humilis*. **C**, corolla X2.5; **D**, anther X10; **E**, indusium X7.5; **F**, habit X2 (**C–F**, G.Gardiner, 12 Oct. 1969, PERTH). **G–J**, *S. ramosissima*. **G**, indusium X7.5; **H**, anther X10; **I**, corolla X2; **J**, habit X0.5 (**G–J**, O.Evans, 29 Oct. 1927, SYD). Drawn by D.Mackay.

51. **Scaevola laciniata** F.M.Bailey, *Queensland Fl.* 3: 910 (1900)

T: between Camooweal and Urandangi, Qld, 3 Aug. 1890, *R.C.Burton*; holo: BRI.

Illustration: F.M.Bailey, *op. cit.*, t. XXXIII.

Erect subshrub to 1 m tall, glabrous or with few hairs. Leaves acute to acuminate, elliptic to cuneate, tapering gradually basally, entire or dentate with narrow, acute teeth; lamina 10–40 mm long, 3–23 mm wide. Flowers in spikes to 13 cm long; bracts leaf-like, narrower; bracteoles lanceolate to linear, 4–11 mm long. Sepals c. 1 mm long, connate basally. Corolla 10–16 mm long, glabrous outside, bearded inside, blue purplish or white; barbulae few, simple; wings sinuate, to 0.7 mm wide. Ovary 2-locular; indusium 1–2 mm long, dense purple beard ±equalling bristles on lip. Fruit obovoid, 3–6 mm long, rugose, glabrous or pubescent.

Occurs in N.T., north of 22°S excluding Arnhem Land, and just over the borders into W.A. and Qld; grows in *Triodia* and *Plectrachne* communities and occasionally in woodland. Flowers chiefly Apr.–Aug. Map 165.

W.A.: c. 112 km E of Halls Creek, *J.S.Beard 5646* (PERTH). N.T.: 16 km S of Victoria Hwy, Kidman Springs road, *M.Parker 465* (DNA); c. 80 km N of Tennant Creek, *R.A.Perry 627* (BRI, CANB); S of Elliot, *J.Must 368* (DNA). Qld: c. 30 km NW of Creswell Downs homestead, *R.C.Carolin 9286* (SYD).

Confused in the past with *S. collina* which however has hairs on the outside of the corolla, and is not known to occur outside S.A. See also *S. graminea*.

52. **Scaevola microphylla** (Vriese) Benth., *Fl. Austral.* 4: 100 (1868)

Molkenboeria microphylla Vriese, *Natuurk. Verh. Holl. Maatsch. Wetensch. Haarlem* ser. 2, 10: 44 (1854); *Lobelia microphylla* (Vriese) Kuntze, *Revis. Gen. Pl.* 2: 378 (1891), *nom. illeg.* T: Swan R., W.A., *J.Drummond*; lecto: K, *fide* R.C.Carolin, *Telopea* 3: 506 (1990); isolecto: L, LD.

Illustration W.H.de Vriese, *op. cit.* t. 9, as *Molkenboeria microphylla*.

Prostrate to decumbent herb, with antrorse, ±appressed hairs; base woody; stems to 50 cm long. Leaves sessile, ±stem-clasping, obovate to elliptic, obtuse or terminally toothed, conspicuously dentate; lamina 8–53 mm long, 3–29 mm wide. Flowers in loose racemes to 15 cm long; bracts leaf-like but smaller; bracteoles oblong, 5–8 mm long. Sepals broadly ovate, to 0.5 mm long, connate basally. Corolla 10–21 mm long, pubescent or glabrous outside, bearded inside, blue; barbulae short, ±papillate apically; wings to 1.5 mm wide. Ovary 2-locular; indusium 1.5–2.5 mm wide, dense brownish beard exceeding bristles on lips. Fruit subglobular, 2–3 mm diam., pubescent, usually 1-seeded.

Occurs in extreme south-western W.A. in and on margins of forests. Flowers Sept.–Jan. Map 166.

W.A.: Ellen Brook on road from Yallingup to Poverty Peak, *P.Weston 198* (SYD); 41 mile peg [c. 64 km] on Perth–Walpole road, 11 Dec. 1965, *F.W.Humphreys* (PERTH); Cowaramup, *R.D.Royce 2441* (PERTH); 20 km E of Pemberton, *G.J.Keighery W214S* (PERTH); c. 6 km from Denmark on Manjimup road, *A.R.Fairall 605* (PERTH).

Peduncles are to 5 cm long. See *S. auriculata* for diagnostic characters.

53. **Scaevola auriculata** Benth., *Fl. Austral.* 4: 99 (1868)

Lobelia auriculata (Benth.) Kuntze, *Revis. Gen. Pl.* 2: 378 (1891). T: south-western W.A., *J.Drummond 3: 153*; lecto: K, *fide* R.C.Carolin, *Telopea* 3: 506 (1990); isolecto: BM, MEL.

[*Molkenboeria semiamplexicaulis auct. non* (DC.) Vriese: W.H.de Vriese, *Natuurk. Verh. Holl. Maatsch.*

Wetensch. Haarlem ser. 2, 10: 41 (1854) *p.p.*]

Illustration: W.H.de Vriese, *op. cit.* t. 5, as *Moelkenboeria semiamplexicaulis*.

Ascending herb with stems to 70 cm long, with loosely, not appressed hairs. Leaves sessile, stem-clasping, elliptic to oblong, dentate; lamina 25–75 mm long, 15–35 mm wide. Flowers in loose racemes to 24 cm long; bracts leaf-like but smaller; bracteoles elliptic, 8–14 mm long. Sepals broadly ovate, to 0.5 mm long, basally connate. Corolla 15–27 mm long, with loose hairs outside, densely bearded inside, blue; barbulae papillate apically; wings 1–2.5 mm wide. Ovary 2-locular; indusium 2–3 mm long, pale brown beard exceeding bristles on lips. Fruit obovoid, 2.5–4 mm long, smooth, pubescent.

Occurs in and around the Porongurup Ra. south-western W.A., in woodland. Flowers Sept.–Jan. Map 167.

W.A.: Devils Slide, Porongurup Ra., *T.E.H.Aplin 2163* (PERTH); c. 50 km N of Albany, Dec. 1927, *W.E.Blackall* (PERTH); Mt Argwin, Porongurup Ra., *E.N.S.Jackson 3302* (AD).

Related to *S. microphylla* which has appressed hairs on the stems and corolla and often smaller bracteoles and floral parts. See also *S. macrophylla* and *S. platyphylla*.

54. Scaevola macrophylla (Vriese) Benth., *Fl. Austral.* 4: 98 (1868)

Molkenboeria macrophylla Vriese, *Natuurk. Verh. Holl. Maatsch. Wetensch. Haarlem* ser. 2, 10: 44 (1854); *Lobelia macrophylla* (Vriese) Kuntze, *Revis. Gen. Pl.* 2: 378 (1891), *nom. illeg.* T: Swan R., W.A., *J.Drummond 5: 362*; holo: K; iso: BM, MEL, P, W.

Illustration: W.H.de Vriese, *op. cit.* t. 8, as *Molkenboeria macrophylla*.

Erect herb to 40 cm tall, woody at base, with not appressed, yellowish hairs. Leaves ±stem-clasping, ovate to oblong, dentate; lamina 20–40 mm long, 10–20 mm wide. Flowers in dense racemes to 5 cm long; bracts leaf-like but smaller and less stem-clasping; bracteoles linear-elliptic, c. 8 mm long. Sepals broadly ovate, c. 0.5 mm long, basally connate. Corolla c. 22 mm long, pubescent outside, with yellowish beard inside, blue; barbulae papillate apically; wings to 2 mm wide. Ovary 2-locular; indusium 3 mm wide, long purplish beard exceeding yellowish bristles on lip. Fruit ovoid, c. 3 mm long, rugose, with few scattered hairs.

Occurs in the Cape Riche area, south-western W.A. Flowering not known. Map 168.

W.A.: Konkoberup promontory, Cape Riche, *coll. unknown* (MEL).

Probably closest to *S. auriculata* which has an ascending habit, loose inflorescences, broader leaves, and longer bracteoles about 1/2 as long as the flower. The beard on the back of the indusium is usually purplish in the present species but brownish in *S. auriculata* and *S. microphylla*.

55. Scaevola platyphylla Lindley, *Sketch Veg. Swan R.* xxvi (1839)

Molkenboeria platyphylla (Lindley) Vriese, *Natuurk. Verh. Holl. Maatsch. Wetensch. Haarlem* ser. 2, 10: 43 (1854); *Lobelia platyphylla* (Lindley) Kuntze, *Revis. Gen. Pl.* 2: 378 (1891). T: no precise locality, *coll. unknown*; holo: CBS; iso: ?K.

S. semiamplexicaulis DC., *Prodr.* 7(2): 509 (1839); *Molkenboeria semiamplexicaulis* (DC.) Vriese, *Natuurk. Verh. Holl. Maatsch. Wetensch. Haarlem* ser. 2, 10: 41 (1854) *p.p.* T: W.A., *J.Drummond 391*; holo: G-DC; iso: BM, K.

S. candollei Vriese in J.G.C.Lehmann, *Pl. Preiss.* 1: 405 (1845). T: Darling Ra., W.A., Sept. 1841, *L.Preiss 1497*; lecto: L, *fide* R.C.Carolin, *Telopea* 3: 507 (1990); isolecto: MEL.

Illustration: W.H.de Vriese, *op. cit.* t. 6, as *Molkenboeria platyphylla;* R.Erickson *et al*, *Fl. Pl. W. Australia*

45, t. 107 (1973).

Spreading shrub to 70 cm tall, usually with long, not appressed hairs. Leaves sessile, ±stem-clasping, elliptic, entire or dentate; lamina 24–63 mm long, 13–42 mm wide. Flowers in terminal racemes to 17 cm; bracts leaf-like but smaller; bracteoles elliptic, 20–30 mm long. Sepals ovate, to 1.2 mm long, basally connate. Corolla 25–40 mm long, with dense coarse closely appressed hairs outside, densely bearded inside, blue; barbulae simple; wings to 4 mm wide. Ovary 2-locular; indusium 2.5–3.5 mm long, almost obscured by dense, white beard at least equalling bristles on lips. Fruit cylindrical, 4–8 mm long, pubescent, smooth. $n = 8$, W.J.Peacock, *Proc. Linn. Soc. New South Wales* 88: 8 (1963).

Occurs in south-western W.A. in the Darling Ra. and nearby areas. Flowers Aug.–Dec. Map 169.

W.A.: Red Hill, Midland to Toodyay road, *T.E.H.Aplin 906* (PERTH); c. 3 km from Wongan Hills to Piawaning road, *F.Lullfitz L1654* (PERTH); Cut Hill, *C.A.Gardner 13575* (PERTH); Glen Forest, *T.E.H.Aplin 1047* (PERTH); Kelmscott, 7 Dec. 1907, *A.Morrison* (BRI).

Indusium compressed laterally. Similar to *S. auriculata* but the hairs on the outside of the corolla of that species are much looser and less dense. The bracteoles are generally longer in the present species being at least half as long as the flower whereas those of *S. auriculata* are less than half as long as the flower. Cytological voucher is *W.J.Peacock 60820.2* (SYD).

Subsect. 2. Pogonanthera

Scaevola sect. **Xerocarpa** subsect. **Pogonanthera** (G.Don) Carolin, *Telopea* 3: 490 (1990).

Scaevola sect. *Pogonanthera* G.Don, *Gen. Hist.* 3: 729 (1834); *Scaevola* sect. *Pogonandra* DC., *Prodr.* 7(2): 511 (1839), *nom. illeg.* T: *S. striata* R.Br.; lecto: *fide* R.C.Carolin, *Telopea* 3: 490 (1990).

Leaves usually well-developed. Flowers in few-flowered, terminal thyrses or racemes; peduncle well-developed; pedicel obsolete. Anthers hairy at apex (except *S. hookeri*). Ovary 2-locular; indusium hairy on back, usually with dense beard not equalling lips.

Seven species endemic in Australia.

56. Scaevola ramosissima (Smith) K.Krause, *Pflanzenr.* 54: 141 (1912)

Goodenia ramosissima Smith, *Spec. Bot. New Holland* 15, 65 (1793); *Trans. Linn. Soc. London, Bot.* 349 (1794). T: New South Wales, 1792, *J.White*, Smithian Herb.; holo: LINN; iso: K.

S. hispida Cav., *Anales Hist. Nat.* 1(2): 99 (1799); *Merkusia hispida* (Cav.) Vriese, *Ned. Kruidk. Arch.* 2: 162 (1851); *Lobelia hispida* (Cav.) Kuntze, *Revis. Gen. Pl.* 2: 378 (1891). T: between Port Jackson and Paramatta, N.S.W., Apr. 1793, L.Née; *n.v.*

S. apterantha F.Muell., *Fragm.* 1: 121 (1859); *Lobelia apterantha* (F.Muell.) Kuntze, *Revis. Gen. Pl.* 2: 378 (1891); *S. ramosissima* var. *apterantha* (F.Muell.) K.Krause, *Pflanzenr.* 54: 14 (1912). T: ranges beyond the Snowy River, [N.S.W.?], Jan. 1855, *F.Mueller*; lecto: MEL, *fide* R.C.Carolin, *Telopea* 3: 493 (1990); isolecto: MEL.

Illustrations: A.J.Cavanilles, *Icon. Pl.* 6: 7, t. 510 (1801); G.R.Cochrane *et al.*, *Fl. Pl. Victoria* 152, fig. 475 (1968).

Decumbent to ascending herb to 40 cm tall, hispid, with simple and glandular hairs. Leaves sessile, linear to oblanceolate, entire or dentate; lamina 2–10 cm long, 2–10 mm wide. Flowers in thyrses or racemes to 30 cm long; bracts leafy; peduncle to 10 cm long; bracteoles linear, to 2.5 cm long, entire, exceeding ovary and usually sepals. Sepals linear,

5 mm long, free. Corolla 1.5–2.5 cm long, hispid outside with hairs at 90°, usually bearded in throat, pale violet to purple; barbulae simple or clavate; wings 1–3 mm wide, not striate. Indusium 2.5–3 mm wide, short white beard above. Fruit ellipsoidal, c. 5 mm long, striate-rugose, hairy $n = 8$, W.J.Peacock, *Proc. Linn. Soc. New South Wales* 88: 8 (1963). Figs 38A, 43G–J.

Occurs in eastern Australia from the Blackdown Tableland, Qld, through N.S.W. to Vic., mostly E of the Great Dividing Ra.; grows in heath and open forest. Flowers chiefly Aug.–Mar. Map 170.

Qld: c. 6.41 km N from Mimosa Ck, Blackdown Tableland, *K.A.W.Williams 74055* (BRI). N.S.W.: Kremnos Ck, c. 13 km N of Glenreagh, *K.Grieves 88635* (NSW); Wahroonga, *H.Salasoo 623* (NSW); Agnes Banks, *R.C.Carolin 3653* (SYD). Vic.: Mallacoota–Wingan coast NE of Bendock River mouth, *A.C.Beauglehole 31079* (MEL).

Differs from *S. pilosa* which has broader, dentate bracteoles and auriculate stem leaves, and occurs in W.A. Cytological voucher is *W.J.Peacock 60950.1* (SYD).

57. **Scaevola hookeri** (Vriese) F.Muell. ex J.D.Hook., *Fl. Tasman.* 1: 231 (1856)

Merkusia hookeri Vriese, *Ned. Kruidk. Arch.* 2: 159 (1851); *Lobelia hookeri* (Vriese) Kuntze, *Revis. Gen. Pl.* 2: 378 (1891). T: Hampshire Hills, Tas., Feb. 1837, *R.C.Gunn 848*; lecto: K, *fide* R.C.Carolin, *Fl. Australia* 35: 333 (1992).

Illustrations: W.H.de Vriese, *Natuurk. Verh. Holl. Maatsch. Wetensch. Haarlem* ser. 2, 10: t. 12 (1854) as *Merkusia hookeri*; J.D.Hooker, *Fl. Tasman.* 1: t. 67 (1861); G.R.Cochrane *et al.*, *Fl. Pl. Victoria* 167, fig. 530 (1968).

Prostrate, stoloniferous herb, often forming mats, with not appressed hairs; stolons to 30 cm long, rooting at nodes. Leaves sessile, ovate to oblong, usually entire; lamina 6–50 mm long, 2–15 mm wide. Flowers in racemes mostly to 10 cm long; bracts leafy; peduncle to 8 mm long; bracteoles elliptic to oblong, 4–6 mm long, entire or dentate. Sepals minute or obsolete. Corolla 5–8 mm long, with not appressed hairs outside, sparsely pubescent inside, white or blue sometimes yellowish in throat; barbulae simple; wings to 1 mm wide, not striate. Indusium to 2 mm wide, with some stiff hairs above and lips ±glabrous. Fruit obovoid, to 2.5 mm long, rugose, pubescent, often 1-seeded. Fig. 56.

Occurs in the eastern highlands from northern N.S.W. to Vic. and Tas., usually at higher altitudes in damp situations. Flowers Dec.–Mar. Map 171.

N.S.W.: Allyn R. to Barrington Tops, Jan. 1928, *M.Fuller* (SYD); c. 3 km above Tuross Falls, *A.Rodd 623* (NSW); Upper Spencers Ck, N of Johnnys Plains, *J.Thompson 406* (NSW). Vic.: Three Mile Ck, near Mt Wellington, Jan. 1949, *J.M.Whaite* (MEL, NSW). Tas.: Lake Dobson, Mt Field Natl Park, *Hj.Eichler 16708* (AD).

Notable in this subsection for its obtuse, glabrous anthers, and smaller flowers. The indusium is depressed-obovate. Sometimes confused with *S. albida* which has spicate inflorescences and an indusium with a short white or purple beard.

58. **Scaevola pilosa** Benth. in S.L.Endlicher *et al.*, *Enum. Pl.* 69 (1837)

Molkenboeria pilosa (Benth.) Vriese, *Natuurk. Verh. Holl. Maatsch. Wetensch. Haarlem* ser. 2, 10: 39 (1854); *Lobelia pilosa* (Benth.) Kuntze, *Revis. Gen. Pl.* 2: 378 (1891). T: Swan River, W.A., *C.A.Hügel*; holo: W; iso: K.

S. membranacea Benth. in S.L.Endlicher, *et al.*, *Enum. Pl.* 69 (1837); *Molkenboeria membranacea* (Benth.) Vriese, *Natuurk. Verh. Holl. Maatsch. Wetensch. Haarlem* ser. 2, 10: 40 (1854); *S. pilosa* var. *membranacea* (Benth.) K.Krause, *Pflanzenr.* 54: 141 (1912). T: Swan River, W.A., *C.A.Hügel*; holo: W.

Illustrations: K.Krause, *Pflanzenr.* 54: 140, fig. 27E–H (1912); W.H.de Vriese, *op. cit.* t. 7, as *Molkenboeria pilosa*.

Ascending to decumbent herb to 70 cm high, hispid, with simple hairs to 1 mm long at 90°, minute, simple hairs and minute, glandular hairs. Lower leaves spathulate, dentate near apex, with lamina 1.5–7.5 cm long, 0.5–3 cm wide; upper leaves sessile, auriculate, smaller. Flowers in racemes to 50 cm long; bracts leafy; peduncle 2–6 cm long; bracteoles ovate to linear-ovate, 1–2.5 cm long, dentate. Sepals orbicular, to 2 mm long, free. Corolla 10–25 mm long, with not appressed, white hairs outside, bearded inside, blue to mauve; barbulae simple; wings to 2 mm wide, not striate. Indusium c. 2.5 mm wide, dense short beard above. Fruit ellipsoidal, c. 5 mm long, smooth, pubescent. $n = 8$, W.J.Peacock, *Proc. Linn. Soc. New South Wales* 88: 8 (1963). Fig. 38B.

Occurs in the south of W.A., from Jurien Bay to Cape Leeuwin, with a record in the central wheatbelt. Flowers chiefly Sept.–Dec. Map 172.

W.A.: Cockleshell Gully, NE of Jurien, *A.E.Orchard 4245* (AD, CANB); between Merredin and Southern Cross, Sept. 1929, *W.E.Blackall* (PERTH); Mundaring, *C.A.Gardner 1198* (PERTH); Kelmscott, *A.Morrison 17099* (PERTH); 10 km NE of Yarloop, *R.J.Cranfield 947* (PERTH).

Similar to *S. ramosissima* and *S. striata*. Cytological voucher is *W.J.Peacock 60884.1* (SYD).

59. Scaevola tenuifolia Carolin, *Telopea* 3: 492 (1990)

T: foot of East Mount Barren, W.A., 26 Nov. 1931, *W.E.Blackall 1417*; holo: PERTH.

Decumbent to prostrate herb, to 1 m diam., hispid with short, stiff, simple hairs at 90°, and small, red, glandular ones. Leaves sessile, linear, revolute; lamina 7–45 mm long, 1–4 mm wide. Flowers in racemes or thyrses to 20 cm long; bracts leafy; peduncle curved, 10–50 mm long; bracteoles linear-triangular, 2–8 mm long, entire. Sepals linear, c. 2 mm long, free. Corolla 12–20 mm long, with long, white hairs at 90° outside, densely bearded inside, blue to mauve; barbulae simple, short; wings triangular, 2 mm wide, not striate. Indusium c. 2 mm wide, with dense beard. Fruit ellipsoidal, c. 4 mm long, ribbed, hispid. Fig. 38C.

Occurs on East Mount Barren and the Thumb Peak range, W.A. Flowers Aug.–Jan. Map 173.

W.A.: SW and E slopes of E Mt Barren, 14 Oct. 1961, *J.H.Willis* (MEL); Thumb Peak range, *A.S.George 7117* (PERTH).

Similar to *S. striata* which has straight peduncles, stem hairs which are at an acute angle, and striate petal wings. The narrower leaves distinguish it from *S. pilosa*.

60. Scaevola striata R.Br., *Prodr.* 586 (1810)

[*S. stricta* R.Br. ex Roemer & Schultes, *Syst. Veg.* 5: 167 (1819), *orth. var.*]; *Molkenboeria striata* (R.Br.) Vriese, *Natuurk. Verh. Holl. Maatsch. Wetensch. Haarlem* ser. 2, 10: 42 (1854); *Lobelia striata* (R.Br.) Kuntze, *Revis. Gen. Pl.* 2: 378 (1891). T: King George Sound, [W.A.], Dec. 1801, *R.Brown*; lecto: BM, *fide* R.C.Carolin, *Telopea* 3: 493 (1990); isolecto: K, MEL.

S. macrodonta DC., *Prodr.* 7(2): 511 (1839). T: King George Sound, [W.A.], *A.Cunningham*; holo: G-DC; iso: K.

[*Baudinia humilis* Lesch. ex DC., *Prodr.* 7: 511 (1839), *nom. inval.*]

S. prostrata Vriese in J.G.C.Lehmann, *Pl. Preiss.* 1: 406 (1845). T: near Middleton Beach, [Albany], W.A., 23 Sept. 1840, *L.Preiss 1490*; lecto: LD, *fide* R.C.Carolin, *Telopea* 3: 493 (1990); isolecto: L.

S. striata var. *depauperata* E.Pritzel in F.L.E.Diels & E.Pritzel, *Bot. Jahrb. Syst.* 35: 569 (1905). T: near Albany, W.A., *L.Diels 2283*; holo: ?B (destroyed) *n.v.*

Ascending to prostrate herb to 20 cm tall, hispid. Leaves sessile, ovate to linear, dentate or serrate; lamina 1–5.5 cm long, 3–20 mm wide, dentate. Flowers in racemes or thyrses to 30 cm long; bracts leafy; peduncle to 8 cm long; bracteoles elliptic to obovate, 10–25 mm long, entire. Sepals linear, 3–8 mm long, free. Corolla 13–27 mm long, with sparse hairs outside, bearded inside, blue to purplish; barbulae simple, short; wings 2–6 mm wide, striate. Indusium c. 3.5 mm wide, with yellowish beard. Fruit ellipsoidal, to 5 mm long, tuberculate, with long, simple, white hairs and minute, glandular ones.

Occurs in southern W.A. The pronounced nerves in the corolla wings and the entire bracteoles of *S. striata* separate it from *S. pilosa*. See also *S. tenuifolia*, *S. calliptera* and *S. phlebopetala*. Two varieties are recognised.

Hairs on stem antrorse to appressed **60a.** var. **striata**

Hairs on stem not appressed, stiff **60b.** var. **arenaria**

60a. Scaevola striata R.Br. var. striata

Stems with antrorse to antrorse-appressed hairs to 1 mm long. Cauline leaves mostly ovate to narrowly oblong. Bracteoles elliptic to narrowly oblong. *Royal Robe*. Figs 38D, 59.

Occurs along the southern coast of W.A. from West Mount Barren westwards to Denmark. Flowers chiefly Aug.–Dec. Map 174.

W.A.: c. 5 km S of Mt Barker, *Hj.Eichler 15980* (AD); Porongurup Ra., *W.E.Blackall* (PERTH); between Middleton Beach and King R., *D.J.E.Whibley 5137* (AD, PERTH); c. 10 km SW of Walpole, *J.Green 960* (PERTH); Denmark, *D.H.Perry* (PERTH).

60b. Scaevola striata var. arenaria E.Pritzel in F.L.E.Diels & E.Pritzel, *Bot. Jahrb. Syst.* 35: 569 (1905)

T: near Warrungup, Stirling Ra., W.A., Oct. 1901, *L.Diels 4945*; holo: ?B (destroyed) *n.v.*; neotype: 1 mile [1.6 km] SE of Kukerin, 16 Oct. 1964, *K.Newbey 1509*; neo: PERTH, *fide* R.C.Carolin, *Telopea* 3: 493 (1990).

Illustration: K.Krause, *Pflanzenr.* 54: 140, fig. 27A–D (1912) as *S. striata*.

Stems densely hispid with long, stiff hairs, dense, minute, simple hairs and minute, glandular hairs. Cauline leaves oblanceolate to obovate, entire. Bracteoles obovate.

Occurs from the Stirling Ra. to the Lake Grace area, south-western W.A., on sand plains. Flowers chiefly Aug. Dec. Map 175.

W.A.: c. 1.5 km SE of Kukerin, *K.Newbey 995* (PERTH); 152 mile peg [c. 244 km], Albany road, *A.R.Fairall 378* (PERTH).

61. Scaevola calliptera Benth. in S.F.L.Endlicher *et al.*, *Enum. Pl.* (1837)

T: Swan River, W.A., *C.A.Hügel*; holo: W; iso: K.

S. macropoda DC., *Prodr.* 7(2): 509 (1839). T: Swan River, W.A., *J.Drummond*; holo: G-DC; iso: K.

S. benthamea Vriese in J.G.C.Lehmann, *Pl. Preiss.* 1: 411 (1845). T: Canning R., near Perth, W.A., 2 Nov. 1839, *L.Preiss 1520*; lecto: W, *fide* R.C.Carolin, *Telopea* 3: 493 (1990); isolecto: K, MEL, P.

[*Scaevola striata* auct. non R.Br.: R.Erickson *et al.*, *Fl. Pl. W. Australia* 44, t. 101 (1973)]

Erect herb to 40 cm tall, with glandular hairs and long or short, simple, hairs at 90°. Leaves sessile, oblanceolate to narrowly oblong, dentate; lamina 2–6 cm long, 4–25 mm wide. Flowers in racemes or thyrses to 30 cm long; bracts leafy; peduncle 2–7.5 cm long;

bracteoles elliptic to narrowly oblong, 10–35 mm long, dentate. Sepals linear, 6–16 mm long, free. Corolla 17–30 mm long, with long hairs outside, densely bearded inside, blue to purple with yellow throat; barbulae simple; wings striate, 3–5 mm wide. Indusium c. 1 mm wide, with short, yellowish beard, with white or occasionally purple hairs below. Fruit ellipsoidal, to 6 mm long, tuberculate, with long and minute hairs. *Royal Robe*. Figs 38E, 61.

Occurs in south-western W.A. from Bullsbrook to the south coast, in forest, woodland and heath. Flowers chiefly Aug.–Dec. Map 176.

W.A.: Parkerville, *F.Lullfitz L1684* (PERTH); near crossing of Aschendon road, Canning R. *A.E.Orchard 4288* (AD, CANB); c. 15 km NE of Perth, Darling Ra. *M.Koch 1574* (AD); 48 km S of Busselton, *D.J.E.Whibley 3270* (AD); Capel to Donnybrook road, *R.J.Cranfield 916* (MEL, PERTH).

This species was previously united with *S. striata* which typically has appressed antrorse hairs on the stems. See also *S. phlebopetala*.

62. Scaevola phlebopetala F.Muell., *Fragm.* 2: 18 (1860)

Lobelia phlebopetala (F.Muell.) Kuntze, *Revis. Gen. Pl.* 2: 378 (1891). T: Murchison R., *A.Oldfield*; lecto: MEL, *fide* R.C.Carolin, *Telopea* 3: 493 (1990); isolecto: K.

S. phlebopetala β *foliosa* E.Pritzel, *loc. cit.* T: near Coorow, W.A., July 1901, *L.Diels 3322*; holo: ?B (destroyed) *n.v.*

S. phlebopetala γ *subaphylla* E.Pritzel, *loc. cit.* T: near Dongara, W.A., Dec. 1901, *L.Diels 5725*; holo: ?B (destroyed) *n.v.*

Illustration: K.Krause, *Pflanzenr.* 54: 140, fig. 27J–M (1912); R.Erickson *et al.*, *Fl. Pl. W. Australia* 98, t. 284 (1973).

Herb, often prostrate with stems to 50 cm long, hispid, sometimes scabrous, hairs at 90°. Leaves sessile, cuneate to linear, sometimes recurved, usually dentate; lamina 1.2–10 cm long, 3–17 mm wide. Flowers in racemes or thyrses to 30 cm long; bracts leafy; peduncle curved, 2–4.5 cm long; bracteoles linear, 5–10 mm long, entire. Sepals linear, 4–14 mm long, free. Corolla mostly 10–27 mm long, with short, white hairs and long, stiff, yellow hairs outside, densely bearded inside, deep purple, yellow in throat; barbulae simple; wings striate, c. 2–3 mm wide. Indusium c. 3 mm wide, with sparse, sometimes purplish beard. Fruit obovoid, 5–6 mm long, striate-tuberculate, hairy. $n = 8$, W.J.Peacock, *Proc. Linn. Soc. New South Wales* 88: 8 (1963). Figs 38F, 60.

Occurs in south-western W.A. from the Murchison R. to the Perth area, in sandy heath. Flowers June–Oct. Map 177.

W.A.: Murchison R., *A.C.Burns 1028* (PERTH); Kalbarri Natl Park, *R.D.Royce 7781* (PERTH); between Northampton and Lynton, *W.E.Blackall 2683* (PERTH); Irwin R., *F.W.Went 233* (PERTH); Brand Hwy, 31 km N of junction with road between Gingin and Wanneroo road, *P.Weston 269* (SYD).

Related to *S. striata* and *S. calliptera* from which it is distinguished by its linear bracteoles and deeper corolla colour. Cytological voucher is *W.J.Peacock 60886.2* (SYD).

Subsect. 3. Parvifoliae

Scaevola sect. **Xerocarpa** subsect. **Parvifoliae** Carolin, *Telopea* 3: 490 (1990).

Type: *S. parvifolia* F.Muell. ex Benth.

Cauline leaves usually reduced. Flowers in terminal thyrses, racemes or spikes, usually pedunculate; pedicel 3.5 mm long to obsolete. Anthers glabrous at tip (except *S.*

depauperata). Ovary 2-locular; indusium often hairy on back, with a beard of stiff basal hairs.

Nine species endemic in Australia.

63. Scaevola hamiltonii K.Krause, *Pflanzenr.* 54: 153 (1912)

T: near Standort, W.A., *A.A.Hamilton*; lecto: NSW, *fide* R.C.Carolin, *Telopea* 3: 493 (1990); isolecto: ?B (destroyed) *n.v.*

Stiff, erect perennial, with many striate stems, to 45 cm tall, with long, stiff, simple hairs at 90° or curved, and minute, glandular hairs. Leaves ±sessile or shortly petiolate, linear to oblanceolate, sometimes dentate; lamina 0.5–8 cm long, 1–10 mm wide. Flowers in distant spikes or thyrses to 15 cm long; bracts leaf-like but smaller; peduncle to 5 mm long; bracteoles linear, 3.5–5 mm long. Sepals triangular, 1 mm long, free. Corolla 7–17 mm long, with not appressed, yellowish brown, simple and glandular hairs outside, bearded inside, white to lilac; barbulae simple; wings 1.5 mm wide. Indusium c. 2 mm wide, glabrous or with few long hairs at base. Fruit globular, c. 3 mm diam., 5-ribbed, pubescent.

Occurs in the south of W.A., from the Murchison R. south to Cadoux and inland to Lake Moore. Flowers Aug.–Dec. Map 178.

W.A.: c. 11 km N of Murchison R. on North West Coastal Hwy, *A.S.George 7872* (PERTH); East Yuna Reserve, NE of Geraldton, *A.C.Burns 64* (PERTH); c. 3 km W of Pindar, *C.A.Gardner 7771* (PERTH); c. 6 km from Mullewa towards Pindar, *M.E.Phillips 038980* (CBG, SYD); Lake Moore, *A.Robinson 282* (PERTH).

Compared with other members of the subsection *Parvifoliae*, the cauline leaves of this species are mostly larger and the basal leaves more persistent. The beard on the indusium is absent or represented by a very few hairs. This species also shows some relationship to members of subsection *Biloculatae* ser. *Pogogynae*, but its yellowish hairs will distinguish it.

64. Scaevola basedowii Carolin, *Telopea* 2: 73 (1980)

T: Mt Unapproachable, S.A., 1 Jul. 1926, *H.Basedow 134*; holo: K (does not include the sheet marked 'Sht 2.').

Stiff, erect perennial, many-stemmed, to 60 cm tall, with minute, glandular hairs and long, stiff, simple hairs; stems striate. Basal leaves sessile, narrowly elliptic, entire, with lamina to 22 mm long, to 6 mm wide; cauline leaves ±triangular, with lamina to 4 mm long. Flowers in distant thyrses or racemes to 15 cm long; bracts as cauline leaves; peduncle 1.7–5.5 cm long; bracteoles ovate-elliptic, 1–2 mm long. Sepals triangular, 0.2–0.35 mm long, basally connate. Corolla 15–25 mm long, with stiff, ±appressed hairs outside, bearded inside, pale mauve with darker lines; barbulae thin, simple; wings c. 1 mm wide. Indusium c. 3 mm wide, dense beard ±equalling bristles on lips. Fruit globular to ellipsoidal, 5 mm long, tuberculate, ribbed, hairy.

Occurs in central Australia in W.A., N.T. and S.A., from the central ranges to the Gibson and Great Victoria Deserts on red sand dunes or desert flats, often with *Triodia basedowii* and *Allocasuarina decaisneana*. Flowers May–Sept. Map 179.

W.A.: c. 80 km NE of Cosmo Newberry, *A.S.George 8114* (PERTH); Great Victoria Desert, Aug. 1960, *A.R.Main* (PERTH). N.T.: c. 44 km W of Mt Olga, *M.Lazarides 8293* (CANB, DNA); 54 km SSW of The Granites, *S.Parker 27886* (AD, DNA, NSW). S.A.: 65 km W of Musgrave Park, *D.J.E.Whibley 971* (AD).

Sepals connate for at least 1/3 their length. Similar to *S. depauperata* which has few long hairs on the ovary and almost glabrous striate stems. Also related to *S. parvifolia* which has longer calyx lobes and mostly simple hairs on the stems. The indumentum of the stems is quite variable in density and in a very few specimens the stems are almost glabrous. R.C.Carolin, *op. cit.* 74, compared this species and its close relatives.

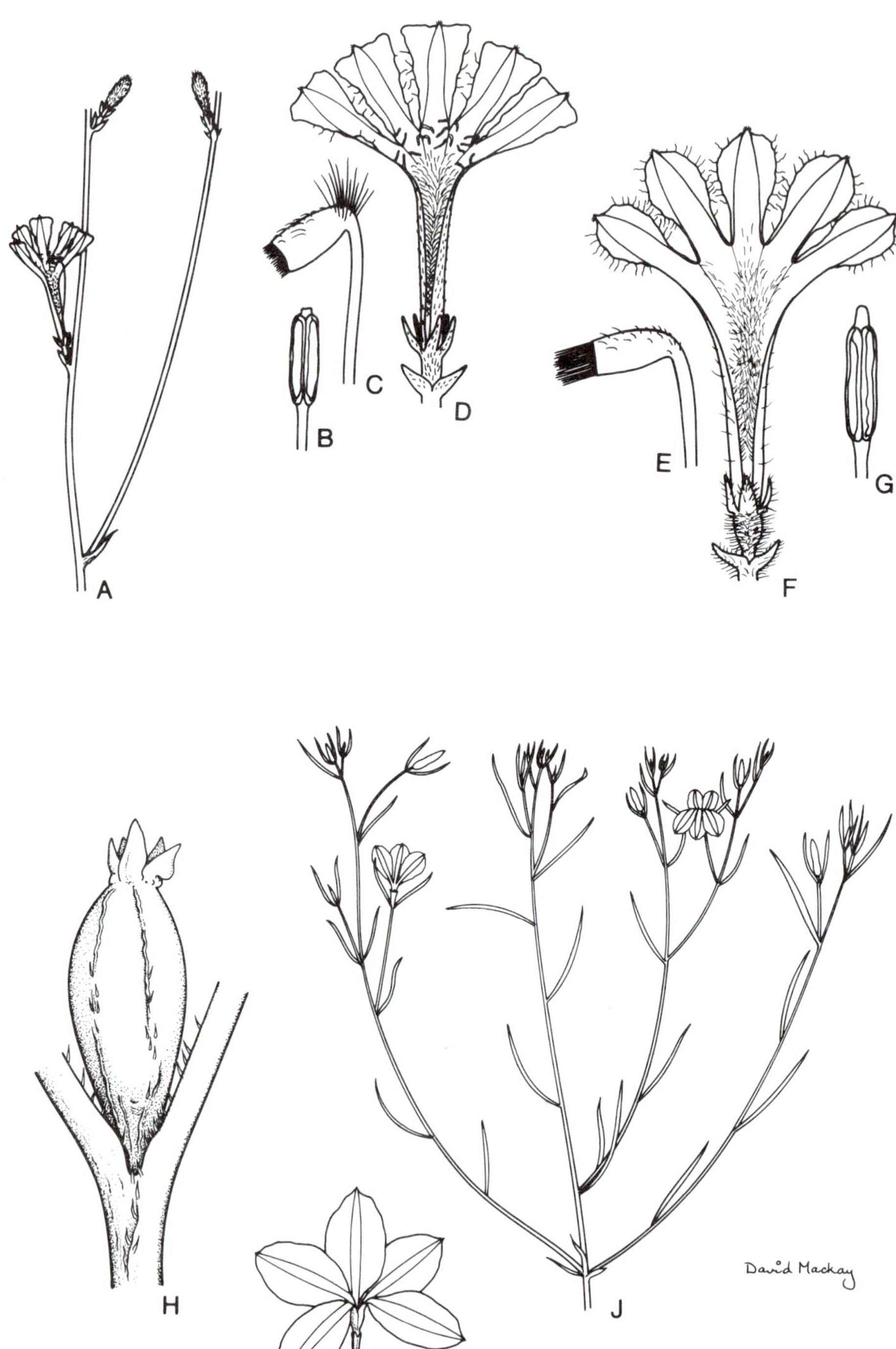

Figure 44. **A–G** *Scaevola*. **A–D**, *S. restiacea* subsp. *restiacea*. **A**, habit X1; **B**, anther X10; **C**, indusium X7.5; **D**, corolla X2.5 (**A–D**, M.Phillips, 10 Sept. 1968, SYD). **E–G**, *S. parvifolia*. **E**, indusium X7.5; **F**, corolla X2.5; **G**, anther X10 (**E–G**, voucher not recorded). **H–J**, *Diaspasis filifolia*. **H**, fruit X10 (S.Carlquist 3987, NSW), **I**, flower X1.5; **J**, habit X0.5 (**I–J**, J.Wrigley, 12 Oct. 1968, SYD). Drawn by D.Mackay.

65. Scaevola parvifolia F.Muell. ex Benth., *Fl. Austral.* 4: 91 (1868)

Lobelia parvifolia (F.Muell. ex Benth.) Kuntze, *Revis. Gen. Pl.* 2: 378 (1891). T: Hooker Creek, [N.T.], *F.Mueller*; holo: K.

S. parvifolia var. *brevifolia* F.Muell., *J. Bot.* 15: 304 (1877). T: [Queen] Victoria Spring, W.A., *Birch*; lecto: MEL, fide R.C.Carolin, *Fl. Australia* 35: 333 (1992).

S. daleana Blakely, *Austral. Naturalist* 11: 11 (1941). T: Connors Well, N.T., *B.A.Dale* per *J.Buckingham*; holo: NSW.

Stiff, erect, many-stemmed perennial to 60 cm tall, with hairs at 90°; stems scarcely striate. Basal leaves sessile, linear to lanceolate, entire, with lamina 18–35 mm long, 3–6 mm wide; cauline leaves ovate to linear, with lamina 1.5–27 mm long. Flowers in thyrses to 40 cm long; bracts leafy; peduncle to 6.5 cm long; bracteoles ovate to linear, 1–4 mm long. Sepals triangular, 1–3 mm long, free or connate basally. Corolla 13–32 mm long, with not appressed hairs outside, bearded inside, blue to white; barbulae simple; wings 0.5–2 mm wide. Indusium c. 2 mm long, with stiff beard exceeding or equalling bristles on lips. Fruit ellipsoidal, 4–8 mm long, striate, tuberculate, hairy. Fig. 44E–G.

Widespread in arid north-western and central Australia. Pedicels to 1.5 mm long. Related to *S. depauperata* which has almost glabrous distinctly striate stems. See also *S. basedowii*. Three subspecies are recognised.

1 Corolla obtuse or acute in bud

2 Sepals with glandular and simple hairs **65a.** subsp. **parvifolia**

2: Sepals with glandular hairs only **65b.** subsp. **pilbarae**

1: Corolla acuminate in bud **65c.** subsp. **acuminata**

65a. Scaevola parvifolia F.Muell. ex Benth. subsp. parvifolia

Stems with tortuous hairs almost at 90°, sometimes with a dense sub-indumentum of glandular and simple hairs. Basal leaves linear to lanceolate, with lamina to 35 mm long; cauline leaves broadly ovate to narrowly elliptic, with lamina to 1.5–15 mm long, 1–4 mm wide. Bracteoles ovate, mostly c. 1 mm long. Sepals 1–2 mm long, with glandular and simple hairs. Corolla with glandular and/or stiff hairs at 90° outside, obtuse or acute in bud.

Occurs in the drier areas of Australia in W.A., N.T., S.A. and Qld, generally between 17°S and 28°S. Flowers chiefly May–Sept. Map 180.

W.A.: 40 km S of Broome, *A.C.Beauglehole 11289* (PERTH, SYD); 37 km S of Halls Creek, *R.C.Carolin 7916* (SYD). N.T.: c. 3 km S of Wauchope, *R.E.Winkworth 83088* (AD, BRI, DNA, NSW); c. 16 km S of Alice Springs, *D.J.Nelson 9434* (AD, DNA). Qld: Oban Stn, *S.L.Everist 1714* (BRI).

Specimens from between Port Hedland and Broome have short, arcuate, harsh hairs while those from a little further south are densely covered with both long and short, simple hairs.

65b. Scaevola parvifolia subsp. pilbarae Carolin, *Telopea* 3: 494 (1990)

T: 50 km S of Rudall River, W.A., 15 Aug. 1971, *P.G.Wilson 10545*; holo: PERTH.

Stems with long, simple hairs at 90° and minute, simple and glandular hairs. Basal leaves lanceolate, with lamina to 2 cm long; cauline ones ovate to lanceolate, with lamina 3–20 mm long. Bracteoles triangular, 1.5–3 mm long. Sepals 2–3 mm long, glandular-hairy. Corolla densely glandular-hairy outside, obtuse or acute in bud.

Occurs in the Pilbara region and into the Great Sandy Desert, W.A. Flowers Mar.–Sept. Map 181.

W.A.: c. 90 km from Port Hedland on Broome road, *R.C.Carolin 7610* (SYD); Hamersley Ra., *M.Cole 5026* (PERTH); S Mt Hodgson, *G.Davies 174* (PERTH); c. 110 km N of Sandstone towards Wiluna, *R.D.Royce 10365* (PERTH).

65c. **Scaevola parvifolia** subsp. **acuminata** Carolin, *Telopea* 3: 495 (1990)

T: 25 km E of Depot Springs, W.A., 27 Aug. 1970, *R.A.Saffrey 1058*; holo: PERTH.

Stems hispid with glandular and simple hairs. Basal leaves linear, 20–35 mm long, with lamina c. 5 mm wide; cauline ones linear to lanceolate, with lamina 8–27 mm long. Bracteoles ovate to lanceolate, 1.5–4 mm long. Sepals 2–3 mm long with minute, glandular and long, simple hairs. Corolla glandular and long-simple hairy, acuminate in bud.

Occurs in drier inland areas of southern W.A. including the Great Victoria Desert. Flowers Mar.–Sept. Map 182.

W.A.: c. 19 km S of Mt Magnet, *H.Demarz 5237* (PERTH); c. 76 km S of Neale Junction, *A.S.George 11948* (PERTH); c. 6 km S of Cundeelee Mission, 13 Oct. 1963, *M.C.George* (PERTH); 50 km E of Southern Cross, Sept. 1927, *C.A.Gardner & W.E.Blackall* (PERTH); c. 150 km from Yelma on Leonora road, *R.C.Carolin 5894* (SYD).

66. **Scaevola depauperata** R.Br. in C.Sturt, *Narr. Exped. C. Australia* 2: *(Bot. Appendix)* 83 (1849)

Merkusia depauperata (R.Br.) Vriese, *Natuurk. Verh. Holl. Maatsch. Wetensch. Haarlem* ser. 2, 10: 74 (1854); *Lobelia depauperata* (R.Br.) Kuntze, *Revis. Gen. Pl.* 2: 378 (1891). T: 'in salt ground, in lat. 26°S', S.A., *C.Sturt*; holo: BM.

S. patens F.Muell., *Fragm.* 3: 33 (1862). T: Coopers Creek, [S.A.?], *W.F.Wheeler*; holo: MEL.

Illustration: G.M.Cunningham *et al.*, *Pl. W. New South Wales* 637 (1981).

Stiff, erect perennial, many-stemmed, to 1 m tall, with minute, glandular hairs and simple hairs; stems striate, almost glabrous. Basal leaves sessile, ovate-spathulate, dentate, with lamina to 60 mm long and 15 mm wide; cauline leaves ±triangular, with lamina to 3 mm long. Flowers in thyrses or racemes to 20 cm long; bracts as cauline leaves; peduncle to 6 cm long; bracteoles triangular, 2–4 mm long. Sepals triangular, 1–4 mm long, connate basally. Corolla 20–30 mm long, puberulous outside, densely bearded inside, cream to blue; barbulae thin, simple; wings to 1 mm wide. Indusium c. 2.5 mm wide, beard ±equalling bristles on lips, hairy below. Fruit ellipsoidal, 5–6 mm long, tuberculate; with minute, glandular hairs and simple hairs, occasionally longer, simple.

Occurs in arid areas south of 20°S of all mainland States except W.A., growing in sandy soil. Flowers chiefly Apr.–Dec. Map 183.

N.T.: c. 24 km N of Andado homestead, *J.R.Maconochie 13476* (AD, DNA). S.A.: Mokari Bore, Simpson Desert, *D.E.Symon 9460* (AD). Qld: c. 75 km NNW of Annadale Stn, 26 June 1939, *R.L.Crocker* (AD). N.S.W.: Olive Downs, *P.L.Milthorpe 879* (AD, NSW). Vic.: Hattah Lakes Natl Park, *A.C.Beauglehole 1143* (MEL).

The pedicels are to 3.5 mm long. Anthers sometimes have hairs at the tip. Distinguished from *S. parvifolia* and *S. basedowii* by the mostly glabrous stems.

67. **Scaevola restiacea** Benth., *Fl. Austral.* 4: 91 (1868)

Lobelia restiacea (Benth.) Kuntze, *Revis. Gen. Pl.* 2: 378 (1891). T: south-western W.A., 1849, *J.Drummond 5: 169*; holo: K.

Erect or divaricate subshrub to 50 cm tall; stems striate usually glabrous or glabrescent, rarely with appressed, simple hairs and glandular hairs at 90°. Basal leaves ephemeral; cauline leaves triangular, with lamina 1–2 mm long. Flowers in distant spikes to 20 cm long; bracts and bracteoles similar to leaves; peduncle ±obsolete. Sepals linear, 1–2 mm long, free or connate basally. Corolla 10–17 mm long, with long, appressed hairs outside, densely bearded inside, white or pale blue; barbulae simple or capitate; wings c. 3.5 mm wide, ±ciliate. Indusium 1.5 mm wide, sparse white beard ±equalling lips. Fruit ovoid, 3–5 mm long, tuberculate; hairs coarse, arcuate, golden.

Occurs in southern inland W.A. Distinguished from other members of this subsection by the coarse golden hairs on the ovary and fruit. Two subspecies are recognised.

Lateral stems not spinescent **67a.** subsp. **restiacea**

Some lateral stems spinescent **67b.** subsp. **divaricata**

67a. **Scaevola restiacea** Benth. subsp. **restiacea**

Erect, many-stemmed shrub, with branches not spinescent. Sepals obtuse. n = 8, W.J.Peacock, *Proc. Linn. Soc. New South Wales* 88: 8 (1963). Fig. 44A–D.

Occurs in southern W.A., between 116°E and 122°E, and 29°S and 33°S. Flowers chiefly Aug.–Dec. Map 184.

W.A.: c. 8 km N of Merredin, *S.Carlquist 3664* (NSW, RSA); 6 km SW of Lake Cronin, *K.Newbey 6220* (PERTH); Bencubbin, *E.Wittwer 1218* (PERTH); near Yellowdine, *C.A.Gardner 8051* (PERTH); Muntadgin, *T.W.Stone 896* (PERTH).

Cytological voucher is *W.J.Peacock 60814.1* (SYD).

67b. **Scaevola restiacea** subsp. **divaricata** Carolin, *Fl. Australia* 35: 333 (1992)

T: 15 miles [c. 24 km] E of Zanthus, W.A., 3 Oct. 1956, *R.D.Royce 5576*; holo: PERTH.

Divaricate subshrub, with many of the lateral branches after 1 or 2 nodes spinescent. Sepals acute.

Occurs between Spargoville and Zanthus, W.A. Flowers chiefly Sept.–Dec. Map 185.

W.A.: Ponton Ck, N of Zanthus, *A.S.George 5989* (PERTH); c. 0.5 km N of Spargoville on Norseman–Coolgardie road, *S.Carlquist 3372* (AD, NSW, RSA); c. 3 km N of Zanthus *R.D.Royce 5458* (PERTH).

Similar to *S. oxyclona*.

68. **Scaevola chrysopogon** Carolin, *Telopea* 3: 494 (1990)

T: 16 miles [c. 25 km] S of Wannoo roadhouse, W.A., North West Coastal Hwy, 9 Sept. 1970, *A.S.George 10367*; holo: PERTH.

Subshrub to 60 cm tall, glabrous or with scattered, appressed hairs particularly at nodes; stems slender, striate. Basal leaves ovate, tapering basally, dentate, with lamina 1–4 cm long, 3–10 mm wide; cauline leaves sessile, narrowly elliptic, entire, with lamina 2.5–10 mm long. Flowers in distant spikes or spike-like thyrses to 25 cm; bracts resembling cauline leaves; peduncle less than 2 mm long; bracteoles linear-triangular, 1–2.5 mm long. Sepals triangular, 1.5–2 mm long, free. Corolla 13–23 mm long, with fine, appressed or arcuate hairs outside, densely bearded inside, cream to white; barbulae simple; wings c. 1 mm wide. Indusium 2 mm wide, dense golden brown beard exceeding lips. Fruit not seen; ovary with short, white hairs at 90°.

Occurs in W.A. around Shark Bay and southwards to Wannoo. Flowers Aug.–Oct. Map 186.

W.A.: Peron Peninsula, Shark Bay, *R.C.Carolin 3318* (SYD); E of Nerren Nerren, *J.S.Beard 7111* (PERTH); c. 23 km S of Wannoo, 17 Sept. 1968, *M.E.Phillips* (AD, BRI, CBG); 54 km S of Denham, *A.S.George 9550* (PERTH).

Similar to *S. restiacea* which has coarse, arcuate, golden hairs on the ovary and fruit, and a sparse, white beard on the indusium.

69. **Scaevola oxyclona** F.Muell., *Fragm.* 10: 58 (1876)

Lobelia oxyclona (F.Muell.) Kuntze, *Revis. Gen. Pl.* 2: 378 (1891). T: Fraser Ra., W.A., *A.Dempster*; holo: MEL; iso: K.

Illustration: R.Erickson *et al.*, *Fl. Pl. W. Australia* 141, t. 444 (1973).

Divaricate subshrub to 1.5 m tall, with closely appressed, white hairs; stems striate; branches spinescent. Basal leaves narrowly obovate, tapering basally, entire or dentate, with lamina 1–2 cm long, 4–6 mm wide; cauline leaves sessile, narrowly obovate to linear, with lamina 3–10 mm long. Flowers in distant spikes to 20 cm long; bracts similar to leaves; peduncle ±obsolete; bracteoles linear-triangular, c. 2.5 mm long, tip acute. Sepals ±orbicular, c. 0.5 mm long, ±free. Corolla 12–18 mm long, with fine, appressed, silky, white hairs outside, densely bearded inside, pale blue; barbulae prominent, papillate apically; wings 0.5–1.5 mm wide. Indusium c. 1.5 mm wide, purplish beard ±equalling lips. Fruit ovoid, c. 2.5 mm long, rugose, with white, silky hairs.

Occurs in southern W.A. between the Fraser and Bremer Ranges and in the Coolgardie region. Flowers chiefly Aug.–Dec. Map 187.

W.A.: c. 5 km NE of Norseman at Jimberlana, *N.N.Donner 4641* (AD, SYD); Granite Ra. E of Norseman, *C.A.Gardner 2917* (PERTH); Mt Day, Bremer Ra., *J.S.Beard 3839* (PERTH); Fraser Ra., 3 Oct. 1891, *R.Helms* (AD, NSW); near Duri, c. 100 km W of Coolgardie, 10 July 1959, *W.S.Stewart* (AD, PERTH).

The spinescent, short, lateral branches distinguish this species from all others in the subsection except *S. restiacea* subsp. *divaricata*, which however, has almost glabrous stems, coarse, golden hairs on the fruit and ovary and a whitish beard on the indusium.

70. **Scaevola angulata** R.Br., *Prodr.* 586 (1810)

Merkusia angulata (R.Br.) Vriese, *Ned. Kruidk. Arch.* 2: 167 (1851); *Lobelia angulata* (R.Br.) Kuntze, *Revis. Gen. Pl.* 2: 378 (1891). T: Carpentaria mainland, opposite Groote Island, [N.T.], 4 Jan. 1801, and Carpentaria Islands, [N.T.], *R.Brown*; lecto: BM, *fide* R.C.Carolin, *Telopea* 3: 495 (1990); isolecto: MEL, P.

Straggling, shrubby perennial to 1 m tall, glabrous to hispid; stems striate. Leaves ±sessile or shortly petiolate, ovate to narrowly oblong, usually entire; lamina mostly 15–30 mm long, 5–10 mm wide. Flowers in spikes to 15 cm long; bracts leafy; peduncle ±obsolete; bracteoles narrowly oblong, to 10 mm long. Sepals triangular, c. 1 mm long, connate, ±campanulate. Corolla 11–20 mm long, glabrous or with short appressed hairs towards base outside, bearded inside, white to pale blue sometimes yellow in throat; barbulae ±simple; wings c. 1 mm wide. Indusium c. 2 mm wide, scattered hairy above. Fruit ellipsoidal, 4 mm long, tuberculate, glabrous or occasionally puberulous.

Occurs in northern N.T. including islands of the Gulf of Carpentaria; mostly in damp soil in open forest on sandstone. Flowers most of the year. Map 188.

N.T.: 9 km NE of Jabiru, *L.G.Adams & P.Richardson 3029* (CANB, DNA); Waterfall Ck, South Alligator R., *N.Byrnes 2448* (BRI, DNA, SYD); South Bay, Bickerton Is., *R.L.Specht 551* (AD, BRI, MEL, NSW); Rose R., *C.R.Dunlop 2711* (BRI, DNA); Katherine Gorge, *N.Byrnes N89* (DNA).

There is considerable variation in the size of the leaves, which are often very small towards the top of the stems. However, the broad well-developed leaves distinguish this species from others with distinctly striate stems.

71. **Scaevola tortuosa** Benth., *Fl. Austral.* 4: 91 (1869)

Lobelia tortuosa (Benth.) Kuntze, *Revis. Gen. Pl.* 2: 378 (1891). T: south-western W.A., 1848, *J.Drummond 4: 191*; holo: K; iso: BM, P.

[*Anthotium humile auct. non* R.Br.: W.H.de Vriese, *Natuurk. Verh. Holl. Maatsch. Wetensch. Haarlem* ser. 2, 10: 189 (1854)]

Ascending perennial to 20 cm tall, with small, appressed hairs; stems striate, tortuous. Basal leaves linear-obovate, recurved, usually entire, with lamina 15–35 mm long, 2–3 mm wide; cauline leaves linear-ovate, with lamina 2–6 mm long. Flowers in racemes or thyrses to 10 cm long; bracts leafy; peduncle to 30 mm long; bracteoles ovate, c. 2 mm long. Sepals narrowly ovate, c. 1.5 mm long, free almost to base, silky hairy. Corolla 10–14 mm long, with fine, appressed hairs outside, bearded inside, colour unknown; barbulae simple; wings to 7 mm wide. Indusium 2 mm wide, white beard ±equalling lips. Fruit not seen; ovary with dense, white hairs at 90°.

Occurs rarely in the central and southern wheatbelt, W.A. Flowering unknown. Map 189.

W.A.: Wyola, 1901, *L.Diels* (MEL); Cummering, 1892, *M.Heal* (MEL); Pingrup, *A.M.Ashby 1274* (AD).

This species differs from *S. parvifolia* and *S. depauperata* in the appressed hairs on the stem. In addition the stems are tortuous.

Unplaced names

Scaevola lanceolata var. *gracilis* E.Pritzel in F.L.E.Diels & E.Pritzel, *Bot. Jahrb. Syst.* 35: 571 (1905)

T: near Chapman R., W.A., Aug. 1901, *L.Diels 3754*; holo: ?B (destroyed) *n.v.*

The Berlin specimen has been destroyed. No other specimen has been located. The name cannot be applied on the basis of the description alone.

Scaevola lyratifolia Vriese in J.G.C.Lehmann, *Pl. Preiss.* 1: 405 (1845)

Merkusia lyratifolia (Vriese) Vriese, *Ned. Kruidk. Arch.* 2: 161, 163 (1851). T: near head of Swan R., W.A., 25 July 1839, *L.Preiss 1485*; lecto: LD, *fide* R.C.Carolin, *Telopea* 3: 508 (1990).

There are no flowers on the specimens and the name cannot be applied as yet.

Scaevola ovalifolia var. *media* DC., *Prodr.* 7(2): 509 (1839)

Merkusia ovalifolia var. *media* (DC.) Vriese, *Ned. Kruidk. Arch.* 2: 154 (1851). T: no precise locality, 'v.s. in h. mus. Par.', *coll. unknown*; *n.v.*

No type specimen has been found. The name cannot be applied satisfactorily on the basis of the description.

Scaevola parviflora K.Krause, *Pflanzenr.* 54: 147 (1912)

T: Arrino, W.A., *Hamilton*; ?holo: ?B (destroyed); iso: NSW *n.v.*

The Berlin type has been destroyed and no specimen has been found in NSW. The taxon may be *S. thesioides* or *S. cunninghamii* but the description is too inadequate to be certain.

[*Scaevola verreauxii* F.Muell., *Bot. Teaching* 65 (1877), cited in error in *Index Kewensis* for '*Scaevola* and *Verreauxia*'].

6. DIASPASIS

R.C.Carolin

Diaspasis R.Br., *Prodr.* 586 (1810)

Type: *D. filifolia* R.Br.

Perennial, sprinkled with papillate, ±appressed hairs; stock often branched. Leaves sessile, alternate. Flowers in loose, terminal racemes, bracteolate; pedicel obsolete. Sepals connate towards base, often unequal. Corolla scarcely bilabiate, tubular, ±dilated basally, not pouched, with dense vesicular hairs towards base of lobes; lobes almost equal, winged, without auricles; rows of enations in throat on darker lines. Anthers connate. Ovary 2-locular; style simple; indusium horizontal, compressed laterally; ovules solitary in each locule. Fruit dry, indehiscent, with hard endocarp. Seeds ovoid, without wing or rim. Embryo terete.

W.H.de Vriese, *Natuurk. Verh. Holl. Maatsch. Wetensch. Haarlem* ser. 2, 10: 177–178 (1845); G.Bentham, *Fl. Austral.* 4: 104–105 (1868); K.Krause, *Pflanzenr.* 54: 116–117 (1912).

A single species endemic in south-western W.A.

Diaspasis filifolia R.Br., *Prodr.* 587 (1810)

T: King George Sound, [W.A.], Dec. 1801, *R.Brown*; *n.v.*

Goodenia glandulifera Vriese, *Natuurk. Verh. Holl. Maatsch. Wetensch. Haarlem* ser. 2, 10: 129 (1854). T: near Albany, W.A., 4 Oct. 1840, *L.Preiss 2032*; lecto: LD, *fide* R.C.Carolin, *Telopea* 3: 566 (1990); isolecto: L.

Scaevola clandestina F.Muell., *Fragm.* 1: 206 (1859). T: south-western W.A., per *A.C.Gregory*; not located.

[*Goodenia armeriaefolia auct. non* DC.: W.H.de Vriese in J.G.C.Lehmann, *Pl. Preiss.* 1: 412 (1845)].

Illustration: K.Krause, *Pflanzenr.* 54: 117, fig. 24 (1912).

Erect or ascending, many-stemmed perennial to 30 cm tall. Leaves linear, acute, sometimes dentate, thick; lamina to 4 cm long, 0.5–2 mm wide. Inflorescence with bracts similar to leaves; peduncles 1–3 cm long; bracteoles linear, 10–20 mm long. Sepals triangular, 1–2 mm long. Corolla 9–16 mm long, white or pink, rarely yellowish or lavender, with dark lines in throat; lobes oblong, to 14 mm long; wings rounded at apex, to 2 mm wide. Style c. 2 mm long, hairy; indusium compressed-hemispherical, 1 mm long, c. 2 mm wide. Fruit ellipsoidal, 3–4 mm long. Figs 23B, 44H–J.

Occurs mainly between Busselton and Cape Riche, but also recorded near Esperance, south-western W.A.; grows in bogs and seasonally wet areas. Flowers chiefly Oct.–Feb. Map 190.

W.A.: 1 km E of Bow Bridge, *J.M.Powell 3116* (NSW); c. 11 km E of Denmark, *S.Carlquist 3987* (NSW, RSA).

F.Mueller (*Fragm.* 2: 17, 181, 1860–1861) referred *Scaevola clandestina* to *Diaspasis filifolia*.

7. GOODENIA

R.C.Carolin

Goodenia Smith, *Spec. Bot. New Holl.* 15 (1793); *Trans. Linn. Soc. London, Bot.* 2: 346 (1794); named for Samuel Goodenough (1743–1827), the Archbishop of Carlisle and a well-known member of the Linnean Society at the time.

Type: *G. ovata* Smith; lecto, *fide* R.C.Carolin, *Fl. Australia* 35: 330 (1992).

Goodenoughia Siebert & Voss, *Vilm. Blumeng.* 1: 559 (1896), *nom. illeg.* as *Goodenouphia.* T: *Goodenoughia ovata* (Smith) Siebert & Voss.

Calogyne R.Br., *Prodr.* 579 (1810). T: *C. pilosa* R.Br.

Balingayum Blanco, *Fl. Filip.* 187 (1837). T: *B. decumbens* Blanco.

Picrophyta F.Muell., *Linnaea* 25: 421 (1853). T: *P. calcarata* F.Muell.

Stekhovia Vriese, *Natuurk. Verh. Holl. Maatsch. Wetensch. Haarlem* ser. 2, 10: 166 (1854). T: *S. scapigera* (R.Br.) Vriese, *fide* R.C.Carolin, *Fl. Australia* 35: 332 (1992).

Aillya Vriese, *Natuurk. Verh. Holl. Maatsch. Wetensch. Haarlem* ser. 2, 10: 75 (1854). T: *A. umbellata* (Vriese) Vriese.

Tetrathylax (G.Don) Vriese, *Natuurk. Verh. Holl. Maatsch. Wetensch. Haarlem* ser. 2, 10: 164 (1854); *Goodenia* sect. *Tetrathylax* G.Don, *Gen. Hist.* 3: 725 (1834). T: *T. quadrilocularis* (R.Br.) Vriese.

Catospermum Benth. in Hooker's *Icon. Pl.* t. 1028 (1868); *Catosperma* Benth., *Fl. Austral.* 4: 83 (1868). T: *C. muelleri* Benth.

Symphyobasis K.Krause, *Pflanzenr.* 54: 40 (1912). T: *S. macroplectra* (F.Muell.) K.Krause.

Neogoodenia C.Gardner & A.S.George, *J. Roy. Soc. W. Australia* 46: 138 (1963). T: *N. minutiflora* C.Gardner & A.S.George.

Herbs or low shrubs. Leaves variable. Flowers in thyrses, racemes, spikes or subumbels, bracteolate or ebracteolate. Sepals ±adnate to ovary or free. Corolla sometimes with pouch or spur, sometimes with enations inside, usually bilabiate; lobes winged, 2 long and adaxial, 3 shorter and abaxial, sometimes adaxial lobes auriculate. Stamens free, epigynous. Ovary usually incompletely 2-locular; style unbranched or 2–4-fid; indusia 1–4. Ovules in 2 rows in each locule or scattered over surface of placentas, rarely solitary. Fruit usually capsular, rarely a small, 1-seeded nut or a large, hard, 4-seeded drupe. Seeds flat, usually winged. Basic number $x = 8$, W.J.Peacock, *Proc. Linn. Soc. New South Wales* 88: 8–27 (1963); *ibid.* 87: 271–277 (1962).

A genus of 179 species; 178 in Australia, most endemic. *Goodenia pilosa* extends to Indonesia, southern China and the Philippines, *G. armstrongiana*, *G. purpurascens* and *G. pumilio* extend to New Guinea and *G. konigsbergeri* (Back.) Back. ex Bold. is endemic in Java.

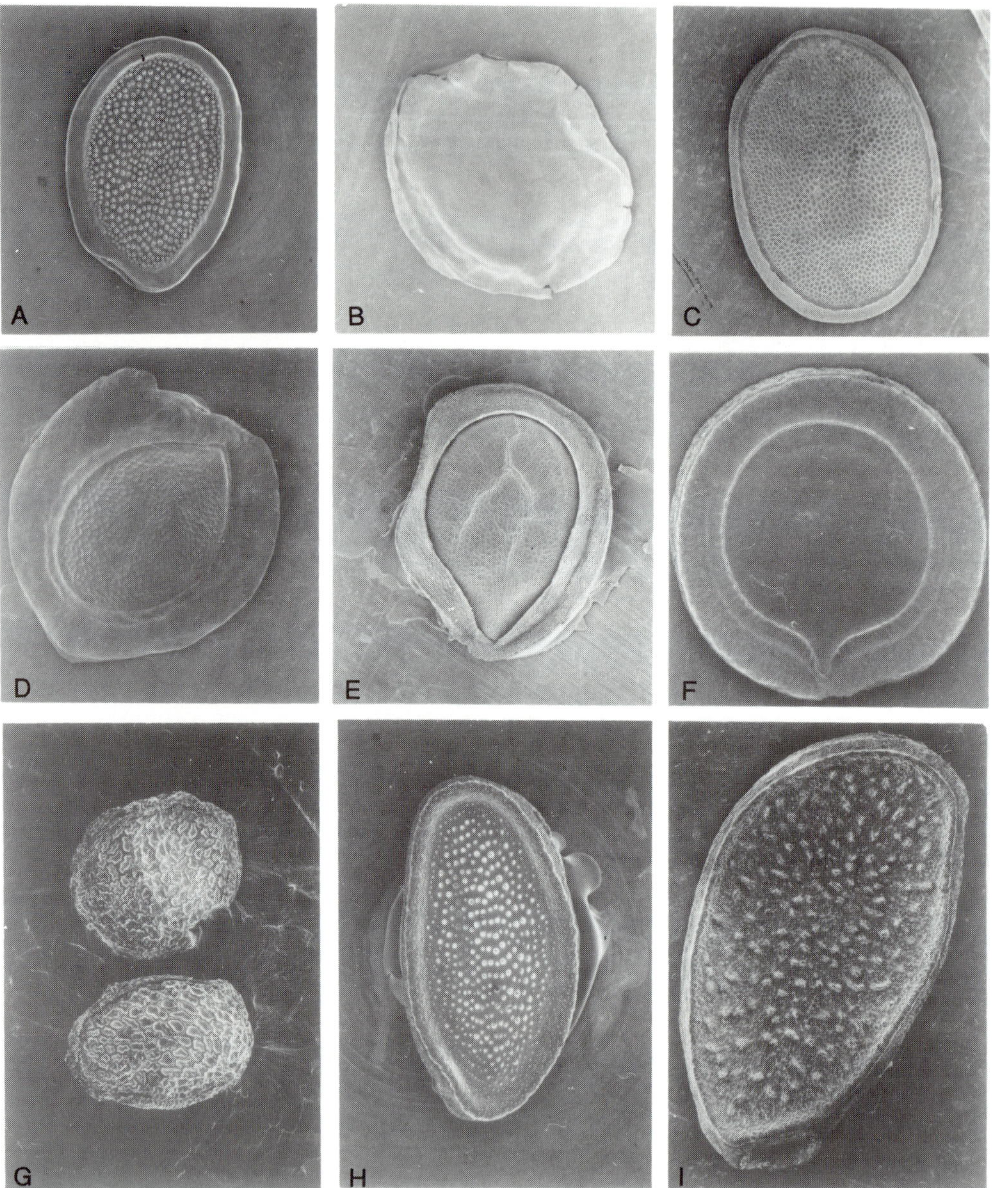

Figure 45. Characteristics of seed surface and wing in *Goodenia*. **A**, *G. calcarata* X8.5, tuberculate; wing narrow (A.Beauglehole 11271, PERTH). **B**, *G. hirsuta* X8.5, smooth, glossy; wing scarcely demarcated from body (R.Carolin 8516, SYD). **C**, *G. scaevolina* X17, colliculate-punctate; wing narrow (A.Beauglehole 11271, PERTH). **D**, *G. fascicularis* X8.5, reticulate-foveate; wing broad (R.Carolin 6368, SYD). **E**, *G. occidentalis* X17, colliculate-smooth, glossy; wing overlapping body of the seed (A.George 8710, SYD). **F**, *G. cycloptera* X8.5, reticulate; wing thick, broad (R.Carolin 4055, SYD). **G**, *G. modesta* X45, reticulate-foveate; wing very narrow (A.Beauglehole 50934, SYD). **H**, *G. centralis* X17, aculeate, cellular projections smaller and more acute than tuberculate; wing narrow (R.Carolin 5852, SYD). **I**, *G. sepalosa* X17, verrucose and granulose; wing almost obsolete (R.Carolin 7421, SYD). Photographs by M.Ricketts & B.Pellow.

G.Bentham, *Goodenia*, *Fl. Austral.* 4: 50–80 (1868); R.C.Carolin, Nomenclatural notes and new taxa in the genus *Goodenia* (Goodeniaceae), *Telopea* 3: 517–570 (1990).

As described in the family treatment, tactile pollinator guides occur in this genus. In sect. *Monochila*, there are long, flat hairs attached to the edge of the corolla wings and projecting into the throat. In other sections, some species have small outgrowths of the corolla throats, here referred to as 'enations'.

The term 'verrucose' is used here in preference to 'verrucate' as used in Carolin, *loc. cit.*

KEY TO INFRAGENERIC TAXA

1 Corolla white, usually with purplish spots at base of lobes or pale pink, rarely yellow or purplish and then style covered with short hairs at 90° — subg. 1. **MONOCHILA**

1: Corolla yellow, cream, blue, purplish or brownish red, rarely white but then without purplish spots; style villous or glabrous — subg. 2. **GOODENIA**

2 Corolla brownish red — sect. 4. **AMPHICHILA**

2: Corolla yellow, cream, blue or purplish

3 Ovules and seeds numerous, in 2 rows on either side of septum — sect. 1. **PORPHYRANTHUS**

3: Ovules and seeds in 2 rows in each locule

4 Corolla blue, often yellowish in throat; pedicels bracteolate; seeds colliculate, reticulate or smooth — sect. 2. **CAERULEAE**

5 Seeds with narrow mucilaginous wing — subsect. 1. **SCAEVOLINA**

5: Seeds with broad membranous wing, notched at hilum — subsect. 2. **CAERULEAE**

4: Corolla usually yellow or cream, rarely blue and then seeds tuberculate or pedicels ebracteolate, rarely white or red — sect. 3. **GOODENIA**

6 Bracteoles present (occasionally absent in *G. calcarata*) — subsect. 1. **GOODENIA**

6: Bracteoles absent (sometimes present in a number of species)

7 Seeds ±orbicular; wing prominent — subsect. 2. **EBRACTEOLATAE**

7: Seeds elliptic; wing very narrow to obsolete — subsect. 3. **BOREALIS**

8 Style unbranched — ser. 1. **BOREALIS**

8: Style 3- or 4-fid — ser. 2. **CALOGYNE**

KEY TO SPECIES

1: Bracteoles absent

2 Style 2–4-branched — **GROUP 10**

2: Style unbranched

3 Wing above auricle on abaxial corolla lobe much narrower than opposite wing — **GROUP 8**

3: Wing above auricle on abaxial corolla lobe equal to opposite one, or both obsolete

4 Cauline leaves auriculate — **GROUP 9**

4: Cauline leaves not auriculate

5 Leaves without simple hairs, or with closely appressed, coarse, simple hairs, sometimes also with glandular ones — **GROUP 6**

5: Leaves with stellate hairs or simple hairs which may be antrorse or retrorse but not closely appressed **GROUP 7**

1: Bracteoles present (although sometimes at base of flower stalk in axil of bract)

6 Corolla various shades of blue through purple, often yellowish in throat **GROUP 1**

6: Corolla (at least adaxial lobes) yellow, white, pink, or rarely greenish

7 Corolla white or pink, rarely greenish, often with purplish markings in throat **GROUP 2**

7: Corolla yellow or yellow-green, sometimes with purplish or brownish markings in throat, and in one species abaxial lobes purplish

8 Leaves glandular-pubescent and/or young plants viscid **GROUP 3**

8: Leaves glabrous, or with stellate, simple, T-shaped or cottony hairs, sometimes minute, glandular hairs present but usually hidden by other hairs; not viscid

9 Mature leaves hairy at least on lower surface **GROUP 4**

9: Mature leaves glabrous or glabrescent **GROUP 5**

GROUP 1

(Flowers bracteate; corolla blue)

1 Plant silvery hairy

2 Corolla lobes almost equal in length **26. G. sericostachya**

2: Corolla lobes very unequal in length

3 Corolla hairy outside **37. G. incana**

3: Corolla glabrous outside **36. G. perryi**

1: Plant not silvery hairy

4 Plant densely glandular-hairy or varnished or viscid

5 Plant varnished when mature **35. G. trichophylla**

5: Plant sometimes viscid but not varnished when mature

6 Glandular hairs long, brownish below their heads

7 No enations inside corolla; leaves to 5 mm wide **30. G. arthrotricha**

7: Enations present inside corolla; leaves mostly more than 5 mm wide

8 Bracteoles linear to linear-lanceolate; W.A. **31. G. eremophila**

8: Bracteoles ovate; northern Australia **29. G. suffrutescens**

6: Glandular hairs small, or short and coarse, white below their heads which are often brownish

9 At least lower leaves lyrate or pinnately lobed

10 Leaf lobes broader than linear; Vic. **57. G. macmillanii**

10: Leaf lobes linear; northern Australia **17. G. gloeophylla**

9: None of the leaves lyrate or pinnately lobed

11 Corolla lobes almost equal in length (adaxial lobes c. 3/4 as long as abaxial lobes) — **27. G. scaevolina**

11: Corolla lobes very unequal in length (adaxial lobes 1/2 to 3/5 as long as abaxial lobes)

12 Herb with basal leaves

13 Corolla to 7 mm long; leaves obovate to lanceolate; northern Australia — **20. G. viscidula**

13: Corolla more than 15 mm long

14 Leaves linear; seed with broad membranous wing; south-western W.A. — **34. G. caerulea**

14: Leaves broader than linear; seed with narrow mucilaginous wing; north-western Australia — **28. G. stobbsiana**

12: Subshrub without basal leaves

15 Leaves cordate or truncate at base; central Australia — **46. G. grandiflora**

15: Leaves tapering very gradually towards base, sessile or with indistinct base; western and north-western Australia

16 Leaves oblanceolate to obovate, mostly more than 20 mm wide — **28. G. stobbsiana**

16: Leaves linear to oblanceolate or narrowly elliptic, mostly to 7 mm wide

17 Indusium oblong; south-western W.A. — **52. G. xanthotricha**

17: Indusium broadly ovate; north-western Australia — **17. G. gloeophylla**

4: Leaves glabrous or simple hairy, or with scattered glandular hairs, not varnished or viscid

18 Cauline leaves ovate-elliptic, stem-clasping, more than 4 mm wide

19 Corolla glabrous outside — **40. G. eatoniana**

19: Corolla cottony hairy outside — **41. G. leptoclada**

18: Cauline leaves linear to oblanceolate, not stem-clasping, less than 4–10 mm wide, or leaves all basal

20 Corolla glabrous outside

21 Sepals obtuse; abaxial lobes of corolla more than 1/2 as long as corolla — **38. G. pterigosperma**

21: Sepals acute; abaxial lobes of corolla less than 1/2 as long as corolla

22 Herb with basal cluster of leaves; seed wing more than 0.2 mm wide — **39. G. glareicola**

22: Shrubby herb or shrub without basal cluster of leaves (*G. hassallii* has a few basal leaves when very young); seed wing less than 0.2 mm wide

23 Flowers with leaf-like bracts — **42. G. hassallii**

23: Flowers on scapes naked except for linear bracts to 5 mm long — **4. G pinifolia**

20: Corolla hairy outside

24 Subshrub or many-stemmed perennial without basal cluster of leaves

25 Leaves glaucous; bracteoles oblanceolate to oblong, more than 4 mm wide **33. G. azurea**

25: Leaves not glaucous; bracteoles linear to lanceolate, less than 4 mm wide **27. G. scaevolina**

24: Herbs with basal cluster of leaves

26 Corolla lobes almost equal in length (adaxial lobes c. 3/4 as long as abaxial lobes); south-western W.A. **3. G. watsonii**

26: Corolla lobes very unequal in length (adaxial lobes less than 3/4 as long as abaxial lobes); northern Australia

27 Corolla more than 8 mm long

28 Basal leaves linear to oblanceolate, to 15 mm wide; enations absent inside corolla **18. G. purpurascens**

28: Basal leaves elliptic-obovate, mostly more than 20 mm wide; enations prominent inside corolla **32. G. ramelii**

27: Corolla 2–5 mm long

29 Corolla mostly white with pale purple markings; corolla wings almost obsolete; indusium not enclosed in auricles **19. G. minutiflora**

29: Corolla mostly bluish purple; corolla wings c. 1 mm wide; indusium enclosed within auricles **21. G. paludicola**

GROUP 2

(Flowers bracteolate; corolla white or pink)

1 Corolla pouch or spur projecting well below base of ovary **56. G. calcarata**

1: Corolla pouch obsolete or as long as ovary

2 Leaves mostly basal on short stock at ground level; herb

3 Corolla 4–8 mm long; south-western W.A. **3. G. watsonii**

3: Corolla less than 4 mm long; northern Australia **19. G. minutiflora**

2: Leaves cauline; plant ±shrubby, at least at base

4 Simple hairs on style usually purplish; fruit an inferior nut; ovary with 1–3 ovules

5 Sepals ovate to elliptic, c. 0.6 mm wide; ovules 1–3 **7. G. stenophylla**

5: Sepals narrowly ovate to linear or deltoid, less than 0.4 mm wide; ovule usually solitary

6 Flowers in condensed leafy spikes or thyrses; bracts greatly exceeding sepals **6. G. fasciculata**

6: Flowers distant in spikes or thyrses; bracts scarcely exceeding sepals

7 Leaves more than 10 mm long **8. G. drummondii**

7: Leaves less than 8 mm long **9. G. helmsii**

4: Simple hairs on style white or yellowish; fruit an inferior capsule; ovary with more than 3 ovules

8 Corolla lobes equal in length or nearly so; adaxial lobes 1/2 as long as corolla, or longer

9 Leaves elliptic to ovate, less than 35 mm long **2. G. decursiva**

9: Leaves linear to narrowly ovate, mostly more than 40 mm long **1. G. scapigera**

8: Corolla lobes very unequal in length; adaxial lobes mostly less than 1/2 as long as corolla

10 Corolla glabrous outside; W.A.

11 Orifice of indusium with minute bristles; fruit 2–3 mm diam. **4. G. pinifolia**

11: Orifice of indusium glabrous; fruit 5–6 mm long **5. G. elderi**

10: Corolla variously hairy outside; eastern, central and southern Australia

12 Leaves linear; Qld **68. G. disperma**

12: Leaves oblanceolate or lanceolate or broader; S.A. and N.T.

13 Leaves mostly sessile (lower sometimes petiolate), glaucous, sometimes with a few glandular hairs near margin **51. G. albiflora**

13: Leaves petiolate, pubescent

14 Corolla with coarse hairs inside on adaxial lobes; plant not viscid **54. G. saccata**

14: Corolla with fine hairs inside on adaxial lobes; plant viscid

15 Wings of corolla lobes with long cilia **47. G. chambersii**

15: Wings of corolla lobes without cilia or with 1 or 2 very short ones **46. G. grandiflora**

GROUP 3

(Flowers bracteolate; corolla yellow; viscid or leaves glandular-hairy)

1 Plant viscid or varnished, particularly in young parts; glandular hairs inconspicuous so that plant may appear to have none

2 Corolla lobes more or less equal **10. G. viscida**

2: Corolla lobes unequal

3 Leaves narrowly elliptic to oblanceolate; corolla sparsely simple-pubescent outside **45. G. vernicosa**

3: Leaves ovate to elliptic, rarely narrower; corolla without simple hairs outside

4 Leaves thick; sepals 2–4 mm long; bracteoles to 1 mm long **44. G. varia**

4: Leaves thin; sepals 3–11 mm long; bracteoles to 6 mm long **43. G. ovata**

1: Plant neither viscid nor varnished, or if viscid then with prominent glandular hairs visible to naked eye

5 Cauline leaves petiolate

6 Leaf or terminal lamina lanceolate, glandular-hairy **53. G. brunnea**

6: Leaf or terminal leaflet ovate to orbicular, or if lanceolate then glabrous or nearly so

7 Leaves with petiolate leaflets attached below main lamina; W.A. **48. G. kingiana**

7: Leaves simple or with small sessile leaflets attached below main lamina; eastern and central Australia

8 Corolla with coarse, simple hairs inside on adaxial lobes **54. G. saccata**

8: Corolla with fine hairs inside on adaxial lobes, or glabrous

9 Wings of corolla lobes with long cilia **47. G. chambersii**

9: Wings of corolla lobes without cilia, or with 1 or 2 very short ones **46. G. grandiflora**

5: Cauline leaves sessile or leaves mostly basal

10 Wing above auricle on abaxial corolla lobes much narrower than opposite one; drier areas of central and northern Australia

11 Stems straight; bracteoles small, inserted just below flower at articulation of pedicel **133. G. glandulosa**

11: Stems divaricate, tangled; bracteoles large, often not opposite, inserted well below flower and articulation of pedicel **138. G. cirrifica**

10: Wings of abaxial corolla lobes almost equal; W.A., S.A., southern Qld and N.S.W.

12 Leaves mostly basal; herb

13 Adaxial lobes of corolla purplish **16. G. bicolor**

13: Adaxial lobes of corolla yellow

14 Leaves linear; corolla c. 20 mm long **65. G. lineata**

14: Leaves broader than linear; corolla 8–14 mm long **12. G. paniculata**

12: Leaves mostly cauline; plant usually ±shrubby

15 Leaves stem-clasping, ovate to elliptic

16 Indusium with straight orifice, or lower lip slightly concave; leaves mostly more than 3 cm long **49. G. amplexans**

16: Indusium deeply notched especially on lower lip; leaves mostly less than 3 cm long **50. G. benthamiana**

15: Leaves not stem-clasping, linear to lanceolate or oblanceolate

17 Pedicels 12–15 mm long; corolla pouch conspicuous **70. G. stephensonii**

17: Pedicels 4–7 mm long; corolla pouch inconspicuous **71. G. heterophylla**

GROUP 4

(Flowers bracteolate; corolla yellow; mature leaves with non-glandular hairs)

1 Upper surface of leaves with stellate hairs

2 Indusium with 2 lateral lobes **85. G. tripartita**

2: Indusium without 2 lateral lobes

3 Hairs on leaves almost all stellate; connate part of corolla 5–7 mm long **88. G. dyeri**

3: Hairs on leaves simple and/or multicellular mixed with stellate; connate part of corolla to 3 mm long

4 Leaves on scapes stem-clasping **87. G. robusta**

4: Leaves on scapes not stem-clasping

5 Outside of corolla without simple hairs amongst the stellate ones, or with a few towards the top; southern Australia E of the Great Australian Bight **86. G. willisiana**

5: Outside of corolla with simple hairs amongst the stellate ones; W.A.

6 Indusium flat, with notched or straight lips **83. G. affinis**

6: Indusium very convex, with convex lips **84. G. convexa**

1: Upper surface of leaves without stellate hairs

7 Mature leaves hairy on lower surface, glabrous on upper surface

8 Indusium bristles inserted behind upper lip; W.A. **76. G. xanthosperma**

8: Indusium bristles inserted on upper lip; southern and eastern Australia

9 Leaves entire; central Australia **79. G. rupestris**

9: Leaves crenate or dentate; southern and eastern Australia

10 Leaves with felted hairs on lower surface **80. G. blackiana**

10: Leaves hairy but not felted on lower surface **75. G. hederacea**

7: Mature leaves hairy on both surfaces

11 Stems prostrate or decumbent or stoloniferous

12 Cauline leaves with single large lobe on one side at base

13 Cauline leaves ovate to oblong, densely hairy **92. G. schwerinensis**

13: Cauline leaves narrowly oblong to narrowly elliptic, glabrous or with few scattered hairs **90. G. glabra**

12: Cauline leaves without single large lobe at base

14 Bracteoles inserted close under ovary **72. G. rotundifolia**

14: Bracteoles inserted well below ovary

15 Leaves lyrate; capsule 4-valved **23. G. lyrata**

15: Leaves rarely lyrate; capsule 2-valved

16 Sepals adnate to lower half of ovary only; W.A. **89. G. wilunensis**

16: Sepals adnate to more than lower half of ovary; central and south-eastern Australia

17 Seed crumpled, with black glossy body grading into yellow wing; central Australia **151. G. hirsuta**

17: Seed usually not crumpled, with brownish body well differentiated from wing; south-eastern Australia, Tas.

18 Flowers solitary in axils of basal leaves **73. G. arenicola**

18: Flowers in racemes 5–40 cm long

19 Corolla 8–10 mm long **58. G. fordiana**

19: Corolla 11–17 mm long

20 Racemes to 5 cm long; indusium strongly folded **81. G. geniculata**

20: Racemes 10–40 cm long; indusium flat or slightly folded

21 Mature leaves glabrous or rarely cobwebbed-pubescent; southern Qld and northern N.S.W. **74. G. delicata**

21: Mature leaves hairy with multicellular, crisped hairs; southern Vic. and Tas. **82. G. lanata**

11: Stems erect to ascending

22 Leaves greyish, glabrous or with T-shaped hairs

23 Leaves covered with grey T-shaped hairs **94. G. mueckeana**

23: Leaves glabrous and grey-glaucous **15. G. lamprosperma**

22: Leaves green or white to brownish; hairs, if present, not T-shaped

24 Hairs on outside of corolla stellate **64. G. glomerata**

24: Hairs on outside of corolla simple and/or glandular

25 Bracteoles inserted close under ovary

26 Flowers mostly sessile

27 Leaves obovate to oblanceolate, 4–10 cm long **66. G. bellidifolia**

27: Leaves linear, 1–5 cm long **69. G. viridula**

26: Flowers stalked

28 Leaves orbicular, to 2 cm long **72. G. rotundifolia**

28: Leaves linear to elliptic, to 12 cm long

29 Indusium strongly folded; south-eastern Australia **81. G. geniculata**

29: Indusium not strongly folded; northern Qld and N.T. **108. G. strangfordii**

25: Bracteoles inserted well below ovary

30 Capsule gaping after dehiscence; valves bifid; south-eastern Australia from southern Qld

31 Cauline leaves indistinctly petiolate; hairs soft; capsule globular **74. G. delicata**

31: Cauline leaves +/-sessile, often lobed near base; hairs hispid; capsule ovoid **71. G. heterophylla**

30: Capsule not gaping after dehiscence; valves usually entire; inland western and central Australia

32 Sepals adnate to lower part of ovary only

33 Peduncles 30–60 mm long; ovules to 14, in 2 rows on each placenta; hairs silky **89. G. wilunensis**

33: Peduncles 5–30 mm long; ovules more than 20, scattered on the placentas; hairs not silky **22. G. berringbinensis**

32: Sepals adnate to ovary almost to top

34 Abaxial corolla lobes purplish; northern Australia **16. G. bicolor**

34: Abaxial corolla lobes yellow

35 Flowers few in dichasia or monochasia in axils of basal leaves; corolla 5–8 mm long **25. G. claytoniacea**

35: Flowers in thyrses or racemes; corolla usually more than 8 mm long

36 Scapes to 10 cm long, usually less than 1.5 times as long as basal leaves

37 Peduncles 5–15 mm long; pedicels 5–12 mm long; indusium flat **14. G. humilis**

37: Peduncles 15–50 mm long; pedicels 20–50 mm long; indusium folded **81. G. geniculata**

36: Scapes 20–35 cm long, mostly more than twice as long as leaves

38 Corolla with simple and glandular hairs outside, without conspicuous pouch; southern and eastern Australia **12. G. paniculata**

38: Corolla without glandular hairs outside, with prominent pouch; central Australia **24. G. modesta**

GROUP 5

(Flowers bracteolate; corolla yellow; mature leaves almost entirely glabrous)

1 Outside of corolla with stellate hairs

2 Leaves basal in a cluster at ground level

3 Inflorescence spike-like; all peduncles, except sometimes the lowest, almost obsolete **63. G. stelligera**

3: Inflorescence an open thyrse; peduncles more than 5 mm long **62. G. dimorpha**

2: Leaves cauline although sometimes more frequent towards base

4 Leaves decurrent on stem **60. G. decurrens**

4: Leaves not decurrent on stem

5 Inflorescence covered with long soft hairs **64. G. glomerata**

5: Inflorescence covered with short hairs **61. G. rostrivalvis**

1: Outside of corolla without stellate hairs

6 Stems prostrate or decumbent

7 Cauline leaves with single large lobe at base

8 Style c. 3 mm long, less than 1/2 as long as corolla; W.A. **91. G. peacockiana**

8: Style 5–6 mm long, c. 1/2 as long as corolla; eastern and central Australia into S.A. **90. G. glabra**

7: Cauline leaves without single lobe at base

9 Ovary and outside of corolla glabrous

10 Bracteoles inserted at articulation of pedicel just under ovary; pedicel obsolete; fruit a capsule **93. G. laevis**

10: Bracteoles inserted well below ovary; pedicel 10–20 mm long, not articulate; fruit a hard, thin-skinned drupe **78. G. goodeniacea**

9: Ovary and outside of corolla hairy

11 Bracteoles inserted close under ovary **72. G. rotundifolia**

11: Bracteoles inserted well below ovary

12 Abaxial sepal longer than others; bristles inserted behind indusium lip **77. G. centralis**

12: Abaxial sepal equal to others; bristles inserted on indusium lip

13 Hairs cottony to felted **75. G. hederacea**

13: Hairs simple, not appressed, soft

14 Basal leaves not lyrate; N.S.W. and southern Qld **74. G. delicata**

14: Basal leaves lyrate; inland western and central Australia **23. G. lyrata**

6: Stems erect or ascending

15 Indusium with dense beard at base **111. G. phillipsiae**

15: Indusium without dense beard at base

16 Indusium tightly folded downwards **11. G. gracilis**

16: Indusium not tightly folded downwards

17 Bracteoles inserted immediately below ovary

18 Leaves linear, terete

19 Herb **96. G. angustifolia**

19: Shrub **69. G. viridula**

18: Leaves flattened, usually broader than linear

20 Indusium furrowed on upper surface **137. G. odonnellii**

20: Indusium not furrowed on upper surface

21 Corolla to 12 mm long

22 Flowers in leafy racemes or thyrses with bracts more than 10 mm long; W.A. **93. G. laevis**

22: Flowers in spikes with bracts mostly less than 10 mm long; eastern Australia **66. G. bellidifolia**

21: Corolla more than 12 mm long

23 Leaves to 20 mm long; inflorescence to 30 cm long; south-eastern Australia **72. G. rotundifolia**

23: Leaves 20–80 mm long; inflorescence to 15 cm long; inland Australia

24 Corolla with large glandular hairs outside; peduncles 5–12 mm long; Barkly Tableland **95. G. nigrescens**

24: Corolla glabrous or with a few simple hairs outside; peduncles 20–50 mm long; inland eastern Australia **97. G. glauca**

17: Bracteoles inserted well below ovary

25 Plants with distinct cluster of basal leaves; scapes arising from it either naked or with reduced leaves

26 Leaves grey-glaucous, conspicuously 3-veined **15. G. lamprosperma**

26: Leaves not grey-glaucous, the veins not conspicuous

27 Corolla more than 15 mm long **65. G. lineata**

27: Corolla to 14 mm long

28 Flowering stems divaricately branched **138. G. cirrifica**

28: Flowering stems not divaricately branched

29 Flowers in a few cymes in axils of basal leaves **25. G. claytoniacea**

29: Flowers in terminal thyrses, racemes or spikes

30 Inflorescence to 10 cm long, to 1.5 times as long as leaves **14. G. humilis**

30: Inflorescence to 30 cm long, usually more than twice as long as leaves

31 Corolla 7–9 mm long **13. G. macbarronii**

31: Corolla 10–14 mm long

32 Sepals attached well below top of ovary and fruit; central Australia **22. G. berringbinensis**

32: Sepals attached near top of ovary and fruit; eastern Australia **12. G. paniculata**

25: Leaves mostly cauline, sometimes also with basal cluster

33 Leaves with decurrent margins

34 Leaves more than 10 mm wide; N.S.W. **60. G. decurrens**

34: Leaves less than 10 mm wide; Qld **67. G. racemosa**

33: Leaves without decurrent margins

35 Leaves entire

36 Corolla with dense beard inside

37 Corolla glabrous outside **111. G. phillipsiae**

37: Corolla glandular-hairy outside **95. G. nigrescens**

36: Corolla without dense beard inside

38 Subshrub **67. G. racemosa**

38: Herb

39 Capsule globular, 2-valved, c. 8 mm long; Qld and northern N.S.W. **74. G. delicata**

39: Capsule ±cylindrical, 4-valved, to 4 mm long; central Australia **24. G. modesta**

35: Leaves dentate to lobed

40 Corolla with mostly glandular hairs outside, or viscid or glabrous

41 Corolla with conspicuous glandular hairs outside; leaves sessile; north-eastern Qld **55. G. stirlingii**

41: Corolla glabrous or viscid outside; leaves petiolate; southern Australia

42 Leaves thick; sepals 2–4 mm long; bracteoles 1 mm long **44. G. varia**

42: Leaves thin; sepals 3–11 mm long; bracteoles to 6 mm long **43. G. ovata**

40: Corolla with ±equal numbers of simple and glandular hairs outside, or all simple

43 Wing above auricle of abaxial corolla lobe almost obsolete **95. G. nigrescens**

43: Wing above auricle of abaxial corolla lobes ±equal to opposite one

44 Corolla with soft hairs outside; leaves tapering towards base, south-western Australia **59. G. quadrilocularis**

44: Corolla with harsh, coarse hairs outside; cauline leaves broad at base, mostly auriculate; northern Australia **163. G. hispida**

GROUP 6

(Flowers not bracteolate; leaves glabrous or appressed simple hairy, not auriculate)

1 Corolla densely glandular-hairy

2 Leaves densely glandular-pubescent, sometimes also with simple hairs

3 Corolla less than 12 mm long **129. G. havilandii**

3: Corolla more than 12 mm long

4 Corolla pouch longer than ovary; sepals adnate to ovary only in lower 1/3 **125. G. quasilibera**

4: Corolla pouch usually 1/2 as long as ovary, rarely as long; sepals adnate to ovary in lower 2/3

5 Corolla usually less than 16 mm long **120. G. lobata**

5: Corolla usually more than 25 mm long **148. G. tenuiloba**

2: Leaves simple hairy with scarcely any glandular hairs, or glabrous

6 Corolla to 6 mm long **126. G. occidentalis**

6: Corolla more than 8 mm long

7 Bristles on lower lip of indusium much longer than those on upper lip **130. G. janamba**

7: Bristles on both lips of indusium ±equal, or upper ones longer

8 Leaves hairy **124. G. concinna**

8: Leaves glabrous or glabrescent

9 Corolla mauve to purple **144. G. pallida**

9: Corolla yellow

10 Corolla with both simple and glandular hairs outside **146. G. muelleriana**

10: Corolla with glandular hairs only outside **143. G. nuda**

1: Corolla with simple hairs or glabrous, sometimes with scattered, minute, glandular hairs

11 Plant glaucous

12 Corolla white, pouch or spur longer than ovary **56. G. calcarata**

12: Corolla yellow, pouch or spur inconspicuous, as long as or shorter than ovary **97. G. glauca**

11: Plant not glaucous

13 Indusium folded, with furrow on upper surface **137. G. odonnellii**

13: Indusium concave, not folded

14 Corolla to 6 mm long

15 Basal leaves oblanceolate, more than 4 mm wide **126. G. occidentalis**

15: Basal leaves linear or with linear lobes, to 1 mm wide

16 Indusium with purplish beard at base; Kimberley region and Arnhem Land **106. G. coronopifolia**

16: Indusium without purplish beard at base; south-eastern Australia **128. G. micrantha**

14: Corolla more than 6 mm long

17 Upper lip of indusium glabrous or with minute bristles

18 Corolla 8–14 mm long; sepals adnate to ovary for at least 1/2 its length **99. G. lunata**

18: Corolla 6–7 mm long; sepals adnate to ovary only near base **121. G. salmoniana**

17: Upper lip of indusium with prominent white bristles

19 Fruit narrowly cylindrical to linear, more than 12 mm long **105. G. corynocarpa**

19: Fruit cylindrical to globular, less than 10 mm long

20 Corolla glabrous outside
21 Corolla to 10 mm long, corolla pouch inconspicuous **106. G. coronopifolia**
21: Corolla more than 12 mm long, corolla pouch conspicuous
22 Plant stoloniferous; leaves entire **112. G. gibbosa**
22: Plant not stoloniferous; leaves pinnatifid or deeply dentate **110. G. pinnatifida**
20: Corolla hairy outside
23 Stems prostrate or decumbent
24 Leaves silvery hairy **127. G. krauseana**
24: Leaves green, glabrous or with scattered hairs
25 Leaf axils with prominent tufts of hairs **145. G. prostrata**
25: Leaf axils glabrous or with almost hidden tufts of small hairs
26 Main stems straight or sinuate; cauline leaves broader than linear **104. G. maideniana**
26: Main stems zig-zag; cauline leaves linear **103. G. anfracta**
23: Stems erect or ascending
27 Leaves linear-terete
28 Cauline leaves numerous; basal leaves ephemeral; central Australia **140. G. virgata**
28: Cauline leaves few; leaves mostly basal; south-western W.A. **123. G. filiformis**
27: Leaves flattened
29 Corolla pouch prominent
30 Sepals free from ovary almost to base; corolla spurred **116. G. macroplectra**
30: Sepals fused to ovary; corolla pouched but not spurred **122. G. pulchella**
29: Corolla pouch obscure
31 Leaves decurrent, mostly to 4 cm long; seeds warty **166. G. armstrongiana**
31: Leaves not decurrent, mostly more than 5 cm long; seeds colliculate to almost smooth
32 Leaves entire
33 Leaves cauline; ovary tapering basally **108. G. strangfordii**
33: Leaves mostly basal; ovary rounded basally **98. G. fascicularis**
32: Leaves dentate or lobed
34 Flowers solitary in axils of basal leaves
35 Axillary hairs dense, conspicuous **102. G. pusilla**
35: Axillary hairs sparse, hidden by silvery indumentum **127. G. krauseana**
34: Flowers in racemes or subumbels
36 Ovary tapering towards base; south-eastern Australia **117. G. elongata**
36: Ovary rounded at base; central and western Australia
37 Corolla 12–23 mm long; central Australia **98. G. fascicularis**
37: Corolla 7–8 mm long; Arnhem Land and W.A.
38 Leaves silvery hairy; W.A. **127. G. krauseana**
38: Leaves glabrous; N.T. **167. G. argillacea**

GROUP 7

(Flowers not bracteolate; leaves with stellate or not appressed, simple hairs, not auriculate)

1 Indusium notched or with furrow on upper surface	
2 Indusium folded with furrow on upper surface	**137. G. odonnellii**
2: Indusium notched, not folded	
3 Corolla 12–25 mm long	**115. G. mimuloides**
3: Corolla 5–7 mm long	**109. G. pusilliflora**
1: Indusium not notched or furrowed	
4 Corolla spur exceeding or equalling ovary in length	
5 Hairs on lower lip of indusium longer than those on upper lip	**130. G. janamba**
5: Hairs on lower lip of indusium shorter than those on upper lip	
6 Leaves mostly glandular-hairy	**116. G. macroplectra**
6: Leaves with mostly simple hairs	
7 Hairs on leaves not appressed	**131. G. cycloptera**
7: Hairs on leaves antrorse	**125. G. quasilibera**
4: Corolla not spurred, or with spur shorter than ovary	
8 Adaxial sepal 4–6 mm long, much longer than others	**136. G. redacta**
8: Adaxial sepal equal to others	
9 Corolla 1–5 mm long	
10 Flowers sessile	
11 Flowers in heads at ground level	**165. G. chthonocephalata**
11: Flowers in spikes	**164. G. subauriculata**
10: Flowers stalked	
12 Leaves sessile	**178. G. kakadu**
12: Leaves petiolate	
13 Hairs on leaves simple; ovule solitary	**153. G. neogoodenia**
13: Hairs on leaves stellate; ovules c. 40	**177. G. pumilio**
9: Corolla more than 5 mm long	
14 Hairs stellate	**152. G. stellata**
14: Hairs simple	
15 Corolla glabrous outside	
16 Indusium with basal beard of purplish hairs; the Kimberley and Arnhem Land	**106. G. coronopifolia**
16: Indusium with white hairs; central and southern Australia	
17 Plant not stoloniferous	**110. G. pinnatifida**
17: Plant stoloniferous	**112. G. gibbosa**
15: Corolla hairy outside	

18 Corolla blue **150. G. vilmorinae**

18: Corolla yellow or almost white

19 Hairs tangled **155. G. arachnoidea**

19: Hairs not appressed, not tangled

20 Leaves lobed or pinnatifid

21 Leaves with 1 or 2 lobes on either side; Qld **118. G. megasepala**

21: Leaves with more than 2 lobes or segments on either side; W.A.

22 Corolla with simple hairs outside; southern W.A. **115. G. mimuloides**

22: Corolla with simple and glandular hairs outside; northern W.A. **148. G. tenuiloba**

20: Leaves entire or dentate (sometimes lobed towards base in *G. iyouta*)

23 Cauline leaves mostly linear even towards base of stem **140. G. virgata**

23: Lower cauline leaves broader than linear

24 Leaf axils with prominent tufts of hairs **145. G. prostrata**

24: Leaf axils glabrous or with tufts of hairs often hidden by general indumentum

25 Flowers in spike-like thyrses or racemes on naked scapes arising from tuft of leaves terminal on a stem covered with dead leaves; bracts linear to triangular, to 5 mm long **149. G. cusackiana**

25: Flowers in spreading thyrses or racemes or subumbels on leafy stems arising from cluster of leaves ±at ground level; bracts more than 5 mm long

26 Bristles on lower lip of indusium much longer than those on upper lip **130. G. janamba**

26: Bristles on lower lip of indusium shorter than those on upper lip, or equal

27 Stems glabrous or glabrescent **143. G. nuda**

27: Stems hairy

28 Glandular hairs minute

29 Pedicels 4–11 cm long

30 Ovary rounded at base; sepals 6–9 mm long; Qld **118. G. megasepala**

30: Ovary tapered at base; sepals 3–4 mm long; south-eastern Australia **117. G. elongata**

29: Pedicels to 4 cm long

31 Leaves hairy all over; seeds black, glossy, grading into a yellow wing, usually crumpled **151. G. hirsuta**

31: Leaves with hairs mostly towards margin and on main vein; seeds warty, almost wingless **166. G. armstrongiana**

28: Glandular hairs conspicuous

32 Leaves and flowers with scattered, simple hairs and dark, glandular hairs **146. G. muelleriana**

32: Glandular hairs pale

33 Simple hairs dense with surface of leaf scarcely visible **147. G. forrestii**

33: Simple hairs less dense with surface of leaf visible

34 Ovary obtuse at base; central Australia **119. G. iyouta**

34: Ovary attenuate at base; south-eastern Australia **117. G. elongata**

GROUP 8

(Flowers not bracteolate; corolla wings unequal on abaxial lobes)

1 Plant with conspicuous glandular hairs, or viscid and varnished on younger parts

2 Leaves linear

3 Base of stem reddish; single-stemmed at base with cauline leaves, distinct internodes and axillary branches **141. G. triodiophila**

3: Base of stem not reddish; stems several from a cluster of basal leaves, internodes lacking

4 Leaves mostly cauline **133. G. glandulosa**

4: Leaves mostly basal

5 Basal leaves glandular-hairy, sometimes with a few scattered simple hairs **139. G. armitiana**

5: Basal leaves pubescent to hirsute with mostly simple hairs **142. G. microptera**

2: Leaves mostly broader than linear

6 Plant stoloniferous **101. G. heteromera**

6: Plant not stoloniferous

7 Stems and leaves with very short hairs or glabrous, often varnished **135. G. faucium**

7: Stems and leaves with conspicuous glandular hairs or sometimes glabrescent, not varnished

8 Stems and leaves with arcuate-antrorse simple hairs; leaves becoming scabrous **133. G. glandulosa**

8: Stems and leaves with soft hairs at 90° **134. G. larapinta**

1: Plant with glandular hairs absent or obscured by simple ones; not viscid nor varnished

9 Corolla glabrous or with a few coarse hairs outside

10 Leaves oblong to oblanceolate, lobed or coarsely dentate **110. G. pinnatifida**

10: Leaves linear, entire **107. G. integerrima**

9: Corolla hairy outside

11 Plant stoloniferous **101. G. heteromera**

11: Plant not stoloniferous

12 Ovary tapering at base; south-eastern Australia **117. G. elongata**

12: Ovary rounded at base; northern Australia, western Australia and drier inland regions

13 Corolla hairs appressed **100. G. pascua**

13: Corolla hairs not appressed

14 Cauline leaves linear

15 Leaves covered with simple and/or multicellular hairs **142. G. microptera**

15: Leaves with minute glandular and sometimes simple hairs **139. G. armitiana**

14: Cauline leaves broader than linear

16 Plant erect; seed with thick wing not thinning towards edge; indusium longer than or as long as broad **132. G. heterochila**

16: Plant decumbent; seed with wing thinning towards edge; indusium broader than long

17 Pedicels 2–4 cm long; seed usually crumpled, glossy, with paler wing scarcely differentiated from body **151. G. hirsuta**

17: Pedicels 5–10 cm long; seed not crumpled, with wing clearly differentiated from body by raised rim **115. G. mimuloides**

GROUP 9

(Flowers not bracteolate; leaves auriculate)

1 Flowers sessile; corolla to 5 mm long **164. G. subauriculata**

1: Flowers pedicellate or if sessile then corolla more than 8 mm long

2 Pedicel with hairs mostly on adaxial side **158. G. durackiana**

2: Pedicel glabrous or with hairs evenly distributed

3 Sepals narrowly elliptic or ovate, 7–12 mm long, more than 1 mm wide **154. G. sepalosa**

3: Sepals narrowly lanceolate to linear, to 6 mm long, to 1.5 mm wide

4 Corolla pink to mauve, often with yellow throat **161. G. malvina**

4: Corolla yellow, sometimes with purplish markings

5 Ovary with mostly glandular hairs

6 Stems with simple hairs **162. G. potamica**

6: Stems with glandular hairs **154. G. sepalosa**

5: Ovary with mostly simple hairs

7 Corolla 7–15 mm long

8 Stem hairs at 90° and retrorse **156. G. brachypoda**

8: Stem hairs at 90° and antrorse, or stems glabrous

9 Leaves oblong to linear, to 6 mm wide **167. G. argillacea**

9: Leaves ovate to lanceolate, mostly more than 6 mm wide **160. G. campestris**

7: Corolla 15–20 mm long

10 Fruit c. 8 mm wide; seeds smooth **157. G. leiosperma**

10: Fruit to 5 mm wide; seeds warty

11 Sepals 0.4–0.6 mm wide, densely hispid; fruit ellipsoidal **163. G. hispida**

11: Sepals 0.8–1 mm wide, with a few marginal bristles; fruit ±globular **159. G. byrnesii**

GROUP 10

(Flowers not bracteolate; style branched)

1 Style 4-branched **176. G. quadrifida**

1: Style 2- or 3-branched

2 Style 2-branched

3 Corolla pouch prominent, almost spurred; sepals 1–1.5 mm wide **114. G. ochracea**

3: Corolla pouch not prominent; sepals narrower **113. G. berardiana**

2: Style 3-branched

4 Corolla purple or purplish brown

5 Corolla with dense, simple, closely appressed hairs and some glandular hairs outside; seeds smooth or colliculate

6 Corolla and pedicel with glandular hairs; corolla 8–12 mm long **174. G. symonii**

6: Corolla and pedicel with very few and inconspicuous glandular hairs; corolla 7–8 mm long **168. G. porphyrea**

5: Corolla glandular-hairy outside; seeds granulate or warty **175. G. purpurea**

4: Corolla yellow at least on wings

7 Bracts and pedicels glabrous or with mostly simple hairs

8 Corolla 8–15 mm long; wings of abaxial corolla lobes equal **169. G. pilosa**

8: Corolla c. 6 mm long; wing above auricle of abaxial corolla lobe narrower than opposite one **171. G. heteroptera**

7: Bracts and pedicels with mostly glandular hairs

9 Corolla with appressed, simple hairs and a few glandular ones outside; seeds smooth or colliculate **170. G. neglecta**

9: Corolla glandular-hairy outside; seeds granulate to warty

10 Corolla less than 8 mm long **171. G. heteroptera**

10: Corolla more than 10 mm long

11 Cauline leaves auriculate, usually with 2 broad teeth at base; median indusium ovate; corolla 14–20 mm long **172. G. holtzeana**

11: Cauline leaves petiolate or sessile, not auriculate; median indusium depressed-ovate; corolla 10–12 mm long **173. G. heppleana**

Subg. 1. Monochila

Goodenia subg. **Monochila** (G.Don) Carolin, *Fl. Australia* 35: 330 (1992)

Goodenia sect. *Monochila* G.Don, *Gen. Hist.* 3: 725 (1834). T: *G. scapigera* R.Br.

Stekhovia Vriese, *Natuurk. Verh. Holl. Maatsch. Wetensch. Haarlem* ser. 2, 10: 166 (1854). T: *S. scapigera* (R.Br.) Vriese, *fide* R.C.Carolin, *Fl. Australia* 35: 330 (1992).

Scaevola ser. *Parviflorae* Benth., *Fl. Austral.* 4: 86, 103 (1868). T: *S. fasciculata* Benth.

Shrubs, subshrubs or herbs. Leaves basal or cauline. Flowers in terminal or axillary thyrses or spikes; pedicel bracteolate, usually not articulate. Sepals wholly adnate to ovary basally, by the midline to near top. Corolla white, cream or yellow, (bluish in *G. watsonii*), often with purplish markings, with ±stiff hairs in throat, without enations or auricles; pocket inconspicuous, to 1/2 ovary length. Ovary with septum c. 2/3 locule length; ovules 2-rowed in each locule. Fruit either capsular, valves 2, persistent, entire; or nutlike, indehiscent. Seeds black, pale or dark brown, smooth or colliculate-punctate; wing white, mucilaginous, less than 0.5 mm wide or obsolete.

The subgenus consists of 10 species in south-western Australia. All species have an oblong indusium, and a corolla without auricles or conspicuous pockets.

1. Goodenia scapigera R.Br., *Prodr.* 578 (1810)

Stekhovia scapigera (R.Br.) Vriese, *Natuurk. Verh. Holl. Maatsch. Wetensch. Haarlem* ser. 2, 10: 167 (1854). T: Bay I [Lucky Bay], [W.A.], 11 Jan. 1802, *R.Brown*; holo: BM.

Scaevola stricta Vriese in J.G.C.Lehmann, *Pl. Preiss.* 1: 408 (1844). T: near Konkoberup Hill [Mt Melville, near Cape Riche], W.A., 19 Nov. 1840, *L.Preiss 1511*; lecto: LD, *fide* R.C.Carolin, *Telopea* 3: 507 (1990); isolecto: G, L, MEL, P.

G. scapigera var. *parviflora* Benth., *Fl. Austral.* 4: 57 (1868). T: Phillips Ra. [i.e. The Barrens], W.A., *G.Maxwell*; lecto: K, *fide* R.C.Carolin, *Fl. Australia* 35: 332 (1992); isolecto: MEL.

Illustration: R.Erickson *et al.*, *Fl. Pl. W. Australia* 84, t. 234 (1973).

Erect perennial to 1 m tall, glabrous. Leaves cauline, ±clustered towards stem apices, linear to narrowly obovate, entire or dentate, thick; lamina 4–6 cm long, 2–10 mm wide. Flowers in thyrses to 20 cm long; bracts linear, c. 5 cm long; peduncle to 20 mm long; bracteoles linear, c. 2 mm long; pedicel 1–3 mm long. Sepals linear to triangular, 7–8 mm long. Corolla 8–18 mm long; lobes ±equal, c. 7–11 mm long; wings c. 1 mm wide. Ovary with septum c. 1/2 its length; ovules 30–40; style with short, simple hairs and glandular hairs. Fruit a capsule, 2-valved, ovoid to ellipsoidal, c. 5–7 mm long. Seeds elliptic, c. 1 mm long. $n = 8$, W.J.Peacock, *Proc. Linn. Soc. New South Wales* 88: 11 (1963).

Occurs in south-western W.A., usually in sandy soil in woodland and heath. Flowers chiefly Sept.–Jan. Map 191.

W.A.: Cut Hill, The Lakes to York road, *F.Lullfitz, L 1691* (PERTH); Pingelly, *C.A.Gardner 1020* (PERTH); Ravensthorpe, *R.C.Carolin 3582* (SYD); Ellen Peak, *A.Morrison 12290* (K); base of Bluff Knoll, *R.D.Royce 6039* (PERTH).

The corolla is white with purplish spots in throat. There is considerable variation in leaf shape in this species, e.g. from almost linear to narrowly obovate, and from entire to serrate. One specimen (near Lake Hope, Sept. 1929, *C.A.Gardner*, PERTH), has very narrow leaves. Cytological vouchers are *W.J.Peacock 6095.6* and *6092.1* (SYD).

2. Goodenia decursiva W.Fitzg., *J. W. Austral. Nat. Hist. Soc.* 2: 25 (1905)

T: Esperance, W.A., Oct. 1903, *C.Andrews*; lecto: PERTH, *fide* R.C.Carolin, *Telopea* 3: 518 (1990); isolecto: K, NSW, PERTH.

G. scapigera β R.Br., *Prodr.* 578 (1810). T: Bay I [Lucky Bay], [W.A.], 2 Jan. 1803, *R.Brown*; holo: BM.

G. scapigera var. *foliosa* F.Muell. ex Benth., *Fl. Austral.* 4: 57 (1868); *G. foliosa* (F.Muell. ex Benth.) Domin, *Vestn. Král. Ceské Spolecn. Nauk, Tr. Mat.-Prír.* 2: 111 (1923). T: Cape Arid and Cape le Grand, W.A., *G.Maxwell*; lecto: K, *fide* R.C.Carolin, *Telopea* 3: 518 (1990); isolecto: MEL.

Illustration: K.Krause, *Pflanzenr.* 54: 43, fig. 12J–M (1912).

Erect shrub to 1 m tall, glabrous. Leaves cauline, dense, ascending, ±obscuring stem, ±stem-clasping, decurrent, elliptic to ovate, dentate; lamina 10–35 mm long, 8–12 mm wide. Flowers in compact thyrses to 10 cm long; bracts linear, to 5 cm long; peduncle to 10 cm long; bracteoles linear-lanceolate, c. 7 mm long; pedicel 1–2 mm long. Sepals linear, 9 mm long, acute, entire, sometimes slightly viscid. Corolla 15–18 mm long lobes

±equal, c. 8 mm long; wings c. 2 mm wide. Ovary with septum almost to top; ovules 30–40; style with simple and glandular hairs. Fruit a cylindrical capsule, 9 mm long. Seeds elliptic, c. 1 mm long. Fig. 46D.

Occurs along the south coast from Esperance to Israelite Bay, W.A., in heath on granite outcrops mostly near the sea. Flowers Sept.–Jan. Map 192.

W.A.: Dempster Head, Esperance, *W.E.Blackall 1079* (PERTH); Duke of Orleans Bay, 30 Nov. 1950, *J.H.Willis* (MEL); Cape le Grand, 1881, *Webb* (MEL); near Cape Arid, 1875, *G.Maxwell* (MEL); Israelite Bay, 1885, *S.Brooke* (MEL).

The corolla is white with purplish spots in throat. Similar to *G. scapigera*, differs in having broad, stem-clasping, decurrent leaves and compact inflorescences.

3. **Goodenia watsonii** F.Muell. & R.Tate, *Bot. Centralbl.* 53: 268 (1893) as *G. watsoni*

T: near Gnarlbine, W.A., 12 Nov. 1891, *R.Helms, Elder Explor. Exped.*; lecto: MEL, *fide* R.C.Carolin, *Telopea* 3: 518 (1990); isolecto: AD, K, MEL, NSW.

Perennial to 50 cm tall, glabrous or glandular-hairy; stout stock. Leaves mostly basal, narrowly obovate to narrowly elliptic, dentate to entire, thick; lamina 3–15 cm long, 4–10 mm wide. Flowers in thyrses to 20 cm long; bracts linear, to 5 mm long; peduncle to c. 4 cm long; bracteoles linear, 1–3 mm long; pedicel 1–2 mm long. Sepals narrowly ovate, 1–2 mm long. Corolla 4–8 mm long; lobes ±equal, 2–4 mm long; wings c. 0.5 mm wide. Ovary with septum more than 1/2 its length; ovules 12–40; style with simple and glandular hairs. Fruit a capsule, ±globular, 4–5 mm long. Seeds ellipsoidal to orbicular, c. 0.5 mm diam.

Occurs in south-western W.A. in heath. Corolla white, cream or bluish. Differs most conspicuously from all other species in the section in the short basal stock with basal leaves. Two subspecies are recognised.

Flowers glabrous outside **3a.** subsp. **watsonii**

Flowers glandular-hairy outside **3b.** subsp. **glandulosa**

3a. **Goodenia watsonii** F.Muell. & R.Tate subsp. **watsonii**

Plant glabrous or nearly so. Corolla usually more than 5 mm long, white. Ovules usually more than 15.

Occurs in drier parts of south-western W.A. Flowers Oct.–Dec. Map 193.

W.A.: Wubin–Kalannie, *W.E.Blackall* (PERTH); Mollerin Rock, *S.B.Rosier 172* (PERTH); c. 25 km S of Tammin, *R.D.Royce 9343* (PERTH); c. 1 km from Lake King towards Ravensthorpe, *E.M.Canning 7341* (CBG, SYD).

3b. **Goodenia watsonii** subsp. **glandulosa** Carolin, *Telopea* 3: 518 (1990)

T: 20 miles [c. 32 km] E of Hyden, W.A., 12. Jan. 1965, *J.S.Beard 3908*; holo: PERTH.

Plant with peduncles, pedicels, ovary, sepals and outside of corolla glandular-pubescent. Corolla usually 4–5 mm long, blue with yellow throat, or whitish. Ovules 12–16.

Restricted to Lake Grace–Hyden area, W.A. Flowers Oct.–Jan. Map 194.

W.A.: c. 13 km W of Lake Grace, *A.S.George 520* (PERTH); c. 26 km W of Lake Grace, *W.E.Blackall 1314* (PERTH); c. 19 km N of Lake Grace, *P.R.Jefferies 641029* (PERTH).

4. Goodenia pinifolia Vriese, *Natuurk. Verh. Holl. Maatsch. Wetensch. Haarlem* ser. 2, 10: 157 (1854)

T: Swan R., 1839, W.A., *J.Drummond s.n.*; lecto: K, *fide* R.C.Carolin, *Telopea* 3: 519 (1990); isolecto: MEL.

Erect or spreading shrub to 1.5 m tall, glabrescent. Leaves cauline, linear to subulate, entire; lamina 2–6 cm long, 1–3 mm wide. Flowers in loose thyrso-racemes to 15 cm long; scape naked; bracts linear, to 5 mm long; peduncle to 6 cm long; bracteoles linear, to 2 mm long; pedicel to 8 cm long. Sepals linear to deltoid, 2–6 mm long. Corolla 9–15 mm long; lobes unequal, abaxial 6–7 mm long; wings 0.5–1 mm wide. Ovary with septum about 1/2 its length; ovules 12–14; style with white hairs. Fruit a subglobular capsule, 2–3 mm diam. Seeds elliptic, c. 1 mm long. $n = 8$, W.J.Peacock, *Proc. Linn. Soc. New South Wales* 88: 11 (1963).

Occurs in south-western W.A. between Perenjori and Ravensthorpe. Flowers Oct.–Dec. Map 195.

W.A.: c. 6 km N of Perenjori, *N.T.Burbidge 4698* (CANB, PERTH); c. 6 km E of Wubin, *F.Lullfitz 102* (PERTH); Wongan Hills, *F.Stoward 867* (BM, K); c. 6 km N of Wyalkatchem, *J.W.Green 834* (PERTH); c. 10 km from Ravensthorpe, 27 Oct. 1968, *J.W.Wrigley* (CANB, SYD).

The corolla is white or pale blue. There is some variation within the species with respect to length of indusium and of leaves. Similar to *G. elderi*, see note under that species. Cytological voucher is *W.J.Peacock 60816.1* (SYD).

5. Goodenia elderi F.Muell. & R.Tate, *Bot. Centralbl.* 53: 268 (1893)

T: near Warangering, W.A., 14 Nov. 1891, *R.Helms*; holo: MEL; iso: AD, K, NSW.

Erect or virgate shrub to 80 cm tall, white-tomentose when young, glabrescent. Leaves cauline, ±fasciculate, linear, entire, thick; lamina 5–20 mm long, 1–1.5 mm wide. Flowers in spreading racemes or thyrses 4–12 cm long; scape slender; bracts linear, to 5 mm long; peduncle to 2 cm long; bracteoles linear, to 2 mm long; pedicel to 12 mm long. Sepals linear, 4–6 mm long. Corolla 9–11 mm long; lobes unequal, abaxial to 7 mm long; wings c. 1 mm wide. Ovary with septum 1/2 its length; ovules 12–16; style with stiff, white hairs and minute, glandular hairs. Fruit an elliptic capsule, 5–6 mm long. Seeds elliptic, c. 1 mm long. Fig. 46B–C.

Occurs between Karalee (W of Coolgardie) and Queen Victoria Spring, W.A., in sandy soil. Flowers chiefly Sept.–Jan. Map 196.

W.A.: Karalee, *J.W.Green 1337* (PERTH); Coolgardie goldfields, *E.Pritzel 856* (AD, NSW); Coolgardie, 1900, *L.C.Webster* (NSW); Kalgoorlie, 13 Dec. 1943, *H.M.Wilson* (PERTH); c. 20 km N of Cundeelee Mission, N of Zanthus, *A.S.George 6001* (PERTH).

The corolla is white with purplish markings, with brownish markings in throat. Similar to *G. pinifolia*, differs in its more elongated fruit, glabrous indusial lips (*G. pinifolia* has bristles on the indusium), and generally smaller flowers and shorter leaves.

6. Goodenia fasciculata (Benth.) Carolin, *Telopea* 3: 519 (1990)

Scaevola fasciculata Benth. in S.L.Endlicher *et al.*, *Enum. Pl.* 68 (1837); *Lobelia fasciculata* (Benth.) Kuntze, *Revis. Gen. Pl.* 2: 378 (1891). T: Swan River, W.A., *C.Hügel*; holo: W (photo SYD).

G. squarrosa Vriese in J.G.C.Lehmann, *Pl. Preiss.* 1: 413 (1844). T: Maddington, W.A., 2 Nov. 1839, *L.Preiss 1467*; lecto: LD, *fide* R.C.Carolin, *Telopea* 3: 519 (1990); isolecto: K, MEL.

Scaevola fasciculata var. *parviflora* E.Pritzel in F.L.E.Diels & E.Pritzel, *Bot. Jahrb. Syst.* 35: 572 (1905). T: near Dandaragan, W.A., Dec. 1901, *F.L.E.Diels 5767*; holo: ?B (destroyed) *n.v.*

Illustration: R.Erickson *et al.*, *Fl. Pl. W. Australia* 43, t. 97 (1973) as *Scaevola fasciculata*.

Ascending shrub to 1 m tall, villous when young. Leaves cauline, fasciculate, narrowly linear; margins entire, revolute; lamina 6–12 mm long, to 0.5 mm wide. Inflorescence a spike, 5–8 cm long; bracts similar to upper leaves, at least 1/2 as long as corolla; bracteoles 5–9 mm long. Sepals narrowly ovate, c. 0.5 mm long, with white hairs at margins. Corolla 5–8 mm long; lobes ±equal, 2–3.5 mm long; wings c. 1 mm wide. Ovary with septum almost obsolete; ovule solitary; style with stiff, purplish hairs and minute, glandular hairs. Fruit a nut, globular, 2 mm diam. Seed elliptic, c. 1 mm long. n = 8, W.J.Peacock, *Proc. Linn. Soc. New South Wales* 88: 14 (1963) as *Scaevola fasciculata*.

Occurs in the Darling Ra. and adjacent areas, south-western W.A. Flowers chiefly Sept.–Dec. Map 197.

W.A.: 65 mile peg [c. 104 km from Perth] on Geraldton Rd, Sept. 1948, *D.H.Perry* (PERTH); Red Hill, *T.E.H.Aplin 318 & 324* (PERTH); Gooseberry Hill, *A.Morrison 800* (K); Darling Ra., *E.Pritzel 51* (K, NSW); Kalamunda, Nov. 1924, *C.A.Gardner* (K, PERTH).

The corolla is white with purplish markings. Differs from other *Goodenia* species with indehiscent fruits, in its longer, leafy bracts and bracteoles, and more compact inflorescence. Cytological voucher is *W.J.Peacock 60875.5* (SYD).

7. **Goodenia stenophylla** F.Muell., *Fragm.* 1: 113 (1859)

Scaevola stenophylla (F.Muell.) Benth., *Fl. Austral.* 4: 104 (1868); *Lobelia stenophylla* (F.Muell.) Kuntze, *Revis. Gen. Pl.* 2: 378 (1891). T: Phillips Ranges [i.e. The Barrens], W.A., ?*G.Maxwell 137*; holo: MEL.

Erect shrub to 1 m tall, glabrous. Leaves cauline, fasciculate, linear to subulate, channelled above, entire; lamina 18–30 mm long, to 0.5 mm wide; axillary hairs few. Inflorescence a spike or spike-like thyrse, to 12 cm long; bracts triangular, not exceeding sepals; bracteoles linear, 1 mm long. Sepals ovate to elliptic, 1 mm long. Corolla 8–10 mm long; lobes ±equal, 4.5–6 mm long; wings c. 0.8 mm wide. Ovary septum obsolete, ovules 1–3; style with stiff, purplish hairs and minute, glandular hairs. Fruit a globular nut, 1.5 mm diam. Seeds elliptic, c. 1 mm long.

Occurs from the Fitzgerald R. Natl Park to Bremer Bay, south-western W.A. Flowers Sept.–Jan. Map 198.

W.A.: Young R., *G.Maxwell* (K); Whoogarup Ra., *A.S.George 7191* (PERTH); gorge of Fitzgerald R., *C.A.Gardner 9236* (PERTH); near Mt Barren, *W.E.Blackall 1444* (PERTH); Bremer Bay, *C.A.Gardner 6555* (PERTH).

The corolla is white with purplish spots in throat. Differs from other *Goodenia* species with indehiscent fruits in its broader sepals and more prominent corolla wings.

8. **Goodenia drummondii** Carolin, *Telopea* 3: 519 (1990)

T: south-western W.A., *J.Drummond s.n.*; holo: MEL; ?iso: MEL (4 sheets).

Erect shrub to 1 m tall, glabrous. Leaves cauline, fasciculate, linear to oblanceolate, entire or dentate, thick; lamina 1–4 cm long, 1–4 mm wide. Inflorescence a spike-like thyrse, to 20 cm long; bracts triangular, 1–2 mm long, not exceeding sepals; bracteoles similar, smaller. Sepals narrowly ovate to linear, 1–1.5 mm long. Corolla c. 6 mm long; lobes ±equal, c. 2.5 mm long; wings 0.3–0.4 mm wide. Ovary with septum obsolete; ovule solitary; style with stiff, purplish hairs and minute, glandular hairs. Fruit a globular nut, c. 1 mm diam. Seed elliptic, c. 1 mm long.

Occurs from Kalbarri Natl Park to Latham, W.A. Flowers Sept.–Jan. Map 199.

W.A.: Latham, *D.A.Sargent 1372* (MEL); c. 3 km along Balla Rd from Geraldton–Carnarvon Hwy, *A.C.Burns 1070* (PERTH); c. 43 km E of Kalbarri, *A.S.George 7926* (PERTH); Moresby Ra., *A.C.Burns 6* (PERTH); without locality, *J.Drummond 365* (K, MEL).

The corolla is white with purplish spots in throat. Similar to *G. stenophylla* and *G. helmsii*, differing in the former having larger flowers with broader corolla wings, and narrower entire leaves, and from the latter having shorter narrower leaves.

9. **Goodenia helmsii** (E.Pritzel) Carolin, *Telopea* 3: 520 (1990)

Scaevola helmsii E.Pritzel in F.L.E.Diels & E.Pritzel, *Bot. Jahrb. Syst.* 35: 572 (1905). T: near Southern Cross, W.A., Nov. 1901, *E.Pritzel 881*; lecto: NSW, *fide* R.C.Carolin, *Telopea* 3: 520 (1990); isolecto: ?B (destroyed) *n.v.*, K.

Erect or ascending shrub to 80 cm tall, glabrous except for copious axillary wool. Leaves cauline, fasciculate, linear-terete, entire, thick; lamina mostly c. 2 mm long but main stem leaves to 5 mm long, to 0.5 mm wide. Inflorescence a spike or spike-like thyrse, to 12 cm long; bracts ±flattened, to 0.7 mm long, not exceeding sepals; bracteoles deltoid, 0.5 mm long. Sepals deltoid, 0.5–0.7 mm long. Corolla 4–5.5 mm long; lobes equal, 1.5–2.5 mm long, wings to 0.3 mm wide. Ovary with septum obsolete; ovule solitary; style with stiff, purplish hairs and minute, glandular hairs. Fruit a nut, subglobular, c. 1 mm diam. Seeds elliptic, c. 0.6 mm long.

Occurs in inland south-western W.A. Flowers chiefly July–Dec. Map 200.

W.A.: c. 11 km S of Pindar, *F.Lullfitz L1177* (PERTH); c. 4 km SW of Kokardine siding, *A.S.George 11165* (PERTH); Wongan Hills, *R.D.Royce 2181* (PERTH); Bruce Rock, *R.C.Carolin 3134* (SYD); 19 km N of Northam, *S.Carlquist 897* (RSA, SYD).

The corolla is white with purplish spots in the throat and glabrous outside. Similar to *G. drummondii* and *G. fasciculata*, see notes under these species for differences. The specimens which occur furthest west (Darling Ra., near Perth), have longer leaves and slightly larger (5.5 mm long) corollas. These specimens overlap in range with *G. fasciculata*, which has longer leaves and corollas. One specimen in particular (*A.S.George 464*, PERTH) has leaves 6 cm long but the inflorescence is clearly not that of *G. fasciculata*, in many ways approaching *G. drummondii*.

10. **Goodenia viscida** R.Br., *Prodr.* 578 (1810)

Stekhovia viscida (R.Br.) Vriese, *Natuurk. Verh. Holl. Maatsch. Wetensch. Haarlem* ser. 2, 10: 168 (1854). T: Bay I [Lucky Bay], [W.A.], 12 Jan. 1802, *R.Brown*; lecto: BM, *fide* R.C.Carolin, *Telopea* 3: 520 (1990); isolecto: K.

G. spicata F.Muell., *Fragm.* 3: 35 (1862). T: south-western Australia, *G.Maxwell*; holo: MEL.

Illustration: K.Krause, *Pflanzenr.* 54: 43, fig. 12E–H (1912).

Erect perennial to 40 cm tall, glandular, viscid; stock woody. Leaves cauline, narrowly oblong to obovate, dentate; lamina 5–20 mm long, 2–5 mm wide. Inflorescence a spike, to 15 cm long: bracts leaf-like; bracteoles narrowly oblong, c. 7 mm long. Sepals triangular, c. 2.5 mm long. Corolla 6–7 mm long, with short hairs inside; lobes equal, 3–4 mm long; wings 0.3–0.4 mm wide. Ovary with septum about 1/2 its length; ovules 12–16; style with stiff, simple, purplish or white hairs and glandular hairs. Fruit a subglobular capsule, 3 mm long. Seeds elliptic, 1.4 mm long, smooth, glossy; wing c. 0.1 mm wide, white, mucilaginous.

Occurs from near Esperance to the Stirling Ra., W.A., in sandy loam in depressions. Flowers Oct.–Jan. Map 201.

Figure 46. *Goodenia*. **A**, *G. gloeophylla,* habit X0.5 (N.Byrnes, 2 May 1969, NSW). **B–C**, *G. elderi*. **B**, habit X0.5; **C**, flower X2 (**B–C**, A.Strid 21318, SYD). **D**, *G. decursiva,* habit X1 (H.Eichler 19893, SYD). **E**, *G. paniculata,* flower X2 (C.Burgess CBG 027336, NSW). **F–G**, *G. gracilis*. F, flower X2; **G**, indusium X10 (**F–G**, P.Ollerenshaw 1485 & D.Kratzing, SYD). Drawn by D.Mackay.

W.A.: Forrestania, 19 Jan. 1929, *C.A.Gardner* (K, PERTH); Young R., 29 Jan. 1935, *C.A.Gardner* (K, PERTH); c. 24 km E of Esperance on Cape le Grand road, *A.S.George 2198* (PERTH); near Whoogarup Ra., *A.S.George 1969* (PERTH); Toolbrunup, *A.Meebold 7059* (AD, K).

Corolla yellow. Difficult to place systematically. It apparently belongs to subg. *Monochila* by virtue of the palmate form of the corolla, the typical indusium shape, (athough this is somewhat obscured by being notched and ±folded), and the form of the hairs on the style. In the rest of the subgenus the yellow corolla is unknown, while viscidness only occurs in one other species, *G. watsonii*.

Subg. 2. Goodenia

Goodenia Smith subg. **Goodenia**

Subshrubs or herbs. Leaves basal or cauline. Flowers in terminal and axillary thyrses or racemes; pedicel bracteolate or ebracteolate, articulate or not. Sepals adnate to ovary for varying distances. Corolla yellow, blue or purplish, sometimes with long, stiff hairs on margins and sometimes in throat, enations often present, auriculate or not; pocket various. Ovary with septum 2/3 as long as locule to very short; ovules 2-rowed in each locule or scattered. Fruit a capsule; valves 2, persistent or deciduous. Seeds various; wing narrow or broad.

The subgenus contains 169 species, 168 Australian, 3 extending to New Guinea, 1 to the Philippines and 1 endemic in Java.

Sect. 1. Porphyranthus

Goodenia subg. **Goodenia** sect. **Porphyranthus** G.Don, *Gen. Hist.* 3: 725 (1834)

Type: *G. purpurascens* R.Br.; lecto, *fide* R.C.Carolin, *Fl. Australia* 35: 330 (1992).

Herbs with basal stock. Leaves mostly basal (except *G. gloeophylla*). Flowers in terminal, spreading, thyrses or racemes; pedicel bracteolate, articulate. Sepals adnate to ovary ±to top. Corolla yellow or blue, glabrous inside or with few, sometimes long and stiff, hairs, enations sometimes present, sometimes auriculate; pocket inconspicuous. Ovary with septum 2/3 to almost as long as locule; ovules more than 30, scattered over placentas. Fruit a capsule; valves 2, sometimes split to base and thus appearing as 4, persistent. Seeds elliptic to orbicular, mostly less than 1 mm wide, reticulate-foveate, glossy; wing c. 0.1 mm wide, mucilaginous.

A section of 15 species, mostly in eastern and northern Australia, extending into the arid central areas and 1 species also in New Guinea.

11. Goodenia gracilis R.Br., *Prodr.* 575 (1810)

T: Broad Sound, [Qld], 15 Sept. 1802, *R.Brown*; lecto: BM, *fide* R.C.Carolin, *Telopea* 3: 533 (1990); isolecto: K.

G. semiteres Domin, *Biblioth. Bot.* 89: 644 (1929). T: near Pentland, Qld, Mar. 1910, *K.Domin 8788*; lecto: PR, *fide* R.C.Carolin, *Fl. Australia* 35: 332 (1992); isolecto: PR.

Illustration: G.M.Cunningham *et al.*, *Pl. W. New South Wales* 633 (1981).

Annual or perennial herb to 50 cm tall, almost glabrous; tap root to 50 cm long. Leaves linear-lanceolate, entire or dentate, thick; lamina 5–17 mm long, 1–5 mm wide. Flowers in racemes 3–20 cm long; bracts linear, to 20 mm long; peduncle mostly 10–25 mm long; bracteoles 1–2 mm long; pedicel mostly 8–20 mm long. Sepals narrowly triangular, 1–1.4 mm long. Corolla 7–10 mm long, with blunt enations, auriculate; abaxial lobes 2.5–4 mm long; wing c. 1 mm wide. Indusium notched and folded, the two sides of the lower lip appressed. Ovules many, scattered. Fruit subglobular to ovoid, 3–6 mm long; valves entire. Seeds brown. Fig. 46F–G.

Widespread in eastern Australia in Qld, N.S.W. and Vic., in damp places in heavy soil. Flowers most of the year; in the south mainly in spring and summer; in the north in autumn and winter. Map 202.

Qld: Camp-Oven Waterhole, *R.C.Carolin 9166* (SYD). N.S.W.: Wee Waa, Oct. 1916, *H.M.R.Rupp* (NSW); c. 48 km SE of Deniliquin, 17 Nov. 1964, *J.H.Leigh* (NSW); Gerogery, *E.J.McBarron 3225* (NSW, SYD). Vic.: near Winton Swamp, 15 Apr. 1959, *H.I.Aston* (MEL).

The corolla is yellow. Both corolla and ovary are glabrous, or simple- or glandular-pubescent outside. The species is distinctive in its closely folded indusium.

12. **Goodenia paniculata** Smith, *Trans. Linn. Soc. London, Bot.* 2: 348 (1794)

T: Port Jackson, N.S.W., 1793, *D.Burton*; holo: LINN (photo SYD); iso: K.

G. flexuosa Vriese, *Natuurk. Verh. Holl. Maatsch. Wetensch. Haarlem* ser. 2, 10: 126 (1854). T: Port Jackson, N.S.W., *F.Bauer*; holo: W.

G. rosulata Domin in F.L.E.Diels, *Biblioth. Bot.* 89: 644 (1929). T: near Jericho, Qld, Mar. 1910, *K.Domin 8783*; lecto: PR, *fide* R.C.Carolin, *Telopea* 3: 533 (1990).

Short-lived herb to 50 cm tall, pubescent or glabrous; root system a short tap-root and numerous adventitious roots. Leaves obovate to oblanceolate, dentate; lamina 14–100 mm long, 6–10 mm wide. Flowers in racemes to 25 cm long; bracts linear to narrowly elliptic, 4–40 mm long, 0.5–2 mm wide; peduncle 7–80 mm long; bracteoles 1.5–2 mm long; pedicel 6–16 mm long, articulate. Sepals triangular to lanceolate, 1–2 mm long. Corolla 10–14 mm long, with or without a very few indistinct enations; abaxial lobes 5–6 mm long; wing c. 1.5 mm wide. Indusium truncate-obtriangular, purplish, villous towards base, concave. Ovules many, scattered. Fruit globular-ovoid, 5–6 mm long; valves entire. Seeds brown. $2n = 32$, $n = 16$, W.J.Peacock, *Proc. Linn. Soc. New South Wales* 88: 13 (1963). Fig. 46E.

Occurs in eastern Australia from northern Qld, through N.S.W. into north-eastern Vic. in damp sandy places. Flowers chiefly Oct.–Apr. Map 203.

Qld: c. 24 km W of Pentland, *R.C.Carolin 8477* (SYD); Mooloolaba, Aug. 1962, *J.Galbraith* (MEL). N.S.W.: c. 11 km E of Murwillumbah, *E.F.Constable 6532* (NSW); Mellong Swamp, *R.C.Carolin 3862* (SYD); Dapto, *E.J.McBarron 4157* (NSW). Vic.: near Maramingo Ck, Genoa, 27 Dec. 1957, *M.Allender* (MEL).

The corolla is yellow. Both corolla and ovary have glandular and simple hairs outside. See also *G. macbarronii*. The specimens from the area where the type of *G. rosulata* was collected have broader leaves and paler glandular hairs than those from further south. The leaf width, however, is very variable throughout the species and there does seem to be a gradation in gland colour.

13. **Goodenia macbarronii** Carolin, *Telopea* 3: 533 (1990)

T: Holbrook, N.S.W., 10 Feb. 1947, *E.J.McBarron 647*; holo: NSW; iso: SYD.

Herb to 30 cm tall, glabrous; roots adventitious and sometimes a persistent taproot. Leaves oblanceolate, dentate, thick; lamina 5–11 cm long, 2–5 mm wide. Flowers in racemes to 25 cm long; bracts linear, 2–20 mm long; peduncle mostly to 20 mm long; bracteoles triangular, 1–1.5 mm long; pedicel 0.5–3 mm long. Sepals linear to elliptic, 1–2 mm long. Corolla 7–9 mm long, with obscure enations, auriculate; abaxial lobes 4–4.5 mm long; wings c. 1 mm wide. Indusium square, purplish towards top, flat. Ovules many, scattered. Fruit ovoid, 3–4 mm long; valves entire. Seeds yellow-brown. n = 8, W.J.Peacock, *Proc. Linn. Soc. New South Wales* 88: 13 (1963), as *G. gracilis*.

Occurs on the western slopes of the Great Dividing Ra., N.S.W. and in north-eastern Vic. in damp, usually sandy places. Flowers chiefly Oct.–Mar. Map 204.

N.S.W.: Tingha, Mar. 1917, *J.L.Boorman* (NSW); Warrumbungle Ra., *E.Betche 60* (MEL); c. 24 km E of Rylstone, *H.S.Mckee 444* (SYD); Khyber Pass, Rylstone, *W.J.Peacock 604.26.1* (SYD). Vic.: Ovens R., 9 Feb. 1853, *F.Mueller* (MEL).

Corolla is yellow, glandular- and ±simple-pubescent outside. Cytological voucher is *W.J.Peacock 604.26.1* (SYD). Previously confused with *G. gracilis* which has a closely folded indusium, and with *G. paniculata* which has thinner wider leaves, larger, obtriangular, entirely purple indusium, larger flowers and different chromosome number. The specimens from Vic. have larger flowers.

14. Goodenia humilis R.Br., *Prodr.* 575 (1810)

T: Arthurs Seat, Port Phillip, [Vic.], 25 Jan. 1804, *R.Brown*; lecto: BM, *fide* R.C.Carolin, *Telopea* 3: 534 (1990); isolecto: K, MEL, P.

G. graminifolia J.D.Hook., *Hooker's London J. Bot.* 4: 265 (1847). T: George Town, Tas., *R.C.Gunn*; holo: K.

G. nana Vriese, *Natuurk. Verh. Holl. Maatsch. Wetensch. Haarlem* ser. 2, 10: 132 (1854). T: Flinders Is., Tas., *R.C.Gunn 1177*; holo: K. (W.H.de Vriese, *loc. cit.* (1854) gives no. *1777* which is apparently a misprint.)

G. humilis var. *alpigena* F.Muell., *Monthly Not. Pap. & Proc. Roy. Soc. Tasmania for 1873 62* (1874). T: near Lake St Clair, Tas., *Th. & B.Gulliver*; holo: MEL.

Illustration: G.R.Cochrane *et al.*, *Fl. Pl. Victoria & Tasmania* 27, fig. 61 (1980).

Weak, perennial herb to 20 cm tall, softly hairy; root system a tap root and adventitious roots. Leaves linear to oblanceolate, dentate or entire, thick, often glabrescent; lamina mostly 4–10 cm long, 4–8 mm wide. Flowers in racemes; scape to 1.5 times leaf length, usually slightly zig-zag; bracts linear, 5–14 mm long; peduncle 5–15 mm long; bracteoles linear, 3–4 mm long; pedicel 5–12 mm long. Sepals linear-lanceolate, 1.5–3 mm long. Corolla 8–12 mm long, without enations, auriculate; abaxial lobes 4–5 mm long; wing c. 1 mm wide. Indusium square to depressed-oblong, purplish, flat or convex. Ovules many, scattered. Fruit ovoid, 3–3.5 mm long; valves entire. Seeds yellowish brown.

Occurs in south-eastern Australia in S.A., Vic. and Tas. in damp places. Flowers chiefly Nov.–Mar. Map 205.

S.A.: 3 km SSW of Tarpeena, *P.Wilson 1244* (AD); c. 10 km W of Naracoorte, *D.Hunt 385* (AD). Vic.: The Grampians, *R.Filson 785* (MEL). Tas.: Flinders Is., *J.Milligan 573* (BM); Macquarie Harbour, *coll. illegible 889* (MEL23369).

The corolla is yellow and sometimes purplish on the auricles. Distinguished from other members of the section by the dense, soft indumentum and the scapes which are usually slightly zig-zag and not as long as the leaves, or only slightly exceeding them.

15. Goodenia lamprosperma F.Muell., *Fragm.* 1: 116 (1859)

G. gracilis var. *lamprosperma* (F.Muell.) Ewart, *Proc. Roy. Soc. Victoria* ser. 2, 38: 86 (1926). T: upper Victoria R., [N.T]., Dec. 1855, *F.Mueller*; lecto: MEL 23395, *fide* R.C.Carolin, *Telopea* 3: 534 (1990); isolecto: K.

Perennial herb to 50 cm tall, glabrous or glabrescent; tap root thick. Leaves usually with 1 or 3 conspicuous nerves, linear-oblanceolate, dentate to entire, ±glaucous; lamina 3–8 cm long, 3–10 mm wide. Flowers in racemes to 35 cm long; bracts linear, 5 cm long; peduncle to 8 cm long; bracteoles linear, 2–5 mm long; pedicel 3–6 mm long. Sepals lanceolate to narrowly oblong, 0.5–2 mm long. Corolla 8–10 mm long, without enations, auriculate; abaxial lobes 1.5–3 mm long; wings 0.5–1.5 mm wide. Indusium oblong, 1 mm long, concave. Ovules many, scattered. Fruit cylindrical, 3–4 mm long; valves entire. Seeds yellow-brown.

Occurs in northern W.A., N.T. and north-western Qld, in seasonally damp areas, usually in sandy soil but sometimes in heavier soil. Flowers chiefly Apr.–Aug. Map 206.

W.A.: c. 32 km S of Wyndham, *N.T.Burns 5796* (CANB); Dampier, *R.C.Carolin 7883* (SYD); Tragedy Paddock, Mount Anderson, *H.F.Broadbent 601* (PERTH). N.T.: Adelaide River–Marraki road, *N.Byrnes NB853* (DNA). Qld: c. 120 km W of Croydon on Normanton road, *R.C.Carolin 8658* (SYD).

The corolla has simple hairs and minute, glandular hairs outside. May have an aquatic stage before flowering, with long-petiolate leaves with floating lamina. Differs from *G. purpurascens* in the colour of the corolla and from *G. gracilis* in its concave indusium. The indumentum on the corolla varies to some extent; the specimens from the Ashburton R. area, W.A., tend to have scarcely any glandular hairs whilst those around Darwin, N.T., have mostly glandular hairs. The species has sometimes been distinguished by the prominent veins in the leaves; this feature will usually separate it from *G. gracilis* but not necessarily from other members of the section.

16. Goodenia bicolor F.Muell. ex Benth., *Fl. Austral.* 4: 80 (1868)

T: between Macadam Ra., and Providence Hill, [N.T.], *F.Mueller*; *n.v.* (the specimen – 'near Table Hill, Oct. 1855, *F.Mueller*' (K) may be a type).

G. propinqua W.Fitzg., *J. & Proc. Roy. Soc. W. Australia* 3: 213 (1918). T: Inglis Gap, King Leopold Ra., W.A., *W.V.Fitzgerald*; lecto: NSW, *fide* R.C.Carolin, *Telopea* 3: 534 (1990).

Annual or ephemeral herb to 40 cm tall, with soft, glandular hairs and simple hairs; roots mostly adventitious. Leaves obovate to lanceolate, dentate, thin; lamina 20–50 mm long, 6–20 mm wide. Flowers in racemes to 30 cm long; bracts narrowly elliptic, to 3 mm long; peduncle 2–5 mm long; bracteoles narrowly elliptic, 2 mm long; pedicel resembling peduncle. Sepals narrowly ovate, 1.5 mm long, 0.7 mm wide. Corolla 5–8 mm long, without enations, auriculate; abaxial lobes c. 3 mm long; wings 0.8 mm wide, unequal on adaxial lobes. Indusium oblong, 11 mm long, villous basally, concave. Ovules many, scattered. Fruit ellipsoidal, 3–4 mm long, 1.5 mm wide; valves usually entire. Seeds brown.

Occurs in the north Kimberley, W.A., and Victoria R. area, N.T.; usually in scrub communities in stony run-off areas. Flowers Mar.–June. Map 207.

W.A.: Kimberley, 24 June 1921, *C.A.Gardner* (PERTH); c. 115 km NNW of Gibb River Stn, *N.H.Speck 4955* (CANB); c. 2.5 km W of Lake Argyle, *A.C.Beauglehole* (MEL); Packsaddle Ck, S of Kununurra, *K.F.Kenneally 1931* (PERTH, SYD).

Has a bicoloured (yellow and purple) corolla unlike all other species in the section.

17. **Goodenia gloeophylla** Carolin, *Telopea* 3: 534 (1990)

T: Longini Landing, Kalumburu Mission, W.A., 29 May 1971, *D.E.Symon 7125*; holo: AD; iso: CANB, PERTH, SYD.

Erect, slightly woody subshrub to 50 cm tall, viscid, glandular-hairy; root a tap root. Leaves cauline, linear to narrowly elliptic, dentate or narrowly lobed; lamina 4–8 cm long, 1–2.5 mm wide. Flowers in racemes to 35 cm long; bracts leaf-like, not exceeding the flower; peduncle to 18 mm long; bracteoles linear, 3–8 mm long; pedicel 2–4 mm long. Sepals narrowly deltoid, 1–1.5 mm long, 0.3–0.5 mm wide. Corolla 10–12 mm long, without enations, auricle indistinct; abaxial lobes 5–6 mm long; wings c. 1 mm wide. Indusium broadly ovate, folded, with a few long hairs towards base. Ovules many, scattered. Fruit cylindrical, 8–10 mm long; valves bifid. Seeds brown. Fig. 46A.

Occurs in the north Kimberley, W.A., and northern N.T., in heath and scrub in sandy soil. Flowers about Apr.–May. Map 208.

N.T.: c. 3 km N of El Sharana, *P.Martensz & R.Schodde AE 565* (DNA); Waterfall Ck, c. 1 km above falls (South Alligator), *N.T.Byrnes NB1518* (DNA); 13°03'S, 132°56'E, *M.Lazarides 8011* (CANB); c. 65 km from Pine Creek to U.D.P. Falls, *C.H.Gittins 2846* (BRI, SYD).

The corolla is pale to deep purple. Differs from other species in the section in its ±woody stem. It also has a prominent pocket on the corolla.

18. **Goodenia purpurascens** R.Br., *Prodr.* 578 (1810)

T: Bay No. 3, Point 2, [Arnhem Bay, N.T.], 3 Mar. 1803, *R.Brown*; lecto: BM, *fide* R.C.Carolin, *Telopea* 3: 535 (1990).

Usually perennial herb, to 50 cm tall, glabrescent; roots adventitious. Leaves mostly basal, ascending, linear to oblanceolate, sparsely and minutely dentate, thick, usually glaucous, often densely hairy when young; lamina 3–20 cm long, 2–15 mm wide. Flowers in spreading thyrses or panicles to 30 cm long; bracts linear, to 20 mm long; peduncle 1–4 cm long; bracteoles linear, 1–2 mm long; pedicel 5–10 mm long. Sepals lanceolate, 1–1.5 mm long. Corolla 8–12 mm long, without enations, auriculate; abaxial lobes 3–5 mm long; wings 1–2 mm wide. Indusium oblong, 1.5 mm long. Ovules many, scattered. Fruit ovoid to subglobular, 2–4 mm long; valves entire. Seeds yellow-brown.

Occurs in the north-eastern Kimberley, W.A., northern N.T. and Qld, in grey cracking soil and in creek beds; also in New Guinea. Flowers chiefly Jan.–May. Map 209.

N.T.: Cahill Crossing, East Alligator R., *R.C.Carolin 6893* (SYD); Canon Hill Airstrip, *P.Martensz AE756* (DNA); Port Darwin, *F.Schultze 463* (MEL); c. 115 km E of Carlton Stn, *R.A.Perry 2607* (CANB). Qld: Settlement Ck, *L.J.Brass 333* (CANB).

The corolla is purple, and glandular- and simple-hairy outside. Similar to *G. paludicola* and *G. minutiflora*, differing in their much smaller corollas. Similar to *G. viscidula*, differing in its definitely glandular-pubescent leaves clammy to the touch. The plants growing in creek beds differ from clay soil plants in having a greater proportion of glandular hairs on the flowers.

19. **Goodenia minutiflora** F.Muell., *Fragm.* 8: 244 (1874)

T: between the Norman and Gilbert Rivers, Qld, 1874, *T.Gulliver 68*; lecto: MEL, *fide* R.C.Carolin, *Telopea* 3: 535 (1990); isolecto: MEL.

G. purpurascens var. *minima* F.Muell. ex Benth., *Fl. Austral.* 4: 78 (1868); *G. minima* (F.Muell. ex Benth.) Domin, *Biblioth. Bot.* 22: 643 (1929). T: upper Victoria R., [N.T.], F.Mueller; holo: K.

[*G. lamprosperma auct. non* F.Muell.: F.M.Bailey, *Syn. Queensl. Fl.* 3rd Suppl., 37 (1888); *Queensl. Fl.* 904 (1900)].

Annual herb to 20 cm tall; roots adventitious. Leaves mostly basal, ascending, linear-oblanceolate, ±entire, glabrous; lamina 2.5–10 cm long, 2–5 mm wide. Flowers in loose, terminal thyrses or racemes to 15 cm long; bracts linear, to 15 mm long; peduncle 4–10 mm long; bracteoles linear, to 1.5 mm long; pedicel 5–7 mm long. Sepals lanceolate, c. 1 mm long. Corolla 2–3 mm long, without enations, auricles indistinct; abaxial lobes 0.5 mm long; wings absent or very narrow. Indusium square, c. 0.5 mm long, ±flat. Ovules many, scattered. Fruit subglobular, c. 3 mm diam.; valves entire. Seeds yellowish.

Occurs mainly along the southern shores of the Gulf of Carpentaria, in N.T. and Qld; in seasonally damp places, particularly in relatively sandy soil. Flowers chiefly Mar.–May. Map 210.

Qld: between Peters and Cliffdale Creeks at turn-off for Camp-Oven Waterhole, *R.C.Carolin 9161* (SYD); Six-mile Ck SE of Westmoreland, *R.C.Carolin 9191* (SYD); c. 22 km NW of Corinda on road to Westmoreland, *R.C.Carolin 9148* (SYD).

The corolla is white suffused purple or purplish outside, and has dark-headed, glandular hairs and long, simple hairs at 90° outside. Similar to *G. purpurascens*, which, however, has much larger flowers and shorter, glandular hairs.

20. **Goodenia viscidula** Carolin, *Telopea* 3: 535 (1990)

T: 166 miles [c. 265 km] from Borroloola on Daly Waters road, N.T., 19 May 1974, *R.C.Carolin 9339*; holo: NSW; iso: SYD.

Erect, annual herb to 25 cm tall, viscid, with soft, glandular and simple hairs; roots adventitious. Leaves mostly basal, ±petiolate, ascending, obovate or oblanceolate, entire or obscurely dentate, sometimes with 2 narrow, basal lobes; lamina 4–11 cm long, 10–20 mm wide. Flowers in loose thyrses or panicles to 20 cm long; bracts linear, to 10 mm long; peduncle to 15 mm long; bracteoles linear, to 5 mm long; pedicel 9–11 mm long. Sepals lanceolate, 1 mm long. Corolla 5–7 mm long, without enations, indistinctly auriculate; abaxial lobes 2 mm long; wings c. 1 mm wide. Indusium square to oblong, c. 1 mm long. Ovules many, scattered. Fruit subglobular, c. 1.5 mm diam.; valves bifid to midmark. Seeds yellowish brown.

Occurs in northern N.T. and near the southern end of the Gulf of Carpentaria, Qld; in seasonally damp situations. Flowers chiefly Mar.–May. Map 211.

N.T.: Camp-Oven Waterhole, *R.C.Carolin 9167* (SYD); Edith Falls, *R.A.Perry 1941* (BRI). Qld: c. 22 km NW of Corinda on road to Westmoreland, *R.C.Carolin 9146* (SYD);

The corolla is bluish purple. Distinguished from other species in the section by its pubescent, clammy or viscid, ±distinctly petiolate leaves.

21. **Goodenia paludicola** Carolin, *Telopea* 3: 536 (1990)

T: 14 miles [c. 22 km] NW of Corinda on road to Westmoreland, Qld, 7 May 1974, *R.C.Carolin 9147*; holo: NSW; iso: SYD.

Ascending, annual herb to 25 cm tall; roots adventitious. Leaves mostly basal, ascending, lanceolate, entire or indistinctly dentate, glabrous; lamina 4–10 cm long, 2–4 mm wide. Flowers in compound thyrses or panicles to 20 cm long; bracts leaf-like but smaller; peduncle to 3 mm long; bracteoles linear, 3–5 mm long; pedicel 7–9 mm long. Sepals ovate to lanceolate, 1.5–1.8 mm long. Corolla usually 4–5 mm long, without enations, distinctly auriculate; abaxial lobes 1–2 mm long; wings c. 1 mm wide. Indusium almost square, 0.5 mm long, purplish. Ovules many, scattered. Fruit subglobular, 1.5–2.5 mm diam.; valves entire. Seeds yellowish.

Extends from Arnhem Land, N.T., to the southern shores of the Gulf of Carpentaria, Qld. Flowers chiefly Mar.–May. Map 212.

N.T.: Tin Camp Ck, c. 32 km S of Nabarlek, *T.G.Hartley 13775* (DNA); Katherine Gorge, *C.L.Gunn 14* (DNA); Ferguson R., *M.Parker 112* (DNA). Qld: c. 6 km N of Maggieville on Myravale road, *R.C.Carolin 8766* (SYD).

The corolla is bluish purple and has dark-headed, glandular hairs and simple hairs at 90° on the outside. Similar to *G. minutiflora* which, however, has a smaller, paler corolla and indusium not enclosed by auricles.

22. **Goodenia berringbinensis** Carolin, *Telopea* 3: 537 (1990)

T: bed of Berringbine Ck, Belele Stn, W.A., 15 Oct. 1945, *C.A.Gardner 7857*; holo: PERTH.

Annual herb to 30 cm tall, softly hairy; root system a thin tap root and adventitious roots. Leaves mostly basal, oblanceolate, entire or dentate, sometimes glabrescent; lamina 3.5–6 cm long, 2–10 mm wide. Flowers in loose, spreading, terminal thyrses to 15 cm long; bracts linear to elliptic, 4–15 mm long; peduncle 5–30 mm long; bracteoles linear-elliptic, 3–4 mm long; pedicel 13–20 mm long. Sepals lanceolate, c. 3 mm long, attached only to lower 1/2 of ovary. Corolla c. 12 mm long, with prominent enations, auriculate; abaxial lobes 4–5 mm long; wings c. 2 mm wide, unequal on adaxial lobe. Indusium obtriangular, 1.5 mm long, lower lip glabrous. Ovules many, scattered. Fruit ±cylindrical, c. 8 mm long; valves entire. Seeds yellow-brown. Fig. 47D–E.

Occurs near Belele Stn, W.A., in red sandy loam. Flowers about Oct. Map 213.

W.A.: Belele Stn, *D.W.Goodall 3275* (PERTH).

The corolla is yellow with purplish auricles. Distinguished from all other members of the section by the lower attachment of the sepals to the ovary (fruit extended c. 4 mm above sepal attachment). Probably has a restricted range.

23. **Goodenia lyrata** Carolin, *Telopea* 3: 538 (1990)

T: 20 miles [c. 32 km] W of Laverton, W.A., 22 Aug. 1961, *A.S.George 2798*; holo: PERTH.

Prostrate herb, with scattered hairs; stems to 13 cm long. Leaves cauline and basal, thick; axillary hairs prominently villous; basal leaves oblanceolate, lyrate, with lamina 7–28 mm long, 2–5 mm wide; cauline leaves smaller. Flowers in terminal racemes to 8 cm long; bracts leaf-like, 5–9 mm long, 2–2.5 mm wide; peduncle 2–3 mm long; bracteoles elliptic, c. 3 mm long; pedicel 3–5 mm long. Sepals lanceolate, 2–3 mm long. Corolla 10–12 mm long, with enations, auriculate; abaxial lobes c. 4 mm long; wings 1.5–2 mm wide, dentate. Indusium broadly oblong, c. 1.2 mm long, purple, villous especially on undersurface. Ovules many, scattered. Fruit ovoid, 5–6 mm long, beaked; valves split to base and appearing as 4. Mature seeds not seen.

Known only from the type collection, from near Laverton, W.A. Grows in red sandy loam with Mulga (*Acacia aneura*). Flowers about Aug. Map 214.

The corolla is yellow with purplish auricles. Distinguished from other members of the section by its lyrate leaves.

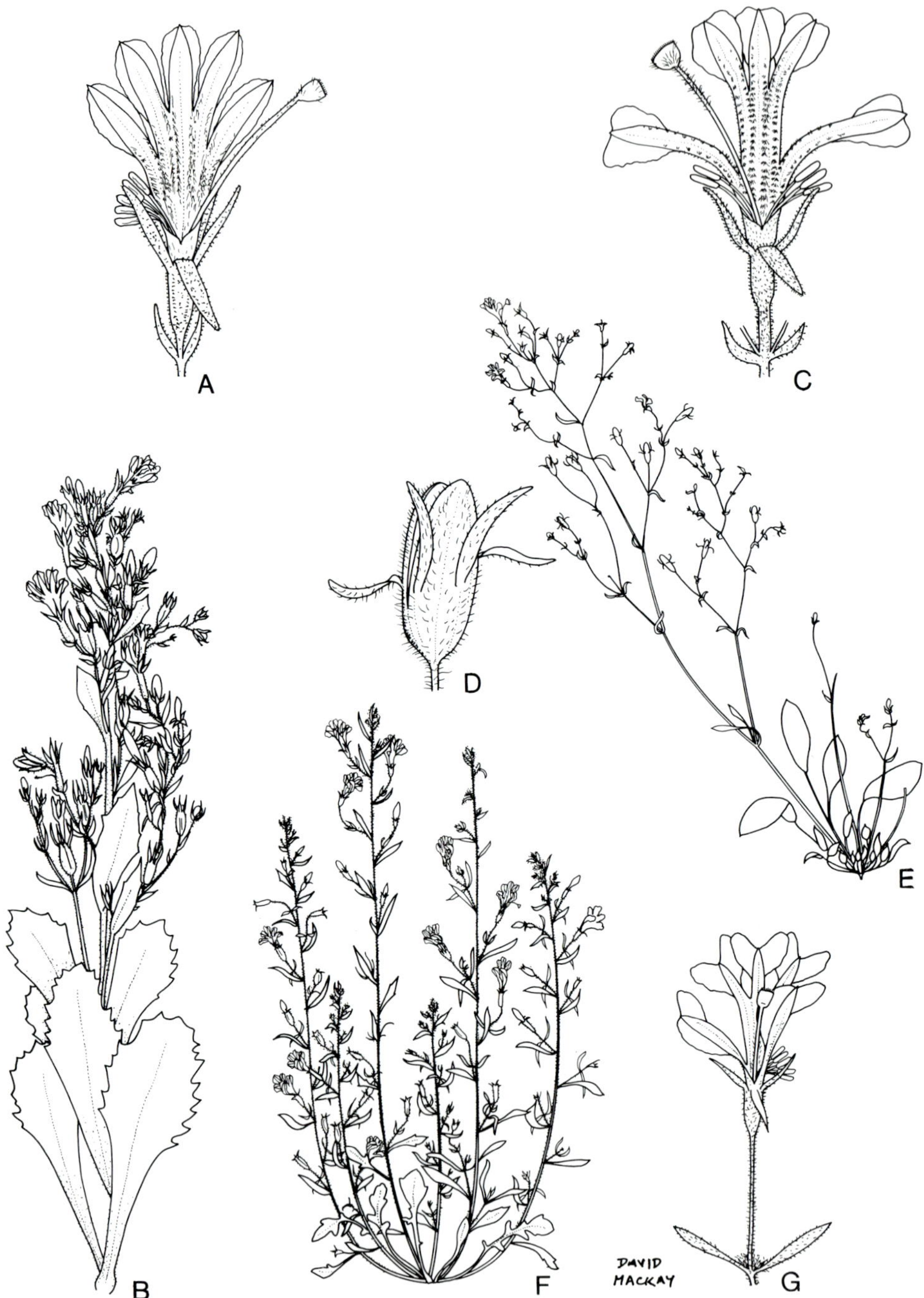

Figure 47. *Goodenia*. **A–B**, *G. scaevolina*. **A**, habit X0.5; **B**, flower X2 (**A–B**, A.Beauglehole 11271, SYD). **C**, *G. stobbsiana*, flower X2 (R.Carolin 7670, SYD). **D–E**, *G. berringbinensis*. **D**, capsule X4, **E**, habit X0.5 (**D–E**, D.Goodall 3275, PERTH). **F–G**, *G. modesta*. **F**, habit X0.5; **G**, flower X2 (**F–G**, A.George 8905, PERTH). Drawn by D.Mackay.

24. **Goodenia modesta** J.M.Black, *Trans. & Proc. Roy. Soc. S. Australia* 36: 172 (1912)

T: Tarcoola, S.A., June 1912, *J.W.Mellor*; holo: AD; iso: K, MEL.

G. erecta Ewart in A.J.Ewart & O.B.Davies, *Fl. N. Terr.* 265 (1917). T: 12 miles [c. 19 km] NW of Nth. Terr. Survey Camp III, 12 June 1911, *G.F.Hill 329*; lecto: MEL, *fide* R.C.Carolin, *Telopea* 3: 539 (1990); isolecto: MEL, NSW.

Erect herb to 50 cm tall, softly hairy to almost glabrous; tap root thin. Basal leaves oblanceolate, lyrato-pinnatifid, sometimes glabrescent; lamina to 8 cm long, 2 cm wide. Cauline leaves dentate to entire, smaller. Flowers in branched thyrses or racemes to 35 cm long; bracts linear to obovate, 5–20 mm long, dentate, thick; peduncle 1–12 mm long; bracteoles linear-oblanceolate, 2–7 mm long; pedicel 3–5 mm long. Sepals linear to lanceolate, 2–3 mm long. Corolla 12–14 mm long, villous inside, without enations, scarcely auriculate; abaxial lobes c. 4 mm long; wings c. 1.5 mm wide. Ovules many, scattered; indusium obtriangular, 1.5 mm long. Fruit ±cylindrical, 4 mm long, prominently beaked; valves split to base and appearing as 4. Seeds brownish. Figs 45G, 47F–G.

Occurs in central Australia in W.A., N.T. and S.A. between 128°E and 135°E. Grows in red loam with Mulga (*Acacia aneura*) and in sand. Flowers most of the year. Map 215.

W.A.: c. 11 km W of Dovers Hills, *A.S.George 9012* (PERTH); Mu Hills, NE of Sir Frederick Ra., *A.S.George 8905* (PERTH). N.T.: Hanson R., 21 Apr. 1962, *R.Rawlins* (DNA); c. 19 km SE of Rabbit Flat, *P.K.Latz 3947* (DNA). S.A.: Commonwealth Hill Stn, *D.E.Symon 3377* (AD, K, NSW).

The corolla is yellow. Distinguished from other members of the section by its many-flowered partial inflorescences. It also has a prominent corolla pocket like *G. gloeophylla*, which, however, has a purple corolla.

25. **Goodenia claytoniacea** F.Muell. ex Benth., *Fl. Austral.* 4: 79 (1868) as *laytoniana*

T: swampy flats of Don River, W.A., *G.Maxwell*; lecto: K, *fide* R.C.Carolin, *Telopea* 3: 539 (1990); isolecto: MEL.

[*G. tenella auct. non* R.Br.: F.Mueller, *Fragm.* 2: 111 (1861)]

Ascending, perennial herb to 20 cm tall, glabrous or slightly pubescent; rootstock thick. Leaves mostly basal, linear to oblanceolate, entire; lamina 2–6 cm long, 2–4 mm wide; axillary hairs a dense tuft. Flowers in dichasia or monochasia to 10 cm long, flowers often solitary in axil of basal leaf; peduncle 2–10 cm long, from basal rosette; bracteoles leaf-like, linear to oblanceolate, mostly 4–15 mm long; pedicel 7–20 mm long. Sepals lanceolate, c. 1 mm long. Corolla 5–8 mm long, indistinctly auriculate; abaxial lobes 2–2.5 mm long; wings 1.2–1.4 mm wide. Indusium oblong, c. 1 mm long. Ovules many, scattered. Fruit ovoid, tapering towards base, c. 5 mm long; valves split to base and appearing as 4. Seeds brown.

Occurs in south-western W.A. in damp situations in sandy soil. Flowers Oct.–Feb. Map 216.

W.A.: Midland Junction, 4 Dec. 1900, *A.Morrison* (BM); Ellen Brook Sanctuary, *N.T.Burbidge 7930* (CANB); no precise locality, *J.Drummond 406* (MEL).

The corolla is yellow, with darker markings. Distinguished from other species in the section by the insertion of the partial inflorescences (cymes) in the axils of the basal leaves. For discussion of the spelling of the specific epithet, see R.C.Carolin, *loc. cit.*

Sect. 2. Caeruleae

Goodenia subg. **Goodenia** sect. **Caeruleae** (Benth.) Carolin, *Fl. Australia* 35: 330 (1992)

Goodenia ser. *Caeruleae* Benth., *Fl. Austral.* 4: 53, 65 (1868). T: *G. caerulea* R.Br.; lecto: *fide* R.C.Carolin, *Fl. Australia* 35: 330 (1992).

Subshrubs or herbs with basal stock. Leaves basal or cauline. Flowers in terminal thyrses, racemes or spikes; pedicel bracteolate, articulate. Sepals adnate to ovary nearly to top. Corolla blue, usually with yellowish throat, usually with enations and hairs inside, scarcely auriculate; pocket various. Ovary with septum at least 2/3 as long as locule; ovules 20–60, in 2 rows in each locule. Fruit a capsule; valves 2, ±persistent. Seeds more than 1.5 mm wide, colliculate, reticulate or smooth; wing narrow and mucilaginous or broad and membranous.

This section contains 17 species distributed mostly in the western half of Australia but extending to the Great Dividing Ra., northern Qld.

Subsect. 1. Scaevolina

Goodenia subg. **Goodenia** sect. **Caeruleae** subsect. **Scaevolina** Carolin, *Fl. Australia* 35: 331 (1992)

T: *G. scaevolina* F.Muell.

Subshrubs or herbs. Leaves basal or cauline. Corolla blue, mostly with prominent enations. Ovules 30–60. Seeds colliculate-punctate, with narrow mucilaginous wing c. 0.1 mm wide.

This subsection contains 8 species distributed over the same areas as the section, mostly in northern Australia.

26. Goodenia sericostachya C.Gardner, *J. Roy. Soc. W. Australia* 47: 63 (1964)

T: 45 km N of Murchison R., W.A., 5 Jan. 1960, *C.A.Gardner 12430*; holo: PERTH.

Erect, short-lived or perennial herb to 40 cm tall, silvery hairy. Leaves mostly basal, obovate to lanceolate, sinuate-lobed to entire; lamina 5–10 cm long, 2–2.5 cm wide. Inflorescence a compact spike or spike-like thyrse, to 15 cm long; bracts oblong to linear, to 10 mm long; bracteoles linear, c. 6 mm long. Sepals linear, 5 mm long, adnate to ovary along midrib only except basally. Corolla c. 15 mm long, with stiff hairs inside towards base; lobes ±equal, c. 8 mm long; wings c. 0.5 mm wide. Indusium oblong, 2 mm long. Ovules to 60. Fruit ±cylindrical, c. 6 mm long; valves entire. Seeds elliptic, c. 1 mm long, black.

Occurs in the Yuna–lower Murchison R. area, W.A., on sand plain, appearing particularly after fire. Flowers Oct.–Jan. Map 217.

W.A.: between 390 and 394 mile pegs [c. 630 km N of Perth] on North West Coastal Hwy, *A.C.Burns 1043* (PERTH); c. 30 km E of Yuna, 28°18'S, 115°17'E, *A.S.George 16407* (PERTH) 425 mile peg [c. 680 km N of Perth] on North West Coastal Hwy, end 1964, *F.Humphreys* (PERTH).

The corolla is blue, to pinkish mauve with a yellow spot at top of throat and an inconspicuous pocket.

27. Goodenia scaevolina F.Muell., *Fragm.* 1: 118 (1859)

T: Depot Creek, [N.T.], Mar. 1856, *F.Mueller*; holo: MEL. The specimen 'Upper Victoria River, 1856, *F.Mueller*' (K) is probably an isotype.

Much-branched, perennial subshrub to 80 cm tall, glandular-viscid. Leaves cauline, obovate to oblanceolate, dentate, sometimes glabrous; lamina 4–7 cm long, 2–3 cm wide. Flowers in thyrses to 35 cm long; bracts leaf-like basally, smaller apically; peduncle 1–5 cm long; bracteoles elliptic to linear, 5–10 mm long; pedicel 1–7 mm long. Sepals lanceolate to narrowly elliptic, 5–10 mm long. Corolla 20–25 mm long, villous-pubescent inside, with enations in rows, scarcely auriculate; lobes ±equal, 10–12 mm long; wings 1.5–3.5 mm wide. Indusium broadly obtriangular, 3 mm long, convex. Ovules 30–50. Fruit ovoid, 8–9 mm long, valves entire. Seed elliptic, 2 mm long, blackish brown. Figs 45C, 47A–B.

Occurs in the Kimberley, W.A., and Arnhem Land, N.T., in scrub communities in shallow stony soil, with one collection from the Great Dividing Ra., northern Qld. Flowers chiefly Mar.–Aug. Map 218.

W.A.: Derby–Broome road intersection, *R.C.Carolin 7468* (SYD). N.T.: c. 80 km SW of Willeroo homestead, 9 May 1960, *G.Chippendale* (DNA); c. 60 km ESE of Limbunya Stn, *R.A.Perry & M.Lazarides 2296* (AD, BRI, CANB, DNA, NSW). Qld: about 22.5 km NNW of Yarrowmere Stn on Great Dividing Ra., *R.J.Henderson, G.P.Guymer & H.A.Dillewaard H2844* (BRI).

The corolla is blue, the pocket is usually prominent, and the corolla enations each have a few coarse white hairs. The specimens from the Victoria R. eastwards show some slight differences from those from the Kimberley, particularly in the longer, looser inflorescence and usually somewhat longer peduncles. Moreover, the specimens from the Kimberley westwards to the De Grey R. have very short glandular hairs on the leaves, some being almost glabrous; there appears to be a complete intergradation in the West Kimberley region. The single collection from the highlands of northern Qld resembles the specimens from the Victoria R. area but has a very prominent corolla pouch.

28. Goodenia stobbsiana F.Muell., *Fragm.* 11: 49 (1878)

T: Yule River, W.A., 1878, *J.Forrest*; holo: MEL; iso: L.

G. stapfiana K.Krause, *Pflanzenr.* 54: 48 (1912). T: between Ashburton and De Grey Rivers, W.A., *E.Clement*; holo: K.

G. clementii K.Krause, *op. cit.* 66 T: between Ashburton and De Grey Rivers, W.A., *E.Clement*; holo: K.

Usually several-stemmed, much-branched, perennial subshrub to 80 cm tall, glandular-viscid. Leaves basal and cauline, obovate to oblanceolate, sometimes dentate, glandular-pubescent to glabrescent; lamina 4–7 cm long, 2–3 cm wide. Flowers in loose thyrses to 40 cm long; bracts leaf-like, smaller apically; peduncle to 5 cm long; bracteoles usually linear, to 10 mm long; pedicel mostly to 4 mm long. Corolla hairy inside; enations in regular rows bearing conspicuous tuft of hairs; lobes unequal to 14 mm long. Otherwise similar to *G. scaevolina*. Figs 47C, 66.

Occurs in the Pilbara region, W.A., appearing particularly after fire and in disturbed ground. Flowers chiefly Mar.–Oct. Map 219.

W.A.: c. 85 km N of Marble Bar, *R.C.Carolin 7664* (SYD); near Mt Herbert, *C.A.Gardner 6299* (PERTH); Hamersley Ra., *C.A.Gardner 6299A* (PERTH); c. 20 km N of Marble Bar, *R.C.Carolin 7654* (SYD); 19 km E of Roy Hill, *J.S.Beard 2805* (PERTH).

The corolla is blue. Similar to *G. scaevolina*, differing in the more unequal corolla lobes, the more numerous simple hairs on the outside of the corolla and the more conspicuous enations inside the corolla.

29. **Goodenia suffrutescens** Carolin, *Telopea* 2: 67 (1980)

T: 30 miles [c. 48 km] S of Halls Creek, Billiluna Stn, W.A., 22 Aug. 1970, *R.C.Carolin 7915*; holo: NSW; iso: PERTH, SYD.

Undershrub to c. 1 m tall, glandular-viscid; stems woody, thick, decumbent. Leaves usually ±clustered at base, oblanceolate to obovate, dentate, pubescent; lamina 8–11 cm long, 3–4 cm wide. Flowers in terminal thyrses to 40 cm long; bracts ovate, broader and smaller than leaves; peduncle to 2 cm long; bracteoles ovate-elliptic, 6–14 mm long; pedicel to 5 mm long. Sepals lanceolate, c. 3 mm long. Corolla 15–20 mm long, stiffly hairy inside, enations prominent, auricles indistinct; lobes unequal, abaxial c. 4 mm long; wings c. 2 mm wide. Indusium oblong, 2 mm long. Ovules c. 40. Fruit ±cylindrical, c. 1 cm long, entire. Mature seeds not seen.

Occurs on the north-western edge of the Tanami Desert in W.A., on lateritic pavement. Flowers c. Aug. Map 220.

W.A.: between Billiluna and Balgo Stns, *J.S.Beard 5566* (PERTH); 19 km from Bloodwood Bore towards Balgo, *C.H.Gittins 02454* (BRI).

Corolla blue. Differs from the following closely related species in the characters indicated; from *G. azurea* which lacks shrubby habit and extensive glandular indumentum; from *G. scaevolina* which has different bract and bracteole shape, finer glandular hairs, and lacks adaxial corolla lobes closing over the indusium; and from *G. eremophila* which lacks the shrubby habit and dense simple hairs on the inside surface of the sepals.

30. **Goodenia arthrotricha** F.Muell. ex Benth., *Fl. Austral.* 4: 62 (1868)

T: south-western W.A., *J.Drummond 4*: *190*; lecto: K, *fide* R.C.Carolin, *Fl. Australia* 35: 331 (1992); isolecto: BM, MEL.

G. bonneyana F.Muell., *Fragm.* 6: 226 (1868). T: south-western W.A., *J.Drummond 190*; lecto: MEL, *fide* R.C.Carolin, *Telopea* 3: 539 (1990); isolecto: BM, K.

Illustration: F.Mueller, *op. cit.* t. 53.

Erect, perennial herb to 40 cm tall, with glandular hairs brownish below head. Basal leaves sessile, linear-oblanceolate, with lamina to 5 cm long, 3–5 mm wide; cauline leaves smaller. Flowers in thyrses to 20 cm long; bracts leaf-like, 1–2 cm long; peduncle 2–6 cm long; bracteoles linear, 1.5–2.5 mm long; pedicel 6–2.5 mm long. Sepals lanceolate, 9–10 mm long. Corolla 14–20 mm long, some hairs inside, without enations, scarcely auriculate; lobes unequal, abaxial c. 5 mm long; wings 2 mm wide. Indusium broadly oblong, 2 mm long. Ovules several. Fruit ovoid, 4–5 mm long; valves divided to base. Seeds elliptic, pale-brown, 2 mm long.

The only authentically located specimen is from Wannamal, W.A. Flowers Oct.–Nov. Map 221.

W.A.: Wannamal, *V.Mann 202 & A.S.George* (K); 15 km N of Moora, *D.Whibley 4874* (AD).

The corolla is blue.

31. **Goodenia eremophila** E.Pritzel, *Bot. Jahrb. Syst.* 35: 558 (1905)

T: S of Menzies, W.A., Oct. 1901, *F.L.E.Diels 51B*; holo: ?B (destroyed) *n.v.*

Illustration: K.Krause, *Pflanzenr.* 54: 65, fig. 13H–K (1912).

Ascending herb to 50 cm tall, with stock, with glandular hairs purplish brown below head. Leaves sessile, linear to elliptic-obovate, dentate or entire; lamina 6–9 cm long, 15–20 mm wide. Flowers in thyrses to 35 cm long; bracts leaf-like; peduncle 10–40 mm long; bracteoles linear-deltoid, 5–20 mm long; pedicel 5–10 mm long. Sepals lanceolate, 3–4 mm long. Corolla to 20 mm long, hairy inside, with prominent enations, auriculate; abaxial lobes 6–7 mm long; wings c. 2 mm wide. Indusium ±square, 2.5–3 mm long. Ovules to 50. Fruit ovoid, 9–12 mm long; valves entire. Seeds flat, elliptic, 1.5 mm long, very dark brown, colliculate.

Occurs from Wiluna to Kalgoorlie, W.A. Flowers Oct.–Dec. Map 222.

W.A.: NE of Wiluna, 14 Oct. 1947, *G.E.Brockway* (PERTH); Mt Keith, Lawlers, 16 Oct. 1931, *J.H.Rogers* (K, PERTH); Comet Vale, *C.A.Gardner 1118* (PERTH); N of Goongarrie, *H.F. & M.Broadbent 1682* (BM, NSW); c. 112 km N of Kalgoorlie, *P.R.Jeffries 64101* (PERTH).

Corolla blue with a yellow throat and a prominent pocket. Similar to *G. azurea* which usually has narrower bracteoles and lacks glandular hairs on the leaves; where glandular hairs do occur in *G. azurea*, i.e., on the floral parts, they are shorter with ±globular rather than ovoid heads. *Goodenia eremophila* resembles *G. ramelii* which has very few glandular hairs and a less prominent pocket on the corolla.

32. **Goodenia ramelii** F.Muell., *Fragm.* 3: 20 (1862)

T: Attack Creek, S.A., *J.Macdouall Stuart*; holo: MEL.

G. basedowii K.Krause, *Pflanzenr.* 54: 49 (1912). T: southern central Australia, *H.J.Basedow 2*; holo: ?B (destroyed) *n.v.*

Illustration: F.Mueller, *Fragm.* 3: t. 17 (1862).

Perennial herb with basal rosette, to 1 m tall, glabrous or sparsely glandular-hairy; tap root thick. Basal leaves elliptic-obovate, dentate, sometimes ±glaucous; lamina 10–25 cm long, 20–45 mm wide. Cauline leaves smaller. Flowers in terminal thyrses or racemes to 40 cm long; bracts lanceolate, 8–15 mm long; peduncle 5–15 mm long; bracteoles lanceolate, 2–7 mm long; pedicel 1–4 mm long. Sepals lanceolate, 1.5–2 mm long. Corolla 12–15 mm long, villous inside, with enations, ±auriculate; abaxial lobes 7–8 mm long; wings 1.5–2 mm wide. Indusium ±square, c. 2 mm long. Ovules 40–50. Fruit ovoid, c. 7 mm long; valves entire. Seeds elliptic, 1.7 mm long, greyish to black, colliculate. Fig. 48F–G.

Occurs in central Australia in W.A., N.T. and Qld, from the Gibson Desert to the Barkly Tableland, in stony soils. Flowers Mar.–Sept. Map 223.

W.A.: Sir Frederick Ra., *D.E.Symon 2257* (AD). N.T.: c. 32 km S of Elliott, *H.F. & M.Broadbent 919* (BM, NSW); Docker R., *R.C.Carolin 5840* (SYD). Qld: c. 45 km NW of Mt Isa, *P.Ollerenshaw & D.Kratzing 1257* (CANB, SYD). S.A.: 0.17 km W of Arckaringa homestead, *E.A.Shaw 512* (AD).

The corolla is blue with a yellow throat, and has glandular hairs and a few simple hairs outside. Similar to *G. azurea* and *G. eremophila*, *q.v.*

33. **Goodenia azurea** F.Muell., *Fragm.* 1: 117 (1859)

T: Sturt Creek, [N.T.], Feb. 1856, *F.Mueller*; holo: MEL; iso: K.

More or less divaricate, many-stemmed, perennial herb to 80 cm tall, glaucous. Leaves obovate, dentate, thick; lamina to 7 cm long, to 15 mm wide. Flowers in thyrses or racemes, to 50 cm long; bracts leaf-like, oblanceolate to orbicular, 10–30 mm long, 4–15 mm wide; peduncle 20–35 mm long; bracteoles oblanceolate to oblong, 5–20 mm long; pedicel mostly 2–3 mm long, articulate. Sepals lanceolate, 3–3.5 mm long. Corolla 12–15 mm long, densely villous towards base inside, with enations, indistinctly auriculate; abaxial lobes 5–6 mm long; wings 1–1.5 mm wide. Indusium oblong-deltoid, c. 3 mm long. Ovules 40–50. Fruit ovoid-cylindrical, 10–12 mm long; valves bifid. Seeds broadly elliptic, 1.8 mm long, black, colliculate. Fig. 48C.

Occurs in northern central Australia, extending into the central deserts of W.A., in sandy soil with lateritic pebbles. Flowers Apr.–Oct. Map 224.

W.A.: c. 10 km S of Googhenama Ck, *R.D.Royce 1791* (PERTH); c. 45 km W of Warburton Ra., *R.C.Carolin 5947* (SYD). N.T.: c. 65 km S of Wave Hill Stn, 11 July 1956, *G.Chippendale* (CANB, DNA, MEL, NSW); c. 29 km W of Inverway, *R.C.Carolin 7371* (SYD); c. 65 km from Mt Doreen on The Granites road, *R.C.Carolin 7953* (SYD).

The corolla is blue with a yellow throat, has scattered, coarse, glandular hairs outside and is often viscid. Similar to *G. ramelii* which, however, has narrower bracts and bracteoles and is not many-stemmed.

Subsect. 2. Caeruleae

Goodenia subg. **Goodenia** sect. **Caeruleae** subsect. **Caeruleae** (Benth.) Carolin, *Fl. Australia* 35: 331 (1992)

Goodenia ser. *Caeruleae* Benth., *Fl. Austral.* 4: 53, 65 (1868). T: *G. caerulea* R.Br.; lecto: *fide* R.C.Carolin, *Fl. Australia* 35: 331 (1992).

Herbs with basal stock. Leaves usually basal. Flowers in terminal racemes, on stems arising from axils of basal leaves; bracts leaf-like, but smaller. Corolla with or without enations, blue. Ovules 20–40. Seeds mostly colliculate, usually with dry, hyaline wing more than 0.1 mm wide.

There are 9 species in this subsection, all occurring in south-western Australia.

34. Goodenia caerulea R.Br., *Prodr.* 578 (1810)

T: Princess Royal Harbour, King George Sound, [W.A.], Dec. 1801, *R.Brown*; lecto: BM, *fide* R.C.Carolin, *Telopea* 3: 540 (1990); isolecto: K, MEL, P.

G. rigida Benth. in S.L.Endlicher, *Enum. Pl. Hügel* 71 (1837). T: Swan River, W.A., *C.A.Hügel*; holo: W; iso: K.

G. barilletii F.Muell., *Fragm.* 3: 140 (1863). T: SW extremity of the Great Bight, W.A., *coll. unknown*; holo: MEL; iso: MEL.

Scaevola tenera Vriese in J.G.C.Lehmann, *Pl. Preiss.* 1: 409 (1845). T: interior, south-western W.A., Feb. 1841, *J.A.L.Preiss 1442*; lecto: LD, *fide* R.C.Carolin, *Telopea* 3: 507 (1990); isolecto: G, L.

Scaevola tenera var. *pauciflora* Vriese in J.G.C.Lehmann, *Pl. Preiss.* 1: 409 (1845). T: Sussex district, W.A., Dec. 1839, *J.A.L.Preiss 1482*; lecto: LD, *fide* R.C.Carolin, *Telopea* 3: 507 (1990); isolecto: G, L, P, W.

G. teretifolia Vriese, *Natuurk. Verh. Holl. Maatsch. Wetensch. Haarlem* ser. 2, 10: 130 (1854). T: none cited.

Illustration: K.Krause, *Pflanzenr.* 54: 70, fig. 14A–C (1912).

Erect to ascending, perennial herb to 50 cm tall, with glandular hairs and often cottony. Basal leaves sessile, linear, entire or with few teeth, thick, sometimes glabrescent; lamina 3–7 cm long, 1–3 mm wide. Raceme mostly to 30 cm long; bracts leaf-like, smaller; peduncles 10–25 mm long; bracteoles linear, 5–5 mm long; pedicel 5–15 mm long. Sepals linear-lanceolate, 3–7 mm long. Corolla 15–25 mm long, few hairs inside, with enations, auriculate; abaxial lobes 5–10 mm long; wings c. 2 mm wide. Indusium obtriangular 2–2.5 mm long. Ovules 30–40. Fruit ovoid, 7–8 mm long; valves ±bifid. Seeds orbicular, 1.5 mm diam., brown. $n = 24$, W.J.Peacock, *Proc. Linn. Soc. New South Wales* 88: 13 (1963) as *G. pterygosperma*. Figs 25I, 72.

Occurs in south-western W.A. from near Shark Bay to the south coast, growing in a variety of habitats. Flowers chiefly Sept.–Jan. Map 225.

W.A.: W of Watheroo, *T.E.H.Aplin 1312* (PERTH); Goomalling–Wyalkatchem road, *F.Lullfitz 3034* (PERTH); Gooseberry Hill, *A.Morrison 18085* (K); East Mt Barren, *A.S.George 555* (PERTH); Tuttanning [Nature] Reserve, *A.S.George 7362* (PERTH).

The corolla is blue. Cytological voucher is *W.J.Peacock 60886.1* (SYD).

A very variable species, especially in the indumentum. Specimens from the Perth area northwards tend to be relatively densely glandular-pubescent, those along the south coast tend to be glabrous except the ovary and sepals. Specimens from intermediate areas down to the Porongurup Ra. have intermediate indumentum. They may be part of a cline. Only one chromosome count is available, a hexaploid: experience in other species of the genus indicates that polyploid series may be the cause of some of the variation. Similar to *G. glareicola*, differing in having glandular hairs. See note under *G. trichophylla*.

35. Goodenia trichophylla Vriese ex Benth., *Fl. Austral.* 4: 67 (1868)

T: south-western W.A., *J.Drummond 3: 158*; lecto: K, *fide* R.C.Carolin, *Telopea* 3: 540 (1990); isolecto: BM, MEL, P.

Erect to ascending herb to 30 cm tall, viscid or varnished, with appressed, peltate hairs. Basal leaves linear, entire, thick; lamina 2–4 cm long, c. 2 mm wide. Raceme to 20 cm long; bracts leaf-like, smaller; peduncles 5–9 mm long; bracteoles linear, 1–1.2 mm long; pedicel 4–5 mm long. Sepals lanceolate, 1–1.5 mm long. Corolla 8–12 mm long, almost glabrous inside, with enations, indistinctly auriculate; abaxial lobes 4–5 mm long; wings c. 1 mm wide. Indusium oblong, 1.5 mm long; lower lip almost glabrous. Ovules c. 20. Fruit ovoid-oblong, 5–6 mm long; valves entire. Seeds orbicular, c. 3 mm diam., brown.

Occurs in the Lake King area, W.A. Flowers chiefly Aug.–Dec. Map 226.

W.A.: c. 24 km E of Lake King on Norseman Rd, *J.W.Green 5200* (PERTH); 15.5 km ENE of Coujinup Hill, 25.3 km NE of rabbit proof fence on Norseman track, *M.A.Burgess & C.Layman 3315* (PERTH).

The corolla is blue. Closely related to *G. caerulea* and *G. glareicola*, differing in having peltate hairs, which are completely hidden by their secretion of viscid varnish and cover the younger leaves and outside of the flowers.

36. Goodenia perryi Gardner ex Carolin, *Telopea* 3: 540 (1990)

T: Bunjil, W.A., 15 Oct. 1961, ex herb. *C.A.Gardner s.n.*; holo: PERTH.

Ascending herb to 25 cm tall, silvery cottony-villous. Basal leaves oblanceolate, entire; lamina 4–5 cm long, 6–10 mm wide. Raceme to 15 cm long; bracts leaf-like, smaller; peduncles to 5 mm long; bracteoles linear, 2–6 mm long; pedicel 2–3 mm long. Sepals linear deltoid, 7 8 mm long. Corolla 15–18 mm long, with few short hairs inside, with enations, auriculate; abaxial lobes 8–9 mm long; wings c. 3 mm wide. Indusium suborbicular, c. 2 mm diam. Ovules c. 40. Fruit and seeds not seen.

Occurs near Bunjil, W.A.; known only from the type collection. Flowers Oct. Map 227.

The corolla is blue, glabrous outside. Similar to *G. incana* which, however, has hairs on the outside of the corolla.

37. **Goodenia incana** R.Br., *Prodr.* 578 (1810)

T: Bay I [Lucky Bay], South Coast, [W.A.], 7 Jan. 1802, [not 1804] *R.Brown*; lecto: BM, *fide* R.C.Carolin, *Telopea* 3: 541 (1990); isolecto: K.

Scaevola pterosperma Vriese in J.G.C.Lehmann, *Pl. Preiss.* 1: 408 (1844). T: near Seven Mile Bridge, W.A., 20 Dec. 1840, *J.A.L.Preiss 1499*; lecto: LD, *fide* R.C.Carolin, *Telopea* 3: 508 (1990); isolecto: G, L, MEL, P, W.

Illustration: K.Krause, *Pflanzenr.* 54: 65, fig. 13a–d (1912).

Ascending herb to 30 cm tall, with silvery white, multicellular hairs. Basal leaves linear to oblanceolate, entire, ±thick; lamina mostly 2–4 cm long, 3–8 mm wide. Raceme to 20 cm long; bracts leaf-like, smaller; peduncles mostly 8–15 mm long; bracteoles linear, 6–12 mm long; pedicel 2–5 mm long. Sepals lanceolate, 4–7 mm long. Corolla 12–15 mm long, glabrous or with few hairs inside, enations indistinct, auriculate; abaxial lobes 6–7 mm long; wings c. 2 mm wide. Indusium broadly oblong, c. 2 mm long. Ovules c. 30. Fruit ovoid, 8–9 mm long; valves bifid to base. Seed orbicular, 2.5 mm diam. Fig. 24A.

Occurs widely in south-western W.A. in sandy soil in heath and open forest. Flowers chiefly Sept.–Jan. Map 228.

W.A.: c. 22 km W of Boorabbin, *A.S.George 6037* (PERTH); Burracoppin, Apr. 1943, *C.A.Gardner* (PERTH); c. 24 km E of Jerramungup, *K.Newbey 1195* (PERTH); Warrungup, 2 Nov. 1943, *C.A.Gardner* (PERTH); Kalgan Plains, Dec. 1909, *J.H.Maiden* (NSW).

The corolla is blue, and the wing of the seed is notched. The year of collection of the type of *G. incana* is 1802, since Brown (on the Flinders expedition) visited Lucky Bay only once.

38. **Goodenia pterigosperma** R.Br., *Prodr.* 578 (1810)

T: Bay I [Lucky Bay], South Coast, [W.A.], 12 Jan. 1802, *R.Brown*; lecto: BM, *fide* R.C.Carolin, *Telopea* 3: 541 (1990); isolecto: K.

G. cyanea F.Muell., *Fragm.* 1: 155 (1859). T: [Phillips Ra.], W.A., *G.Maxwell*; lecto: K, *fide* R.C.Carolin, *Telopea* 3: 541 (1990).

Illustration: K.Krause, *Pflanzenr.* 54: 70, fig. 14G–M (1912) as *pterygosperma*.

Erect perennial to 40 cm tall, glabrous. Basal leaves linear to oblanceolate, ±incurved, with few teeth, thick; lamina 2–5 cm long, 3–8 mm wide. Raceme to 20 cm long; bracts leaf-like, smaller; peduncles upcurved, mostly 15–20 mm long; bracteoles linear, 1.5–2 mm long; pedicel mostly 1–5 mm long. Sepals oblong, 1–2.5 mm long, obtuse. Corolla 12–14 mm long, villous towards base inside, without enations, auriculate; lobes subequal, abaxial lobes 8–9 mm long; wings 2–2.5 mm wide. Indusium ±oblong, 2 mm long. Ovules 20–30. Fruit ovoid, 6 mm long, sometimes glaucous; valves deeply bifid. Seeds elliptic, 3 mm long, colliculate, yellow-brown. n = 8, W.J.Peacock, *Proc. Linn. Soc. New South Wales* 88: 13 (1963). Fig. 48E.

Occurs near the south coast of south-western W.A. from Jerramungup to Israelite Bay, in heath and open woodland in sandy soil. Flowers Sept.–Jan. Map 229.

W.A.: c. 14 km from Lake Newdegate, 6 Nov. 1968, *J.W.Wrigley* (CBG, SYD); Grasspatch, *C.A.Gardner 2837* (K, PERTH); c. 5 km N of Truslove, *C.A.Gardner & W.E.Blackall 1125* (PERTH); NE of Jerramungup, *A.S.George 7039* (PERTH); Gordon Inlet, *T.E.H.Aplin 2771c* (PERTH).

The corolla is dark blue, and the wing of the seed is notched. Cytological voucher is *W.J.Peacock 60957* (SYD); the other specimen cited by Peacock is *G. caerulea.* Distinguished from other members of the subsection by its obtuse sepals and bracteoles, its shorter pedicels, and its subequal corolla lobes.

Figure 48. *Goodenia*. **A–B**, *G. kingiana*. **A**, flower X2; **B**, habit X0.5 (**A–B**, J.Peacock 608581, SYD). **C**, *G. azurea*, flower X2 (A.Beauglehole 60170 & E.Errey 3870, SYD). **D**, *G. glareicola,* flower X2 (K.Kenneally 7163, SYD). **E**, *G. pterigosperma*, flower X2 (N.Donner 3048, SYD). **F–G**, *G. ramelii*. **F**, flower X2; **G**, habit X0.5 (**F–G**, N.Donner 4457, SYD). Drawn by D.Mackay.

39. Goodenia glareicola Carolin, *Telopea* 3: 541 (1990)

T: near Newdegate, W.A., 18 Nov. 1931, *W.E.Blackall 1364*; holo: PERTH.

Illustration: R.Erickson *et al.*, *Fl. Pl. W. Australia* 126, t. 392 (1973) as *G. brevisepala, nom. nud.*

Erect, perennial herb to 30 cm tall, glabrous or glaucous. Basal leaves linear to oblanceolate, thick, with lamina 20–30 mm long, 2–4 mm wide; upper leaves smaller. Raceme to 20 cm long; bracts leaf-like, smaller; peduncles 8–20 mm long; bracteoles linear-lanceolate, 2.5–3 mm long; pedicel mostly 5–10 mm long. Sepals lanceolate, 3–4 mm long, acute. Corolla c. 15 mm long, almost glabrous inside, with obscure enations, indistinctly auriculate; abaxial lobes 6–7 mm long; wings 2–2.5 mm wide. Indusium broadly obovate, 1.5 mm long. Ovules c. 60. Fruit ovoid, 8–9 mm long; valves often splitting to base. Seeds elliptic, 2 mm long, colliculate, brown. Fig. 48D.

Occurs in south-western W.A. from Mullewa to Lake Grace, in gravelly and sandy soil. Flowers Oct.–Jan. Map 230.

W.A.: Bendering, *C.A.Gardner 2026* (PERTH); c. 11 km E of Hyden, *F.Lullfitz L3821* (PERTH); c. 25 km N of Lake Biddy, *W.E.Blackall 1364* (PERTH); Tarin Rock, *J.S.Beard 2150* (PERTH); c. 15 km W of Lake Grace, 29 Oct. 1962, *M.E.Phillips* (CBG, PERTH).

The corolla is blue. Distinguished from most other members of the subsection by the lack of hairs on the external parts, except in the leaf axils. Close to *G. pterigosperma* which has wider, obtuse sepals; and from *G. trichophylla* which has viscid varnish on the young leaves.

40. Goodenia eatoniana F.Muell., *Fragm.* 8: 186 (1874)

T: Blackwood R., W.A., *J.Forrest*; lecto: MEL, *fide* R.C.Carolin, *Telopea* 3: 542 (1990); isolecto: K.

Illustration: K.Krause, *Pflanzenr.* 54: 65, fig. 13E–G (1912).

Ascending, perennial herb to 40 cm tall, glabrous, ±glaucous. Basal leaves oblanceolate, tapering basally, entire to dentate; lamina 10–50 mm long, to 15 mm wide. Cauline leaves stem-clasping, ovate-elliptic, smaller. Raceme to 20 cm long; bracts leaf-like, smaller; peduncles 10–20 mm long; bracteoles lanceolate to linear, 3–7 mm long; pedicel 5–12 mm long. Sepals lanceolate, c. 4 mm long. Corolla 14–18 mm long, with scattered hairs inside, enations ±obsolete, auriculate; abaxial lobes 10–11 mm long; wings 2–2.5 mm wide. Indusium broadly oblong, 1.5 mm long. Ovules to 25. Fruit sub-globular, c. 3 mm diam.; valves entire. Seed elliptic, 1.8 mm long, yellow-brown, colliculate.

Occurs in the extreme SW of W.A. Flowers Oct.–Jan. Map 231.

W.A.: Collie R. and Preston R., 8 Dec. 1877, *F.Mueller* (MEL); Mullalyup, Dec. 1904, *C.Andrews* (PERTH); c. 9 km S of Manjimup, *T.E.H.Aplin 1380* (PERTH); near Augusta, Jan. 1936, *H.J.Hall* (PERTH); Darradup, *R.D.Royce 8312* (PERTH).

The corolla is blue, and the wing of the seed is obscure. Similar to *G. leptoclada*, see note under that species.

41. Goodenia leptoclada Benth., *Fl. Austral.* 4: 67 (1868)

T: south-western W.A., *J.Drummond 188*; holo: K; iso: MEL, P.

Ascending, perennial herb to 30 cm tall, becoming scabrous, with scattered, multicellular hairs. Basal leaves oblanceolate to obovate, with blunt teeth or entire; lamina 5–25 mm long, 1–3 mm wide. Cauline leaves stem-clasping, ovate to oblong; lamina 7–12 mm long.

Raceme to 15 cm long; bracts leaf-like, smaller; peduncles 10–25 mm long; bracteoles linear-lanceolate, 2.5–3.5 mm long; pedicel 0–5 mm long. Sepals lanceolate, 3 mm long, acute. Corolla 10–13 mm long, hairs scattered inside, enations obscure, auriculate; abaxial lobes 6–7 mm long; wings 2–2.5 mm wide. Indusium broadly deltoid, 1.5 mm long. Ovules 4 or 5. Fruit subglobular, c. 2 mm diam.; valves entire. Seeds elliptic, 1 mm long, greyish brown, reticulate.

Occurs in the extreme SW of W.A. Flowers Nov.–Jan. Map 232.

W.A.: Scott R., 17 Jan. 1945, *R.D.Royce* (PERTH); E end of Porongurup Ra., *K.Newbey 1642* (PERTH); Wilson Inlet, 22 Dec. 1877, *F.Mueller* (MEL); Boggy Lake, 27 Dec. 1957, *D.Churchill* (PERTH); no precise locality, *J.Drummond 188* (BM).

Corolla blue and villous outside; seed wing almost obsolete. Similar to *G. eatoniana* which, however, has larger and more numerous seeds in each capsule, and is glabrous.

42. **Goodenia hassallii** F.Muell., *Fragm.* 6: 10 (1867)

T: south-western W.A., *J.Drummond s.n.*; lecto: MEL, *fide* R.C.Carolin, *Telopea* 3: 542 (1990); isolecto: K, MEL.

Illustrations: F.Mueller, *op. cit.* t. 51 (1867); K.Krause, *Pflanzenr.* 54: 70, fig. 14D–F (1912).

Erect, shrubby perennial to 50 cm tall, glabrous. Basal leaves linear to oblanceolate, entire or with coarse teeth, thick; lamina 5–8 cm long, 1.5–10 mm wide. Raceme to 20 cm long; bracts leaf-like, smaller; peduncles 5–15 mm long; bracteoles narrowly elliptic, 2.5–3 mm long; pedicel 2–3 mm long. Sepals narrowly oblong, 5 mm long. Corolla 13–15 mm long, with some hairs inside, with enations, auriculate; abaxial lobes 5–7 mm long; wings c. 2 mm wide. Indusium square, 2 mm long. Ovules to 30. Fruit ovoid, 6 mm long; valves usually entire. Seeds orbicular, 1.5 mm diam., yellow-brown, colliculate-punctate to reticulate.

Occurs in W.A., west of 118°E and between 25°S and 29°S. Flowers Oct.–Jan. Map 233.

W.A.: Morawa to Three Springs road, c. 5 km from Three Springs, *J.Beard & F.Lullfitz 1961* (PERTH); Wongan Hills, *F.Stoward 826* (K); Mogumber, *F.Meebold 6676* (AD).

The corolla is blue. The seed wing is almost obsolete.

In this subsection, *G. leptoclada* and *G. eatoniana* also have seeds with very narrow or obsolete wings, but both those species have stem-clasping leaves, and *G. eatoniana* is not woody at the base. *G. hassallii* is also similar to *G. azurea* in subsect. *Scaevolina*, but *G. azurea* has large, glandular hairs on the flower.

Sect. 3. Goodenia

Goodenia Smith subg. **Goodenia** sect. **Goodenia**

Goodenia sect. *Ochrosanthus* G.Don, *Gen. Hist.* 3: 724 (1834); *Goodenia* sect. *Eugoodenia* Benth., *Fl. Austral.* 4: 51 (1868). T: *G. ovata* Smith, *fide* R.C.Carolin, *Fl. Australia* 35: 330 (1992).

Goodenia sect. *Tetrathylax* G.Don, *op. cit.* 725; [*Goodenia* sect. *Tetraphylax* DC., *Prodr.* 7: 515 (1839) *orth. var.*]. T: *G. quadrilocularis* R.Br.

Subshrubs or herbs. Leaves basal and cauline or mostly basal; cauline leaves smaller, narrower. Flowers in leafy terminal thyrses, racemes or subumbels; pedicel bracteolate or not, articulate or not. Sepals variously adnate to ovary. Corolla pubescent to glabrous inside, sometimes with long, stiff hairs marginally and in throat, sometimes with enations, sometimes auriculate, usually pouched, usually yellow to purplish. Ovary septum 2/3 locule length, to much shorter. Ovules less than 20, 2-rowed in each locule. Fruit capsular or rarely indehiscent; valves 2, persistent or deciduous. Seeds more than 1.5 mm wide, colliculate to tuberculate; wing obsolete or 0.1–0.2 mm wide, mucilaginous or membranous.

The section contains 135 species, mostly Australian; 1 extends to New Guinea, 1 to the Philippines and 1 endemic in Java.

Subsect. 1. Goodenia

Goodenia Smith subg. **Goodenia** sect. **Goodenia** subsect. **Goodenia**

Goodenia ser. *Racemosae* Benth., *Fl. Austral.* 4: 51, 57 (1868). T: *G. decurrens* R.Br., *fide* R.C.Carolin, *Fl. Australia* 35: 331 (1992).

Goodenia ser. *Bracteolatae* Benth., *Fl. Austral.* 4: 52, 59 (1868); *Goodenia* subsect. *Bracteolatae* (Benth.) K.Krause, *Pflanzenr.* 54: 46 (1912); *Goodenia* ser. *Suffruticosae* K.Krause, *loc. cit.* T: *G. ovata* Smith, *fide* R.C.Carolin, *Fl. Australia* 35: 330 (1992).

Goodenia ser. *Rosulatae* K.Krause, *loc. cit.* T: *G. geniculata* R.Br., *fide* R.C.Carolin, *Fl. Australia* 35: 331 (1992).

Subshrubs or herbs. Leaves basal and cauline. Flowers in thyrses or racemes; pedicel bracteolate, not conspicuously articulate. Corolla usually with scattered hairs inside, often with enations, not auriculate, with pouch, yellow or rarely purplish. Fruit a capsule; valves 2, entire or bifid, persistent. Seeds with a narrow mucilaginous wing.

A subsection of 53 species; 52 in Australia, most occurring in southern and central Australia, with 2 in W.A., 2 in northern Qld and 1 endemic in Java.

43. Goodenia ovata Smith, *Trans. Linn. Soc. London, Bot.* 2: 347 (1794)

Goodenoughia ovata (Smith) Siebert & Voss, *Vilm. Blumengärtn.* 1: 559 (1896). T: Sion Garden, Dec. 1793, *Mr. Hoy*; holo: LINN, Smith. Herb. 325; iso: K?

Goodenia acuminata R.Br., *Prodr.* 575 (1810). T: Grose River, W.A., 1803, *R.Brown*; lecto: BM, *fide* R.C.Carolin, *Telopea* 3: 520 (1990); isolecto: K, MEL.

Goodenia ovata var. *latifolia* Schldl., *Linnaea* 20: 601 (1847). T: near Torrens and Oncaparinga, S.A., *H.H.Behr 83*; *n.v.*

Goodenia ovata var. *cordata* F.Muell., *Fragm.* 6: 15 (1867). T: near Port Jackson; *n.v.*

Goodenia ovata var. *lanceolata* F.Muell., *Fragm.* 6: 15 (1867). T: near Newcastle, N.S.W., *Leichhardt*; holo: MEL.

Illustrations: A.J.Cavanilles, *Icon.* 6: 4, t. 506 (1801); E.P.Ventenat, *Jard. Cels.* t. 3 (1800); G.R.Cochrane, *et. al.*, *Fl. Pl. Victoria & Tasmania* 81, fig. 381 (1980).

Erect, ascending to prostrate shrub to 2 m tall, viscid, often varnished. Leaves ovate to elliptic, dentate; lamina 3–8 cm long, 1–4 cm wide; petiole to 3 cm long. Flowers in thyrses or racemes mostly to 35 cm long; bracts leaf-like; peduncle to 3 cm long; bracteoles linear, to 6 mm long; pedicel to 8 mm long, articulate. Sepals linear to lanceolate, 3–11 mm long. Corolla 10–19 mm long, pubescent inside, with enations, auriculate; abaxial lobes 5–7 mm long; wings to 2.5 mm wide. Indusium depressed-obovate. Ovules to 40. Fruit cylindrical, 8–12 mm long; valves entire. Seeds elliptic, c. 2 mm long, pale brown to whitish, aculeate. $n = 8$, W.J.Peacock, *Proc. Linn. Soc. New South Wales* 88: 12 (1963). Figs 23A, 25C, 70.

Extends from the Flinders Ra., S.A., through Vic. to the N.S.W. coast and northwards to the Blackdown Tableland, Qld, and in Tas. Occurs in forest and woodland and sometimes in exposed rocky situations near the sea. Flowers all year but mostly Oct.–Mar. Map 234.

S.A.: Harrisons Ck, c. 3 km S of Palmer, 18 Mar. 1961, *N.N.Donner* (AD). Qld: Byron Ck, *S.L.Everist 7567* (BRI). N.S.W.: Huskisson, *F.A.Rodway 539* (NSW). Vic.: Briggs Bluff, *T.B.Muir 2727* (MEL). Tas.: Roaring Beach, NE of Nubeena, *R.C.Carolin 1809* (SYD).

Corolla yellow. Cytological voucher is *W.J.Peacock 6012.5.3* (SYD). An extremely variable species particularly in the relative length and breadth of the sepals, leaf shape and corolla length. The corolla sometimes has simple hairs outside. The thinner more flexible leaves with a more distinct petiole separate this species from *G. varia* and *G. vernicosa.* The latter is much more viscid and has more numerous simple hairs on the outside of the corolla, while *G. varia* has shorter sepals and bracteoles than *G. ovata*.

44. **Goodenia varia** R.Br., *Prodr.* 576 (1810)

T: Bay III, S Coast [Fowler Bay, S.A.] *R.Brown Iter Australiense 2517*; lecto: BM, *fide* R.C.Carolin, *Telopea* 3: 520 (1990); isolecto: K, MEL.

Ascending to prostrate shrub to 1 m tall, usually viscid when young, glabrous. Leaves elliptic to orbicular, usually dentate; lamina 2–4 cm long, 1–2 cm wide; petiole to 5 mm long. Flowers in thyrses or racemes to 15 cm long; bracts leaf-like; peduncle to 10 mm long; bracteoles triangular, 1 mm long; pedicel to 3 mm long, articulate. Sepals linear, 2–3.5 mm long. Corolla 8–18 mm long, few hairs inside, ±auriculate; abaxial lobes 5–8 mm long; wings c. 1.5 mm wide. Indusium broadly obovate. Ovules to 16. Fruit cylindrical, 8–10 mm long; valves entire. Seeds oblong, 2–2.5 mm long, pale brown, aculeate or reticulate. $n = 8$, W.J.Peacock, *Proc. Linn. Soc. New South Wales* 88: 12 (1963)

Occurs in southern Australia extending from Eucla, W.A., through Eyre Peninsula, S.A., to far western Vic. Grows in spinifex grassland, mallee and coastal communities. Flowers most of the year. Map 235.

S.A.: Buckleboo, *K.D.Rohrlach 111* (AD); 4 km S of Corny Point, *H.Eichler 14043* (AD); S of Port Moorowie, *D.N.Kraehenbuehl 1993* (AD). Vic.: Nhill, Oct. 1922, *H.B.Williamson* (MEL); Ouyen, *A.Morris 1558* (NSW).

Corolla yellow, sometimes with brownish lines, usually viscid and varnished outside. Cytological voucher is *W.J.Peacock 6081.1* (SYD). Similar to *G. vernicosa*, differing in the less viscid leaves and stems; and to *G. ovata*, differing in the thicker, stiffer leaves, shorter sepals and shorter bracteoles.

This is an extremely variable species with regard to leaf shape, leaf margin, habit and persistence of leaf bases. R.Brown recognised this and distinguished three forms. These intergrade with each other and, since he used only Greek letters to distinguish them, they need not be considered from a nomenclatural point of view. It seems that there are several ecotypes, in particular the ±prostrate coastal form with broad leaves.

45. **Goodenia vernicosa** J.Black, *Trans. & Proc. Roy. Soc. S. Australia.* 43: 41 (1919)

T: Mount Patawarta, near Moolooloo, S.A., *E.H.Ising*; holo: AD; iso: K, MEL, NSW.

Erect shrub to 1 m tall, viscid, becoming varnished, glabrous. Leaves shortly petiolate, narrowly elliptic to oblanceolate, entire to dentate; lamina 1.5–4 cm long, 5–20 mm wide. Flowers in thyrses or racemes to 20 cm long; bracts leaf-like; peduncle 3–7 mm long;

bracteoles linear, 2–5 mm long; pedicel 3–8 mm long, articulate. Sepals linear, 6–10 mm long. Corolla 12–15 mm long, sparsely pubescent inside, viscid outside, not auriculate; abaxial lobes 6–7 mm long; wings triangular, to 3 mm wide. Indusium broadly oblong. Ovules 30–40. Fruit obovoid to cylindrical, c. 7–9 mm long; valves entire. Seeds elliptic, c. 2.5 mm long, pale yellow-brown.

Occurs in the Flinders Ra., S.A. Flowers chiefly Sept.–Jan. Map 236.

S.A.: W face of Hans Heysen Ra., N of Brachina Gorge, *E.N.S.Jackson 1877* (AD, NSW); Wonoka Hill, *K.Hill & L.A.S.Johnson*, 1 Nov. 1986 (AD, CANB, MEL, NSW).

The corolla is yellow, simple-pubescent, viscid and varnished outside.

46. Goodenia grandiflora Sims, *Bot. Mag.* 23: t. 890 (1805)

Goodenoughia grandiflora (Sims) Siebert & Voss, *Vilm. Blumengärtn.* 1: 559 (1896). T: 'from Mr. WHITLEY of Old-Brompton who raised it from seeds from New South Wales'; not found: illustration in J.Sims, *loc. cit.*, here taken as the type.

G. appendiculata Jacq., *Fragm.* 62, t. 92 (?1809). T: illustration in N.Jacquin, *loc. cit.*

G. mollis R.Br., *Prodr.* 577 (1810). T: near Broad Sound, [Qld], *R.Brown*; lecto: BM, *fide* R.C.Carolin, *Telopea* 3: 520 (1990); isolecto: K, MEL.

G. horniana F.Muell. & Tate ex Tate, *Rep. Horn. Exped.* 3: 169, 189 (1896). T: George Gills Range, N.T., June 1894, *R.Tate*; lecto: AD, *fide* R.C.Carolin, *Telopea* 3: 520 (1990); isolecto: K, MEL, NSW.

[*G. chambersii auct. non* F.Muell.: F.M.Bailey, *Syn. Queensland Fl.* 37 (1888)]

Erect subshrub to c. 1.5 m high, viscid, with glandular and simple hairs. Leaves ovate to orbicular, obtuse to cordate at base, dentate; lamina 2.5–5.0 cm long, to 6 cm wide; petiole to 5 cm long. Flowers in racemes or thyrses to 25 cm long; bracts leaf-like; peduncle to 3 mm long; bracetoles linear, to 4 mm long; pedicel to 5 cm long, articulate. Sepals lanceolate, to 6 mm long. Corolla to 23 mm long, villous inside, with enations, not auriculate; abaxial lobes 8–9 mm long; wings c. 3 mm wide. Indusium depressed-obovate. Ovules 30–50. Fruit obovoid, 10–13 mm long; valves sometimes 2-fid. Seeds orbicular to elliptic, c. 3 mm long, brownish, colliculate to tuberculate. Fig. 68.

Occurs in mountainous areas of central Australia in W.A., N.T., ranges in N Qld and the E coast to the Deua Natl Park, N.S.W., in rocky situations. Flowers May–Nov. Map 237.

W.A.: E end of Schwerin Mural Crescent, *R.C.Carolin 6229* (SYD). N.T.: c. 8 km NW of Aileron, *D.J.Nelson 406* (DNA, K, SYD). Qld: Zamia Ra., *R.W.Johnson 1406* (BRI, CANB); Mt Byron, *S.L.Everist 7601* (BRI). N.S.W.: N slope of Gloucester Bucketts, 7 Sept. 1967, *R.Coveny* (NSW).

The corolla is yellow, white or purplish, and the petiole sometimes carries small, sessile, leafy appendages.

There is considerable variation within this species. *Goodenia appendiculata* Jacq. was based on a specimen showing definite lateral pinnae on the petiole: however, all gradations are found from this, through very narrow small pinnae, to none at all. The type of *G. mollis* shows a more prominent simple pubescence than other specimens. Some specimens are densely villous inside the corolla. The specimens from Standley Chasm have been separated as *G. horniana*, primarily on the purple corolla. However, the differences shown by the populations from the central Australian mountains, while consistent in one locality, are not large, e.g., the corolla is purple at Standley Chasm, and yellow elsewhere. There are other differences between these populations, such as the number of teeth on the leaves, which are, presumably, due to their isolation from each other.

47. Goodenia chambersii F.Muell., *Fragm.* 1: 204 (1859)

G. grandiflora var. *chambersii* (F.Muell.) K.Krause, *Pflanzenr.* 54: 75 (1912). T: north-western S.A., *J.Macdouall Stuart*; lecto: MEL, *fide* R.C.Carolin, *Telopea* 3: 520 (1990); isolecto: K.

G. helenae Ising, *Trans. & Proc. Roy. Soc. S. Australia* 78: 117 (1955). T: Evelyn Downs, c. 144 km SW of Oodnadatta, S.A., 15 Sept. 1950, *E.H.Ising 3626*; holo: AD; iso: K, L, MEL, NSW, UC.

Ascending shrub to 50 cm tall, ±viscid, glandular- and simple-pubescent. Leaves orbicular to broadly ovate, dentate, rarely lyrate; lamina 15–30 mm long, 10–30 mm wide; petiole to 15 mm long. Flowers in racemes or thyrses to 20 cm long; bracts leaf-like; peduncle 2–5 mm long; bracteoles linear, 5–7 mm long; pedicel 5–10 mm long, articulate. Sepals lanceolate, 5–6 mm long. Corolla 18–25 mm long, villous inside, with prominent enations, scarcely auriculate; abaxial lobes c. 10 mm long; wings c. 1.5 mm wide, ciliate. Indusium depressed-obovate. Ovules 30–40. Fruit ovoid, c. 12 mm long; valves sometimes 2-fid. Seeds elliptic-oblong, 4 mm long. Fig. 63.

Occurs on ranges near the Lake Eyre Basin, S.A., in rocky situations. Flowers Sept.–Oct. Map 238.

S.A.: c. 22 km S of Oodnadatta, Aug. 1970, *H.Stace* (SYD).

Corolla yellow or cream. Distinguished from other members of the group by long cilia on the petal wings, more prominently tuberculate seeds and a less villous to almost glabrous style.

48. Goodenia kingiana Carolin, *Telopea* 3: 521 (1990)

T: 16 miles [25.6 km] E of Kalli, W.A., 22 July 1958, *N.H.Speck 1052*; holo: CANB; iso: AD, BRI, K.

[*G. grandiflora auct. non* Sims: C.A.Gardner, *Enum. Pl. Austral. Occid.* 124 (1930)].

Erect shrub to 1.5 m tall, viscid, with long and short, glandular hairs but few simple hairs. Leaves lyrato-pinnatifid; terminal lamina 2–4 cm long, 1–2.5 cm wide; petiole 3–7 cm long. Flowers in racemes or thyrses to 12 cm long; bracts leaf-like; peduncle 3–10 mm long; bracteoles linear to oblanceolate, pinnatifid, 5–10 mm long; pedicel 15–20 mm long, articulate. Sepals lanceolate, 4–5 mm long. Corolla 25–30 mm long, villous inside towards base, with enations, not auriculate; abaxial lobes 14–16 mm long; wings 3–4 mm wide. Indusium depressed-ovate. Ovules to 50. Fruit ovoid, 8 mm long; valves ±bifid. Only immature seeds seen. $n = 8$, W.J.Peacock, *Proc. Linn. Soc. New South Wales* 88: 12 (1963) as '*G. Sp.*' Fig. 48A–B.

Occurs near Cue, W.A., on cliff lines of lateritic 'breakaways'. Flowers July–Oct. Map 239.

W.A.: c. 69 km SE of Mileura, *N.H.Speck 696* (CANB); 60 km NW of Cue, *P.G.Wilson* 9901 (PERTH); Tching Ra., *W.J.Peacock 60858.1* (SYD); Cue, June 1902, *W.D.Campbell* (K).

The corolla is yellow. Cytological voucher is *W.J.Peacock 60858.3* (SYD). On the leaves the terminal lamina is ovate, cordate & dentate; while lateral pinnae are smaller, ovate to linear, and when ovate are petiolulate; the bracteoles sometimes have a few lobes at base, and are usually ±petiolate.

Differs from other members of the section in having longer, glandular hairs, larger flowers with broader wings making a more acute angle at the apex, and petiolulate segments below the main lamina of the leaf.

49. **Goodenia amplexans** F.Muell., *Trans. & Proc. Philos. Inst. Victoria* 2: 70 (1857)

T: foot of the Mount Lofty Ranges, S.A., *F.Mueller*; lecto: MEL, *fide* R.C.Carolin, *Telopea* 3: 522 (1990).

G. amplexans var. *angustifolia* K.Krause, *Pflanzenr.* 54: 62 (1912). T: Western R., Kangaroo Is., S.A., Sept. 1908, *Rogers*; lecto: NSW, *fide* R.C.Carolin, *Fl. Australia* 35: 331 (1992); isolecto: B (destroyed).

Erect undershrub or shrubby herb to 1 m tall, viscid, glandular-pubescent, with long and short, glandular hairs and fine, simple hairs, aromatic. Leaves sessile, stem-clasping, ovate to oblong, denticulate; lamina 3–9 cm long, 5–30 mm wide. Inflorescence a spike-like raceme, to 15 cm long; bracts leaf-like; peduncles 1–2 mm long; bracteoles linear-lanceolate, 1–2 mm long; pedicels 2–3 mm long, indistinctly articulate. Sepals lanceolate, 2.5–3.5 mm long. Corolla 12–15 mm long, ±glabrous inside, with enations, inconspicuously auriculate; abaxial lobes 3.5–7.5 mm long; wings 1–2.5 mm wide. Indusium broadly oblong. Ovules 20–40. Fruit ellipsoidal, 6 mm long; valves sometimes bifid. Seeds elliptic, 3 mm long, yellow-brown, colliculate to tuberculate.

Occurs in the southern Flinders Ra. and the Mt Lofty Ra. and on Kangaroo Is., S.A., in open forest and on sea cliffs. Flowers chiefly Aug.–Feb. Map 240.

S.A.: Alligator Gorge, 20 Nov. 1957, *H.M.Cooper* (AD); Mambray Ck, Oct. 1955, *A.R.R.Higginson* (AD); Hallet Cove, 30 Sept.1893, *R.Brummit* (AD); Slape Gully, 9 km SE of Adelaide, *P.G.Wilson 1040* (AD); Atholstone, *R.Filson 1539* (MEL).

The corolla is yellow. The leaves are somewhat variable in shape and the narrow-leaved form has been given varietal status by K.Krause. At first it was thought to be confined to Kangaroo Is., but there are specimens from the mainland which approach the more broad-leaved Kangaroo Is. specimens.

50. **Goodenia benthamiana** Carolin in J.P.Jessop, *Fl. S. Australia* 4th edn, 3: 1392 (1986)

G. amplexans var. *parvifolia* Benth., *Fl. Austral.* 4: 60 (1868). T: Wimmera, Vic., *J. Dallachy*; holo: K; iso: MEL.

Erect undershrub to 40 cm tall, viscid, with long and short, glandular hairs and few simple hairs, aromatic. Leaves sessile, ±stem-clasping, ovate to broadly elliptic, dentate; lamina 8–25 mm long, 4–20 mm wide. Inflorescence a spike-like raceme to 10 cm long; bracts leaf-like; peduncles 0.5–1 mm long; bracteoles ovate to lanceolate, 2.5–3 mm long; pedicels 0.5–1 mm long, articulate. Sepals ovate to lanceolate, 2–3 mm long. Corolla 10–12 mm long, pubescent and with enations inside, indistinctly auriculate; abaxial lobes 4 mm long; wings c. 1.5 mm wide. Indusium broadly elliptic. Ovules c. 15. Fruit and seeds not seen.

Occurs from Eyre Peninsula, S.A., to western Vic., in mallee and forest usually over alkaline soil. Flowers most of the year. Map 241.

S.A.: Jamieson, 20 km S of Kimba, *K.D.Rohrlach 84* (AD). Vic.: c. 16 km NNE of Huntley, 27 May 1944, *J.H.Willis* (MEL); near Nhill, Oct. 1889, *C.Walter* (NSW).

The corolla is yellow. The indusium is folded on the upper surface and deeply notched especially on the lower lip. Similar to *G. amplexans*, differing in leaf shape, shorter hairs, less numerous ovules and a notched indusium.

51. **Goodenia albiflora** Schldl., *Linnaea* 20: 599 (1847)

Picrophyta albiflora F.Muell., *Linnaea* 25: 421 (1853); *Goodenia grandiflora* var. *albiflora* (Schldl.) K.Krause, *Pflanzenr.* 54: 75 (1912). T: South Australia, *H.Behr 82*; *n.v.*

Erect shrub to 70 cm tall, glaucous, with few large, glandular hairs when young, otherwise almost glabrous except flowers. Leaves almost sessile, elliptic to ovate, toothed, rarely entire, thick; lamina 3–7 cm long, 15–40 mm wide. Flowers in racemes to 10 cm long; bracts lanceolate, 8–15 mm long; peduncle 2–4 mm long; bracteoles lanceolate, c. 3 mm long; pedicel 1–2 mm long, indistinctly articulate. Sepals lanceolate, 5–6 mm long. Corolla 15–20 mm long, with coarse hairs inside, with enations, indistinctly auriculate; abaxial lobes c. 10 mm long; wings 2.5–3 mm wide. Indusium broadly ovate, folded. Ovules c. 40. Fruit ovoid, c. 1 cm long; valves bifid. Seeds elliptic, 3 mm long, light brown, tuberculate.

Occurs in S.A. from Eyre Peninsula to Flinders Ranges and Mt Lofty Ra. Flowers chiefly Oct.–Jan. Map 242.

S.A.: Horrocks Pass between Stirling North and Wilmington, 4 Oct. 1954, *J.B.Cleland* (AD); Baroota, 18 Sept. 1961, *J.C.Cleland* (AD); Wirrabara, Nov. 1882, *J.H.Love* (MEL); Flinders Ra., *M.Koch 572* (NSW).

The corolla is white, softly glandular and simple hairy outside. Similar to *G. calcarata*, which is also glaucous, but which has more dissected leaves, a prominent spur on the corolla and less shrubby habit.

52. **Goodenia xanthotricha** Vriese, *Natuurk. Verh. Holl. Maatsch. Wetensch. Haarlem* ser. 2, 10: 155 (1854)

T: [Australia], *J.P.Verreaux*; holo: L.

G. leptotheca F.Muell., *Fragm.* 6: 13 (1867). T: south-western W.A., *J.Drummond 4: 194*; holo: MEL.

G. hilliana C.Gardner, *J. & Proc. Roy. Soc. W. Australia* 27: 197 (1942). T: Hill River, W.A., Feb. 1940, *C.A.Gardner s.n.*; holo: PERTH.

Undershrub or shrubby herb to 50 cm tall, viscid, glandular-hairy. Leaves almost sessile, linear to oblanceolate, slightly recurved, dentate; lamina 15–35 mm long, 2–7 mm wide. Flowers in racemes to 6 cm long; bracts leaf-like, oblanceolate to obovate, c. 5 mm long; peduncle 2–4 mm long; bracteoles narrowly obovate, 4–10 mm long; pedicel c. 1 mm long, not articulate. Sepals narrowly oblong, 4–5 mm long. Corolla c. 14 mm long, hairs scattered inside, with enations, indistinctly auriculate; abaxial lobes c. 5 mm long; wings to 2 mm wide. Indusium oblong. Ovules to 15. Fruit cylindrical to ovoid, 5 to 6 mm long; valves split eventually to base. Seeds elliptic, 1.8 mm long, blackish brown, aculeate.

Restricted to the Hill R. region, south-western W.A., on gravelly hills. Flowers chiefly Nov.–Feb. Map 243.

W.A.: Hill R., *C.A.Gardner 12078* (PERTH); Mt Lesueur, *R.D.Royce 7725* (PERTH).

The corolla is blue. K.Krause, (*Pflanzenr.* 64: 44, 1912) placed this in his sect. *Monochila* but it has none of the features associated with that group. It is a species difficult to place. Since the seeds are aculeate it does not appear to belong to sect. *Caeruleae*, despite the blue corolla. C.A.Gardner, when describing *G. hilliana*, associated it with *G. quadrilocularis*, from which it differs in the aculeate seeds and flower colour. In describing *G. leptotheca*, Mueller considered the corolla to be possibly yellow, but he based this on dried material and it is almost certainly in error.

53. Goodenia brunnea Carolin in J.P.Jessop, *Fl. S. Australia* 4th edn, 3: 1393 (1986)

T: 1 mile [c. 1.6 km] east of Mount Woodroffe, S.A., 30 June 1958, *R.Hill & T.R.N.Lothian 711*; holo: AD; iso: AAU, TL.

[*G. grandiflora auct. non* Sims: E.Robertson in J.M.Black, *Fl. S. Australia* 2nd edn, 3: 822 (1957) *p.p.*].

Erect or ascending shrub to 80 cm tall, viscid, glandular-hairy, with few simple hairs. Leaves lanceolate, dentate, usually lyrate; lamina 3–6 cm long, 10–20 mm wide; petiole to 3 cm long. Flowers in racemes or thyrses to 25 cm long; bracts leaf-like; peduncle ±obsolete; bracteoles lanceolate, 2–3 mm long; pedicel 5–10 mm long, articulate. Sepals lanceolate, 4–5 mm long. Corolla to 20 mm long, with long, fine, white hairs ±in rows inside, with enations, scarcely auriculate; abaxial lobes c. 9 mm long; wings c. 2.5 mm wide, dentate. Indusium depressed-obovate. Ovules c. 40. Fruit ovoid, 10 mm long; valves sometimes bifid. Seeds elliptic-oblong, 3 mm long, dark brown, colliculate.

Occurs in the Musgrave Ra., S.A., in rocky situations. Flowers chiefly June–Nov. Map 244.

S.A.: Mt Morris, Musgrave Ra., *H.Eichler 17324* (AD); Alalka Gorge, Ernabella, *J.R.Maconochie 834* (DNA); Ernabella, Sept. 1945, *J.B.Cleland* (AD).

The corolla is yellow. Differs from other members of the section in its narrower leaves which are glandular-viscid.

54. Goodenia saccata Carolin in J.P.Jessop, *Fl. S. Australia* 4th edn, 3: 1403 (1986)

T: Gammon Ra., S.A., 17 Sept. 1956, *H.Eichler 12676*; holo: AD; iso: B, K, L, MEL, NSW, P, TI, UC, US.

[*G. grandiflora auct. non* Sims; E.Robertson in J.M.Black, *Fl. S. Australia* 2nd edn, 3: 822 (1957) *p.p.*].

Erect, branching shrub to c. 1 m tall, with glandular and simple hairs, glabrescent. Leaves ovate, obtuse at base, dentate; lamina 2–5 cm long, 2–3 cm wide; petiole 1.5–5 cm long. Flowers in racemes or thyrses, to 20 cm long; bracts leaf-like; peduncle less than 1 mm long; bracteoles linear, c. 1.5 mm long; pedicel 4–6 mm long, articulate. Sepals lanceolate, 5–6 mm long. Corolla to 17 mm long, with short, coarse hairs inside, with enations, indistinctly auriculate; abaxial lobes c. 8 mm long; wings c. 2 mm wide, often laciniate. Indusium deltoid, slightly folded. Ovules to 25. Fruit ovoid, c. 10 mm long, valves usually bifid. Seeds broadly elliptic, 3 mm long, brown, dull, colliculate.

Occurs in the northern Flinders Ra. and ranges of the Lake Torrens basin, S.A., in rocky situations. Flowers chiefly Sept.–Nov. Map 245.

S.A.: Umberatana homestead, *T.R.N.Lothian 3318* (AD); Arcoona Ck, *K.D.Rohrlach 643* (AD); near junction of Moolooloo and Parachilna roads, *D.Kraehenbuehl 23* (AD); Aroona Dam, *H.Eichler 12997* (AD).

The corolla is yellow or 'off white', the sepals are c. 1 mm wide, and adnate to the ovary to the top. Differs from *G. albiflora* in having stalked, densely hairy leaves. Differs from *G. grandiflora* especially in having coarse hairs inside the corolla.

55. Goodenia stirlingii F.M.Bailey, *Queensland Agric. J.* 15: 492 (1904) as *G. stirlingi*

T: Herberton, Qld, *J.Stirling*; holo: BRI; iso: K.

Erect undershrub to 30 cm tall, glabrous except flowers. Leaves sessile, narrowly elliptic, tapering towards base, dentate; lamina 2–4 cm long, 5–15 mm wide. Flowers in short racemes to 10 cm long; bracts leaf-like; peduncle to 5 mm long; bracteoles linear-lanceolate, c. 5 mm long; pedicel 9–12 mm long, articulate. Sepals lanceolate, 7–8 mm long. Corolla c. 22 mm long, villous inside, with enations, scarcely auriculate; abaxial lobes 6–8 mm long; wings 1–1.5 mm wide. Indusium broadly oblong. Ovules c. 20. Fruit ovoid-obloid, 12 mm long; valves bifid. Seeds elliptic, 3 mm long, yellowish, minutely reticulate.

Occurs in north-eastern Qld. Flowers chiefly Mar.–July. Map 246.

Qld: near Herberton, June 1899, *Dixon* (NSW); Stannary Hills, *C.H.Gittins 531* (BRI, NSW); Batch Garnet Railway Line, 2 Mar. 1939, *H.R.Thurston* (BRI).

The corolla is yellow, and glandular-puberulous outside. It is distinguished from others in this section by its glabrous, sessile leaves, glabrous pedicels and from *G. albiflora* by its lack of coarse hairs on the inside of the corolla.

56. **Goodenia calcarata** (F.Muell.) F.Muell., *Fragm.* 6: 14 (1867)

Picrophyta calcarata F.Muell., *Linnaea* 25: 422 (1853). T: NE of Cudnaka (Kanyaka), S.A., 1851, *F.Mueller*; holo: MEL; iso: K, P.

Illustration: G.M.Cunningham *et al.*, *Pl. W. New South Wales* 631 (1981).

Erect, annual herb to 60 cm tall, glaucous, glabrous or coarsely glandular-hairy. Leaves obovate to oblong, dentate and/or lyrate; lamina 2–5 cm long, 10–35 mm wide; petiole to 20 cm long. Flowers in racemes, to 15 cm long; lower bracts leaf-like, upper ones linear, smaller; peduncle 2–7 mm long; bracteoles absent or deltoid, to 1 mm long; pedicel 10–25 mm long, indistinctly articulate. Sepals narrowly elliptic, c. 5 mm long. Corolla 10–15 mm long, with few, short, often purplish hairs inside, with enations, scarcely auriculate, spurred; abaxial lobes 6–9 mm long; wings 1.5–2 mm wide. Indusium broadly elliptic. Ovules to 35. Fruit ovoid, 12–15 mm long; valves bifid. Seeds elliptic, 3.5 mm long, yellowish, tuberculate. Figs 26E, 45A, 49A–C.

Occurs in the drier areas of eastern S.A., south-eastern Qld and north-western N.S.W.; usually in stony situations. Flowers June–Dec. Map 247.

S.A.: 11 km west of Arckaringa homestead, *E.A.Shaw 511* (AD); Gammon Ranges, *H.Eichler 12644* (AD, K); Aroona Reservoir, *T.R.N.Lothian 1408* (AD). Qld: Quilpie, 20 Oct. 1950, *D.Haplin* (BRI); N.S.W.: c. 22 km west of Yandama, Aug. 1941, *N.C.W.Beadle* (SYD).

The corolla is white with brownish markings, has a spur to twice as long as ovary, and the indusium has purplish hairs towards the base. Differs from all other members of the section in the long corolla spur. See note under *G. albiflora*.

57. **Goodenia macmillanii** F.Muell., *Fragm.* 1: 119 (1859)

G. grandiflora var. *macmillanii* (F.Muell.) K.Krause, *Pflanzenr.* 54: 75 (1912). T: near Macalister R., Gippsland, Vic., *F.Mueller*; lecto: MEL, *fide* R.C.Carolin, *Fl. Australia* 35: 332 (1992); isolecto: ?BM, K, MEL.

Erect shrub to 50 cm tall, -glandular pubescent, with few simple hairs. Leaves ovate to elliptic in outline, pinnately lobed to lyrate; lamina 4–8 cm long, 2–5 cm wide; petiole to 3 cm long. Flowers in racemes to 20 cm long; bracts leaf-like; peduncle 0–1 cm long; bracteoles linear, 2 mm long; pedicel 15–21 mm long, indistinctly articulate. Sepals lanceolate, 6–8 mm long. Corolla 18–22 mm long, densely hairy inside, with enations, scarcely auriculate; abaxial lobes 9–10 mm long; wings 2–2.5 mm wide. Indusium broadly elliptic. Ovules 40–50. Fruit cylindrical to ovoid; 10–12 mm long; valves bifid. Seeds broadly elliptic, 2–5 mm long, yellow-brown, tuberculate.

Occurs in the forests of E Gippsland, Vic. Flowers chiefly Nov.–Feb. Map 248.

Vic.: Upper Snowy R., c. 25 km N of Buchan, 13 Feb. 1948, *L.Hodge* (MEL); Blanket Hill, Macalister R., Dec. 1890, *A.H.S.Lucas* (NSW)

Corolla bluish purple, which distinguishes it from all other members of the section, except some races of *G. grandiflora* which has dentate, not lyrate–pinnate, leaves.

58. **Goodenia fordiana** Carolin, *Telopea* 3: 522 (1990)

T: Bushman Ra. near Coramba, N.S.W., Oct. 1913, *W.Heron*; holo: NSW.

Prostrate herb with stock, to 20 cm long, softly pubescent with simple and multicellular hairs. Basal leaves elliptic to broadly elliptic, dentate to entire; lamina 10–25 mm long, 8–14 mm wide; petiole 1–3 cm long. Flowers distant in racemes to 20 cm long; bracts leaf-like; peduncle 1–2 cm long; bracteoles linear, 2 mm long; pedicel 1–1.5 cm long, not articulate. Sepals linear, 2.5 mm long. Corolla 8–10 mm long, pubescent inside, with enations, auriculate; abaxial lobes c. 3.5 mm wide; wings c. 2 mm wide. Indusium broadly elliptic. Ovules not seen. Fruit and seeds not known.

Occurs in forests in the ranges of north-eastern N.S.W. N of Bulahdelah. Flowers Oct.–Dec. Map 249.

N.S.W.: Lookout Rd, Bellangry State Forest, 23 Oct. 1958, *J.C.Cousins* (NSW); Bulahdelah, 12 Nov. 1923, *H.M.R.Rupp* (NSW); Ellenborough Falls, *R.Coveny & L.Haegi 9883* (NSW).

The corolla is yellow. Similar to *G. rotundifolia*, differing in its softer pubescence, longer pedicel, slender scapes with distant flowers and almost entire leaves. Its systematic position remains doubtful at present.

59. **Goodenia quadrilocularis** R.Br., *Prodr.* 578 (1810)

Tetrathylax quadrilocularis (R.Br.) Vriese, *Natuurk. Verh. Holl. Maatsch. Wetensch. Haarlem* ser. 2, 10: 165 (1854). T: Lucky Bay, [W.A.], Jan. 1801, *R.Brown*; holo: BM.

G. taylori F.Muell., *Fragm.* 3: 141 (1863). T: Duke of Orleans Bay, W.A., (*G.Maxwell*?); lecto: MEL, *fide* R.C.Carolin, *Telopea* 3: 523 (1990); isolecto: K?

Erect, perennial herb to 50 cm tall, glabrous except flowers. Leaves mostly towards base, ovate to oblanceolate, tapering basally, dentate, thick; lamina 3–5 cm long, 10–17 mm wide. Cauline leaves smaller. Flowers in racemes to 30 cm long; bracts lanceolate, 5–20 mm long; peduncle 4–20 mm long; bracteoles lanceolate, 2–5 mm long; pedicel 2–5 mm long, indistinctly articulate. Sepals lanceolate, c. 5 mm long. Corolla c. 20 mm long, softly pubescent outside, villous inside, with enations, indistinctly auriculate; abaxial lobes c. 7 mm long; wings 2–2.5 mm wide. Indusium broadly oblong. Ovules 60–65. Fruit cylindrical, 15 mm long; valves ±bifid. Seeds broadly elliptic, 1.5 mm long, black or dark brown, reticulate. Fig. 49D.

Restricted to the Cape le Grand area, W.A., on sand dunes and granite slopes. Flowers Sept.–Dec. Map 250.

W.A.: Cape le Grand, Sept. 1952, *N.H.Speck* (PERTH); near Cape le Grand, *P.G.Wilson 5553* (PERTH); Mt le Grand, *A.S.George 11011* (PERTH).

The corolla is yellow and has soft, simple hairs and glandular hairs outside.

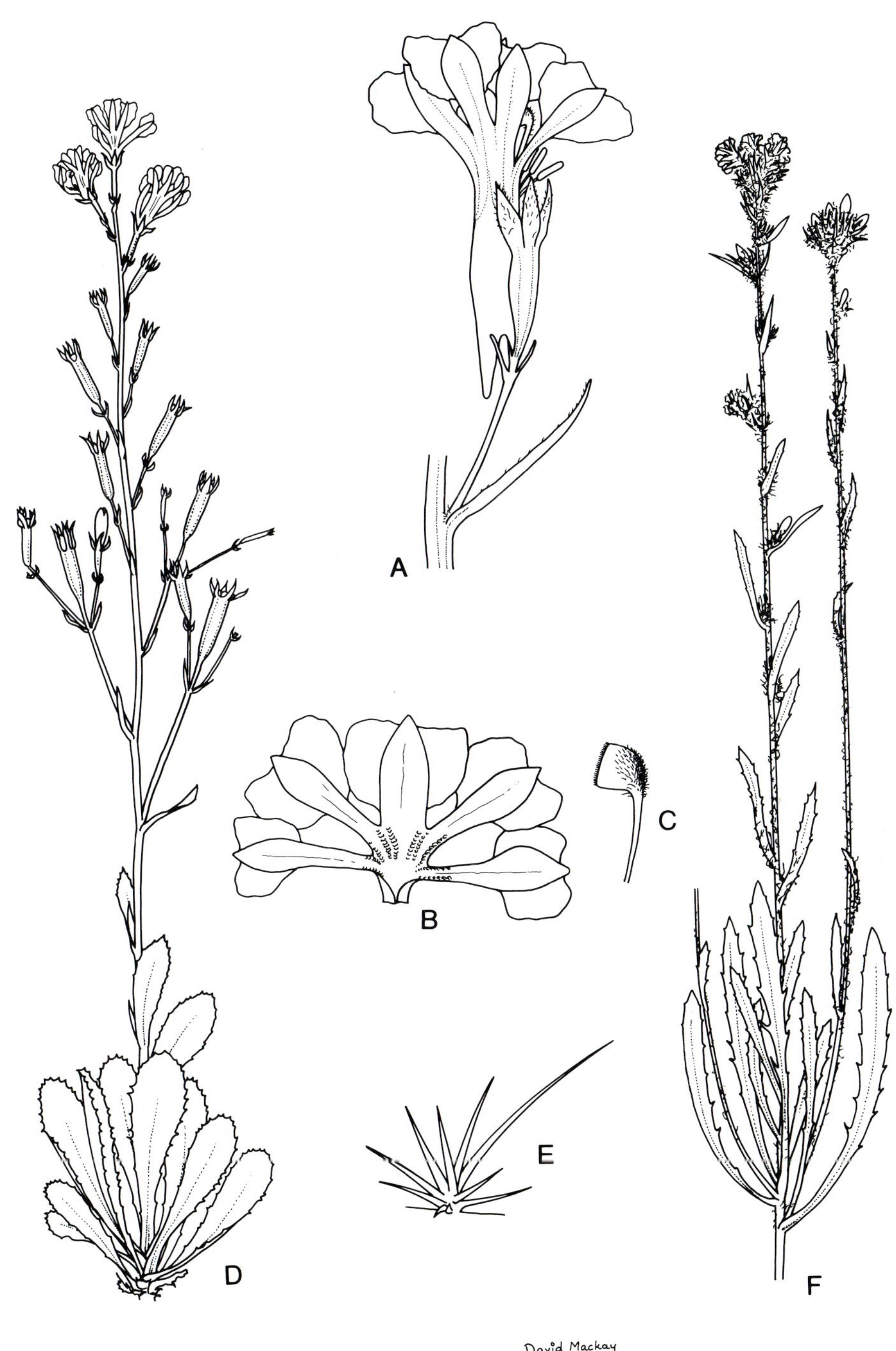

Figure 49. *Goodenia*. **A–C**, *G. calcarata*. **A**, flower X2; **B**, corolla X2; **C**, indusium X2 (**A–C**, R.Pearce 136, AD). **D**, *G. quadrilocularis*, habit X0.5 (J.Wrigley, 30 Oct. 1968, SYD). **E–F**, *G. glomerata*, **E**, hair X50; **F**, habit X0.25 (**E–F**, I.Telford BR258A, NSW). Drawn by D.Mackay.

60. Goodenia decurrens R.Br., *Prodr.* 575 (1810)

T: banks of Grose River, N.S.W., 1803, *R.Brown*; lecto: BM, *fide* R.C.Carolin, *Telopea* 3: 523 (1990); isolecto: K, MEL, NSW, P.

Erect undershrub to 80 cm tall, glabrous. Leaves sessile, decurrent, lanceolate to elliptic, dentate, thick; lamina 5–10 cm long, 10–25 mm wide. Flowers in thyrses or racemes to 45 cm long; bracts linear to lanceolate, usually to 10 mm long; peduncle to 15 mm long; bracteoles similar to bracts, smaller; pedicel to 10 mm long, articulate. Sepals narrowly ovate, 4–7 mm long. Corolla 15–18 mm long, villous inside, with enations, indistinctly auriculate; abaxial lobes 7–8 mm long; wings 2–3 mm wide. Indusium depressed-obovate. Ovules c. 50. Fruit ovoid, c. 9 mm long; valves bifid. Seeds elliptic to orbicular, c. 1.8 mm long, pale brown, tuberculate-granulate. n = 16, W.J.Peacock, *Proc. Linn. Soc. New South Wales* 88: 11 (1963). Fig. 25B.

Occurs in the Blue Mtns, N.S.W., on sandstone cliffs. Flowers chiefly Oct.–Mar. Map 251.

N.S.W.: Lees Pinch, Wollar, 10 Aug. 1950, *L.A.S.Johnson & E.F.Constable* (NSW); Mt Wilson road, *S.Carlquist 1308* (RSA, SYD); Burragorang, *R.H.Cambage 3226* (SYD); Upper Colo, 6 Oct. 1963, *C.Burgess* (NSW); near waterfall opposite Milsons Is., 20 Dec. 1909, *J.B.Cleland* (NSW).

The corolla is yellow, usually has simple and glandular and occasionally stellate hairs outside. There is usually a line of hairs through each internode above the leaf axil. Cytological voucher is *W.J.Peacock 6011.26.3* (SYD).

61. Goodenia rostrivalvis Domin, *Biblioth. Bot.* 22: 639 (1929)

T: near Katoomba, Blue Mountains, N.S.W., Apr. 1910, *K.Domin*; lecto: PR, *fide* R.C.Carolin, *Telopea* 3: 523 (1990).

Ascending subshrub, to c. 60 cm tall, glabrous, ±glossy. Leaves obovate to oblanceolate, tapering basally, dentate; lamina 3.5–7 cm long, 8–20 mm wide. Flowers in thyrses or racemes to 40 cm long; bracts linear, 0.5–2 cm long; peduncle mostly 0.5–2 cm long; bracteoles similar to bracts, smaller; pedicel 0.5–3 cm long, not articulate. Sepals linear to lanceolate, 5–6 mm long. Corolla 14–18 mm long, with few, short hairs inside, with enations, ±auriculate; abaxial lobes 7–8 mm long; wings c. 2.5 mm wide. Indusium depressed-obovate. Ovules c. 20. Fruit ovoid, 10 mm long; valves bifid to base. Seeds elliptic, 1.5–1.7 mm long, brown, glossy, reticulate-foveate.

Occurs in the Blue Mtns, N.S.W., on damp south facing sandstone cliffs. Flowers Oct.–Apr. Map 252.

N.S.W.: Mt Tomah, 20 Feb. 1962, *C.Burgess* (CBG); Wentworth Falls, 22 Jan. 1915, *J.McLuckie* (SYD); Wentworth Falls, 6 Jan. 1943, *O.D.Evans* (SYD); National Pass, Wentworth Falls, *R.Coveny 4780* (NSW); Lawson, 3 Jan. 1959, *B.P.Fitzgerald* (NSW).

The corolla is yellow, covered with mostly stellate, yellowish hairs and some cottony, white hairs. Similar to *G. dimorpha* and *G. decurrens*, differing in the former not having the elongated basal part of the stem bearing most of the leaves, and in the latter having leaves which are decurrent and not glossy.

62. **Goodenia dimorpha** Maiden & Betche, *Proc. Linn. Soc. New South Wales* ser. 2, 28: 907 (1904)

T: Woodford, Blue Mtns, N.S.W., Jan. 1899, *J.H.Maiden*; lecto: NSW, *fide* R.C.Carolin, *Telopea* 3: 523 (1990); isolecto: MEL.

Erect herb to 50 cm tall, with adventitious roots, glabrous. Leaves mostly basal from stock, linear to obovate, narrowing basally, entire or slightly dentate, thick; lamina 1–5 cm long, 2–11 mm wide. Flowers in spreading thyrse-like panicles to 40 cm long; bracts linear, 0.5–3 cm long; peduncle to 14 cm long; bracteoles similar to bracts but smaller; pedicel c. 3 mm long, obscurely articulate. Sepals narrow to very narrowly obovate, 2–6 mm long. Corolla 12–15 mm long, pubescent inside, with enations, auriculate; abaxial lobes c. 6 mm long; wings 3 mm wide. Indusium broadly obovate. Ovules c. 25. Fruit narrowly cylindrical to obovoid, 8–10 mm long; valves ±bifid. Seeds elliptic, 1.5 mm long, dark brown, glossy, reticulate-foveate.

Occurs in the Sydney area and Blue Mtns, N.S.W. The corolla is yellow, with stellate and cottony hairs outside. Similar to *G. bellidifolia* and *G. stelligera*, differing from the former in the fruit and ovary shape, and in the spreading inflorescence, and from the latter in the spreading inflorescence. Two varieties are recognised.

Basal leaves obovate **62a**. var. **dimorpha**

Basal leaves linear **62b**. var. **angustifolia**

62a. **Goodenia dimorpha** Maiden & Betche var. **dimorpha**

Basal leaves obovate. $n = 8$, W.J.Peacock, *Proc. Linn. Soc. New South Wales* 88: 11 (1963) as var. *angustifolia*.

Occurs in the Blue Mtns and adjacent coastal regions of N.S.W., in swampy places on sandstone. Flowers chiefly Oct.–Mar. Map 253.

N.S.W.: Cowan track, 10 Mar. 1918, *W.F.Blakely* (NSW); Wentworth Falls, Dec. 1926, *O.D.Evans* (SYD); Leura, Apr. 1914, *A.A.Hamilton* (NSW); Linden, 4 Feb. 1959, *E.F.Constable* (NSW); Hazelbrook, Feb. 1963, *M.Pearce* (NSW).

Cytological voucher is *W.J.Peacock 611.18.1* (SYD).

62b. **Goodenia dimorpha** var. **angustifolia** Maiden & Betche, *Proc. Linn. Soc. New South Wales* ser. 2, 28: 907 (1904)

T: National Park, N.S.W., Jan. 1903, *J.L.Boorman*; lecto: NSW, *fide* R.C.Carolin, *Telopea* 3: 523 (1990); isolecto: BM, SYD.

Leaves linear.

Grows in swamps of the coastal sandstone plateaus near Sydney, N.S.W. Flowers chiefly Nov.–June. Map 254.

N.S.W.: Wondabyne, 22 Feb. 1925, *W.F.Blakely* (NSW); Loftus, 1894, *J.H.Camfield* (NSW); Uloola track, Royal Natl Park, *R.Coveny 1045a* (NSW); Waterfall, 27 Feb. 1887, *J.J.Fletcher* (NSW).

63. **Goodenia stelligera** R.Br., *Prodr.* 575 (1810)

T: Port Jackson (Sydney & South Head), N.S.W., May–June 1802, *R.Brown*; lecto: BM, *fide* R.C.Carolin, *Telopea* 3: 523 (1990); isolecto: K, MEL, P.

G. armeriifolia Sieber ex Sprengel, *Syst. Veg. 4 Cur. Post.* 276 (1827) as *armerifolia*. T: Australia, probably near Sydney, N.S.W., *F.W.Sieber*; *n.v.* [type is almost certainly from the same collection as the type of *G. armeriifolia* Sieber ex DC.].

G. armeriifolia Sieber ex DC., *Prodr.* 7: 513 (1839) as armeriaefolia. T: Port Jackson, N.S.W., *F.W.Sieber 229*; holo: G; iso: BM, K, MEL.

G. longifolia Vriese, *Natuurk. Verh. Holl. Maatsch. Wetensch. Haarlem* ser. 2, 10: 127 (1854); *G. stelligera* var. *longifolia* (Vriese) Domin, *Biblioth. Bot.* 22: 640 (1929). T: Sydney, N.S.W., date unknown, *coll. unknown*; holo: K.

Illustration: E.R.Rotherham, *et al.*, *Fl. Pl. New South Wales & S. Queensland* 43, fig. 97 (1975).

Erect herb to 60 cm tall, with adventitious roots, glabrous. Leaves mostly basal from stock, sessile, linear to oblanceolate, entire or denticulate, thick, usually glossy; lamina 5–25 cm long, 1–12 mm wide. Flowers in thyrses or spikes to 60 cm long; bracts linear-lanceolate, mostly 1–2.5 cm long; peduncle mostly obsolete, lowermost to 6 cm long; bracteoles linear-lanceolate, 3–6 mm long; pedicel 0–4 mm long, not articulate. Sepals linear, 4–5 cm long. Corolla 13–15 mm long, ±glabrous inside, with enations, indistinctly auriculate; abaxial lobes 7–8 mm long; wings 1.5–2.5 mm wide. Indusium broadly oblong. Ovules 30–40. Fruit obovoid, 5–9 mm long; valves bifid. Seeds elliptic, 1.5 mm long, brown, reticulate-foveate. n = 8, W.J.Peacock, *Proc. Linn. Soc. New South Wales* 88: 11 (1963). Fig. 23E.

Occurs in coastal areas of southern Qld and N.S.W., in swamps on sandstone. Flowers chiefly Aug.–Feb. Map 255.

Qld: Fraser Is., Oct. 1921, *C.T.White* (BRI); Coolum Beach E of Yandina, *C.L.Wilson 613, 215* (BRI). N.S.W.: Hat Head, 18 Jan. 1953, *E.F.Constable* (K, NSW); North Head, Sydney, *F.A.Rodway 1647* (NSW); Currockbilly Mtn, near Braidwood, Dec. 1915, *J.L.Boorman* (NSW).

The corolla is yellow, with whitish, cottony hairs and yellow, stellate hairs outside. Cytological vouchers are *W.J.Peacock 6012.2.1 & 5810* (SYD).

The leaf width and shape are particularly variable in this species. *Goodenia longifolia*, for instance, represents a form with spathulate leaves, often with numerous cauline ones as well. The narrow-leaved forms do not occur on the higher plateaus, even near the sea.

64. **Goodenia glomerata** Maiden & Betche, *Proc. Linn. Soc. New South Wales* ser. 2, 24: 646 (1900)

T: Braidwood, N.S.W., Dec. 1884, *W.Bäuerlen*; holo: NSW.

Erect herb to 50 cm tall, woody at base, yellowish to grey villous, glabrescent. Leaves mostly towards base, elliptic to oblanceolate, narrowing basally, dentate to lyrate; lamina mostly 6–20 cm long, 10–22 mm wide. Flowers in compact spikes to 12 cm long to 30 cm long; bracts lanceolate, 12–20 mm long; bracteoles linear-lanceolate, 8–15 mm long. Sepals linear-lanceolate, 8–10 mm long. Corolla 20–30 mm long, pubescent inside, with enations, ±auriculate; abaxial lobes 9–11 mm long; wings c. 2 mm wide. Indusium broadly oblong. Ovules 20–25. Fruit cylindrical, 8–9 mm long; valves split ±to base. Seeds elliptic to orbicular, 1.5 mm long, brown, reticulate-foveate. Fig. 49E–F.

Occurs on the Budawang Ra. and nearby areas, N.S.W., in swamps and seepages on sandstone and conglomerate. Flowers chiefly Oct.–Jan. Map 256.

N.S.W.: The Castle, Budawang Ra., *R.C.Carolin 4414* (SYD); Newhaven Ck, Budawang Ra., 22 Oct. 1971, *I.R.Telford* (CBG, SYD); track from Yadboro House to The Castle, Budawang Ra., *R.C.Carolin 6965* (SYD).

The corolla is yellow, with cottony hairs and yellow, stellate hairs outside.

65. **Goodenia lineata** J.H.Willis, *Muelleria* 1: 151 (1967)

T: summit of Mt William, The Grampians, Vic., 3 Jan. 1964, *J.H.Willis*; holo: MEL; iso: K, MEL, NSW.

Erect, perennial herb to 50 cm tall, glandular pubescent or glabrescent. Leaves mostly basal from stock, oblanceolate, tapering basally, ±dentate, thick; lamina 3–8 cm long, 3–9 mm wide. Inflorescence a few-flowered raceme to 20 cm long; bracts linear, 3–9 mm long; peduncles 4–8 cm long; bracteoles triangular, 3–4 mm long, pedicels 3–6 mm long, articulate. Sepals lanceolate, 4–7 mm long. Corolla c. 20 mm long, with few hairs inside, with enations, auriculate; abaxial lobes 8–10 mm long; wings c. 2 mm wide. Ovules 15–20; indusium broadly elliptic. Fruit and seeds unknown.

Restricted to The Grampians, Vic., in heath in sandy soil. Flowers chiefly Nov.–Feb. Map 257.

Vic.: Stachans Crossing, Glenelg R., 11 Dec. 1966, *J.H.Willis & A.C.Beauglehole* (MEL); Mafeking, Mt William Ra., 18 Dec. 1965, *E.M.Tucker* (MEL); Victoria Ra., *P.E.Finck & A.C.Beauglehole 4059* (MEL); Kalimna Falls, Mt William, 29 Oct. 1961, *D.Dawson* (MEL); Kentruck Heath, 22 Nov. 1970, *M.Beck* (MEL).

The corolla is yellow. In the absence of seeds the systematic position of this species remains uncertain. The presence of bracteoles seems to indicate differences from the rest of the *Pedicellosae* of K.Krause (subsect. *Ebracteolatae p.p.* of this treatment) with which J.H.Willis, *loc. cit.*, allies it: such an affinity is possible, however, since some of the species grouped around *G. glauca* of subsect. *Ebracteolatae* also have bracteoles. The inflorescence is not strictly a paniculate cyme, as Willis says, but a simple raceme; reduction of the flower number may result in a superficial resemblance to a cyme.

66. **Goodenia bellidifolia** Smith, *Trans. Linn. Soc. London, Bot.* 2: 349 (1794)

T: Port Jackson, N.S.W., *J.White*; holo: LINN, Smith. Herb. 325.7; iso: K?

G. spathulata Vriese, *Natuurk. Verh. Holl. Maatsch. Wetensch. Haarlem* ser. 2, 10: 123 (1854). T: Botany Bay, N.S.W., *Hügel post. spec. cult.*; holo: W.

G. bellidifolia var. *ramosissima* K.Krause, *Pflanzenr.* 54: 50 (1912). T: without precise locality, N.S.W., *G.Caley*; lecto: BM, *fide* R.C.Carolin, *Telopea* 3: 523 (1990); isolecto: W.

Erect herb to 60 cm tall, glabrous to cottony. Leaves usually basal on stock, oblanceolate to obovate, narrowing basally, entire to dentate; lamina 4–9.5 cm long, 5–20 mm wide. Flowers in narrow thyrses or racemes to 40 cm long; bracts linear to lanceolate, to 10 mm long; peduncle to 2.5 cm long; bracteoles linear, c. 3 mm long; pedicel to 1 mm long, ±articulate. Sepals linear, 1.5–2.0 mm long. Corolla to 12 mm long, villous inside, with enations, ±auriculate; abaxial lobes 2.5–4.5 mm long; wings 1–1.5 mm wide. Indusium oblong to square. Ovules 15–20. Fruit subglobular to obovoid, c. 4 mm long; valves entire. Seeds ±orbicular, 1 mm diam., blackish, reticulate-foveate.

Occurs from south-eastern Qld to eastern Vic. The corolla is lemon to orange, with cottony, white hairs and often appressed, yellow hairs outside. Similar to *G. stelligera*, differing in its simple, not stellate, hairs on the outer surface of the corolla. Two subspecies are recognised.

Corolla always with numerous, coarse, appressed, yellowish, simple hairs outside **66a.** subsp. **bellidifolia**

Corolla with mostly white hairs outside **66b.** subsp. **argentea**

66a. **Goodenia bellidifolia** Smith subsp. **bellidifolia**

Leaves towards stem base, usually oblanceolate, narrowing very gradually towards base. Corolla always with numerous, coarse, appressed, yellowish hairs outside, orange to yellow. Indusium depressed-oblong, c. 0.8 mm long and 1.2 mm wide, ±concave below; bristles usually 0.3 mm long. Fruit obovoid, usually c. 4 mm long. Seeds usually brown. *n* = 8, 16, 24, W.J.Peacock, *Proc. Linn. Soc. New South Wales* 87: 388–396 (1962).

Extends from the Hunter Valley, N.S.W., to eastern Vic., along the coast and on adjacent highlands, in dry heath, scrub and woodland. Flowers chiefly Aug.–Mar. Map 258.

N.S.W.: Khyber Pass, Rylstone, *W.J.Peacock 6011.26.7* (SYD); Bundanoon, *C.W.E.Moore 3592* (CANB); between Turramurra and Bobbin Head, *R.H.Goode 201* (NSW); Yerranderie, 2 Dec. 1911, *R.H.Cambage* (SYD). Vic.: Genoa Ck, Dec. 1916, *A.J.Mahe* (MEL).

66b. **Goodenia bellidifolia** subsp. **argentea** Carolin, *Telopea* 3: 523 (1990)

T: 18 miles [c. 29 km] west of Emmaville, N.S.W., *W.J.Peacock 6111.23.1*; holo: NSW.

Leaves basal, usually ovate, narrowing more abruptly at base than in subsp. *bellidifolia*. Outside of corolla with very few, short, appressed, pale yellow hairs, scarcely visible without magnification and usually obscured by numerous, white, cottony hairs, usually lemon yellow. Indusium usually oblong, 1.0 mm long, with a double groove and a single usually villous ridge on undersurface; bristles usually less than 0.2 mm long. Fruit subglobular, c. 2.0 mm diam. Seeds very dark brown to black. *n* = 8, 16, 24, 32, W.J.Peacock, *Proc. Linn. Soc. New South Wales* 87: 388–396 (1962).

Occurs in south-eastern Qld, and north of the Hunter R., N.S.W. Flowers chiefly Mar.–Jan. Map 259.

Qld: Caboolture, *R.C.Carolin 565* (SYD); Whites Hill, *C.E.Hubbard 4345* (BRI, K). N.S.W.: Wallangarra, Nov. 1912, *J.L.Boorman* (NSW); c. 5 km N of Bundarra, *W.J.Peacock 6111.14.1* (SYD); Bulahdelah, *W.J.Peacock 611.3.2* (SYD).

67. **Goodenia racemosa** F.Muell., *Fragm.* 1: 114 (1859)

T: Burnett R., Qld, Nov. 1856, *F.Mueller*; lecto: MEL, *fide* R.C.Carolin, *Telopea* 3: 524 (1990); isolecto: K, MEL.

Erect undershrub to 120 cm tall, mostly glabrous. Leaves cauline, sessile, decurrent, linear to narrowly oblong, entire; lamina 3.5–6.0 cm long, 2–10 mm wide. Flowers in thyrses or racemes to 15 cm long; bracts linear, mostly c. 10 mm long; peduncle c. 10 mm long; bracteoles linear, 2.5–3.5 mm long; pedicel c. 3 mm long, not articulate. Sepals linear, c. 5 mm long. Corolla 12–15 mm long, almost glabrous inside, with enations, ±auriculate; abaxial lobes 5–7 mm long; wings 1.2–1.5 mm wide. Indusium oblong to obovate. Ovules c. 15. Fruit globular to ovoid, 4–5 mm long; valves bifid. Seeds orbicular, c. 1.8 mm diam., pale brown, reticulate-foveate.

Occurs in the Carnarvon Ra., Blackdown Tableland and Isla Gorge, Qld. The corolla is vivid yellow, glabrous or sparsely, cottony hairy outside; peduncles have 2 lines of hairs; and sepals are adnate to lower 3/4s of ovary. This species resembles *G. decurrens* which, however, has leaves 10–25 mm wide. Two varieties are recognised.

Leaves to 4 mm wide ..... **67a.** var. **racemosa**

Leaves 7–10 mm wide ..... **67b.** var. **latifolia**

67a. **Goodenia racemosa** F.Muell. var. **racemosa**

Leaves linear, to 4 mm wide. Corolla wings obtuse, c. 1.2 mm wide; auricles narrow, slightly pubescent; abaxial lobes c. 7 mm long. Indusium oblong to square.

Occurs in the Carnarvon Ra. and Blackdown Tableland, Qld. Flowers Sept.–Dec. Map 260.

Qld: Mt Playfair, *H.S.Biddulph 38* (MEL); Carnarvon Ra., *C.T.White 11356* (BRI); Blackdown Tableland, *R.J.Henderson et al. 1095* (BRI).

67b. **Goodenia racemosa** var. **latifolia** Carolin, *Telopea* 3: 524 (1990)

T: Isla Gorge, 16 miles [c. 25 km] SSW of Theodore, Qld, 1 Sept. 1963, *F.D.Hockings 11*; holo: BRI.

Leaves narrowly oblong, 7–10 mm wide. Corolla wings with ±acute angles, c. 1.5 mm wide; auricles conspicuous, pubescent; abaxial lobes c. 5 mm long. Indusium depressed-obovate.

Restricted to Isla Gorge, Qld. Flowers Sept. Map 261.

Qld: Isla Gorge, *S.L.Everist 8060* (BRI).

68. **Goodenia disperma** F.Muell., *Fragm.* 1: 113 (1859)

T: Dawson [River], Qld, *coll. unknown*; lecto: MEL, *fide* R.C.Carolin, *Telopea* 3: 524 (1990); isolecto: K.

G. sessiliflora F.Muell., *Fragm.* 4: 145 (1864). T: Cape River, Qld, *E.M.Bowman 231*; lecto: MEL, *fide* R.C.Carolin, *Telopea* 3: 525 (1990).

Erect undershrub to 30 cm tall, sparsely branched from erect stock, with brownish, cottony hairs, glabrescent. Leaves cauline, linear, narrowing basally, recurved, entire; lamina 3–50 mm long, 1.5–2.0 mm wide. Flowers in racemes to 15 cm long; bracts leaf-like; peduncle 1–5 mm long; bracteoles linear, c. 1 mm long; pedicel 1–2 mm long, pedicels articulate. Sepals linear to oblanceolate, 4–5 mm long. Corolla c. 9 mm long, villous inside, with enations, auriculate; abaxial lobes 3–5 mm long; wings c. 0.5 mm wide. Indusium depressed-obovate. Ovules 4–10. Fruit ovoid, c. 8 mm long; valves entire. Seeds elliptic-oblong, c. 3 mm long, yellow-brown, aculeate. Fig. 78A–B.

Occurs in the Great Dividing Ra., Qld, from W of Townsville to W of Bundaberg, in sclerophyll forest and woodland. Flowers Nov.–May. Map 262.

Qld: Herbert Ck, *E.M.Bowman 187* (MEL); c. 22 km N of Lynd on Mt Garnet Rd, *R.C.Carolin 8564* (SYD); c. 17 km W of Pentland, *R.C.Carolin 8315* (SYD); Boatman Stn, *S.L.Everist 4299* (BRI).

The corolla is white, with white hairs outside. The indusium has an upper lip considerably overlapping the lower one. Similar to *G. racemosa*, differing in *G. racemosa* having a yellow corolla which is almost glabrous inside, lacking brownish cottony hairs on the ovary, and having bifid capsule valves.

69. **Goodenia viridula** Carolin, *Telopea* 3: 525 (1990)

T: about 5 miles [c. 8 km] east of Jericho, Qld, Oct. 1940, *L.S.Smith & S.L.Everist 97842*; holo: BRI; iso: CANB, K, MEL.

Asending undershrub to 40 cm tall, with white or yellowish, cottony hairs, glabrescent. Leaves linear, revolute, entire; lamina 1.5–5 cm long, 0.5–1.1 mm wide. Inflorescence a spike, to 10 cm long; bracts leaf-like; bracteoles linear, c. 1 mm long. Sepals ovate, 0.8–1 mm long. Corolla 4–9 mm long, densely villous inside, enations hidden, auriculate; abaxial lobes 2–4 mm long; wings c. 0.5 mm wide. Indusium oblong. Ovules 4–6. Fruit subglobular, c. 3 mm diam.; valves entire. Seeds oblong-elliptic, c. 3 mm long, yellow-brown, reticulate.

Restricted to near Jericho, Qld. Flowers Nov.–May. Map 263.

Qld: Jericho, Apr. 1946, *M.S.Clemens* (BRI).

The corolla is greenish yellow. Related to *G. disperma* which has glabrescent stems and recurved leaves.

70. Goodenia stephensonii F.Muell., *Victorian Naturalist* 3: 138 (1887) as *stephensoni*

T: upper Hunter R., N.S.W., Nov. 1886, *L.Stephenson*; holo: MEL.

Erect, ±woody herb to 80 cm tall, glandular-hairy, ±viscid. Leaves cauline, linear to oblong, narrowing basally, flat or slightly recurved, ±dentate or entire; lamina 3–6 cm long, 3.5 mm wide. Flowers in racemes to 20 cm long; bracts leaf-like; peduncle 8–12 mm long; bracteoles linear, 2–3 mm long; pedicel 12–15 mm long, articulate. Sepals linear-lanceolate, c. 2 mm long. Corolla 10–15 mm long, hairy inside, with enations, auriculate; abaxial lobes 3–4 mm long; wings c. 0.5 mm wide. Indusium depressed-obovate. Ovules 10–14. Fruit cylindrical, 6 mm long; valves separated to midmark, usually entire. Seeds broadly elliptic, 2 mm long, pale yellow, aculeate. $n = 8$, W.J.Peacock, *Proc. Linn. Soc. New South Wales* 88: 13 (1963). Fig. 25G.

Occurs in the Denman–Baerami region in the Hunter Valley, N.S.W., in sclerophyll forest. Flowers Oct.–Jan. Map 264.

N.S.W.: Baerami Ck shale mine, *R.H.Cambage 2696* (NSW, SYD); Coxs Gap, *H.W.McKee 358* (SYD); Murrumbo Gap, 49 km W of Denman, *H.Streimann 052997* (CBG).

The corolla is yellow and has a conspicuous pouch. Similar to *G. heterophylla*, differing in its conspicuous corolla pouch, and in its larger leaves. Cytological voucher is *W.J.Peacock 6012.6.1* (SYD).

71. Goodenia heterophylla Smith, *Trans. Linn. Soc. London. Bot.* 2: 349 (1794)

T: Port Jackson, N.S.W., *J.White* holo: LINN Smith. Herb. 325.1; iso: ?K.

Procumbent to erect plant to 40 cm tall, ±woody; hairs glandular or simple. Leaves mostly cauline, ±sessile, linear to ovate, entire to lobed; lamina 1–4 cm long, 2–10 mm wide; basal leaves ephemeral. Flowers in thyrses or racemes to 30 cm long; bracts leaf-like; peduncle slender, 6–15 mm long; bracteoles linear, 2–4 mm long; pedicel 4–7 mm long, not articulate. Sepals linear to lanceolate, 2–4 mm long. Corolla to 12 mm long, glabrous or with few hairs inside, auriculate; abaxial lobes to 5 mm long; wings to 1.5 mm wide. Indusium oblong, often grooved above and below. Ovules 4–12. Fruit broadly ovoid, to 3 mm long; valves entire, spreading. Seeds oblong, 2 mm long, pale yellow, minutely aculeate.

Occurs from south-eastern Qld to central-coastal N.S.W. The corolla is yellow. This species differs from others in the group in its ephemeral rosette, mixed indumentum and slender pedicels and peduncles. Four subspecies are recognised.

1 Hairs mostly glandular

2 Leaves mostly with 2 basal lobes **71a.** subsp. **heterophylla**

2: Leaves never with 2 basal lobes **71c.** subsp. **teucriifolia**

1: Hairs mostly eglandular

3 Leaves 2–3 mm wide, revolute **71d.** subsp. **montana**

3: Leaves more than 3 mm wide, flat **71b.** subsp. **eglandulosa**

71a. Goodenia heterophylla Smith subsp. heterophylla

Ascending to erect herb, with simple, glandular, and multicellular hairs. Leaves usually ovate, sometimes linear, sometimes recurved, very rarely entire, usually serrate or lobed with acute scarcely acuminate teeth, with 2 large lobes or teeth near base; lamina 1.5–3 cm long, 3–8 mm wide.

Occurs in coastal and tableland regions of central N.S.W., in forest and woodland on Triassic sandstone. Flowers chiefly Aug.–May. Map 265.

N.S.W.: c. 14 km S of Cessnock, *R.Story 6722* (CANB); c. 5 km N of Putty, *H.Salasoo 2892* (NSW); Cameron track, *R.C.Carolin 2032* (SYD); Mona Vale Rd, Frenchs Forest, *J.Hufton 36* (SYD); Wentworth Falls, *E.J.McBarron 8703* (NSW).

71b. Goodenia heterophylla subsp. eglandulosa Carolin, *Telopea* 3: 526 (1990)

T: south end of Jervis Bay, N.S.W., Oct. 1954, *F.A.Rodway*; holo: NSW.

Ascending to erect herb, with simple and multicellular hairs, sometimes glabrescent. Leaves ovate, with recurved margins, serrate, each tooth ±mucronate, sometimes with 2 basal lobes; lamina 1–3 cm long, 3–8 mm wide.

Occurs in coastal and tableland regions, N.S.W., on Quaternary sand, Nowra sandstone, and on unknown formations in the Upper Williams valley. Flowers chiefly Aug.–May. Map 266.

N.S.W.: Wallis Lake, *H.Salasoo 3312* (NSW); Boolambyte Lake, *J.Hufton 11* (SYD); Norah Heads, Nov. 1960, *A.R.H.Martin* (SYD); Kogarah, Nov. 1896, *J.H.Camfield* (NSW); lower Budgong road, c. 5 km N of Shoalhaven R., 24 Nov. 1960, *L.A.S.Johnson & E.F.Constable* (NSW).

Listed as *G. heterophylla* subsp. B by S.W.L.Jacobs & J.Pickard in *Pl. New South Wales* 133 (1981).

71c. Goodenia heterophylla subsp. teucriifolia (F.Muell.) Carolin, *Telopea* 3: 526 (1990)

G. teucriifolia F.Muell., *Trans. & Proc. Philos. Inst. Victoria* 2: 70 (1858). T: Glass House Mtns, [near] Moreton Bay, Qld, *coll. unknown*; lecto: MEL, *fide* R.C.Carolin, *Telopea* 3: 527 (1990); isolecto: K, MEL.

Decumbent to procumbent herb, with simple, glandular, and multicellular hairs. Leaves ovate, flat, serrate with the teeth shortly mucronate, never with 2 basal lobes; lamina 1–4 cm long, 5–15 mm wide.

Restricted to the Glass House Mtns region, Qld, in sclerophyll forest. Flowers c. Oct. Map 267.

Qld: Pearsons Falls, *F.M.Bailey* (BRI); Highfields, Oct. 1877, *F.M.Bailey* (BRI); Mt Ngungun, *C.E.Hubbard 3344* (K).

71d. **Goodenia heterophylla** subsp. **montana** Carolin, *Telopea* 3: 527 (1990)

T: Mt Colong, N.S.W., 4 Mar. 1948, *E.F.Constable*; holo: NSW.

Erect plant, ±shrubby, villous-cottony when young, scabrid with short hairs when mature. Leaves linear to narrowly oblong, revolute, entire or nearly so; lamina 1.5–2.5 cm long, usually 2–3 mm wide. Fig. 78C.

Occurs in the Blue Mtns and S towards Nerriga, N.S.W., in sclerophyll forest. Flowers Sept.–Mar. Map 268.

N.S.W.: Big Hill Stn, Upper Burragorang Valley, 4 Oct. 1956, *E.F.Constable* (NSW); Yerranderie, *R.H.Cambage 3131* (NSW, SYD); Wingello State Forest, 21 Jan. 1956, *E.F.Constable* (NSW); Bluebush Ridge–Kiaramba Ridge, 28 Mar. 1948, *L.A.S.Johnson* (NSW); Tolwong turnoff, Nerriga–Jervis Bay road, *R.Pullen 2025* (NSW).

This subspecies is listed as *G. heterophylla* subsp. C by S.W.L.Jacobs & J.Pickard in *Pl. New South Wales* 133 (1981). J.G.Hufton (Hons Thesis, Univ. of Sydney, 1972) studied the 3 more southerly subspecies using computer techniques and made extensive observations on their ecological preferences. She demonstrated that they showed different ecological preferences and had distinct geographic distributions. Where the distributions overlap there are occasional intermediates.

72. **Goodenia rotundifolia** R.Br., *Prodr.* 576 (1810)

G. rotundifolia var. *glaberrima* R.Br., *loc. cit.*; [*G. rotundifolia* var. *typica* Domin, *Biblioth. Bot.* 22: 641 (1929)]. T: Shoalwater Bay Passage, [Qld], 26 Sept. 1802, *R.Brown*; lecto: BM, *fide* R.C.Carolin, *Telopea* 3: 528 (1990); isolecto: K, MEL, P.

G. rotundifolia var. *pubescens* R.Br., *loc. cit.* T: Paterson River, [N.S.W.], *R.Brown*; holo: BM.

G. strongylophylla F.Muell., *Fragm.* 6: 12 (1867). T: Princhester Creek, Qld, *E.M.Bowman 57*; holo: MEL; iso: K.

G. rotundifolia var. *hirsuta* Domin, *Biblioth. Bot.* 22: 641 (1929). T: Tamborine Mt, Qld, Mar. 1910, *K.Domin*; holo: PR.

Prostrate to erect, perennial herb to 50 cm tall, hirsute or glabrous. Leaves mostly cauline, orbicular to obovate, dentate to crenate; lamina 8–20 mm long, 5–20 mm wide; petiole to 10 mm long; basal leaves ephemeral. Flowers in racemes to 30 cm long; bracts leaf-like; peduncle 8–23 mm long; bracteoles linear, c. 5 mm long; pedicel 0–2 mm long, not articulate. Sepals linear to lanceolate, 5–7.5 mm long. Corolla 12–16 mm long, with few hairs inside, with enations, auriculate; abaxial lobes 5–6 mm long; wings 0.8–1.5 mm wide. Indusium broadly elliptic. Ovules 12–14. Fruit subglobular, 5–7 mm long; valves deeply bifid. Seeds oblong, 2–2.5 mm, pale brown, aculeate. $n = 8$, W.J.Peacock, *Proc. Linn. Soc. New South Wales* 88: 12, 13 (1963), some as *G. heterophylla*. Fig. 71.

Distributed throughout the coastal and tableland regions of southern Qld and N.S.W., in sclerophyll woodland and forest. Flowers chiefly Sept.–May. Map 269.

Qld: Blackdown Tableland, *R.W.Johnson 1040* (BRI); Fraser Is.,*C.E.Hubbard 4524* (BRI, K); Gympie, *S.L.Everist 7896* (BRI). N.S.W.: Drake, Oct. 1901, *J.L.Boorman* (NSW); East Pilliga State Forest, *R.C.Carolin 9570* (SYD).

Corolla yellow, sometimes with brownish markings, and bracteoles usually close under flower, so that the pedicel is obsolete or very short. Cytological vouchers are *W.J.Peacock 6012.2.3*, *6012.6.3* and *611.22.2* (SYD). A very variable species particularly in indumentum and habit; may be prostrate, erect or stoloniferous. Distinguished from all other species of the section by its coarse, simple indumentum, its bracteoles being close under the flower and the very obtuse angle between the wings at the apex of the corolla lobes. The indumentum of *G. heterophylla* subsp. *eglandulosa* is sometimes similar but the bracteole position and wing form will distinguish them; furthermore the seeds of *G. heterophylla* are not distinctly aculeate.

73. **Goodenia arenicola** Carolin, *Telopea* 3: 528 (1990)

T: Stradbroke Island, Qld, *coll. unknown HT 13903*; holo: NSW.

Stoloniferous or rhizomatous herb, softly pubescent. Leaves mostly clustered at ends of short stems, oblanceolate, narrowing basally, entire or dentate; lamina 4–10 cm long, 8–14 mm wide. Flowers solitary in axil of basal leaves; peduncle 3–4 cm long; bracteoles linear, 4–6 mm long; pedicel 3–4 cm long. Sepals linear, 5–6 mm long. Corolla 15–17 mm long, pubescent inside, with enations, auriculate; abaxial lobes 7–8 mm long; wings c. 2.5 mm wide. Indusium broadly oblong, folded. Ovules 30–50. Fruit and seeds not seen.

Known only from the type collection; grows on stabilised sand dunes. Flowering time unknown. Map 270.

Corolla yellow. Differs from other stoloniferous species in this section, in having minute, soft indumentum and an attenuate ovary.

74. **Goodenia delicata** Carolin, *Telopea* 3: 525 (1990)

T: 13 miles [c. 20 km] W of Westmar, Qld, 19 Oct. 1959, *L.Pedley 506*; holo: BRI.

G. hederacea f. *angustior* F.Muell., *Fragm.* 6: 15 (1867). T: Bockhara Creek, N.S.W., *L.Leichhardt*; holo: MEL.

[*G. geniculata auct. non* R.Br.: F.M.Bailey, *Queensland Fl.* 898 (1900)]

Decumbent to ascending herb to 50 cm long, cobwebbed-pubescent, glabrescent. Leaves mostly towards base, linear to narrowly elliptic, narrowing basally, entire or dentate; 2–6 cm long, 2–13 mm wide. Flowers in racemes to 40 cm long; bracts leaf-like; peduncle slender, mostly 5–20 mm long; bracteoles linear, 1–2 mm long; pedicel 5–10 mm long, not articulate. Sepals linear-ovate, 2–2.5 mm long. Corolla 10–13 mm long, sparsely pubescent inside, without enations, auriculate; abaxial lobes 4–4.5 mm long; wings c. 1 mm wide. Indusium broadly oblong. Ovules c. 12. Fruit globular, c. 8 mm diam.; valves bifid to midpoint. Seeds elliptic-oblong, 2 mm long, aculeate, pale yellow.

Occurs on the tablelands of southern Qld and northern N.S.W., in forest and woodland. Flowers Oct.–June. Map 271.

Qld: Eidsvold, *H.S.McKee 10230* (CANB); Biggenden, *C.T.White 7280* (BRI); Gayndah, *G.J.E.Schoneveld 249* (BRI); c. 5 km N of Glenmorgan, *W.J.Peacock 6111.36.2* (SYD).

Corolla yellow. Differs from *G. hederacea* in its narrower, often almost entire leaves, its sparse indumentum (when present) of very short, soft hairs at 90°, and its globular capsule.

75. **Goodenia hederacea** Smith, *Trans. Linn. Soc. London, Bot.* 2: 349 (1794)

T: Port Jackson, N.S.W., *J.White*; holo: LINN; iso: ?K.

Prostrate to ascending herb, with basal stock, to 80 cm long, cottony. Leaves orbicular to narrowly oblong, lobed to entire, glabrous above, glabrous or cottony-tomentose below; lamina 1–12 cm long, 3–25 mm wide; petiole to 4 cm long. Flowers in racemes to 80 cm long; bracts leaf-like; peduncle 10–30 mm long, slender; bracteoles linear, 3–5 mm long; pedicel 5–20 mm long, not articulate. Sepals linear to lanceolate, 3–5 mm long. Corolla 8–15 mm long, pubescent inside, rarely with small enations, auriculate; abaxial lobes 3.5–7 mm long; wings to 2 mm wide. Indusium depressed-obovate. Ovules 8–20. Fruit ovoid, 5–9 mm long; valves spreading, bifid. Seeds elliptic-oblong, 2.5 mm long, aculeate, yellowish to brownish.

Occurs from south-eastern Qld to eastern Vic. Corolla yellow. Abaxial sepal may be larger than the rest. Distinguished from its relatives by its indumentum of distorted multicellular hairs. *Goodenia blackiana* may approach it in some forms but the undersurface of the leaf in that species is very densely tomentose and also has some simple hairs, which effectively distinguishes them. Two subspecies are recognised.

Stems not rooting at nodes **75a.** subsp. **hederacea**

Stems rooting at nodes **75b.** subsp. **alpestris**

75a. Goodenia hederacea Smith subsp. hederacea

G. hederacea var. *hypotephra* F.Muell., *Fragm.* 6 (1867) 15. T: head of Gwydir R., N.S.W., *L.Leichhardt*; holo: MEL.

G. hederacea var. *hartmannii* F.Muell., *Fragm.* 10: 110 (1877). T: near Towamba, N.S.W., *Hartmann*; syn: MEL.

G. boormannii K.Krause, *Pflanzenr.* 54: 56 (1912). T: Dubbo, N.S.W., Jan. 1898, *J.Boorman*; lecto: W, *fide* R.C.Carolin, *Fl. Australia* 35: 331 (1992); isolecto: ?B (destroyed) *n.v.*, NSW.

Illustration: E.R.Rotherham *et. al.*, *Fl. Pl. New South Wales & S. Queensland* 72, fig. 206 (1975).

Stems not rooting at nodes. Cauline leaves orbicular to ovate or narrowly oblong, obscurely dentate to entire, pubescent on lower surface, sparsely pubescent on both surfaces, or glabrescent; lamina 1.4–10 mm long, 5–8 mm wide; petiole 3–10 mm long. Pedicels mostly 5–15 mm long. Corolla 10–14 mm long. Seeds obscurely papillose, pale yellow. $n = 8$, W.J.Peacock, *Proc. Linn. Soc. New South Wales* 88: 12 (1963) as var. *alpestris*. Fig. 24B.

Extends from southern Qld through N.S.W. to Vic. Flowers chiefly Aug.–Mar. Map 272.

Qld: without precise locality, *B.Scortechini* (NSW 100846). N.S.W.: N Minore Rd, 18 km NW of Dubbo, on Mitchell Hwy, *P.J.Myerscough & A.Denham 140* (SYD); 2 km from Torrington, towards Silent Grove, *B.Pellow & D.Keith 71* (SYD); Flemington, 3 Dec. 1926, *O.D.Evans* (SYD). Vic.: Beechworth, *A.Meebold 21623* (MEL).

Cytological vouchers are *W.J.Peacock 6012.7.2* and *6111.50.1* (SYD). Specimens of this subspecies from the northern tablelands of N.S.W. resemble subsp. *alpestris* in the cauline leaf shape, indumentum and margin. The stems, however, do not root at the nodes and the basal leaves are more persistent and more elongate.

75b. Goodenia hederacea subsp. alpestris (K.Krause) Carolin, *Telopea* 3: 529 (1990)

G. hederacea var. *alpestris* K.Krause, *Pflanzenr.* 54: 56 (1912). T: Mt Buller, Vic., *F.Mueller*; lecto: K, *fide* R.C.Carolin, *Telopea* 3: 529 (1990).

G. hederacea var. *cordifolia* Ewart, *Fl. Victoria* 1073 (1931). T: none designated.

Illustration: A.B.Costin, *et al.*, *Kosciusko Alpine Fl.* t. 290 (1980).

Stems prostrate, usually rooting at nodes. Leaves orbicular to broadly ovate, obtuse at base and apex, crenate, usually densely tomentose on undersurface; lamina usually to 30 mm long, to 25 mm wide; petiole usually 15–30 mm long. Pedicels 10–20 mm long. Corolla 8–10 mm long. Seeds strongly papillate to aculeate, yellow-brown. Fig. 64.

Occurs in subalpine areas of south-eastern N.S.W. and eastern Vic., in grassland and woodland. Flowers Nov.–Apr. Map 273.

N.S.W.: Tumut Ponds to Tooma Dam, *E.J.McBarron 7445* (NSW); Happy Jacks Plain, 17 Jan. 1958, *H.J.Wier* (NSW). A.C.T.: Tidbinbilla, *R.H.Cambage 3045* (NSW). Vic.: Bogong High Plains, *H.I.Aston 215* (MEL); Mt Buller, *R.C.Carolin 1094* (SYD).

Listed as *G. hederacea* subsp. 'C' by S.W.L.Jacobs & J.Pickard, *Pl. New South Wales* 133 (1981).

76. Goodenia xanthosperma F.Muell., *Fragm.* 10: 12 (1876)

T: Queen Victoria Spring, W.A., 27 Sept.–6 Oct. 1875, *J.Young*; holo: MEL.

G. discolor K.Krause, *Pflanzenr.* 54: 57 (1912). T: Camp 44, Great Victoria Desert, W.A., 7 Sept. 1891, *R.Helms*; holo: K; iso: MEL, NSW.

G. mooreana K.Krause, *loc. cit.* T: Coolgardie district, W.A., *W.Webster*; lecto: BM, *fide* R.C.Carolin, *Telopea* 3: 529 (1990); isolecto: ?B (destroyed) *n.v.*

[*G. geniculata* var. *primulacea auct. non* (Schldl.) Benth.: E.Pritzel, *Bot. Jahrb. Syst.* 35: 559 (1904)]

Prostrate herb, with basal stock, tomentose; stems to 80 cm long. Basal leaves elliptic to ovate, narrowing basally, dentate to lobed, glabrescent above; lamina 3–6 cm long, 1.3–3 cm wide; cauline leaves smaller. Flowers in racemes to 80 cm long; bracts leaf-like; peduncle 1–3 cm long; bracteoles linear, 4–5 mm long; pedicel 2–25 mm long, not articulate. Sepals lanceolate, 5 mm long. Corolla c. 15 mm long, pubescent inside, with enations, auriculate; abaxial lobes 5–7 mm long; wings c. 2 mm wide. Indusium broadly oblong. Ovules 40–60. Fruit ovoid to ellipsoidal, c. 10 mm long; valves bifid. Seeds oblong, 2.5 mm long, aculeate to setose, dark yellow. $n = 8$, W.J.Peacock, *Proc. Linn. Soc. New South Wales* 88: 12 (1963) as *G. discolor*.

Found in drier parts of southern W.A., in various communities in sandy soil. Flowers chiefly May–Oct. Map 274.

W.A.: near Cosmo Newberry, *R.C.Carolin 5929* (SYD); N of Cundeelee Mission, *A.S.George 5832* (PERTH); Boorabbin, *A.Morrison 16056* (K); c. 32 km N of Bencubbin, *W.E.Blackall 3333* (PERTH); c. 35 km from Coolgardie, *N.T.Burbidge 2642* (CANB).

Cytological voucher is *W.J.Peacock 60911.1* (SYD). Corolla yellow, often with purplish markings, hairy towards base outside; indusium with bristles inserted behind lips.

Differs from *G. centralis* in the undersurface of the leaves remaining hairy. Differs from other members of the group which have the indusial bristles inserted on the lips. In the eastern parts of the distribution the pedicels and peduncles tend to be shorter (the bracteoles frequently being inserted just under the ovary), and the outside of the corolla and peduncles etc., tend to be more densely tomentose. K.Krause (*Pflanzenr.* 54: 57, 1912) used some of these characters to distinguish *G. discolor* from *G. mooreana*.

W.E.Blackall & B.J.Grieve (*How to Know W. Austral. Wildfl.* 423, 1954) used the name *G. mooreana* for a member of the *G. affinis* complex. Their description is in key form and does not provide enough detail to determine to which species it has been misapplied. C.A.Gardner, *Enum. Pl. Austral. Occid.* 124 (1931), gave no description but from his identifications it seems probable that he was referring to *G. peacockiana* when he used *G. xanthosperma*.

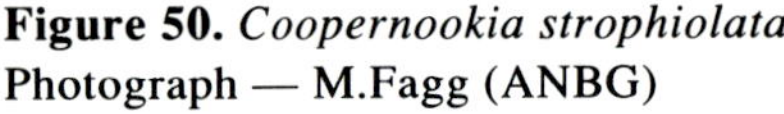

Figure 50. *Coopernookia strophiolata*
Photograph — M.Fagg (ANBG)

Figure 51. *Scaevola enantophylla*
Photograph — M.Fagg

Figure 52. *Scaevola taccada*
Photograph — T.Low

Figure 53. *Scaevola spinescens*
Photograph — A.George

Figure 54. *Scaevola crassifolia*
Photograph — T.Low

Figure 55. *Scaevola collaris*
Photograph — A.George

Figure 56. *Scaevola hookeri*
Photograph — A.George

Figure 57. *Scaevola calendulacea*
Photograph — T.Low

Figure 58. *Scaevola albida*
Photograph — M.Fagg

Figure 59. *Scaevola striata* var. *striata*
Photograph — A.George

Figure 60. *Scaevola phlebopetala*
Photograph — A.George

Figure 61. *Scaevola calliptera*
Photograph — A.George

Figure 62. *Goodenia tripartita*
Photograph — M.Fagg

Figure 63. *Goodenia chambersii*
Photograph — A.George

Figure 64. *Goodenia hederacea* subsp. *alpestris*
Photograph — A.George

Figure 65. *Goodenia vilmorinae*
Photograph — A.George

Figure 66. *Goodenia stobbsiana*
Photograph — A.George

Figure 67. *Goodenia neogoodenia*
Photograph — A.George

Figure 68. *Goodenia grandiflora*
Photograph — M.Fagg (ANBG)

Figure 69. *Goodenia glabra*
Photograph — M.Fagg

Figure 70. *Goodenia ovata*
Photograph — T.Low

Figure 71. *Goodenia rotundifolia*
Photograph — M.Fagg (ANBG)

Figure 72. *Goodenia caerulea*
Photograph — A.George

Figure 73. *Velleia panduriformis*
Photograph — A.George

Figure 74. *Velleia paradoxa*
Photograph — M.Fagg

Figure 75. *Selliera radicans*
Photograph — M.Fagg

Figure 76. *Velleia rosea*
Photograph — A.George

Figure 77. *Verreauxia villosa*
Photograph — M.Fagg

77. **Goodenia centralis** Carolin, *Telopea* 2: 66 (1980)

T: 1.5 miles [c. 2.5 km] E of Irving Creek, Petermann Ra., S.A., 24 June 1958, *G.Chippendale*; holo: DNA; iso: NSW.

Prostrate, annual herb to 80 cm long, cottony. Leaves basal and cauline, obovate to spathulate, narrowing basally, dentate, ±glabrescent; lamina 3–10 cm long, 15–30 mm wide. Flowers in racemes to 60 cm long; bracts leaf-like; peduncle 1–4 cm long; bracteoles linear, 1–2 mm long; pedicel mostly 1–2 cm long, not articulate. Sepals linear-deltoid, c. 2 mm long. Corolla 12–15 mm long, sparsely pubescent inside, with prominent enations, auriculate; abaxial lobes c. 5 mm long; wings c. 1.5 mm wide. Indusium depressed-obovate. Ovules 28–30. Fruit ±ellipsoidal, c. 8 mm long; valves bifid to midmark. Seeds oblong, 2.5 mm long, minutely aculeate, yellow-brown. n = 8, W.J.Peacock, *pers. comm.* Fig. 45H.

Occurs in the deserts of central-eastern W.A., south-western N.T. and northern S.A., in grassland and scrub in sandy and stony soil. Flowers chiefly June–Sept. Map 275.

W.A.: near southern end of Walter James Ra., *R.C.Carolin 6254* (SYD); c. 1.5 km E of Todd Ra., Gunbarrel Hwy, *A.S.George 5343(a)* (PERTH). N.T.: Docker R., *R.C.Carolin 5349* (SYD); Petermann Ra. area, *N.Henry 426* (DNA); 5 km SW of Ayers Rock, *N.N.Donner 4383* (DNA).

Corolla yellow, sometimes with purplish markings, the abaxial sepal is ±larger, and the indusium has bristles inserted slightly behind the lip. Cytological voucher is *R.C.Carolin 5313* (SYD). Similar to *G. hederacea* but with a very different habitat, and distinguished by the prominent enations on the inside of the corolla and the bristles clearly inserted back from the lips of the indusium. Previously confused with both *G. hirsuta* and *G. xanthosperma*. The former species has both leaf surfaces hirsute; and the latter species has a persistent, dense, cottony tomentum on the lower surface of the leaf.

78. **Goodenia goodeniacea** (F.Muell.) Carolin, *Telopea* 3: 529 (1990)

Scaevola goodeniacea F.Muell., *Fragm.* 1: 121 (1858); *Catospermum goodeniaceum* (F.Muell.) Baillon, *Hist. Pl.* 8: 370 (1885). T: Sturts Creek, [N.T.], Mar. 1856, *F.Mueller*; holo: MEL.

Catospermum muelleri Benth., *Hooker's Icon. Pl.* t. 1028 (1868). T: illustration in W.J.Hooker, *Icon. Pl.* t. 1028 (1868).

Illustration: J.P.Jessop (ed.), *Fl. Centr. Australia* 359, fig. 457 (1981), as *Catosperma goodeniacea.*

Prostrate herb, ±perennial; stems to 50 cm long. Basal leaves elliptic to ovate, petiolate, dentate, glabrous or with few cottony hairs; lamina 5–6 cm long, 3–4 cm wide; cauline leaves similar, smaller. Flowers in thyrses to 40 cm long; bracts leaf-like; peduncle 5–15 mm long; bracteoles linear-deltoid, to 2 mm long; pedicel 10–20 mm long, not articulate. Sepals linear, 3 mm long. Corolla 13–15 mm long, with few scattered hairs in throat, without enations, auriculate; abaxial lobes 5 7 mm long; wings c. 1.5 mm wide. Indusium obovate, c. 1 mm long. Ovules 4. Fruit subglobular, to 10 mm diam., 10-ribbed, indehiscent. Seeds orbicular, c. 3 mm diam., aculeate. Fig. 78D–E.

Occurs in scattered localities from Tennant Creek to Sturt Ck, N.T., and in Qld; there is an unauthenticated record from N of Mundiwindi, W.A. Flowers May–Aug. Map 276.

N.T.: Tennant Creek, *H.I.Aston 307* (AD, MEL); c. 69 km SW of Barrow Creek, *M.Lazarides 5801* (AD, CANB, MEL). Qld: Boatman Stn, *S.L.Everist 2825* (BRI, K).

Corolla yellow with purplish lines, glabrous or with a few, weak, cottony hairs outside. The ovary is ±cottony-hairy or glabrescent, the style is curved and c. 5 mm long, and the seed wing is narrow and obtuse.

Figure 78. *Goodenia*. **A–B**, *G. disperma*. **A**, flower X4, **B**, habit X1 (**A–B**, R.Carolin 8564, SYD). **C**, *G. heterophylla* subsp. *montana*, habit X2 (H.Cambage, Dec. 1911, NSW). **D–E**, *G. goodeniacea*. **D**, fruit T.S. X4; **E**, habit X0.5 (**D–E**, I.Olsen 458, NSW). Drawn by D.Mackay.

79. Goodenia rupestris Carolin, *Telopea* 2: 64 (1980)

T: unnamed pass between the Hull and Docker Rivers, Petermann Ra., N.T., 21 Aug. 1966, *R.C.Carolin 5333*; holo: NSW; iso: SYD.

Ascending or pendulous, perennial herb, to 20 cm long, cottony-tomentose. Leaves crowded on thick stock, elliptic to oblanceolate, narrowing basally, entire, ±glabrescent; lamina 2–6 cm long, c. 10 mm wide; cauline leaves smaller. Flowers in racemes to 15 cm long; bracts leaf-like; peduncle 7–10 mm long; bracteoles linear, 4–5 mm long; pedicel 5–7 mm long, not articulate. Sepals linear, 4–5 mm long. Corolla 14–16 mm long, pubescent inside, with enations, auriculate, yellow; abaxial lobes 7–8 mm long; wings c. 1.5 mm wide. Indusium depressed-obovate. Ovules 16–20. Fruit ellipsoidal, c. 6 mm long; valves bifid to midmark. Seeds not seen. $n = 8$, W.J.Peacock, *pers. comm.*

Restricted to the Petermann Ra., N.T., in crevices in cliffs. Flowers c. Aug. Map 277.

N.T.: Mannana Ra., Petermann Reserve, *P.K.Latz 8016* (DNA).

Corolla yellow. Similar to *G. xanthosperma* which has dentate or lobed leaves, no thick stock, fruit c. 10 mm long, and indusial bristles inserted behind the lips. Cytological voucher is *R.C.Carolin 5333* (SYD).

80. Goodenia blackiana Carolin in J.Jessop & H.R.Toelken, *Fl. S. Australia* 4th edn, 1393 (1986)

T: Teddy Bear Gap, The Grampians, Vic., 24 Oct. 1953, *T. & J.Whaite 1470*; holo: NSW.

[*G. primulacea auct. non* Schldl.: K.Krause, *Pflanzenr.* 54: 53 (1912); E.Robertson in J.M.Black, *Fl. S. Australia* 2nd edn, 819 (1957)].

[*G. geniculata auct. non* R.Br.: F.Mueller, *Key Syst. Victorian Pl.* 355 (1888) *p.p.*]

Prostrate to ascending herb to 20 cm long, cottony-pubescent; stems usually stoloniferous. Basal leaves obovate to oblanceolate, narrowing basally, dentate, glabrescent above, cottony-tomentose and with few short hairs, sometimes ±glabrescent below; lamina 2–6 cm long, 5–15 mm wide; cauline leaves smaller. Flowers distant in racemes to 5 cm long, or solitary in leaf axils; bracts leaf-like; peduncle 3–5 cm long; bracteoles linear, 5–12 mm long; pedicel 3–5 cm long, not articulate. Sepals narrowly oblong to ovate, 4–5 mm long. Corolla 13–14 mm long, pubescent inside, with small enations, auriculate; abaxial lobes 5–6 mm long; wings c. 2 mm wide. Indusium broadly oblong. Ovules 12–18. Fruit ovoid, c. 10 mm long; valves bifid at apex. Seeds elliptic, 3 mm long, aculeate, yellowish. Figs 24D, 79C–D.

Extends from Eyre Peninsula, S.A., to central Vic., in grassland, woodland and mallee. Flowers chiefly Sept.–Jan. Map 278.

S.A.: Spring Gully, Clare Hills, *D.N.Kraehenbuehl 450* (AD); between Corny Point and Spencer, *H.Eichler 13997* (AD); Crafers, *H.Eichler 13371* (AD). Vic.: W of Dergholm, *A.C.Beauglehole 8444* (SYD); NW slopes of Mt Difficult, *T.B.Muir 2652* (MEL).

The corolla is yellow, and the basal leaves rarely reach 12 cm in length. Similar to *G. geniculata*, differing in having longer leaves which are not hirsute on both surfaces, enations inside corolla, a concave (not folded) indusium, and longer bracteoles. Similar also to *G. lanata*, differing in having longer bracteoles, and the hairy but not densely tomentose undersurface of leaves.

81. **Goodenia geniculata** R.Br., *Prodr.* 557 (1810)

T: Port Phillip, [Vic.], *R.Brown*; lecto: BM, *fide* R.C.Carolin, *Telopea* 3: 529 (1990); isolecto: K, P.

G. primulacea Schldl., *Linnaea* 20: 601 (1847); *G. geniculata* var. *primulacea* (Schldl.) Benth., *Fl. Austral.* 4: 63 (1868). T: near Bethany, S.A., *H.Behr 84*; holo: HAL; iso: K.

Decumbent to ascending herb, hirsute; stems to 25 cm long, often stoloniferous. Leaves basal on stock and cauline, linear to oblanceolate, narrowing basally, dentate, usually hirsute on both surfaces; lamina 3–10 cm long, 3–10 mm wide. Flowers in racemes to 5 cm long, or solitary in leaf axils; bracts leaf-like; peduncle 1.5–5 cm long; bracteoles linear, c. 5 mm long; pedicel 2–5 cm long, not articulate. Sepals oblong to narrowly oblong, 4–5 mm long. Corolla 14–16 mm long, pubescent inside, without enations, ±auriculate; abaxial lobes 7–8 mm long; wings to 3 mm wide. Indusium broadly obovate, folded. Ovules 14–16. Fruit obovoid, c. 10 mm long; valves shortly bifid. Seeds elliptic, c. 3.5 mm long, aculeate, yellow-brown. n = 8, W.J.Peacock, *Proc. Linn. Soc. New South Wales* 88: 12 (1963) as *G. lanata*. Fig. 79B.

Extends from Eyre Peninsula, S.A., to southern Vic. and Tas., in woodland and forest. Flowers Sept.–Jan. Map 279.

S.A.: Mt Compass, *D.E.Symon 2817* (AD); Mylor, *N.N.Donner 439* (AD). Vic.: Wimmera, *W.E.Matthews 25* (MEL); c. 5 km W of Halls Gap, *T.B.Muir 2636* (MEL). Tas.: Hobart, 1805, *G.Caley* (BM).

Corolla yellow, with stiff, yellowish hairs and sometimes, cottony hairs outside; pedicels geniculate at the bracteoles; indusium folded so that the opposite sides almost meet. Cytological voucher is *W.J.Peacock 6011.12.45* (SYD). Similar to *G. blackiana* and *G. lanata*, but distinguished by its strongly folded indusium. The leaf shape varies considerably and there are a number of specimens, from apparently swampy situations, with glabrous leaves.

82. **Goodenia lanata** R.Br., *Prodr.* 577 (1810)

G. geniculata var. *lanata* (R.Br.) Rodway, *Tasman. Fl.* 102 (1903). T: Port Dalrymple, Tas., 5–6 Jan. 1804, *R.Brown*; lecto: BM, *fide* R.C.Carolin, *Telopea* 3: 529 (1990); isolecto: K, MEL.

[*G. hederacea auct. non* Smith: J.D.Hooker, *Fl. Tasman.* 1: 232 (1856)]

Prostrate or decumbent, perennial herb, with soft, silvery grey, multicellular, crisped, simple hairs; stems to 30 cm long. Leaves basal on stock, obovate, narrowing basally, dentate to lyrate, hairy on both surfaces; lamina 1–8 cm long, 5–20 mm wide; cauline leaves smaller. Flowers in racemes to 20 cm long; bracts leaf-like; peduncle 2–6 cm long; bracteoles linear, 3–5 mm long; pedicel 1–4 cm long, not articulate. Sepals narrowly oblong, 5–6 mm long. Corolla 10–15 mm long, pubescent inside, usually without enations, indistinctly auriculate; abaxial lobes 5–6 mm long; wings c. 2 mm wide. Indusium broadly oblong, slightly folded. Ovules 10–12. Fruit ovoid to cylindrical, 6–7 mm long; valves deeply bifid. Seeds elliptic, 3 mm long, pale yellow.

Occurs in southern Vic. and in Tas., in woodland and forest. Flowers chiefly Sept.–Mar. Map 280.

Vic.: near Killiwarra, *T.B.Muir 1697* (MEL); Smythesdale, *E.Simper 2479* (NSW); Creswick, 31 Mar. 1928, *J.H.Willis* (MEL). Tas.: c. 22 km E of Launceston, *N.T.Burbidge 2958* (CANB); Cambridge, *F.H.Long 643* (CANB).

Corolla yellow; pedicels geniculate at the bracteoles. Similar to *G. geniculata* which has coarser hairs and a strongly folded indusium; and to *G. blackiana* which has longer bracteoles and a thicker, uneven indumentum. *Goodenia lanata* also has longer and generally more prostrate scapes than either of these.

Figure 79. *Goodenia*. **A**, *G. tripartita*, flower X2 (W.Peacock 60887.3, SYD). **B**, *G. geniculata*, flower X2 (R.Carolin, 23 Sept. 1983, SYD). **C–D**, *G. blackiana*. **C**, habit X0.5; **D**, flower X2 (**C–D**, A.Beauglehole 8444, SYD). **E**, *G. convexa*, flower X2 (R.Carolin 3125, SYD). **F**, *G. affinis*, flower X2 (A.Orchard 1133, SYD). **G**, *G. wilunensis*, habit X0.5 (W.Peacock 60861.1, SYD). Drawn by D.Mackay.

83. **Goodenia affinis** Vriese, *Natuurk. Verh. Holl. Maatsch. Wetensch. Haarlem* ser. 2, 10: 137 (1854)

Scaevola geniculata Vriese in J.G.C.Lehmann, *Pl. Preiss.* 1: 404 (1845); *G. geniculata* var. *eriophylla* Benth., *Fl. Austral.* 4: 63 (1868). T: near Cape Riche, W.A., 20 Nov. 1840, *J.A.L.Preiss 1503*; lecto: LD, *fide* R.C.Carolin, *Telopea* 3: 508 (1990); isolecto: MEL.

Decumbent to ascending, usually perennial herb, with simple, silky multicellular, and stellate hairs; stems to 20 cm long. Leaves mostly basal, ovate to oblong, narrowing basally, dentate, hairy on both surfaces; lamina 2–4 cm long, 5–12 mm wide. Flowers in racemes to 5 cm long, or solitary in leaf axils; bracts leaf-like; peduncle 10–20 mm long; bracteoles linear, 4–8 mm long; pedicel 10–20 mm long, not articulate. Sepals narrowly oblong to ovate, 2–3 mm long. Corolla 12–15 mm long, with few hairs inside, enations absent, auriculate; abaxial lobes 7 or 8 mm long; wings c. 1.5 mm wide. Indusium semi-orbicular. Ovules c. 20. Fruit oblong, 10–12 mm long; valves separating to sepal attachment, bifid. Seeds elliptic, 2 mm long, brown, smooth. Fig. 79F.

Extends from King George Sound to the western end of the Great Australian Bight, W.A., in mallee. Flowers chiefly July–Dec. Map 281.

W.A.: c. 14 km S of Cocklebiddy Motel, *A.S.George 8549* (PERTH); c. 3 km E of Ravensthorpe, *K.Newbey 261* (PERTH); Neridup, c. 3 km NE of Howick Hill, *A.E.Orchard 1133* (AD, SYD); 80 km W of Esperance, *P.G.Wilson 7946* (AD, PERTH); c. 30 km N of the coast at Stokes Inlet, *A.E.Orchard 1455* (PERTH).

Corolla yellow, with simple, multicellular, and stellate hairs outside. The pedicels are often geniculate at the bracteoles, while the indusium is 1–2 mm long, arched or ±straight, with the orifice notched or entire.

84. **Goodenia convexa** Carolin, *Telopea* 3: 529 (1990)

T: Swan River, W.A., *J.Drummond 405*; holo: K; iso: MEL.

[*G. robusta auct. non* (Benth.) K.Krause: W.E.Blackall & B.J.Grieve, *How to Know W. Austral. Wildfl.* 423 (1956)]

Decumbent herb to 50 cm long, ±hirsute with mostly stellate and yellowish, multicellular hairs, some simple hairs. Leaves mostly basal, obovate to lanceolate, narrowing basally, usually dentate, hairy on both surfaces; lamina 3–12 cm long, 4–25 mm wide. Flowers in racemes to 5 cm long, or solitary; bracts leaf-like; peduncle 1–6 cm long; bracteoles linear to lanceolate, 4–15 mm long; pedicel 3–5 cm long, not articulate. Sepals narrowly elliptic, 5–8 mm long. Corolla 18–25 mm long, glabrous inside, without enations, auriculate; abaxial lobes 15–18 mm long; wings 2–3 mm wide. Indusium broadly obovate, slightly folded. Ovules c. 30. Fruit ovoid to cylindrical, 10–15 mm long; valves separating to sepal attachment, bifid. Seeds elliptic, 1.8–2.5 mm long, aculeate, pale brown. n = 8, W.J.Peacock, *Proc. Linn. Soc. New South Wales* 88: 12 (1963) as *G. affinis*. Fig. 79E.

Occurs in south-western W.A. from the Hill R. to the Tone R. area and inland to Cowcowing, growing in heath and woodland in sandy soil. Flowers chiefly Aug.–Nov. Map 282.

W.A.: S of junction of Red Gully road and Brand Hwy, *A.S.George 16272* (PERTH); Jurien Bay road, *W.E.Blackall 3658* (PERTH); c. 24 km W of Mogumber, *A.S.George 6429* (PERTH); Taylors Well near Pingelly, *R.C.Carolin 3166* (SYD); Cowcowing, *M.Koch 1573a* (BRI, NSW).

Corolla pale orange, purplish in throat, with both simple and stellate hairs outside. The pedicels are geniculate at the bracteoles, and the indusium, which is c. 3 mm long, has a very convex orifice. Cytological vouchers are *W.J.Peacock 60887.3* and *6095.11* (SYD).

Similar to *G. affinis* which has stellate, simple and silky hairs, and a notched or entire indusial orifice which is arched or straight and has short bristles. Similar also to *G. tripartita* which has a trilobed indusium; and to the eastern *G. robusta* which has cauline leaves and stem-clasping bracts.

85. **Goodenia tripartita** Carolin, *Telopea* 3: 530 (1990)

T: Fitzgerald River, Ongerup–Ravensthorpe [road], W.A., 11 Sept. 1961, *R.C.Carolin 3569*; holo: NSW.

Decumbent to prostrate herb, whitish villous, sometimes felted with stellate, simple and multicellular hairs; stems to 15 cm long. Leaves mostly basal, narrowly elliptic to obovate, narrowing abruptly at base, dentate, hairy on both surfaces; lamina 4–7 cm long, 10–20 mm wide. Flowers in racemes to 8 cm long; bracts leaf-like; peduncle 10–30 mm long; bracteoles linear, c. 4 mm long; pedicel 10–20 mm long, not articulate. Sepals linear, 4–5 mm long. Corolla c. 18 mm long, few hairs inside, without enations, indistinctly auriculate; abaxial lobes 8–9 mm long; wings c. 2.5 mm wide. Indusium broadly ovate, folded. Ovules c. 30. Fruit ovoid, c. 13 mm long; valves separated to midmark, shortly bifid. Seeds elliptic, c. 2 mm long, aculeate, yellowish. $n = 8$, W.J.Peacock, *Proc. Linn. Soc. New South Wales* 88: 12 (1963). Figs 62, 79A.

Occurs in south-western W.A. from near Morowa to the Stirling Ra. and E to the Newdegate area, in heath in sandy soil. Flowers Aug.–Oct. Map 283.

W.A.: Waddouring, *W.B.Alexander 1242* (PERTH); Muntadgin, *E.T.Bailey 954* (PERTH); c. 3 km W of Lake Grace, *W.J.Peacock 6098* (SYD); Needilup, *R.C.Carolin 3558* (SYD); Warrungup, *A.Morrison 12279* (K, PERTH).

Corolla bright yellow; bracteoles obscured by hairs; pedicels geniculate at the bracteoles, and the indusium has 2 lateral, curved, linear arms folded in front. The 3-partite indusium distinguishes it from all other species in the section. Cytological voucher is *W.J.Peacock 60887.3* (SYD).

86. **Goodenia willisiana** Carolin, *Telopea* 3: 531 (1990)

T: Redcliffs, NW Vic., Aug. 1937, *J.H.Willis*; holo: MEL.

Illustration: E.R.Rotherham *et al.*, *Fl. Pl. New South Wales & S. Queensland* 136, fig. 436 (1975) as *G. affinis*.

Erect or ascending herb to 20 cm tall, whitish tomentose, with stellate and multicellular hairs, felted. Leaves crowded basally, narrowing basally, narrowly elliptic to lanceolate, entire or sinuate, hairy on both surfaces; lamina 4–9 cm long, 4–18 mm wide. Flowers solitary in leaf axils; peduncle to 5 cm long; bracteoles linear, c. 2 mm long; pedicel to 4 cm long, not articulate. Sepals narrowly oblong, 3.5 cm long. Corolla 15 mm long, ±glabrous inside, without enations, ±auriculate; abaxial lobes 5–6 mm long; wings c. 2.5 mm wide. Indusium broadly elliptic, folded. Ovules 28–30. Fruit ovoid, c. 10 mm long; valves separated to midmark, bifid. Seeds elliptic, 2 mm long, aculeate, yellow.

Extends from Eyre Peninsula and Flinders Ra., S.A., to south-western N.S.W. and western Vic., in mallee and dry communities. Flowers chiefly Aug.–Feb. Map 284.

S.A.: Gawler Ra., 13 Sept. 1946, *E.H.Ising* (AD); c. 6 km S of Whyalla, Aug. 1970, *M.Tindale* (SYD). N.S.W.: Rankins Springs road, *G.Althofer 8a* (NSW). Vic.: Hattah, Sept. 1940, *J.H.Willis* (MEL); Sea Lake, *W.W.Watts 435* (NSW).

Corolla yellow, with few, hairs at the top outside; pedicels geniculate at the bracteoles; indusium orifice is slightly convex. Previously treated under and similar to *G. affinis*, but occurring in eastern not western Australia. Similar to the eastern *G. robusta* which has yellowish villous indumentum, wider sepals, and stem-clasping leaves.

87. **Goodenia robusta** (Benth.) K.Krause, *Pflanzenr.* 54: 53 (1912)

G. geniculata var. *robusta* Benth., *Fl. Austral.* 4: 63 (1868). T: Marble Ra., S.A., *J.F.C.Wilhelmi*; lecto: K, *fide* R.C.Carolin, *Telopea* 3: 532 (1990); isolecto: MEL, P.

Erect to ascending perennial herb, to 40 cm tall, yellowish villous with mostly stellate and multicellular hairs. Leaves crowded basally, elliptic to narrowly oblong, narrowing basally, entire or sinuate, hairy on both surfaces; lamina 4–12 cm long, 8–20 mm wide; cauline leaves stem-clasping. Flowers in racemes to 25 cm long; bracts leaf-like; peduncle 1.5–5 cm long; bracteoles c. 5 mm long; pedicel mostly 10–30 mm long, not articulate. Sepals lanceolate, c. 5 mm long. Corolla c. 15 mm long, ±glabrous inside, without enations, auriculate; abaxial lobes 5 to 6 mm long; wings c. 2 mm wide. Indusium broadly elliptic, folded. Ovules c. 30. Fruit ellipsoidal, c. 10 mm long; valves often bifid. Seeds flattened, ±orbicular, c. 2.5 mm diam., aculeate, brown.

Extends from the W.A./S.A. border to the Flinders Ra., S.A., and north-western Vic. Flowers chiefly Sept.–Jan. Map 285.

S.A.: Pinkawillinie, *K.D.Rohrlach 177* (AD); Flora and Fauna Reserve NE of Lock, *R.L.Specht 2437* (AD); c. 1.5 km from Wanilla towards Spields, *M.E.Phillips 720* (CBG, SYD). Vic.: Tempy, 2 Oct. 1966, *T.Henshall* (NSW); Nhill, Oct. 1892, *C.Watson* (NSW).

Corolla yellow; pedicels not geniculate at bracteoles; sepals connate at base; indusium orifice very convex. Distinguished from all other species in this section by its stem-clasping leaves on the scapes.

88. **Goodenia dyeri** K.Krause, *Pflanzenr.* 54: 56 (1912)

T: railway between Cunderdin and Dedari, W.A., Oct. 1903, *D.H.Thistleton-Dyer 103*; holo: K.

Ascending herb to 20 cm tall, mostly stellate pubescent. Leaves crowded basally, obovate, narrowing basally, dentate to lyrate; lamina 3–5 cm long, 10–15 mm wide. Flowers solitary in axil of basal leaves; peduncle 5–10 mm long, bracteoles linear, c. 6 mm long; pedicel 2–5 mm long, not articulate. Sepals linear to lanceolate, c. 6 mm long. Corolla 12–14 mm long, ±glabrous inside, without enations, obscurely auriculate; abaxial lobes 4–4.5 mm long; wings c. 1.5 mm wide. Indusium broadly elliptic, ±folded. Ovules c. 20. Fruit subglobular, 5–6 mm diam.; valves separate to sepal insertion, sometimes bifid. Seeds elliptic, 2 mm long, aculeate, yellow.

Occurs between Cowcowing and Kalgoorlie, south-western W.A. Flowers Aug.–Nov. Map 286.

W.A.: Black Flag, Sept. 1898, *W.V.Fitzgerald* (PERTH); Cowcowing, *M.Koch 1173* (BRI, NSW, PERTH); Merredin, *R.D.Royce 8393* (PERTH).

Corolla yellow; pedicels geniculate at bracteoles; abaxial lobes of corolla slightly longer than adaxial lobes. Distinguished from other western Australian members of this section by its very few, simple hairs, short pedicels and peduncles and the abaxial lobes of the corolla being only slightly longer than the adaxial lobes.

89. **Goodenia wilunensis** Carolin, *Telopea* 2: 65 (1980)

T: 25 miles [c. 40 km] W of Wiluna, N.T., 23 July 1931, *C.A.Gardner 2382*; holo: PERTH; iso: BM.

Ascending to prostrate, annual herb, silky, with simple and multicellular hairs; stems to 30 cm long. Leaves mostly basal, elliptic to obovate, narrowing basally, dentate; lamina 4–6 cm long, 8–15 mm wide; cauline leaves smaller. Flowers in racemes to 30 cm long; bracts leaf-like; peduncle 3–6 cm long; bracteoles lanceolate, c. 4 mm long; pedicel 1–2.5 cm long, not articulate. Sepals narrowly ovate, 6 mm long. Corolla c. 18 mm long, pubescent inside, without enations, auriculate; abaxial lobes 8–9 mm long; wings c. 4 mm wide. Indusium broadly elliptic, folded. Ovules 12–14. Fruit subglobular, 7 mm diam; valves separating to sepal insertion, ±entire. Seeds elliptic, 3 mm long, aculeate, yellow. Fig. 79G.

Occurs in the Wiluna–Meekathara–Carnegie area, W.A., in sandy soil, often in creek beds. Flowers July–Oct. Map 287.

W.A.: 120 km N of Mundiwindi, *J.S.Beard 2730* (PERTH); c. 115 km N of Meekatharra, *A.S.George 937* (PERTH); Mileura Stn, *W.J.Peacock 60861.1.17* (SYD); Gascoyne R., 1882, *J.Forrest* (MEL); Carnegie homestead, *A.S.George 5555* (PERTH).

Corolla yellow with purplish markings; sepals adnate to lower 1/2 of ovary. Indusium folded, densely villous in the fold, with the orifice ±straight, and some bristles on the lower lip reflexed over the upper lip. Distinguished from other species in the section by its silky indumentum and the sepals which are adnate to only the lower 1/2 of the ovary.

90. **Goodenia glabra** R.Br., *Prodr.* 577 (1810)

T: Port 1, [between Curtis and Facing Islands, Qld], *R.Brown*; lecto: BM, *fide* R.C.Carolin, *Telopea* 3: 533 (1990); isolecto: K.

G. flagellifera Vriese in T.L.Mitchell, *J. Exped. Trop. Austral.* 378 (1848). T: near Balonne and Maranoa Rivers, Qld, 3 Nov., *T.L.Mitchell*; holo: K.

G. unilobata J.Black, *Trans. & Proc. Roy. Soc. S. Australia* 51: 383 (1927). T: Ooldea, S.A., *coll. unknown*, in Herb. Tate; holo: AD.

Illustration: E.R.Rotherham *et al., Fl. Pl. New South Wales & S. Queensland* 129, fig. 413 (1975).

Prostrate to decumbent herb, ±glabrous or cottony-hairy; stems to 30 cm long. Basal leaves narrowing basally, narrowly oblong to obovate, sinuate to lobed, with lamina 3–9 cm long, 7–11 mm wide; cauline leaves smaller, broader, not narrowing, one side with basal lobe. Flowers in racemes to 25 cm long; bracts leaf-like; peduncle 8–15 mm long; bracteoles linear, c. 3 mm long; pedicel 8–20 mm long, not articulate. Sepals linear-lanceolate, 3–4 mm long. Corolla 10–18 mm long, ±glabrous inside, with enations, auriculate; abaxial lobes 3–6.5 mm long; wings 1.5–2.5 mm wide. Indusium depressed-obovate, folded. Ovules 20–30. Fruit ovoid, 7–10 mm long; valves separating to base, thick, entire. Seeds oblong to elliptic, c. 3 mm long, aculeate, brownish. $n = 8$, W.J.Peacock, *Proc. Linn. Soc. New South Wales* 88: 12 (1963). Fig. 69.

Found mainly in east central Qld and tablelands of N.S.W., with disjunctions in central N.T. and S.A. Grows in a variety of communities. Flowers most of the year. Map 288.

N.T.: c. 70 km SW of Alice Springs, *R.E.Winkworth 753* (DNA). S.A.: Appatina Well, *E.A.Shaw 425* (AD). Qld: Boatman Stn, *S.L.Everist 2830* (BRI). N.S.W.: Carouga Peak, c. 11 km W of Byrock, *C.W.E.Moore 5260* (CANB); c. 45 km E of Nyngan, 19 Aug. 1939, *I.M.Pidgeon & J.W.Vickery* (SYD).

Corolla yellow with purplish markings; indusium tightly folded with the orifice very convex. Cytological voucher is *W.J.Peacock 6111.31.1* (SYD). Distinguished from most species in the section by the single enlarged lobe at the base of the cauline leaves. *Goodenia schwerinensis* and *G. peacockiana* also have this feature, but these species can be separated by their distribution.

91. **Goodenia peacockiana** Carolin, *Telopea* 2: 64 (1980)

T: 23 miles [c. 37 km] from Yelma on Leonora road, W.A., 27 July 1967, *R.C.Carolin 5911*; holo: NSW; iso: SYD.

[*Goodenia xanthosperma auct. non* F.Muell.: W.E.Blackall & B.J.Grieve, *How to Know W. Austral. Wildfl.* 424 (1956)].

Prostrate to decumbent, annual herb, ±glabrous or cottony-hairy; stems to 25 cm long. Leaves mostly basal, elliptic to oblanceolate, narrowing basally, dentate to lobed, with lamina 2–6 cm long, to 12 mm wide; cauline leaves smaller, broad at base, with a basal lobe on one side. Flowers in racemes to 20 cm long; bracts leaf-like; peduncle 5–10 mm long; bracteoles linear 3–4 mm long; pedicel 12–15 mm long, not articulate. Sepals lanceolate, 6 mm long. Corolla c. 15 mm long, densely pubescent inside, with enations, auriculate; abaxial lobes 8–11 mm long; wings c. 2.5 mm wide. Indusium broadly ovate, folded. Ovules c. 35. Fruit ovoid to cylindrical, c. 10 mm long; valves usually entire, spreading. Seeds elliptic, 3 mm long, aculeate to setose, dark brown. n = 8, W.J.Peacock, *Proc. Linn. Soc. New South Wales* 88: 13 (1963).

Occurs in semi-arid areas of central W.A., in scrub. Flowers June–Oct. Map 289.

W.A.: c. 20 km W of Agnew on Sandstone road, *A.S.George 8014* (PERTH); c. 100 km N of Agnew on Wiluna road, *T.E.H.Aplin 6396* (PERTH); Yackabindi, *F.Lullfitz 1601* (PERTH); near Cosmo Newberry, *R.C.Carolin 5942* (SYD); c. 18 km E of Laverton, *A.S.George 4494* (PERTH).

Corolla yellow, often with darker markings. Similar to *G. xanthosperma* which does not have the single basal lobe on the cauline leaves. *Goodenia glabra* and *G. peacockiana* both have a single basal leaf lobe, but *G. glabra* has a much longer style. Also similar to *G. schwerinensis*, which is pubescent. Moreover *G. glabra* occurs in eastern Australia, *G. schwerinensis* in central Australia, and *G. peacockiana* in W.A. Cytological voucher is *W.J.Peacock 60866.2*.

92. **Goodenia schwerinensis** Carolin, *Telopea* 2: 66 (1980)

T: 4 miles [c. 6 km] N of N end of Schwerin Mural Crescent, W.A., 2 Aug. 1962, *D.E.Symon 2431*; holo: PERTH; iso: AD.

Prostrate to ascending herb, with soft, simple and multicellular hairs; stems to 40 cm long. Basal leaves narrowly elliptic to oblanceolate, narrowing basally, dentate to lyrate, with lamina 5–12 cm long, 10–20 mm wide; cauline leaves smaller, not narrowing, with basal lobe on one side. Flowers in racemes to 30 cm long; bracts leaf-like; peduncle 5–15 mm long; bracteoles linear, 4–5 mm long; pedicel 5–20 mm long, not articulate. Sepals linear-triangular, 6–7 mm long. Corolla 12–15 mm long, pubescent inside, with enations, auriculate; abaxial lobes 6–7 mm long; wings c. 1.5 mm wide. Indusium depressed-ovate. Ovules c. 30. Fruit cylindrical, c. 10 mm long; valves separating to midpoint, thick, entire. Seeds elliptic, 3 mm long, aculeate, brown.

Occurs in central eastern W.A., in scrub and spinifex grassland in sandy soil. Flowers May–Sept. Map 290.

W.A.: Lightning Rock, eastern end of Warburton Ra., *R.C.Carolin 5983* (SYD); c. 108 km SW of Warburton Mission, *A.S.George 8160* (PERTH); c. 0.5 km E of Todd Ra., *A.S.George 5340* (PERTH); NE of Wiluna towards Everard Junction, May 1968, *M.de Graaf* (PERTH).

Corolla yellow. Similar to *G. glabra*, differing in its soft indumentum and distribution.

93. Goodenia laevis Benth., *Fl. Austral.* 4: 61 (1868)

T: Phillips Range, W.A., *G.Maxwell*; holo: K.

Prostrate or ascending, ±woody subshrub, with stock, glabrous; stems to 70 cm long. Leaves oblanceolate to oblong, narrowing basally, entire or with 2 teeth towards apex, thick; lamina 1–3 cm long, 3–6 mm wide. Flowers in racemes or thyrses to 60 cm long; bracts leaf-like; peduncle 5–12 mm long; bracteoles linear, c. 2.5 mm long; pedicel obsolete. Sepals linear, 3–4 mm long. Corolla 10–12 mm long, with few stiff hairs inside, with enations; auriculate; abaxial lobes 3–3.5 mm long; wings 0.5–1 mm wide. Indusium depressed-ovate. Ovules c. 35. Fruit obovoid to cylindrical, 6 mm long; valves separate to midpoint or lower, entire. Seeds elliptic, 1 mm long, colliculate to tuberculate, brown. *n* = 8, W.J.Peacock, *Proc. Linn. Soc. New South Wales* 88: 13 (1963).

Restricted to southern inland W.A. Flowers chiefly Aug.–Dec. Map 291.

W.A.: c. 32 km E of Dumbleyung, *W.E.Blackall 1343* (PERTH); near Jerdacuttup R., *T.E.H.Aplin 2688* (PERTH); Ravensthorpe, *R.C.Carolin 3583* (SYD); c. 16 km S of Ravensthorpe, *W.J.Peacock 6095.8* (SYD); Frank Hann Natl Park, *R.D.Royce 10235* (PERTH).

Corolla yellow with purplish markings. Cytological voucher is *W.J.Peacock 6095.8* (SYD).

94. Goodenia mueckeana F.Muell., *Fragm.* 8: 56 (1873)

T: Camps 21–25, Betro Mt, Starr & Gills Ranges, N.T., 1872, *E.Giles*; holo: MEL.

Ascending, ±perennial herb to 30 cm tall, greyish with T-shaped hairs. Leaves linear to ovate, narrowing basally, dentate to pinnatisect; lamina 3–7 cm long, 4–10 mm wide. Flowers in thyrses or racemes to 25 cm long; bracts leaf-like; peduncle 5–15 mm long; bracteoles linear, c. 2 mm long; pedicel 0–0.3 mm long, not articulate. Sepals linear, 3 mm long. Corolla 9–12 mm long, pubescent inside, with enations, auriculate; abaxial lobes 4–5 mm long; wings to 2 mm wide. Indusium broadly obovate. Ovules to 26. Fruit ovoid, 10 mm long; valves usually split to sepals. Seeds elliptic, 2.5 mm long, aculeate, dark brown. *n* = 8, W.J.Peacock, *Proc. Linn. Soc. New South Wales* 88: 13 (1963). Fig. 23G.

Occurs in central Australia and into central W.A., mostly in hummock grassland in sandy soil. Flowers chiefly May–Sept. Map 292.

W.A.: c. 100 km N of Agnew on road to Wiluna, *T.E.H.Aplin 2395* (PERTH); c. 13 km W of Mt Nossiter, *A.S.George 5469* (PERTH); Flint Hill, Rawlinson Ra., *R.C.Carolin 6199* (SYD). N.T.: c. 41 km E of Armstrong R., 25 June 1958, *G.Chippendale* (DNA, NSW); Campbell Ra., *P.K.Latz 2082* (DNA).

Corolla yellow; indusium c. 1.3 mm long. Easily recognised by the appressed, grey pubescence of T-shaped hairs, which are generally quite dense. Cytological voucher is *W.J.Peacock 60867.1* (SYD).

Subsect. 2. Ebracteolatae

Goodenia subg. **Goodenia** sect. **Goodenia** subsect. **Ebracteolatae** K.Krause, *Pflanzenr.* 54: 46 (1912)

Goodenia ser. *Pedicellosae* Benth., *Fl. Austral.* 4: 54, 73 (1868). T: *G. cycloptera* R.Br., *fide* R.C.Carolin, *Fl. Australia* 35: 331 (1992).

Goodenia ser. *Foliosae* Benth., *Fl. Austral.* 4: 53, 69 (1868). T: *G. strangfordii* F.Muell., *fide* R.C.Carolin, *Fl. Australia* 35: 331 (1992).

Subshrubs or herbs. Basal leaves sometimes ephemeral, those on flowering stems usually somewhat smaller. Flowers usually in leafy racemes, sometimes in subumbels; pedicel usually ebracteolate, usually articulate. Corolla yellow, mauve or brownish purple, usually without enations, usually auriculate; hairs inside arranged in rows often becoming confluent towards base; pouch usually inconspicuous. Fruit a capsule; valves 2, entire, deciduous or persistent. Seeds various; wing prominent, usually mucilaginous to some extent.

The subsection contains 59 species, mostly occurring in the drier areas of Australia; 1 species extending to New Guinea.

95. **Goodenia nigrescens** Carolin, *Telopea* 3: 542 (1990)

T: 28 miles [c. 45 km] NE of Banka Banka, N.T., 16 June 1960, *G.Chippendale*; holo: NT.

Erect herb to 20 cm tall, becoming dark brown or black when dried, glabrous or with few glandular hairs, ±glaucous. Leaves oblong to oblanceolate, usually dentate or lobed; lamina 2–5 cm long, 3–5 mm wide. Flowers in racemes to 10 cm long, or subumbels; bracts leaf-like; peduncle 5–12 mm long; bracteoles linear, 2–3 mm long; pedicel 1–2 mm long, articulate. Sepals lanceolate to elliptic, 4–5 mm long. Corolla c. 18 mm long, hairs inside in rows confluent basally, with enations, auriculate; abaxial lobes c. 7 mm long; wings 2.5–3 mm wide, narrower on adaxial lobe above auricle. Indusium oblong-ovate, 2.5 mm long. Ovules c. 30. Fruit ovoid, ±compressed, 10–15 mm long. Seeds orbicular, c. 4 mm diam., reticulate-foveate, brown; wing 0.2–0.3 mm wide.

Occurs on the Barkly Tableland, N.T., in grey and black soils. Flowers chiefly May–Aug. Map 293.

N.T.: c. 11 km N of Brunchilly, 17 June 1960, *G.Chippendale* (DNA); c. 27 km NW of Brunette Downs homestead, *G.Chippendale* (DNA); c. 45 km NE of Alexandria Stn homestead, *C.S.Christian 1545* (CANB).

Corolla orange-yellow, strigose and glandular-hairy outside; sepals adnate to ovary for 3/4 its length; lower wings on adaxial lobes of corolla obsolete. Distinguished from other bracteolate species in the section by the narrower indusium, the glandular hairs on the corolla, the lack of a lower wing on the adaxial corolla lobes and the narrow wing on the seed.

96. **Goodenia angustifolia** Carolin, *Telopea* 2: 67 (1980)

T: Nockatunga, Qld, Aug. 1964, *R.C.Carolin 4159*; holo: NSW.

Ascending herb to 25 cm tall, glabrous, glaucous. Basal leaves linear, terete, channelled, entire, thick; lamina 5–8 cm long, c. 0.5 mm wide; cauline leaves sometimes clustered. Flowers in racemes to 10 cm long, or subumbels; bracts leaf-like; peduncle 5–10 mm long; bracteoles linear-deltoid, 1–1.5 mm long; pedicel 0.7–1.2 mm long, articulate. Sepals lanceolate, 2.5–3 mm long. Corolla 10–12 mm long, with simple and glandular hairs inside, without enations, auriculate; abaxial lobes 6–7 mm long; wings c. 2 mm wide. Indusium broadly oblong, 1.5 mm long. Ovules c. 25. Fruit and seeds not seen.

Occurs in south-western Qld, on stony downs; known only from the type collection. Flowers Aug. Map 294.

Corolla bright yellow; ovary septum c. 1/3 as long as locule; indusium with tufts of hairs on both surfaces. Differs from all other members of the subsection in its linear, terete, channelled leaves.

97. Goodenia glauca F.Muell., *Trans. & Proc. Victorian Inst. Advancem. Sci.* 40 (1855)

T: banks of the Avoca R., Vic., *coll. unknown*; lecto: MEL, *fide* R.C.Carolin, *Telopea* 3: 544 (1990).

Illustration: G.M.Cunningham *et al.*, *Pl. W. New South Wales* 633 (1981).

Erect herb to 20 cm tall, glaucous, glabrous or sparsely strigose. Basal leaves usually distinctly 3-veined, lanceolate to elliptic, entire or dentate, never lobed; lamina 3–8 cm long, 5–10 mm wide. Flowers in racemes to 15 cm long, or subumbels; bracts leaf-like; pedicel 2–5 cm long, articulate; bracteoles usually absent, rarely deltoid, to c. 1 mm long, close under ovary. Sepals lanceolate, 2.5–4 mm long. Corolla 12–17 mm long, hairs inside in rows confluent below, with enations, auriculate; abaxial lobes 5.5–6.5 mm long; wings 2.5–3 mm wide. Indusium hemispherical, 1.5–2 mm long. Ovules to 30. Fruit sub globular, c. 6 mm diam. Seeds broadly elliptic, c. 4 mm long; wing thick, c. 1 mm wide.

Occurs in inland Australia in S.A., Qld, N.S.W. and Vic., in woodland and grassland, usually in heavy soil. Flowers most of the year. Map 295.

S.A.: Lock 6, Murray R., *R.Filson* (MEL 24135 *p.p.*). Qld: Yapunyah, near Thargomindah, Apr. 1885, *Spencer* (MEL). N.S.W.: Goodooga, 14 Sept. 1944, *Alderson* (NSW); c. 8 km ENE of Bourke, *E.F.Constable 4476* (NSW). Vic.: Kings Billabong, near Mildura, *T.Henshall 531* (DNA).

Corolla pale yellow, strigose, hairy outside; ovary septum 1/2–2/3 as long as locule. This species is listed in W.A. by C.A.Gardner, *Enum. Pl. Austral. Occid.* 124 (1930), but this is undoubtedly an error. The locality makes it doubtful that this specimen was *G. glauca*. There appears to be some intergradation between this species and *G. fascicularis*, and some specimens are difficult to assign to one or other species. *Goodenia glauca* has glaucous leaves with few hairs or none, the corolla is a paler yellow and the wing of the seed is considerably thicker whilst in *G. fascicularis* the leaves are green and quite strigose, the corolla is a deeper yellow and the wing of the seed thinner.

98. Goodenia fascicularis F.Muell. & Tate, *Trans. & Proc. Roy. Soc. S. Australia* 13: 108 (1890)

T: Basedow Ranges, N.T., 1889, *W.H.Tietkins*; holo: MEL.

G. subintegra F.Muell. ex J.Black, *Trans. & Proc. Roy. Soc. S. Australia* 51: 383 (1927). T: Darling R., N.S.W., *J.Dallachy*; lecto: K, *fide* R.C.Carolin, *Telopea* 3: 544 (1990); ?isolecto: MEL.

G. glauca var. *sericea* Benth., *Fl. Austral.* 4: 77 (1868). T: Darling R., N.S.W., *J.Dallachy*; lecto: K, *fide* R.C.Carolin, *Telopea* 3: 544 (1990).

[*G. glauca auct. non* F.Muell.: F.M.Bailey, *Syn. Queensl. Fl.* 276 (1883); J.H.Moore & E.Betche, *Handb. Fl. New South Wales* 308 (1893)].

Illustration: G.M.Cunningham *et al.*, *Pl. W. New South Wales* 632 (1981).

Ascending, perennial herb to 20 cm tall, strigose and sometimes cottony- hairy. Basal leaves linear to ovate in outline, dentate, lobed or entire; lamina 2–14 cm long, mostly 4–25 mm wide. Flowers in racemes to 10 cm long, or subumbels, articulate; bracteoles absent. Sepals lanceolate, 2.5–4 mm long. Corolla 12–23 mm long, hairs inside in rows confluent below, without enations, auriculate; abaxial lobes mostly 5–10 mm long; wings 1.2–2.5 mm wide. Indusium broadly oblong, 1–1.5 mm long. Ovules to 24. Fruit ±globular, 5–8 mm diam. Seeds orbicular, c. 5 mm diam.; wing c. 1 mm wide. $n = 8$, W.J.Peacock, *Proc. Linn. Soc. New South Wales* 88: 13 (1963) as *G. subintegra* and *G. glauca*. Figs 23D, 45D.

Widespread in inland areas of the eastern two-thirds of the continent in a large number of communities in a wide variety of soils. Flowers most of the year. Map 296.

N.T.: c. 185 km SW of Calvert Hills, *R.C.Carolin 9284* (SYD). S.A.: c. 6 km N of Oodnadatta, *T.R.N.Lothian 20066* (DNA). Qld: c. 3 km N of Camooweal, *R.A.Perry 936* (CANB, DNA). N.S.W.: Cuttaburra Ck, *R.C.Carolin 3999* (SYD). Vic.: Murray R., near Swan Hill, *G.Morton 14* (MEL).

Corolla yellow; ovary septum almost as long as locule; indusium with long bristles on the lips. Cytological vouchers are *W.J.Peacock 6111.37.1* and *6111.363* (SYD). Confused with a number of species in the past, but these can be recognised by the following character combinations. In *G. pinnatifida* the corolla is glabrous outside and heavily bearded inside; *G. lunata* has very short or no hairs on the upper lip of the indusium; *G. strangfordii* has an attenuate base to the ovary and fruit, and a semi-circular indusium; *G. glauca* is glaucous, has paler yellow flowers and is glabrous or has many fewer hairs.

99. **Goodenia lunata** J.M.Black, *Trans. & Proc. Roy. Soc. S. Australia* 51: 384 (1927)

T: Cordillo Downs, S.A., 27 May 1924, *J.B.Cleland*; lecto: AD, *fide* R.C.Carolin, *Telopea* 3: 544 (1990).

G. argentea J.M.Black, *loc. cit.* T: Strangeways Springs, S.A., *W.L.Cleland*; lecto: AD, *fide* R.C.Carolin, *Telopea* 3: 544 (1990); isolecto: K.

Illustration: J.Jessop (ed.), *Fl. Centr. Australia* 356, 357 (1981).

Ascending or decumbent herb to 25 cm tall, strigose. Basal leaves linear to ovate in outline, dentate to pinnatifid; lamina 4–12 cm long, 6–30 mm wide, including lobes. Flowers in racemes to 15 cm long, or subumbels; bracts leaf-like; pedicel 1–4 mm long, articulate; bracteoles absent. Sepals lanceolate, 2.5–3.5 mm long. Corolla 8–14 mm long, hairs inside in rows confluent below, without enations, auriculate; abaxial lobes 4–5 mm long; wings to 3 mm wide. Indusium broadly oblong, 1.5 mm long. Ovules 10–20. Fruit ±globular, 9–10 mm long. Seeds orbicular, c. 7 mm diam.; wing 1–1.3 mm wide. n = 8, W.J.Peacock *pers. comm.*

Occurs in inland areas of N.T., S.A., Qld and N.S.W., with isolated records near Halls Creek, W.A., and north-western Vic. Flowers chiefly Mar.–Sept. Map 297.

W.A.: 20 km S of Halls Creek, *A.C.Beauglehole* (AD). N.T.: c. 50 km SE of Ranken, 20 June 1960, *G.Chippendale* (DNA). S.A.: c. 17 km W of Arckaringa homestead, *E.A.Shaw 512* (AD). Qld: Currawilla, *S.L.Everist 3946* (BRI). N.S.W.: c. 1.5 km S of Bourke, *S.W.L.Jacobs 47* (NSW).

Corolla yellow; ovary septum c. 1/2 as long as locule; indusium with very short hairs or glabrous on upper lip. Cytological voucher is *R.C.Carolin 5150* (SYD). There is very considerable variation in density of indumentum and width of leaves, the narrowly leaved, densely strigose forms having been separated by J.M.Black as *G. argentea*. There are, however, numerous intergrading specimens. Distinguished from *G. fascicularis* by the indusium which in that species has long bristles on the upper lip. Noted by E.Giles (annotation on MEL 25797) that 'the natives chew this plant instead of tobacco'.

100. **Goodenia pascua** Carolin, *Telopea* 3: 544 (1990)

T: 11 miles [c. 17 km] from Roeburne on Port Hedland road, W.A., 14 Aug. 1970, *R.C.Carolin 7894*; holo: NSW.

Ascending to erect herb to 50 cm tall, strigose. Basal leaves elliptic to oblanceolate, entire or dentate; lamina 4–8 cm long, 5–10 mm wide. Flowers in racemes to 20 cm long, or subumbels; bracts leaf-like; pedicel 2–4 cm long, articulate; bracteoles absent. Sepals lanceolate, c. 2 mm long. Corolla 8–10 mm long, hairs inside in rows confluent below, without enations, auriculate; abaxial lobes 4–4.5 mm long; wings c. 1.5 mm wide, narrower on adaxial lobe above auricle. Ovules c. 20; indusium square to broadly oblong, 1.5–1.7 mm long. Fruit ovoid-ellipsoidal, 7–8 mm long, attenuate at base. Seeds elliptic,

2.8–3 mm long, reticulate, brown; wing 0.3–0.4 mm wide.

Occurs in near-coastal parts of the Pilbara region, W.A., on black soil plains. Flowers May–Aug. Map 298.

W.A.: c. 51 km from Dampier on Onslow road, *R.C.Carolin 7886* (SYD); c. 203 km from Onslow on Roeburne road, *R.C.Carolin 7846* (SYD); c. 80 km from Onslow on Roeburne road, *R.C.Carolin 7859* (AD, SYD); c. 35 km N of Sandy Creek, Rabbit Proof Fence, *R.D.Royce 1687* (PERTH).

Corolla yellow; wings of adaxial corolla lobes narrower above auricle. The ovary septum reaches 1/2 length of ovary. The indumentum on this species shows a clear resemblance to that of the *G. glauca* group of species but *G. pascua* differs in the size and shape of the fruit and indusium and particularly in the unequal wings on the adaxial corolla lobes. The wing on the seed is also narrower.

101. Goodenia heteromera F.Muell., *Fragm.* 1: 115 (1859)

T: Murray R., ?Vic., *coll. unknown*; lecto: MEL, *fide* R.C.Carolin, *Telopea* 3: 545 (1990).

G. heteromera var. *deminuta* J.Black, *Trans. & Proc. Roy. Soc. S. Australia* 51: 395 (1927). T: banks of Wilson R. between Broken Hill and Cordillo Downs, Qld, Sept. 1922, *W.D.K.MacGillivray*; holo: AD.

Illustration: G.M.Cunningham *et al.*, *Pl. W. New South Wales* 634 (1981).

Perennial or annual herb; stems ±stoloniferous, to 20 cm long, strigose, cottony-hairy when young. Leaves tufted on stolons, oblanceolate to obovate, entire or dentate, with dense axillary hairs; lamina 1–10 cm long, 3–10 mm wide. Flowers in very short subumbels or solitary in axils of tufted leaves; pedicel 1–6 cm long, articulate; bracteoles absent. Sepals lanceolate, 2–3 mm long. Corolla 6–11 mm long, hairs inside in rows confluent below; without enations, auriculate; abaxial lobes 4–5 mm long; wings to 2 mm wide, narrower on adaxial lobe above auricle. Indusium oblong to square, 1.2–1.4 mm long. Ovules 20–30. Fruit ovoid, 6–7 mm long. Seeds glossy, orbicular, 1.5–2.5 mm diam., reticulate, brown; wing 0.2–0.5 mm wide.

Occurs in the Murray-Darling River system, in Qld, N.S.W., Vic. and S.A., in heavy soil, mostly in *Eucalyptus camaldulensis* and *E. largiflorens* woodlands. Flowers May–Nov. Map 299.

Qld: Yandilla, Darling Downs, *H.Lau* (MEL). N.S.W.: North Bourke, *R.C.Carolin 3941* (SYD); c. 1.5 km N of Hay, *T. & J.Whaite 1731* (NSW). Vic.: Mildura, Oct. 1923, *H.B.Williamson* (CANB); Minyip, 6 Dec. 1905, *J.R.Weir* (MEL).

Corolla yellow with brownish markings; wings of adaxial corolla lobes obsolete above auricle. Ovary tapering basally, the septum reaching 3/4 the length of the ovary. Hairs dense in leaf axils, where flowers arise. Pedicels sometimes flexuose or geniculate. Distinguished from other members of the section by its narrower indusium, its tapering capsule and its reduction of the wing on the lower margin of the adaxial corolla lobes, to an auricle.

102. Goodenia pusilla (Vriese) Vriese, *Natuurk. Verh. Holl. Maatsch. Wetensch. Haarlem* ser. 2, 10: 131 (1854)

Scaevola pusilla Vriese in J.G.C.Lehmann, *Pl. Preiss.* 1: 412 (1844). T: near Eight-mile Bridge, [c. 13 km N of Albany], W.A., Feb. 1841, *J.A.L.Preiss 1470*; lecto: LD, *fide* R.C.Carolin, *Telopea* 3: 507 (1990); isolecto: G, L.

G. tenella R.Br., *Prodr.* 577 (1810), *nom. illeg.* non Andrews. T: King George Sound, [W.A.], Dec. 1801, *F.Bauer*; holo: BM; iso: K.

G. tenella var. *major* Benth., *Fl. Austral.* 4: 74 (1868). T: Don River, W.A., *G.Maxwell*; lecto: K, *fide*

R.C.Carolin, *Telopea* 3: 545 (1990); isolecto: MEL.

Illustration: K.Krause, *Pflanzenr.* 54: 15, fig. 5A (1912) as *G. tenella*.

Ascending herb to 25 cm tall, strigose; stems sometimes ±stoloniferous. Basal leaves ±petiolate, thin, lanceolate to broadly obovate, entire or with coarse teeth; lamina 2–5 cm long, 4–14 mm wide; axillary hairs dense. Flowers solitary in leaf axils; pedicel c. 2.5 cm long, articulate; bracteoles absent. Sepals lanceolate, c. 2 mm long. Corolla 7–10 mm long, with long, villous hairs inside, without enations, auriculate; abaxial lobes c. 3.5 mm long; wings to 2 mm wide. Indusium depressed-obovate, 0.7 mm long. Ovules 6–10. Fruit ellipsoidal to obovoid, 3.5–4 mm long. Seeds elliptic, 4 mm long, reticulate, brown; wing 0.8 mm wide.

Occurs in south-western W.A. from Gingin to Albany, in moist places in forest, woodland and swamps. Flowers Oct.–Dec. Map 300.

W.A.: head of Gingin Brook, *A.S.George 11144* (PERTH); Blackwood R., *J.Forrest* (MEL); Middleton Beach, *C.Andrews 553* (K).

Corolla yellow with brownish markings; hairs dense in leaf axils; ovary tapering basally; septum reaching 1/2 length of ovary. Specimens from the Albany district are somewhat smaller and, when the stems are elongated, tend to be ±procumbent. However, specimens collected further westwards are larger and more erect; these formed the basis for *G. tenella* var. *major*, but there appear to be intergrades between the forms. Similar to *G. pulchella* which has thicker leaves, a longer corolla pouch often exceeding the ovary, and the ovary base not attenuate.

103. Goodenia anfracta J.M.Black, *Trans & Proc. Roy. Soc. S. Australia* 51: 385 (1927)

T: Cootanoorina, S.A., May 1891, *R.Helms*; holo: AD.

Decumbent herb, strigose; branches zig-zag, to 10 cm long. Basal leaves lanceolate to obovate, entire or dentate, thick, with lamina 1–3 cm long, 2–4 mm wide; cauline leaves terete, sometimes in axillary clusters. Flowers in racemes to 10 cm long, or subumbels; bracts leaf-like but much smaller; pedicel 8–16 mm long, articulate; bracteoles usually absent. Sepals lanceolate to elliptic, 3 mm long. Corolla 8–9 mm long, hairy towards base inside, without enations, auriculate; abaxial lobes c. 3 mm long; wings to 1.5 mm wide. Indusium depressed-oblong, 1.2 mm long. Ovules unknown. Fruit subglobular. Mature seeds not seen, immature ones winged.

Occurs in north-western S.A. and central-eastern W.A. Flowers c. August. Map 301.

W.A.: W end of Hopkins Lake, *D.E.Symon 2343* (AD). S.A.: Strangways Springs, *D.E.Symon 11137* (AD).

Corolla yellow; pedicels sometimes with 2 minute bracteoles; ovary septum less than 1/2 length of ovary.

There is some similarity with the type of *G. fascicularis* but in that species the leaves are strigose or cottony, dentate or lobed, linear to ovate, and the stems not zig-zag. In addition both species bear fascicles of leaves. Those of *G. anfracta* are formed in the axils of main leaves, while in *G. fascicularis* they are much more pronounced and appear to be the result of condensation of the main stem.

104. Goodenia maideniana W.Fitzg., *W. Austral. Nat. Hist. Soc.* 1: 25 (1904)

T: Nannine, W.A., Sept. 1903, *W.V.Fitzgerald*; holo: NSW.

Prostrate or decumbent herb, strigose; stems to 40 cm long. Basal leaves ±petiolate, obovate, dentate, thick; lamina 1–5 cm long, 4–15 mm wide. Flowers in racemes to 35 cm long, or subumbels; bracts leaf-like; pedicel 1.5–3.5 cm long, articulate; bracteoles absent. Sepals oblong to elliptic, 3–4 mm long. Corolla 12–15 mm long, hairy toward base inside, without enations, auriculate; abaxial lobes 5–6 mm long; wings to 3 mm wide. Indusium broadly oblong, 1 mm long. Ovules to 30. Fruit cylindrical to obovoid, to 5 mm long. Seeds broadly elliptic, 3.8 mm long, reticulate, black; wings c. 1 mm wide.

Occurs along the southern borders of the Tanami, Gibson and Great Sandy Deserts in W.A. and N.T., usually on the margins of salt lakes. Flowers June–Oct. Map 302.

W.A.: c. 9 km W of Yelma, *R.C.Carolin 5857* (SYD); c. 95 km W of Wiluna, *R.C.Carolin 5864* (SYD). N.T.: c. 1.5 km W of Central Mt Wedge, *A.O.Nicholls 807* (AD, DNA); Stirling Stn, *A.Mitchell 039* (DNA).

Corolla yellow. Sepals c. 1 mm wide, adnate to ovary for c. 1/2 its length, their margins usually ±free below that. Ovary septum c. 3/4 length of ovary. Similar to *G. fascicularis* and *G. anfracta*, the former being distinctive in its erect habit and longer, narrower, thinner leaves, the latter differing in its narrower leaves and zig-zag stem.

105. Goodenia corynocarpa F.Muell., *Fragm.* 2: 16 (1860)

T: Murchison R., W.A., *A.F.Oldfield*; holo: MEL; iso: K.

G. corynocarpa var. *macrocarpa* Benth., *Fl. Austral.* 4: 73 (1868). T: between Moore and Murchison Rivers, W.A., *A.F.Oldfield*; holo: K; iso: BM, MEL, P.

Illustration: K.Krause, *Pflanzenr.* 54: 73, fig. 15A–D (1912).

Erect herb to 60 cm tall, strigose. Basal leaves linear to elliptic; lamina 6–9 cm long, mostly 4–10 mm wide. Flowers in racemes to 30 cm long, or subumbels; bracts leaf-like; pedicel to 30 mm long, articulate; bracteoles absent. Sepals lanceolate, 3–4 mm long. Corolla 12–15 mm long, almost glabrous inside, without enations, auriculate; abaxial lobes 4–6 mm long; wings to 2 mm wide. Indusium depressed-obovate, 1 mm long. Ovules to 30. Fruit narrowly cylindrical, 12–26 mm long. Seeds flat, elliptic, 2.5 mm long, reticulate, black; wing c. 0.1 mm wide.

Occurs in near-coastal areas from Onslow to the Murchison R., W.A., in grassy plains on heavy soil. Flowers c. Aug. Map 303.

W.A.: c. 25 km from Onslow on Roebourne road, *R.C.Carolin 7830* (AD, SYD); c. 35 km S of Onslow, *I.Olsen 543* (NSW).

Corolla yellow; ovary septum c. 3/4 length of ovary. Easily distinguished from all other species by its long, narrow capsule.

106. Goodenia coronopifolia R.Br., *Prodr.* 577 (1810)

T: Carpentaria Island 5 [Morgans Island, N.T.], 20–21 Jan. 1803, *R.Brown*; lecto: BM, *fide* R.C.Carolin, *Telopea* 3: 546 (1990); isolecto: K, MEL, P.

Prostrate or ascending herb to 40 cm long, often glabrous. Basal leaves linear to oblong-elliptic in outline, dentate to pinnatisect with narrow, acute lobes or rarely entire, with lamina 2–10 cm long, 1–10 mm wide (including lobes); cauline leaves linear, entire or nearly so. Flowers in racemes to 30 cm long, or subumbels; bracts leaf-like; pedicel 10–30 mm long, articulate; bracteoles absent. Sepals lanceolate to oblong, 1–2 mm long. Corolla 5–9 mm long, hairs inside in rows confluent basally, without enations, auriculate; abaxial lobes 2–3 mm long; wings c. 0.5 mm wide. Indusium broadly oblong, 1 mm long. Ovules 4–8. Fruit globular, 4 mm diam. Seeds orbicular, c. 2.5 mm diam., glossy, blackish brown, reticulate; wing c. 0.5 mm wide.

Occurs in the north Kimberley, W.A., and Arnhem Land, N.T. Flowers chiefly May–Oct. Map 304.

W.A.: c. 2.5 km W of Lake Argyle turnoff, Kununurra–Timber Creek road, *A.C.Beauglehole* (AD); c. 80 km N of Turkey Creek Police Stn, *R.A.Perry & M.Lazarides 2518* (AD, CANB, DNA, MEL). N.T.: Mudginberri Stn, c. 3 km W of homestead, *R.C.Carolin 6905* (SYD); c. 16 km W of Timber Creek, *N.Byrnes NB747* (DNA); c. 48 km E of Newry, 10 May 1959, *G.Chippendale* (DNA).

Corolla yellow with purplish lines, and glabrous outside; ovary septum c. 1/4 length of ovary; indusium with long purplish hairs on undersurface. Similar to *G. pinnatifida*, differing in the narrower leaf lobes, and the lack of a dense beard inside the corolla.

107. **Goodenia integerrima** Carolin, *Telopea* 3: 546 (1990)

T: Lake King, W.A., 3 Nov. 1965, *A.S.George 7291*; holo: PERTH.

Decumbent to ascending herb, to c. 9 cm long, glabrous; stock perennial. Cauline leaves clustered, linear, involute, narrowly channelled above, thick, with lamina to 7 cm long, 1–2 mm wide; basal leaves not seen. Flowers in subumbels to c. 8 cm long; bracts leaf-like; pedicel c. 5 mm long, articulate; bracteoles absent. Sepals narrowly deltoid, c. 3 mm long. Corolla c. 7 mm long, hairy inside in throat, without enations, auriculate; abaxial lobes c. 2 mm long; wings c. 1 mm wide, narrower on adaxial lobe above auricle. Indusium broadly oblong, 0.8 mm long. Ovules 6–8. Fruit globular, c. 2.5 mm diam. Seeds orbicular, c. 1.5 mm diam., dark greyish brown, reticulate; wing c. 0.3 mm wide. Fig. 86A–B.

Known only from the type collection in southern W.A., on sandy island in a salt lake. Flowers c. Nov. Map 305.

Corolla yellow with a brownish throat, glabrous outside; wings of adaxial corolla lobes are almost obsolete above auricle; ovary septum as long as locule.

108. **Goodenia strangfordii** F.Muell., *Fragm.* 6: 11 (1867)

T: upper Victoria R., N.T., *F.Mueller*; lecto: MEL, *fide* G.Bentham, *Fl. Austral.* 4: 71 (1869) by implication.

G. strangfordii var. *grandiflora* Benth., *Fl. Austral.* 4: 71 (1868). T: Lara, Flinders R., Qld, *E.B.C.Kennedy*; holo: MEL.

Illustration: F.Mueller, *Fragm.* 6: t. 52 (1867).

Erect, diffuse herb to 30 cm tall, often ±woody basally, strigose. Leaves cauline, narrowly elliptic to oblanceolate, entire; lamina 5–12 cm long, 3–20 mm wide; marginal veins prominent. Flowers in racemes to 10 cm long, or subumbels; bracts leaf-like; pedicel 4–7 cm long, articulate; bracteoles rarely present immediately below ovary. Sepals lanceolate, 4–5 mm long. Corolla 15–19 mm long, hairy inside denser basally, without enations, auriculate; abaxial lobes 6–7 mm long; wings to 3.5 mm wide. Indusium semi-circular, 1.5 mm long. Ovules to 40. Fruit obovoid, 10–12 mm long. Seeds thick, orbicular, 2.5 mm diam., yellowish brown, reticulate; wing c. 0.3 mm wide.

Extends from central Australia into northern N.T. and Qld, in heavy and seasonally wet soil. Flowers chiefly May–Oct. Map 306.

N.T.: Roper Ck, c. 5 km along Mainoru road from Stuart Hwy, *R.C.Carolin 9342* (SYD); c. 40 km E of Stuart Hwy on Borroloola road, *J.Must 453* (DNA); c. 50 km S of Nicholson Stn, *R.A.Perry 2429* (CANB, DNA). Qld: c. 15 km NE of Blackall, *L.S.Smith & S.L.Everist 904* (BRI, MEL); Julia Creek, *M.S.Lorimer A44* (BRI).

Corolla yellow with brownish lines. Bracteoles rarely present immediately below ovary. Ovary tapering basally, with a septum c. 3/4 length of ovary. There is some variation in the density of the indumentum, the eastern specimens being noticeably more pubescent with a less well-developed stem than those from N.T. There are intermediates, e.g., *R.A.Perry 1518* (CANB) and the lectotype itself. Similar to, but differing from *G. fascicularis* and related species, as they have a rounded base to the ovary and a condensed stock with a cluster of basal leaves.

109. **Goodenia pusilliflora** F.Muell., *Victorian Naturalist* 5: 11 (1888)

T: junction of Murray and Darling Rivers, ?N.S.W., Aug. 1887, *C.E.Holding*; lecto: MEL, *fide* R.C.Carolin, *Telopea* 3: 548 (1990).

G. calogynoides E.Pritzel in F.L.E.Diels & E.Pritzel, *Bot. Jahrb. Syst.* 35: 560 (1905). T: near Newcastle, [Toodyay], W.A., *E.Pritzel 550*; lecto: BM, *fide* R.C.Carolin, *Telopea* 3: 548 (1990); isolecto: NSW.

Illustrations: K.Krause, *Pflanzenr.* 54: 87, fig. 16k–m (1912); G.M.Cunningham *et al.*, *Pl. W. New South Wales* 635 (1981).

Decumbent to ascending herb to 20 cm long, with long, not appressed hairs. Basal leaves oblong to obovate or narrowly oblanceolate, dentate to lyrate; lamina 1.5–8 cm long, 5–15 mm wide. Flowers in racemes to 20 cm long, or subumbels; bracts leaf-like; pedicel 1–7 cm long, articulate; bracteoles absent. Sepals elliptic, 1.8–2.2 mm long. Corolla 5–7 mm long, hairy inside towards base, without enations, auriculate; abaxial lobes c. 2 mm long; wings 1–1.5 mm wide, short. Indusium depressed-obovate, 1 mm long, notched. Ovules 4–6. Fruit subglobular, compressed, 5–6 mm long. Seeds orbicular to elliptic, 3.5–4 mm long, black, reticulate; wing brownish white, c. 1 mm wide. $n = 8$, W.J.Peacock, *Proc. Linn. Soc. New South Wales* 88: 13 (1963) as *G. calogynoides*.

Widespread over the drier parts of southern Australia in W.A., S.A., N.S.W. and Vic. Flowers chiefly July–Oct. Map 307.

W.A.: c. 32 km N of Kalgoorlie, *R.C.Carolin 5751* (SYD); W of Burracoppin, *R.D.Royce 1024* (PERTH). S.A.: c. 25 km S of Blinman, *R.H.Kuchel 979* (AD). N.S.W.: c. 6 km from Jerilderie, *E.D'Arnay 391a* (CANB). Vic.: near Lake Hindmarsh, Oct. 1892, *S.E.D'Alton* (MEL).

Corolla yellow; ovary septum very short. Can be distinguished from most others by the notched or cleft indusium. *Goodenia mimuloides* also shows this feature but has much larger flowers (corolla 12–25 mm long). *Goodenia pinnatifida* also shows some similarity but lacks the indusium character, and has a corolla with a mostly glabrous outer surface and a distinct beard inside. Cytological voucher is *W.J.Peacock 60843.1* (SYD).

110. **Goodenia pinnatifida** Schldl., *Linnaea* 21: 450 (1848)

T: South Australia, *H.H.Behr 214*: holo: HAL, photo SYD.

G. pinnatifida var. *minor* F.Muell. & Tate, *Trans. Roy. Soc. S. Australia* 16: 372 (1896). T: Fraser Ra., W.A., *coll. unknown*; *n.v.*

?*G. glabriflora* K.Krause, *Pflanzenr.* 54: 86 (1912). T: near Deniliquin (erroneously given as 'Victoria'), N.S.W., *coll. unknown*; holo: B, destroyed.

Illustration: G.M.Cunningham *et al.*, *Pl. W. New South Wales* 634 (1981).

Decumbent to ascending herb to 40 cm long, glabrous or with crisped hairs. Basal leaves oblong to oblanceolate, dentate to pinnatisect with oblong to linear lobes; lamina mostly 5–8 cm long, 3–20 mm wide. Flowers in racemes to 8 cm long, often in subumbels; bracts leaf-like; pedicel 2–12 cm long, articulate; bracteoles absent. Sepals lanceolate, 2.5–5 mm long. Corolla 8–19 mm long, densely bearded inside, without enations, auriculate; abaxial lobes 3.5–8 mm long; wings to 2.5 mm wide, narrower on adaxial lobe above auricle. Indusium depressed-ovate, 1–1.5 mm long. Ovules 20–35. Fruit globular, c. 8 mm long. Seeds ±orbicular, 5 mm long, glossy, black, almost smooth; wing c. 0.5 mm wide, white. *n*

= 16, W.J.Peacock, *Proc. Linn. Soc. New South Wales* 88: 12 (1963). Figs 23C, 80F–G.

Widespread in the drier parts of southern Australia in W.A., N.T., S.A., N.S.W., Vic. and Tas.; in a variety of habitats. Flowers chiefly May–Nov. Map 308.

W.A.: c. 100 km NE of Wubin, *E.M.Scrymgeour 2104* (PERTH). S.A.: Hambridge Flora and Fauna Res., *C.R.Alcock 1112* (AD). Qld: c. 5 km N of Boggabilla, *W.J.Peacock 6111.29.1* (SYD). N.S.W.: c. 8 km from Junee on Wagga road, *M.Gray 3555* (CANB). Vic.: c. 6 km S of Boweya, *T.B.Muir 1721* (MEL). Tas.: Port Dalrymple, 1804, *R.Brown* (BM, MEL).

Corolla yellow, glabrous or with a few crisped hairs outside, but densely bearded inside; wings of adaxial corolla lobes are narrower above auricle; ovary septum reaching 2/3 length of ovary; indusium bearded beneath. A very widely distributed species showing some variation in size of flower, degree of dissection of leaf, and density of indumentum. Some of the morphological variation appears to have a geographical basis, for instance the form with deeply dissected leaves (often almost linear lobes), small flowers and very little beard on the style, found in the Riverina region, seems to have been the basis for *G. glabriflora*. Cytological vouchers are *W.J.Peacock 6111.10.1* and *6110.1.1* (SYD). The species has some affinities with *G. mimuloides*, but that species has many more hairs that are not crisped, and the corolla is mostly hairy outside. Also confused with *G. fascicularis*, which is strigose, and has the corolla mostly hairy outside and the indusium with no beard. The crisped hairs and bearded indusium of *G. pinnatifida* also distinguish it from *G. elongata* which has a narrower seed wing and an attenuate ovary. To date, all chromosome counts on *G. mimuloides* are $n = 8$, while those of *G. pinnatifida* may show a polyploid series, $n = 8, 16, 24$. *Goodenia glabriflora* is placed in synonymy on the basis of the description in the protologue.

111. **Goodenia phillipsiae** Carolin, *Telopea* 3: 548 (1990)

T: 19 miles [c. 30 km] E of Ravensthorpe, W.A., 3 Nov. 1962, *M.E.Phillips*; holo: PERTH.

Erect to spreading, shrubby perennial to 30 cm high, glabrous. Basal leaves not seen; cauline leaves oblong to linear, entire, thick, with lamina to 4 cm long, to 3 mm wide. Flowers in thyrses to c. 20 cm long; bracts smaller than cauline leaves; peduncle similar to pedicel, shorter; bracteoles linear, to 10 mm long, c. 0.5 mm wide; pedicel 15–30 mm long, arcuate, articulate. Sepals elliptic-lanceolate, 2 mm long. Corolla 10–12 mm long, densely bearded inside, without enations, auriculate; abaxial lobes c. 6 mm long; wings to 1.5 mm wide. Indusium depressed-obovate, 1 mm long. Ovules to 50. Fruit and seeds unknown.

Known only from the type collection, in southern W.A. Flowers c. Nov. Map 309.

Corolla yellow; ovary septum reaching almost to top of locule; indusium bearded at base. The bracteolate inflorescence with arcuate peduncles and pedicels appears to be diagnostic. A species of uncertain affinity. The dense beard inside the corolla points to a relationship with *G. pinnatifida* but the presence of bracteoles is unusual in the close relatives of that species.

112. **Goodenia gibbosa** Carolin, *Telopea* 2: 68 (1980)

T: south end of Dean Ra., W.A., 5 Aug. 1967, *R.C.Carolin 6071*; holo: NSW.

Prostrate to decumbent herb, with soft, sometimes appressed hairs; stems often stoloniferous, to 40 cm long. Leaves basal and at ends of stolons, elliptic to oblanceolate, dentate to entire; lamina mostly 5–8 cm long, 4–14 mm wide. Flowers usually in short subumbels, rarely solitary in axils of tufted leaves; bracts leaf-like; pedicel 4–8 cm long, articulate; bracteoles absent. Sepals lanceolate to ovate, 2–3 mm long. Corolla 12–17 mm long, hairy inside, without enations, auriculate; abaxial lobes 5–6 mm long; wings 2–2.5 mm wide. Indusium broadly oblong, 1.5 mm long. Ovules c. 20. Fruit ±globular, 5–6 mm diam. Seeds ±orbicular, c. 3 mm diam., black, reticulate; wing greyish, c. 3 mm wide. *n* = 8, W.J.Peacock *pers. comm.* Figs 26F, 80C–E.

Common and widely distributed in central Australia in W.A. and N.T.; in sandy soil. Flowers chiefly May–Oct. Map 310.

W.A.: E end of Schwerin Mural Crescent, *R.C.Carolin 6213* (SYD); near Pass of Abencerrages, *A.S.George 8780* (PERTH, SYD). N.T.: Neutral Junction Stn, *A.Mitchell 060* (DNA); Kings Canyon, *P.K.Latz 306* (DNA); c. 16 km W of Mt Olga, *R.C.Carolin 5265* (SYD).

Corolla yellow, ±glabrous outside, with a conspicuous pouch; pedicels often persistent with septum; indusium with a dense beard beneath. Similar to *G. pinnatifida*, both having a beard on the lower surface of the indusium, differing in its less-divided leaves, more prominent corolla pouch, and in often being stoloniferous. Cytological voucher is *R.C.Carolin 5221* (SYD).

113. **Goodenia berardiana** (Gaudich.) Carolin, *Brunonia* 2: 16 (1979)

Distylis berardiana Gaudich. in H.L.C.de Freycinet, *Voy. Bot.* 460 (1829); *Calogyne berardiana* (Gaudich.) F.Muell. ex Benth., *Fl. Austral.* 4: 81 (1868). T: near Shark Bay, [W.A.], 1819–20, *C.Gaudichaud-Beaupré;* holo: P; iso: BM.

Calogyne distylis F.Muell., *Fragm.* 6: 6 (1867). T: near Oldfield R., W.A., *G.Maxwell*; *n.v.* (no specimen with this exact locality has been located).

Goodenia glauca var. *glandulosa* Benth., *Fl. Austral.* 4: 77 (1868). T: Darling R., N.S.W., *J.Dallachy*; lecto: K, *fide* R.C.Carolin, *Telopea* 3: 549 (1990).

Calogyne berardiana var. *major* E.Pritzel in F.L.E.Diels & E.Pritzel, *Bot. Jahrb. Syst.* 35: 563 (1905). T: near Mingenew, W.A., *F.L.E.Diels 3584*; holo: ?B (destroyed) *n.v.*

Calogyne linearis S.Moore, *J. Linn. Soc., Bot.* 45: 185 (1920). T: Kununoppin, W.A., *F.Stoward 307*; lecto: BM, *fide* R.C.Carolin, *Telopea* 3: 549 (1990).

Illustrations: C.Gaudichaud-Beaupré in H.L.C.de Freycinet, *op. cit.* t. 80; K.Krause, *Pflanzenr.* 54: 96, fig. 17C–G (1912) as *Calogyne berardiana*.

Erect, annual herb to 40 cm tall, with not appressed, simple hairs and glandular hairs. Basal leaves linear to obovate, lyrato-pinnatifid to dentate; lamina 2–16 cm long, 2–12 mm wide. Flowers in racemes to 20 cm long, or subumbels; bracts leaf-like; pedicel 2–7 mm long, indistinctly articulate; bracteoles absent. Sepals narrowly elliptic, 1.5–2.5 mm long. Corolla 6–15 mm long, hairy inside, without enations, auriculate; abaxial lobes 2.5–7 mm long; wings 1–2.5 mm wide. Indusium and style bifid; each half-indusium oblong, 1.5 mm long. Ovules 20–30. Fruit ellipsoidal, 8–12 mm long. Seeds orbicular, 4 mm diam., glossy black to dark brown, almost smooth; wing c. 0.5 mm wide, yellowish. *n* = 8, W.J.Peacock, *Proc. Linn. Soc. New South Wales* 88: 10 (1963) as *Goodenia* sp.; and W.J.Peacock *pers. comm.*

Widespread throughout the drier parts of western and southern W.A. and central Australia in N.T., S.A., Qld and N.S.W., and with isoliated occurrences in south-eastern S.A. Grows in a variety of habitats. Flowers chiefly May–Oct. Map 311.

W.A.: c. 32 km S of Shark Bay, *R.C.Carolin 3214* (SYD). N.T.: James Ra., *P.K.Latz 4927* (DNA). S.A.: Tarcoola, *E.H.Ising 1736a* (AD, BRI, MEL, NSW). Qld: c. 95 km E of Windorah, *R.C.Carolin 6337* (SYD). N.S.W.: c. 5 km from Broken Hill, Nov. 1919, *A.Morris* (BRI).

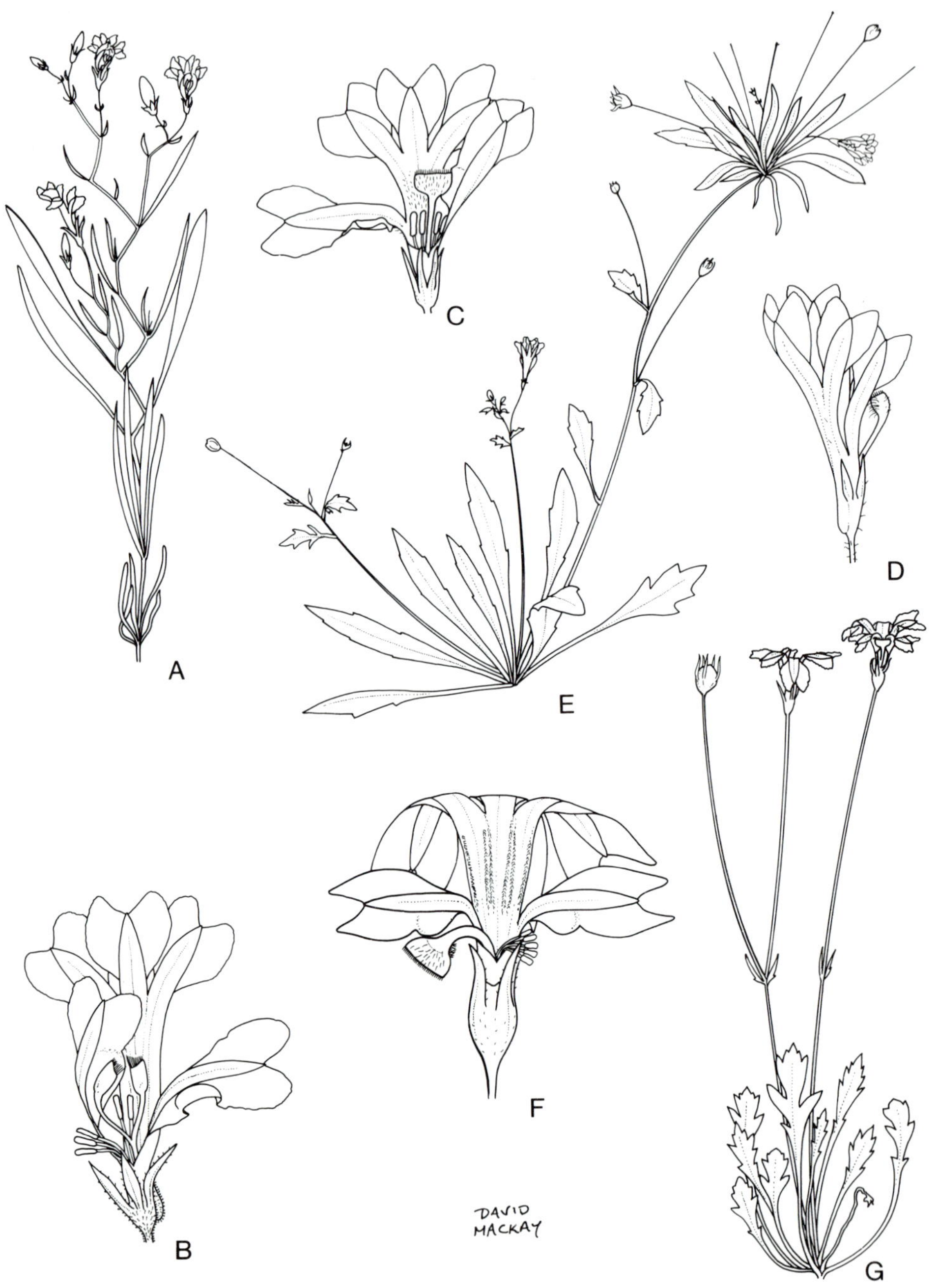

Figure 80. *Goodenia*. **A–B**, *G. ochracea*. **A,** habit X0.5 (A.George 1015a, SYD), **B**, flower X2 (A.George 11508, PERTH). **C–E**, *G. gibbosa*. **C**, flower X2 (A.Beauglehole 60112 & E.Errey 3812, SYD), **D**, flower X2, **E**, habit X0.5 (**D–E**, R & R.Belcher 102, SYD). **F–G**, *G. pinnatifida*. **F**, flower X2; **G**, habit X0.5 (**F–G**, W.Peacock 6111.5.1, SYD). Drawn by D.Mackay.

Corolla yellow with brownish lines. Cytological vouchers are *W.J.Peacock 608502*, *60842.1*, *60861.3* (SYD), *R.C.Carolin 5255* and *5161* (SYD). As is common with such a wide-ranging species there is considerable variation. Several authors have been impressed by the difference in flower (and plant) size, e.g., E.Pritzel who described *Calogyne berardiana* var. *major* and S.Moore who described *Calogyne linearis*, based on small-flowered short forms. The large-flowered forms occur in the more northerly parts of the range whilst the small-flowered forms also occur in the east and south-east. In the Albany–Stirling Ra. region, a form with intermediate-sized flowers occurs. Where the small- and large-flowered forms occur together, other characteristics of the two forms are the same. For example specimens from Taillifer Isthmus, Shark Bay, have glabrous stems; and at the mouth of the Murchison R. they are similar in indumentum. At these two localities the two forms grow in the same populations. The implication is that the large and small-flowered states occur in the same population either as a phenotypic expression or as a genetic polymorphism. Only large-flowered forms have been used for chromosome counts to date, all being $n = 8$.

The indumentum is another feature that shows considerable variation. All specimens have both simple and glandular hairs but the proportions vary, best observed on the undersurface of the leaves. In the eastern specimens almost all the hairs in this position are glandular, while the proportion of simple hairs increases towards the west reaching its highest value in specimens from parts of the Goldfields region of W.A.

114. Goodenia ochracea Carolin, *Telopea* 3: 549 (1990)

T: Shark Bay, W.A., Aug. 1932, *C.A.Gardner*; holo: K.

Decumbent herb, with mostly glandular hairs; stems stoloniferous, to 15 cm long. Leaves clustered near base of scapes, oblanceolate, dentate; lamina 3–6 cm long, 4–8 mm wide. Flowers in racemes to 8 cm long, or subumbels; scape often zig-zag; bracts leaf-like; pedicel 8–15 mm long, indistinctly articulate; bracteoles absent. Sepals narrowly oblong, 4–5 mm long. Corolla 15–16 mm long, ±glabrous inside, without enations, auriculate; abaxial lobes 8–9 mm long; wings c. 2 mm wide. Indusium and style bifid; each half-indusium oblong, 1.5 mm long. Ovules c. 20. Fruit globular, c. 6 mm diam. Seeds orbicular, c. 2 mm diam., black, reticulate; wing 0.1 mm wide, brownish. Fig. 80A–B.

Occurs in the Carnarvon–Shark Bay area, W.A., in sandy soil. Flowers June–Oct. Map 312.

W.A.: c. 11 km N of Quobba homestead, *A.S.George 10159* (PERTH); c. 4 km N of Herald Bay Outcamp, Dirk Hartog Is., *A.S.George 11508* (PERTH).

Corolla deep yellow, with a prominent pocket; sepals 1–1.5 mm wide. Similar to *G. berardiana*, differing in the broader sepals, prominent corolla pouch, and narrower seed wing.

115. Goodenia mimuloides S.Moore, *J. Linn. Soc., Bot.* 35: 167 (1897)

T: Gibraltar, Eastern Goldfields, W.A., Sept. 1895, *S.Moore*; holo: K; iso: BM.

[*G. pinnatifida auct. non* Schldl.: E.Pritzel in F.L.E.Diels & E.Pritzel, *Bot. Jahrb. Syst.* 35: 560 (1905); W.E.Blackall & B.J.Grieve, *How to Know W. Austral. Wildfl.* 430 (1954) *p.p.*].

Figure 81. *Goodenia*. **A–B**, *G. macroplectra*. **A**, habit X0.5; **B**, flower X2.5 (**A–B**, N.Donner 4516, SYD). **C–D**, *G. mimuloides*. **C**, flower X2; **D**, habit X0.5 (**C–D**, R.Saffrey 716, PERTH). Drawn by D.Mackay.

Decumbent to ascending herb, to 40 cm long, with dense, tomentose hairs. Basal leaves narrowly obovate, dentate to pinnatisect; lamina 4–10 cm long, 4–20 mm wide. Flowers in racemes to 8 cm long, or subumbels; bracts leaf-like; pedicel mostly 5–10 cm long, indistinctly articulate; bracteole absent. Sepals narrowly oblong, 2–5 mm long. Corolla 12–25 mm long, hairy inside, without enations, auriculate; abaxial lobes 4–7 mm long; wings 2–3.5 mm wide, narrower on adaxial lobe above auricle. Indusium depressed-ovate, to 2 mm long, notched or straight. Ovules 16–25. Fruit subglobular, 4–7 mm diam. Seeds elliptic, 4–5 mm long, black, reticulate; wings c. 1 mm wide, whitish. n = 8, W.J.Peacock, *Proc. Linn. Soc. New South Wales* 88: 12 (1963) as *Goodenia* sp. and as *G. pinnatifida*. Fig. 81C–D.

Occurs in W.A., from the Goldfields region to the west coast and northwards to Carnarvon, in mallee and open woodland in sandy soil. Flowers June–Oct. Map 313.

W.A.: 64 km E of Mount Magnet, *H.F. & M.Broadbent 1730* (NSW); Bardoc, 11 Sept. 1927, *C.A.Gardner 2114* (K, PERTH); c. 51 km S of Carnarvon, *C.H.Gittins 1523* (NSW); Laverton, Sept. 1909, *J.H.Maiden* (NSW); Great Eastern Hwy, c. 17 km W of Coolgardie, 19 Sept. 1963, *J.H.Willis* (K, MEL).

Cytological vouchers are *W.J.Peacock 60871.1*, *60863.3*, *60855.1* and *60857.2* (SYD). *W.J.Peacock 60860.3* is also given as n = 8 under G. *pinnatifida*, but there are 3 elements in this collection only one of which is *G. mimuloides*.

Corolla yellow, with not appressed hairs outside; wings of adaxial corolla lobe narrower above auricle. Differs from *G. pinnatifida* which has a corolla that is almost glabrous outside, with a dense beard inside.

116. Goodenia macroplectra (F.Muell.) Carolin, *Telopea* 3: 551 (1990)

Velleia macroplectra F.Muell., *Fragm.* 12: 22 (1882); *Symphyobasis macroplectra* (F.Muell.) K.Krause, *Pflanzenr.* 54: 41 (1912). T: Gascoyne R., W.A., *J.Forrest*; lecto: MEL, *fide* R.C.Carolin, *Telopea* 3: 551 (1990).

Illustration: K.Krause, *Pflanzenr.* 54: 41, fig. 11 (1912) as *Symphyobasis macroplectra*.

Erect herb to 20 cm tall, with glandular hairs and simple hairs at 90°. Basal leaves oblanceolate, dentate, with lamina 3–6 cm long, 4–10 mm wide; cauline leaves linear to linear-elliptic, smaller. Flowers in racemes to 10 cm long, or subumbels; bracts leaf-like; pedicel divergent, to 2.5 cm long, articulate; bracteoles absent. Sepals lanceolate, 4 mm long. Corolla 0.8–1.3 cm long, scattered hairy inside, without enations, auriculate; abaxial lobes 3–6 mm long; wings c. 0.5 mm wide. Indusium broadly oblong, c. 1 mm long. Ovules 16–20. Fruit compressed-globular, 6–7 mm diam. Seeds orbicular, 2–3 mm diam., dark brown, minutely reticulate; wing less than 0.3 mm wide. Fig. 81A–B.

Occurs from the Gascoyne R. to Leonora, W.A. Flowers c. Aug.–Sept. Map 314.

W.A.: Mt Leonora, Leonora, *C.A.Gardner 2111* (BM, PERTH); Paynes Find–Sandstone road, *P.G.Wilson 8884* (NSW).

Corolla dark yellow, with a pouch or spur twice as long as the ovary. Sepals free from ovary almost to base.

117. Goodenia elongata Labill., *Nov. Holl. Pl.* 1: 52, t. 75 (1804)

T: [Tasmania], *J.J.H.de Labillardière*; lecto: P *fide* R.C.Carolin, *Telopea* 3: 550 (1990); isolecto: FI, P.

Illustration: J.J.H.de Labillardière, *loc. cit.*

Erect or ascending herb to 40 cm tall, glabrous or with sometimes ±appressed hairs. Leaves usually cauline, oblanceolate, dentate to entire; lamina mostly 5–9 cm long, 4–20 mm wide. Flowers in racemes to 10 cm long, or subumbels; bracts leaf-like; pedicel 4–11 cm long, indistinctly articulate; bracteole absent. Sepals lanceolate, 3.5–4 mm long. Corolla 12–20 mm long, with few hairs towards base inside, without enations, auriculate; abaxial lobes 4–6.5 mm long; wings 1.5–2 mm wide, narrower on adaxial lobe above auricle. Indusium broadly oblong, 1–1.5 mm long. Ovules 25–35. Fruit ovoid, 8–9 mm long. Seeds elliptic, 3.5 mm long, brown, reticulate; wing c. 0.2 mm wide, brownish. *n* = 24, W.J.Peacock, *Proc. Linn. Soc. New South Wales* 88: 13 (1963).

Occurs in south-eastern Australia in S.A., N.S.W., Vic. and Tas.; in open forest, often in damper sites. Flowers Oct.–Feb. Map 315.

S.A.: near Cape Jervis, 1894, *L.Stephenson* (MEL 25565). N.S.W.: Holbrook, *W.J.Peacock 611.1.2, 5* (SYD). Vic.: Mt Buffalo, 19 Feb. 1963, *J.H.Willis* (MEL); Thurra R., *T.B.Muir 1930* (MEL). Tas.: Swanport, *R.C.Carolin 1848* (SYD); D'Entrecasteau R., *R.C.Carolin 1500* (SYD).

Corolla yellow; wings of adaxial corolla lobes slightly narrower above auricle; ovary tapering basally. A somewhat variable species in flower size and indumentum density. Probably close to *G. pinnatifida*, but that species has crisped hairs, and is densely bearded inside the corolla. Cytological voucher is *W.J.Peacock 611.1.2* (SYD); *n* = 8, vouchers not traced.

118. **Goodenia megasepala** Carolin, *Telopea* 2: 69 (1980)

T: Beale Ra., Qld, Aug. 1978, *K.A.W.Williams 78202*; holo: BRI.

Prostrate to decumbent herb, with long, tomentose hairs; stems to 25 cm long, often terminating in leaf cluster. Leaves oblanceolate to narrowly elliptic in outline, dentate or pinnately lobed; lamina 4–8 cm long, 5–25 mm wide. Flowers in racemes to 5 cm long, often in subumbels; bracts leaf-like; pedicel 4–6 cm long, indistinctly articulate; bracteoles absent. Sepals lanceolate, 6–8 mm long. Corolla c. 18 mm long, pubescent inside, without enations, auriculate; abaxial lobes 6–8 mm long; wings c. 2 mm wide. Indusium broadly oblong, 1.5 mm long. Ovules to 3. Fruit globular, c. 6 mm diam. Seeds (mature seeds not seen) flat, ±orbicular; wing broad. Fig. 82A–B.

Occurs in central-western Qld, in red, sandy soil. Flowers c. Aug. Map 316.

Qld: c. 150 km W of Windorah, *H.Stace 93 p.p.* (SYD); c. 95 km E of Windorah, *R.C.Carolin 6336* (SYD); c. 65 km W of Windorah, *R.C.Carolin 6398* (K, SYD).

Corolla yellow; wings of adaxial corolla lobes toothed or cut; capsule tardy in splitting. Similar to *G. fascicularis* which has narrower and shorter sepals, entire wings on the adaxial corolla lobes, closely appressed hairs, and a more readily dehiscent capsule; and to *G. cycloptera* which has a prominent spur to the corolla and narrower sepals.

119. **Goodenia iyouta** Carolin, *Telopea* 2: 69 (1980)

T: near Notabilis Hill, Gunbarrel Hwy, W.A., 25 July 1965, *E.Bettenay*; holo: PERTH.

Prostrate herb, hirsute with not appressed, simple and glandular hairs; stems to 120 cm long. Basal leaves not seen; cauline leaves ovate to elliptic, dentate to almost lobed at base. Flowers in racemes to 20 cm long, or subumbels; bracts leaf-like; pedicel to 4 cm long, articulate; bracteoles absent. Sepals narrowly oblong, c. 3.5 mm long. Corolla 11–15 mm long, pubescent in throat, without enations, auriculate; abaxial lobes 4–5 mm long; wings c. 2.5 mm wide. Indusium broadly oblong, 1.5 mm long. Ovules 12–15. Fruit subglobular, 5–6 mm diam. Seeds ±orbicular, 3.5 mm diam., dark brown to black, reticulate; wing c. 0.5 mm wide, pale brown. Fig. 82C–D.

Occurs in the northern Gibson Desert and near Roebourne, W.A. Flowers c. June–Sept. Map 317.

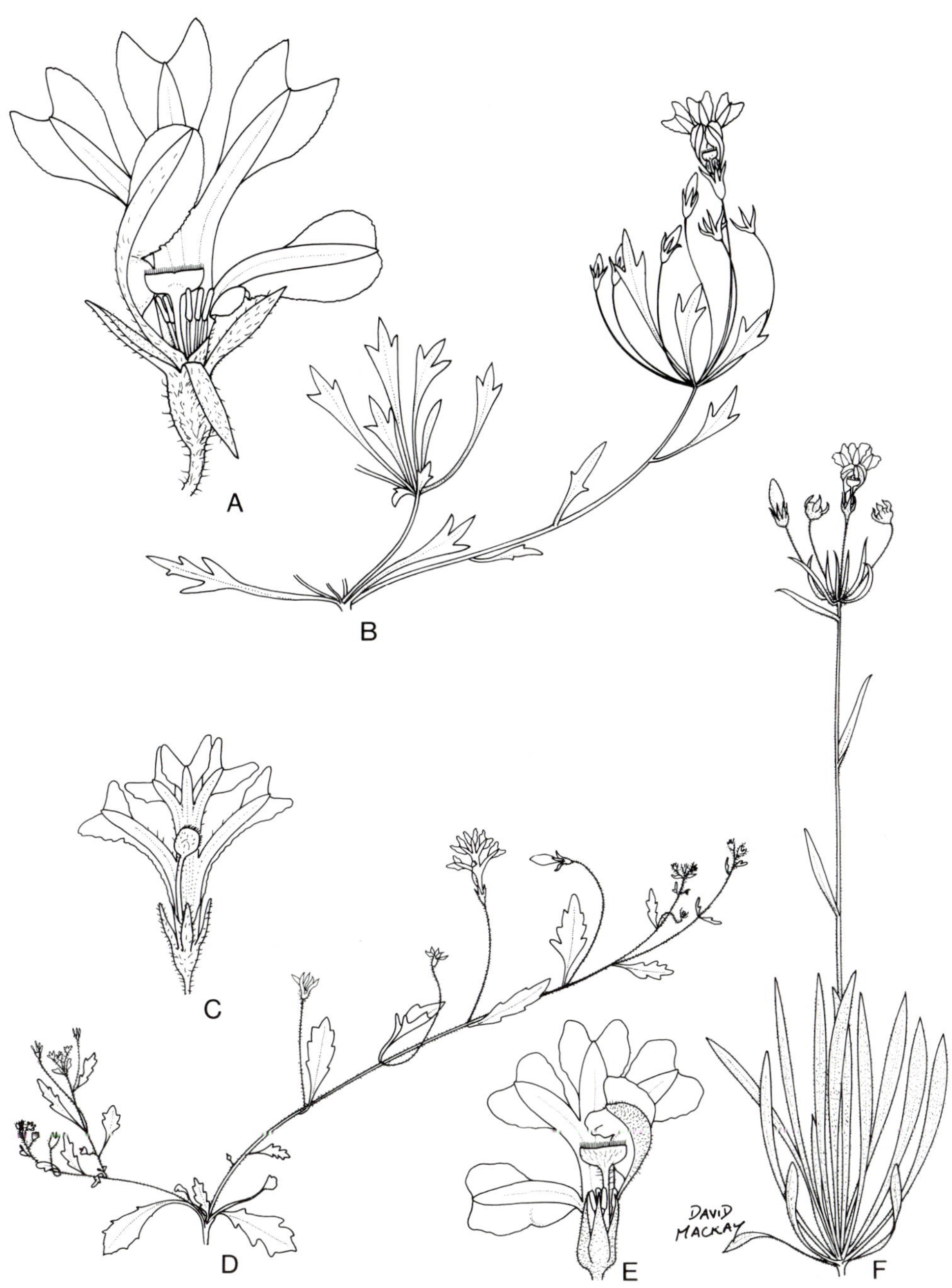

Figure 82. *Goodenia*. **A–B**, *G. megasepala*. **A**, flower X2; **B**, habit X0.5 (**A–B**, K.Williams 78702, SYD). **C–D**, *G. iyouta*. **C**, flower X2; **D**, habit X0.5 (**C–D**, A.George 5539, SYD). **E–F**, *G. quasilibera*. **E**, flower X2; **F**, habit X0.5 (**E–F**, A.George 5985, SYD). Drawn by D.Mackay.

W.A.: c. 35 km W of Browne Ra., Gunbarrel Hwy, *A.S.George 5420* (PERTH); Peawah Ck, Mundabullangana, *A.S.George 3420* (PERTH); c. 4 km E of Carnegie homestead, *A.S.George 5539* (PERTH); near Warburton Ra., 25 June 1960, *J.B.Cleland* (PERTH).

Corolla dark yellow with a distinct pouch; wings of adaxial corolla lobes dentate. Possibly most easily confused with *G. gibbosa* which is considerably less hairy, and has entire corolla wings; and similar to *G. megasepala* which has a very obscure corolla pouch.

120. **Goodenia lobata** Ising, *Trans. Roy. Soc. S. Australia* 81: 169 (1958)

T: Evelyn Downs, S.A., 22 Oct. 1955, *E.H.Ising 3923*; holo: AD; iso: K, MEL, NSW.

Ascending to decumbent herb, strigose, with dense, glandular hairs; stems to 20 cm long. Basal leaves linear to lanceolate, dentate to entire; lamina 4–7 cm long, 2–4 mm wide. Flowers in racemes to 5 cm long, in subumbels, or sometimes in axils of basal leaves; bracts leaf-like; pedicel 1–3.5 cm long, articulate; bracteoles absent. Sepals lanceolate, 3–4 mm long. Corolla 14–16 mm long, villous inside, without enations, auriculate; abaxial lobes c. 3 mm long; wings to 1.5 mm wide. Indusium depressed-obovate, 1.2 mm long. Ovules 25–30. Fruit globular, c. 5 mm diam. Seeds elliptic, 2.2 mm long, glossy, black; wing 0.2 – 0.3 mm wide, hyaline, not overlapping the body at maturity.

Occurs in the Lake Eyre Basin, S.A. Flowers July–Sept. Map 318.

S.A.: Evelyn Downs, 2 Oct. 1955, *E.H.Ising* (AD).

Corolla yellow with a dark throat. Similar to *G. fascicularis* and its relatives, differing in its glandular pubescence, and darker seeds with a narrower wing. Similar also to *G. havilandii* and its relatives, differing in not having the seed wing overlapping the seed at maturity.

121. **Goodenia salmoniana** (F.Muell.) Carolin, *Telopea* 3: 550 (1990)

Velleia salmoniana F.Muell., *Victorian Naturalist* 9: 127 (1892). T: Gascoyne R., W.A., 1889, *M.Forrest*; holo: MEL.

Erect to ascending herb to c. 30 cm tall, ±glabrescent, with few, soft hairs. Basal leaves not seen; cauline leaves linear, ±terete, with lamina 1–3.5 cm long, 0.5–1 mm wide. Flowers in racemes to 10 cm long, or subumbels; bracts leaf-like; pedicel 1–3 cm long, slender, divergent in fruit, articulate; bracteoles absent. Sepals lanceolate, c. 2 mm long. Corolla 6–7 mm long, with few long hairs inside, without enations, auriculate; abaxial lobes c. 3 mm long; wing c. 0.5 mm wide. Indusium broadly deltoid, 1.5 mm long. Ovules not known. Fruit ±globular, 3–4 mm diam. Mature seeds not seen; immature seeds orbicular, reticulate; wing broad.

Known only from the type, from an unspecified locality near the Gascoyne R., W.A. Flowering unknown. Map 319.

Corolla dark yellow; sepals adnate to ovary for less than 1/2 its length; indusium is glabrous on lips. A very imperfectly known species but distinguished from all others by these last two characters.

122. **Goodenia pulchella** Benth. in S.L.Endlicher, *Enum. Pl.* 71 (1837)

G. filiformis var. *pulchella* (Benth.) Benth., *Fl. Austral.* 4: 77 (1868). T: Swan River, W.A., *C.A.Hügel*; holo: W; iso: K.

?*Velleia lanceolata* Lindley, *Sketch Veg. Swan R.* xxvi (1840). T: none cited; *n.v.*

Scaevola umbellata Vriese in J.G.C.Lehmann, *Pl. Preiss.* 1: 411 (1845); *Aillya umbellata* (Vriese) Vriese, *Natuurk. Verh. Holl. Maatsch. Wetensch. Haarlem* ser. 2, 10: 76 (1854); *Goodenia aillya* F.Muell., *Fragm.* 2: 16 (1860), *nom. illeg.* T: near Perth, W.A., 25 Mar. 1839, *J.A.L.Preiss 1435a*; lecto: LD, *fide* R.C.Carolin, *Telopea* 3: 507 (1990); isolecto: W.

Scaevola umbellata var. α *procumbens* Vriese in J.G.C.Lehmann, *Pl. Preiss.* 1: 412 (1845); *Aillya umbellata* var. *procumbens* (Vriese) Vriese, *Natuurk. Verh. Holl. Maatsch. Wetensch. Haarlem* ser. 2, 10: 76 (1854). T: near Perth, W.A., 2 Feb. 1839, *J.A.L.Preiss 1451*; lecto: LD, *fide* R.C.Carolin, *Telopea* 3: 507 (1990).

Scaevola umbellata var. β *denticulata* Vriese in J.G.C.Lehmann, *Pl. Preiss.* 1: 412 (1845); *Aillya umbellata* var. β *denticulata* (Vriese) Vriese, *Natuurk Verh. Holl. Maatsch. Wetensch. Haarlem.* ser. 2, 10: 76 (1854). T: near Fremantle, W.A., 26 Dec. 1838, *J.A.L.Preiss 1428*; lecto: LD; isolecto: G, L, W, *fide* R.C.Carolin, *Telopea* 3: 507 (1990).

Scaevola umbellata var. δ *spathulata* Vriese in J.G.C.Lehmann, *Pl. Preiss.* 1: 412 (1845); *Aillya umbellata* var. δ *spathulata* (Vriese) Vriese, *Natuurk. Verh. Holl. Maatsch. Wetensch. Haarlem* ser. 2, 10: 76 (1854). T: near Perth, W.A., 15 May 1839, *J.A.L.Preiss 1430*; lecto: LD; isolecto: G, L, W, *fide* R.C.Carolin, *Telopea* 3: 507 (1990).

Lobelia longiscapa Vriese in J.G.C.Lehmann, *Pl. Preiss.* 1: 398 (1845). T: interior of south-western W.A., 1 Nov. 1839, *J.A.L.Preiss 1435*; lecto: LD, *fide* R.C.Carolin, *Fl. Australia* 35: 332 (1992); isolecto: G.

G. filiformis var. *glaucoides* E.Pritzel in F.L.E.Diels & E.Pritzel, *Bot. Jahrb. Syst.* 35: 563 (1905). T: Avon district, W.A., Sept. 1903, *F.L.E.Diels 5747*; holo: ?B (destroyed) *n.v.*

G. filiformis var. *hirsuta* K.Krause, *Pflanzenr.* 54: 86 (1912). T: no precise locality, W.A., *J.Drummond 185*; lecto: K, *fide* R.C.Carolin, *Telopea* 3: 550 (1990); isolecto: BM, MEL, P.

[? *G. elongata auct. non* Labill.: W.H.Vriese in J.G.C.Lehmann, *Pl. Preiss.* 1: 412 (1844)].

Illustration: W.H.de Vriese, *Natuurk. Verh. Holl. Maatsch. Wetensch. Haarlem* ser. 2, 10: t. 13 (1854).

Erect to ascending herb to 35 cm tall, strigose. Basal leaves lanceolate, entire or crenate-dentate, often thick; lamina 4–8 cm long, 1–11 mm wide. Flowers in racemes to 15 cm long, or subumbels; bracts leaf-like; pedicel 1–4 cm long, indistinctly articulate; bracteoles absent. Sepals narrowly oblong, 2–3 mm long. Corolla 8–14 mm long, with scattered hairs inside, without enations, auriculate; abaxial lobes 2.5–4 mm long; wings to 1.5 mm wide. Indusium broadly oblong, 0.8–1 mm long. Ovules 20–30. Fruit globular, 2–3 mm diam. Seeds orbicular, 1–1.5 mm diam., dark brown, reticulate; wing c. 0.5 mm wide. n = 16, W.J.Peacock, *Proc. Linn. Soc. New South Wales* 88: 13 (1963).

Occurs in southern W.A. from Kalbarri to the Great Australian Bight. Flowers Sept.–Jan. Map 320.

W.A.: c. 16 km W of Gingin, *R.D.Royce 4738* (PERTH); Needilup, *R.C.Carolin 3554* (SYD); Mt Barker, *B.T.Goadby B2009* (PERTH); Mt Chudalup, *P.G.Wilson 3983* (PERTH); Twin Swamps Wildlife Sanctuary, *N.T.Burbidge 7948* (CANB).

Corolla yellow, with a prominent pouch often longer than the ovary; indusium sometimes bearded below. There is some considerable variation in the indumentum, with the more northerly specimens more pubescent and the hairs more spreading. The hair-type, however, appears to be the same. Cytological voucher is *W.J.Peacock 6094.3* (SYD). Very close to *G. filiformis*, which differs in having terete or linear leaves, a broader more villous indusium which has a straight orifice, and generally more hairy axils of the basal leaves. The true structure of the wing of the seed is similar to that of *G. fascicularis* although thicker. However, the wing swells so much when mature that it appears to overlap the body of the seed as in e.g. *G. havilandii*. *Velleia lanceolata* is placed in synonymy following Krause, *Pflanzenr.* 54: 86 (1912).

123. Goodenia filiformis R.Br., *Prodr.* 578 (1810)

T: between Princess Royal Harbour and [West] Cape Howe, near King George Sound, [W.A.], 18 Dec. 1801, *R.Brown*; lecto: BM, *fide* R.C.Carolin, *Telopea* 3: 550 (1990); isolecto: K.

Erect to ascending herb to 25 cm tall, strigose. Basal leaves terete or narrowly linear, entire; lamina 4–9 cm long, 0.5–1 mm long. Flowers in racemes to 8 cm long, or subumbels; bracts leaf-like; pedicel 10–40 mm long, articulate; bracteoles absent. Sepals ovate, c. 1.5 mm long. Corolla c. 10 mm long, villous inside, without enations, auriculate; abaxial lobes c. 4 mm long; wings to 1.5 mm wide. Indusium broadly oblong, c. 0.7 mm long. Ovules few. Mature fruit and seeds unknown.

Occurs near King George Sound, W.A. Flowers Nov.–Jan. Map 321.

Known only from the type.

Corolla yellow, with a prominent pouch; ovary septum scarcely 1/2 as long as locule. The axils of the basal leaves are ±hairy. See note under *G. pulchella*.

124. Goodenia concinna Benth., *Fl. Austral.* 4: 76 (1868)

T: Eyre's Relief, W.A., *G.Maxwell*; lecto: K, *fide* R.C.Carolin, *Telopea* 3: 551 (1990); isolecto: MEL.

Illustration: K.Krause, *Pflanzenr.* 54: 87, fig. 16a–c (1912).

Erect to ascending, perennial herb to 40 cm tall; stock thick, often much-branched. Basal leaves linear to lanceolate, entire, thick, strigose; lamina 3–9 cm long, 1–8 mm wide. Flowers in racemes to 8 cm long, or subumbels; bracts leaf-like; pedicel 10–20 mm long, indistinctly articulate; bracteoles absent. Sepals lanceolate to narrowly elliptic, 1.5–2.5 mm long. Corolla 10–14 mm long, ±glabrous inside, without enations, auriculate; abaxial lobes 4.5–5.5 mm long; wings c. 1.5 mm wide. Indusium depressed-obovate, 1 mm long. Ovules c. 40. Fruit subglobular, 5–6 mm long. Seeds orbicular, 1.3–1.5 mm diam., glossy, black; wing to 0.3 mm wide, white, mucilaginous, overlapping body. n = 8, W.J.Peacock, *Proc. Linn. Soc. New South Wales* 88: 12 (1963). Fig. 25D.

Occurs near the south coast of W.A., in sand, in heath and forest. Flowers Aug.–Dec. Map 322.

W.A.: Eucla, 1893, *W.Webb* (MEL25763); SW corner of Mt Madden Reserve, *R.A.Saffrey 283* (PERTH); c. 36 km S of Ravensthorpe, *A.S.George 5760* (PERTH); Cape Arid, *R.D.Royce 9815* (PERTH); foot of Mt Trio, *R.C.Carolin 3534* (SYD).

Corolla yellow with a brownish throat, glandular-hairy outside, with a prominent pouch. See notes under *G. quasilibera* and *G. havilandii*. Cytological vouchers are *W.J.Peacock 6092.3* and *6095.10* (SYD).

125. Goodenia quasilibera Carolin in J.P.Jessop & H.R.Toelken, *Fl. S. Australia* 4th edn, 3: 1403 (1986)

T: Thomas R. valley, N of homestead, W.A., 10 Dec. 1960, *A.S.George 2171*; holo: PERTH.

Ascending to erect herb to 30 cm tall, with long, appressed, simple hairs and pale-headed, glandular hairs. Basal leaves lanceolate, entire or dentate, with lamina 3–6 cm long, 3–7 mm wide; cauline leaves smaller, often in fascicles below flowers. Flowers in racemes to 5 cm long, or subumbels; bracts leaf-like; pedicel 1.5–5 cm long, indistinctly articulate; bracteoles absent. Sepals elliptic, 3–5 mm long. Corolla 12–17 mm long, pubescent inside, with enations, auriculate; abaxial lobes usually 6–7 mm long; wings to 2.5 mm wide. Indusium depressed-obovate, 1 mm long. Ovules to 20. Fruit globular, 3–4 mm diam. Seeds orbicular, 1.5 mm diam., glossy, black; wing c. 0.3 mm wide, white, mucilaginous, overlapping. Fig. 82E–F.

Extends disjunctly from Meekatharra to Cape Arid, W.A., and E to Eyre Peninsula, S.A. Flowers chiefly Aug.–Jan. Map 323.

W.A.: c. 16 km W of Meekatharra, *C.A.Gardner 7856* (PERTH); c. 1.5 km N of Cundeelee Mission, 6 Mar. 1963, *M.C.George* (PERTH); Newman Peak, Balladonia, 26 Dec. 1931, *C.A.Gardner* (PERTH). S.A.: Jamieson, *K.D.Rohrlach 79* (AD); Pinkawillinie, *K.D.Rohrlach 156* (AD).

Corolla yellow with brownish lines and with a prominent pouch; sepals adnate to ovary for up to 1/3 its length. The closely related *G. concinna* has dark-headed glandular hairs on the pedicel and flower and smaller flowers.

126. Goodenia occidentalis Carolin, *Telopea* 2: 70 (1980)

T: 140 miles [c. 225 km] W of the Warburton Ra. on the road to Laverton, W.A., 27 July 1964, *R.C.Carolin 5936*; holo: NSW.

Prostrate to decumbent, short-lived herb; stems to 40 cm long. Basal leaves oblanceolate, lyrate to entire, ±thick, strigose; lamina 3–6 cm long, 4–12 mm wide. Flowers in ±secund racemes to 18 cm long; bracts smaller and narrower than leaves; pedicel 6–11 mm long, articulate; bracteoles absent. Sepals mostly ovate, c. 1–1.6 mm long. Corolla 4–6 mm long, villous-pubescent inside, without enations, auriculate; abaxial lobes c. 1.5 mm long; wings c. 0.2 mm wide. Indusium depressed-obovate, c. 0.6 mm long. Ovules to 20. Fruit globular, 2.5–3 mm diam.; valves often deciduous, septum persistent. Seeds orbicular, c. 2 mm diam., glossy, black; wing c. 0.5 mm wide, white, mucilaginous, overlapping body. $n = 8$, W.J.Peacock, *Proc. Linn. Soc. New South Wales* 88: 12 (1963) as *Goodenia* sp. Fig. 45E.

Extends across the drier parts of southern Australia from near the central W coast, W.A., through S.A., to central-western N.S.W., in mallee and *Acacia* scrub. Flowers chiefly June–Sept. Map 324.

W.A.: c. 1.5 km W of Murchison R. bridge on North West Coastal Hwy, *R.C.Carolin 3221* (SYD); c. 225 km W of Warburton Ra. Mission, *R.C.Carolin 5954* (SYD); Cape Arid, 5 Dec. 1971, *R.D.Royce* (PERTH). S.A.: c. 80 km E of Emu, *N.Forde 486* (CANB). N.S.W.: c. 15 km SE of Louth, *C.W.E.Moore 4193* (CANB).

Corolla yellow with a brownish throat, glandular-hairy outside; sepals adnate to ovary for c. 2/3 its length. Similar to *G. micrantha* which, however, has smaller flowers (corolla 1.5–2.5 mm long), narrower sepals, narrower, ±glabrous leaves, and a straight indusium. Cytological voucher is *W.J.Peacock 60856.5* (SYD).

127. Goodenia krauseana Carolin, *Telopea* 2: 64 (1980)

G. nana K.Krause, *Pflanzenr.* 54: 80 (1912) *non* Vriese (1854). T: Camp 53, Elder Explor. Exped., W.A., Sept. 1891, *R.Helms*; holo: K.

Prostrate to ascending herb, densely silvery strigose; stems to 20 cm long. Basal leaves lanceolate, entire or with 2–4 teeth towards top; lamina 3–7 cm long, 3–10 mm wide. Flowers in subumbels to 10 cm long, or solitary in axils of basal leaves; bracts leaf-like; pedicel 8–60 mm long, articulate; bracteoles absent. Sepals linear-elliptic, 2–2.5 mm long. Corolla 7–8 mm long, with long hairs in throat, without enations, auriculate; abaxial lobes c. 2 mm long; wings to 2 mm wide. Indusium broadly oblong, 1 mm long. Ovules c. 10. Fruit globular, 3–4 mm diam. Seeds orbicular; wing overlapping body (mature seeds not seen). Fig. 17C

Occurs on the Eastern Goldfields and in the Great Victoria Desert, W.A. Flowers c. Oct.–Dec. Map 325.

W.A.: Fraser Ra., *A.S.George 8594* (PERTH); c. 8 km W of Cundeelee Mission, N of Zanthus, *A.S.George 5988* (PERTH).

Corolla yellow with a brownish throat. Distinguished from other members of this section with few glandular hairs by the denser indumentum and the large teeth or lobes usually present on the leaf. It also resembles *G. fascicularis* but in that species the seed does not have an overlapping wing.

128. Goodenia micrantha Hemsley ex Carolin, *Telopea* 3: 551 (1990); C.Christensen & C.E.H.Ostenfeld in C.E.H.Ostenfeld, *Biol. Meddel. Kongel. Danske Vidensk. Selsk.* 3: 124, *nom. inval.* in obs. (1921)

G. filiformis var. *minutiflora* F.Muell. ex K.Krause, *Pflanzenr.* 54: 86 (1912). T: cultivated at Kew, ex Western Australia, 12 June 1891, *coll. unknown*; holo: K; iso: MEL.

Prostrate to ascending annual herb; stems to 25 cm long. Basal leaves linear, entire, sparsely strigose or glabrous; lamina 1–5 cm long, 0.4–1 mm wide. Flowers in secund racemes to 25 cm long; bracts leaf-like; pedicel 8–15 mm long, articulate; bracteoles absent. Sepals oblong to lanceolate, c. 1 mm long. Corolla 1.5–2.5 mm long, sparsely hairy inside, without enations, auriculate; abaxial lobes to 1 mm long; wings triangular, short, 0.1–0.2 mm wide. Indusium broadly obovate, 0.3–0.4 mm long. Ovules 12–18. Fruit globular, 2.5–3 mm diam. Seeds orbicular, c. 1 mm diam., glossy, black; wing c. 0.2 mm wide, white, mucilaginous, overlapping body. Fig. 83A–B.

Widespread but of scattered occurrence in southern W.A. from Geraldton to Cape Arid. Flowers chiefly Sept.–Dec. Map 326.

W.A.: N of Arrowsmith Lake, *A.S.George 9768* (PERTH); c. 11 km E of Wannamal, *A.S.George 5932* (PERTH); Cape Arid, *R.D.Royce 9978* (PERTH); Midland Junction, *A.Morrison 1281* (K).

Corolla yellow with a brownish throat, with a few glandular hairs outside. Confused in the past with *G. filiformis* which has larger flowers and subumbellate, never secund, inflorescences. Similar to *G. occidentalis* which has larger flowers and wider, dentate, hairy leaves.

129. Goodenia havilandii Maiden & Betche, *Proc. Linn. Soc. New South Wales* 38: 250 (1913) as *G. havilandi*

T: Cobar, N.S.W., Sept. 1911, *F.E.Haviland*; lecto: NSW, *fide* R.C.Carolin, *Telopea* 3: 551 (1990).

Symphyobasis alsinoides S.Moore, *J. Linn. Soc., Bot.* 45: 184 (1920). T: Mulline, W.A., 1916, *J.E.C.Maryon*; holo: BM.

G. havilandii var. *pauperata* J.M.Black, *Trans. & Proc. Roy. Soc. S. Australia* 51: 385 (1927). T: near Ooldea, S.A., *D.M.Bates*; holo: AD.

Prostrate to ascending, short-lived herb, densely glandular-viscid, sparsely strigose; stems to 40 cm long. Basal leaves linear to lanceolate, dentate to entire; lamina 2–9 cm long, 2–15 mm wide. Flowers in secund racemes to 20 cm long; bracts leaf-like; pedicel 8–15 mm long, arcuate-divergent, articulate; bracteoles absent. Sepals elliptic, c. 1.5 mm long. Corolla 3–12 mm long, with few hairs inside, without enations, auriculate; abaxial lobes 1–3 mm long; wings triangular, 0.5–1 mm wide. Indusium semi-elliptic, 0.5–1 mm long. Ovules 15–20. Fruit globular, 4–5 mm diam. Seeds orbicular, 2 mm diam., glossy, black; wing c. 0.3 mm wide, white, mucilaginous, overlapping body.

Occurs in drier parts of southern Australia in W.A., N.T., S.A., Qld and N.S.W.; in scrub in sandy and loamy soils. Flowers most of the year. Map 327.

W.A.: Miss Gibson Hill, *A.S.George 4092* (PERTH); Yelma Stn, *R.C.Carolin 5869* (SYD). S.A.: Gawler Ra., *R.L.Specht & B.B.Carrodus 34* (SYD). Qld: Tarka, c. 55 km SW of Eulo, *M.Law 57* (BRI). N.S.W.: Tundulya, *C.W.E.Moore 4039* (CANB).

Corolla yellow with a brownish throat. Specimens from Queen Victoria Spring, W.A., have a prominent beard on the lower lip of the indusium. Similar species are *G. quasilibera* and *G. concinna*, which also have dense glandular pubescence. *Goodenia havilandii* can be distinguished from them both by its smaller flowers.

130. **Goodenia janamba** Carolin, *Telopea* 3: 547 (1990)

T: c. 10 miles [c. 16 km] E of S Alligator River on Oenpelli road, N.T., 16 May 1968, *R.C.Carolin 6817*; holo: NSW; iso: SYD.

Erect herb to 60 cm tall, almost glabrous. Basal leaves narrowly oblong to oblanceolate, with few blunt teeth; lamina 5–16 cm long, 2–7 mm wide. Flowers in racemes to 15 cm long, or subumbels; bracts linear, 4–7 mm long, much smaller than leaves; pedicel 2–5 cm long, articulate; bracteoles absent. Sepals deltoid, c. 0.5 mm long. Corolla c. 10–15 mm long, with scattered hairs inside, without enations, auriculate; abaxial lobes 3–6 mm long; wings c. 1 mm wide. Indusium obtriangular, c. 1 mm long; bristles on lower lip much longer than those on upper. Ovules 10–12. Fruit compressed-globular, c. 4 mm diam. Seeds orbicular, c. 4 mm diam., smooth, shining, yellowish; wing 1 mm wide, whitish. Fig. 83F.

Occurs in the Kimberley, W.A., Arnhem Land, N.T. and adjacent Qld; in open woodland. Flowers chiefly Apr.–July. Map 328.

W.A.: just S of Dunham R. crossing, *D.E.Symon 12011* (AD, SYD). N.T.: c. 16 km E of S Alligator R., *R.C.Carolin 6817* (SYD); c. 1.5 km N of Adelaide River township, *N.Byrnes NB650* (DNA, SYD); c. 48 km E of Berwick homestead, *R.C.Carolin 9362* (SYD); c. 35 km from Borroloola on road to Daly Waters, *R.C.Carolin* 9310 (SYD). Qld: c. 88 km E of Croydon on Georgetown road, *R.C.Carolin 8612* (SYD).

Corolla yellow, with simple hairs at 90° and glandular hairs outside; ovary septum is c. 1/4 the length of the ovary.

131. **Goodenia cycloptera** R.Br. in C.Sturt, *Exped. Centr. Australia* 2: app. 83 (1849)

G. nicholsonii F.Muell., *Fragm.* 1: 203 (1859); *G. grandiflora* var. *nicholsonii* (F.Muell.) K.Krause, *Pflanzenr.* 54: 75 (1912). T: NW interior of South Australia, S.A., *J.Macdouall Stuart*; lecto: MEL, *fide* R.C.Carolin, *Telopea* 3: 551 (1990); iso: K.

G. mitchellii Benth., *Fl. Austral.* 4: 71 (1868). T: in the interior, Qld, *T.L.Mitchell*; holo: K; iso: BM.

G. mitchellii var. *typica* Domin, *Biblioth. Bot.* 22: 642 (1929). T: no locality given, *C.Sturt*; holo: BM.

[?*G. heterochila auct. non* F.Muell.: J.H.Maiden & E.Betche, *Census New South Wales Pl.* 191 (1916)].

Illustration: G.M.Cunningham *et al.*, *Pl. W. New South Wales* 631 (1981).

Ascending to decumbent perennial or annual herb, softly hairy; stems to 30 cm long. Leaves basal or cauline, crenate to dentate or almost pinnatifid; basal leaves often petiolate, obovate to spathulate, with lamina mostly 4–10 cm long, 10–15 mm wide; cauline leaves smaller with petiole shorter. Flowers in racemes to 20 cm long, or subumbels; bracts leaf-like; pedicel 15–50 mm long, ±articulate; bracteoles absent. Sepals linear-lanceolate, 2.5–3 mm long. Corolla 10–15 mm long, pubescent in throat, without enations, auriculate; abaxial lobes elliptic to oblong, c. 4 mm long; wings 2–3 mm wide, ±equal. Indusium broadly oblong, c. 1 mm long. Ovules 6–10. Fruit subglobular, 6–7 mm diam. Seeds orbicular, 6 mm diam., reticulate, dark brown to black; wing thick, 1–1.2 mm wide, brown. Fig. 45F.

Occurs in drier parts of inland Australia in W.A., N.T., S.A., Qld and N.S.W., S of 20°S and E of 125°E. Flowers most of the year. Map 329.

W.A.: c. 40 km SE of Giles Meteorological Stn, *A.S.George 4869* (PERTH). N.T.: c. 17 km W of Hermannsburg, 24 Aug. 1956, *G.Chippendale* (CANB, DNA). S.A.: c. 56 km W of Oodnadatta, *T.R.N.Lothian 1967* (DNA). Qld: Curracunya, S of Thargomindah, *R.C.Carolin 4055* (SYD). N.S.W.: Nyngan, Aug. 1903, *J.L.Boorman* (NSW).

Corolla yellow; corolla pouch has a spur at least as long as ovary. Northerly specimens have a more ascending habit, narrower leaves, coarser simple hairs and longer glandular ones, and the hairs on the inside of the corolla extend higher towards the base of the abaxial lobes than in the southern specimens. Similar to *G. heterochila* which lacks the very prominent spurred corolla pocket and has unequal wings on the adaxial corolla lobes. The thick seed-wing and prominent corolla-spur are also absent in *G. odonnellii* and *G. megasepala*.

132. Goodenia heterochila F.Muell., *Fragm.* 3: 142 (1863)

T: Newcastle Waters, [N.T.], 1862, *F.Waterhouse*; holo: MEL.

Illustration: K.Krause, *Pflanzenr.* 54: 73, fig. 15E–H (1912).

Erect or ascending, perennial herb to 40 cm tall, softly hairy. Leaves mostly cauline, oblanceolate to obovate, dentate; lamina 1–5 cm long, 4–14 mm wide. Flowers in racemes to 20 cm long; bracts leaf-like; pedicel 5–10 mm long, articulate; bracteoles absent. Sepals linear-lanceolate, 1.5–2.5 mm long. Corolla 8–12 mm long, hairy inside, without enations, auriculate; abaxial lobes 4–5 mm long; wings c. 1.5 mm wide, narrower above auricle. Indusium oblong, c. 1 mm long. Ovules 6–10. Fruit globular, c. 5 mm diam. Seeds orbicular, 5 mm diam., reticulate, blackish; wing thick, brown, c. 0.6 mm wide. Fig. 24F.

Occurs in arid areas of eastern Australia in N.T., S.A. and Qld with one record from the western Great Sandy Desert, W.A. Usually in sandy soil but usually not on dunes. Flowers most of the year. Map 330.

W.A.: no. 36 Well, S of Jigalong, *R.D.Royce 1992* (PERTH). N.T.: c. 105 km S of Hookers Creek Settlement, 13 July 1956, *G.Chippendale* (DNA, MEL); c. 1.5 km N of Barrow Creek, *P.K.Latz 176* (DNA). S.A.: Mt Carmeena, *H.Eichler 17538* (AD). Qld: c. 64 km W of Windorah, *R.C.Carolin 6392* (SYD).

Corolla yellow with a brownish throat, with a pouch scarcely as long as the ovary. On the adaxial corolla lobes the lower wing is narrower or obsolete. This narrow lower wing and the narrow indusium are distinguishing features of this species.

133. Goodenia glandulosa K.Krause, *Pflanzenr.* 54: 75 (1912)

T: Barrow Ra., W.A., 4 Aug. 1891, *R.Helms*; holo: K; iso: MEL.

Erect, perennial herb to 50 cm tall, with coarse, glandular hairs and arcuate-antrorse, simple hairs, becoming scabrous. Mature leaves oblanceolate to linear, entire, with lamina 6–12 mm long, 1–3 mm wide; younger leaves lobed. Flowers in racemes to 23 cm long; bracts leaf-like; pedicel 5–12 mm long, articulate; bracteoles absent. Sepals ovate-elliptic, 2–3 mm long. Corolla 8–11 mm long, with hairs in lines inside, without enations, auriculate; abaxial lobes c. 3.5 mm long; wings c. 1 mm wide, narrower above auricle. Indusium square, c. 1 mm long. Ovules c. 10. Fruit subglobular, 3–4 mm diam. Seeds elliptic, 3 mm long, reticulate; wing 0.1 mm wide, white.

Occurs in deserts of W.A. and S.A., usually on rocky hillsides. Flowers chiefly July–Oct. Map 331.

W.A.: c. 5 km W of Walter James Ra., *A.S.George 8346* (PERTH); c. 25 km E of Terhan Rockhole, *A.S.George 8135* (PERTH); c. 13 km S of Queen Victoria Spring, *R.D.Royce 5529* (PERTH); S of Cundeelee Mission, *A.S.George 5906* (PERTH). S.A.: 3 km SW of Inila Rock Waters, 11 Oct. 1987, *NPWS Yelabinna Survey 339* (AD).

Corolla yellow; wings of adaxial lobes narrower above auricle. Bracteoles sometimes present below the articulation of the pedicels, but minute. Similar to *G. larapinta* which, however, has soft hairs and dentate leaves not becoming scabrid with age. Considered an indicator of high concentrations of copper in the soil at Warburton, W.A.

134. **Goodenia larapinta** Tate, *Rep. Horn. Exped.* 3: 169, 189 (1896)

T: Glen Edith, N.T., June 1894, *R.Tate*; lecto: AD, *fide* R.C.Carolin, *Telopea* 3: 551 (1990); isolecto: K, MEL.

Erect, perennial herb to 50 cm tall; hairs soft, glandular and simple. Leaves mostly cauline, elliptic-oblong to lanceolate, dentate; lamina 1.5–4 cm long, 5–12 mm wide. Flowers in racemes to 20 cm long; bracts leaf-like; pedicel 6–15 mm long, articulate; bracteoles absent. Sepals lanceolate to narrowly oblong, 2–2.5 mm long. Corolla 12–15 mm long, with scattered hairs inside, without enations, auriculate; abaxial lobes 4–5 mm long; wings 1–2 mm wide, narrower above auricle. Indusium square or broadly ovate, c. 1.5 mm long. Ovules c. 20. Fruit cylindrical, 8–9 mm long; valves rarely bifid. Seeds elliptic, c. 3 mm long, glossy, black; wing 0.2–0.3 mm wide, brownish.

Extends from mountains in central Australia to the Barkly Tableland, N.T., in rocky situations. Flowers chiefly Apr.–Sept. Map 332.

N.T.: c. 85 km N of Tennant Creek, *R.A.Perry 624* (CANB); Mt Ziel, *A.C.Beauglehole 45556* (AD); Yuendumu, 25 Aug. 1950, *J.B.Cleland* (AD); Mt Sonder West, 20 July 1966, *J.H.Willis* (MEL).

Corolla yellow; wings of adaxial lobes narrower above auricle. The long, soft, glandular hairs distinguish this species from *G. glandulosa* and *G. faucium.*

135. **Goodenia faucium** Carolin, *Telopea* 2: 70 (1980)

T: gorge near Mt Liebig, N.T., 23 July 1957, *G.Chippendale 3556*; holo: DNA; iso: CANB, MEL.

Shrubby perennial to 40 cm tall, viscid with minute, peltate hairs. Leaves obovate to narrowly elliptic, dentate; lamina 1.5–3.5 cm long, 6–10 mm wide. Flowers in racemes to 20 cm long; bracts leaf-like; pedicel 8–15 mm long, articulate; bracteoles absent. Sepals narrowly elliptic to lanceolate, 4 mm long. Corolla 14–15 mm long, with short, scattered hairs inside, without enations, auriculate; abaxial lobes c. 4 mm long; wings c. 2 mm wide, narrower above auricle. Indusium broadly oblong, 1 mm long, with villous beard below. Ovules c. 40. Fruit subglobular, 4.5 mm diam. Seeds elliptic, 3 mm long, reticulate; wing c. 0.2 mm wide, hyaline.

Occurs in a few mountainous areas of southern N.T., in rock crevices. Flowers c. July. Map 333.

N.T.: Mt Liebig, 23 July 1966, *J.H.Willis* (MEL); Talla Putta Spring, 25 July 1966, *J.H.Willis* (MEL).

Corolla yellow; wings of adaxial corolla lobes narrower above auricle. Similar to *G. larapinta*, differing in having viscid, peltate hairs which are almost obscured by varnish.

136. **Goodenia redacta** Carolin, *Telopea* 3: 551 (1990)

T: Prince Regent R., W.A., 1891, *J.Bradshaw & W.T.Allen*; holo: MEL.

Prostrate to decumbent herb, with soft, antrorse hairs; stems to 15 cm long. Basal leaves obovate, dentate; lamina 2.5–5 cm long, 10–15 mm wide. Flowers in racemes to 15 cm long, or subumbels; bracts leaf-like; pedicel 5–10 mm long, articulate; bracteoles absent. Sepals 1–6 mm long, unequal. Corolla 6–7 mm long, pubescent towards base inside, without enations, auriculate; abaxial lobes c. 2.5 mm long; wings 0–1 mm wide, obsolete on adaxial lobes. Indusium very broadly obovate, 0.8 mm long. Ovules 6–10. Fruit subglobular, 3–6 mm diam. Seeds elliptic, 3 mm long, reticulate, greyish brown; wing 0.2–0.4 mm wide, white, membranous. Fig. 83E.

Occurs in the Kimberley, W.A., and Arnhem Land, N.T. Flowers c. Apr.–May. Map 334.

W.A.: Bindelong Ck, *E.M.Bennett 1777* (PERTH); 9 km N of Drysdale R. Stn, *D.E.Symon 7090* (AD, PERTH, SYD). N.T.: Katherine Gorge Natl Park, *N.Byrnes NB1477* (DNA).

Corolla yellow with a brownish throat; wings of adaxial lobes obsolete. Sepals are unequal in size; the adaxial and lateral sepals are elliptic-oblong, 1.5–2 mm long, while the abaxial sepal is elliptic, and 2–3 times as long. Related to *G. odonnellii* but distinguished from it and most other species by its enlarged posterior sepal and reduced adaxial corolla lobes which bear an auricle but no wings.

137. Goodenia odonnellii F.Muell., *Austral. J. Pharm.* 1: 278 (1886)

T: Ord R., W.A., *H.T.O'Donnell*; lecto: MEL, *fide* R.C.Carolin, *Telopea* 3: 551 (1990).

G. heterochila var. *foliosa* Benth., *Fl. Austral.* 4: 71 (1868). T: Victoria R., N.T., Dec. 1855, *F.Mueller*; holo: MEL; iso: MEL.

G. heterochila var. *runcinata* Benth., *Fl. Austral.* 4: 71 (1868). T: Arnhem Land, N.T., *F.Mueller*; ?holo: K; iso: MEL.

Erect to decumbent herb, with arcuate hairs or glabrous; stems to 40 cm long. Leaves mostly cauline, oblong to ovate, dentate to lyrate; lamina to 6 cm long, to 18 mm wide. Flowers in racemes to 30 cm long, or subumbels; bracts leaf-like; pedicel 10–40 mm long, articulate; bracteoles absent. Sepals linear to lanceolate, 1–2 mm long. Corolla 5–7 mm long, pubescent in throat and with dense hairs in tube, without enations, auriculate; abaxial lobes c.1.5 mm long; wings 0.5–0.8 mm wide. Indusium oblong, 1.2–1.4 mm long, folded and grooved above. Ovules 6–10. Fruit globular, to 3.8 mm diam.; valves spreading, not separating to base. Seeds elliptic, 3 mm long, aculeate, reticulate; wing 0.1–0.2 mm wide, yellowish. Fig. 83C–D.

Extends from the Kimberley, W.A., to Arnhem Land, N.T., and the Barkly Tableland, Qld, in woodland, often on river flats. Flowers Jan.–July. Map 335.

W.A.: c. 5 km SE of Bedford Downs Stn, *M.Lazarides 6376* (DNA, NSW). N.T.: c. 16 km E of Victoria R. crossing, *N.Byrnes NB737* (DNA, NSW); Armstrong R., Top Springs, *R.C.Carolin 6735* (AD, SYD). Qld: Settlement Ck, *L.J.Brass 331* (CANB); c. 11 km W of Cliffdale Ck, *D.E.Symon 5037* (AD, SYD).

Corolla yellow. Bracteoles rarely present close under flower. There is considerable variation in density of the hairs, with specimens from the eastern part of the range tending to be much less hairy. Some specimens from the Kimberley have much coarser hairs, usually narrower leaves and a more erect habit than elsewhere. The folded indusium distinguishes *G. odonnellii* from most other species in the subsection. Previously confused with *G. heterochila* which has a wider and thicker seed wing, and unequal wings on the adaxial lobes of the corolla.

Figure 83. *Goodenia*. **A–B**, *G. micrantha*. **A**, habit X0.5; **B**, flower X4 (**A–B**, H.Eichler 20309, SYD). **C–D**, *G. odonnellii*. **C**, flower X4; **D**, habit X0.5 (**C–D**, R.Carolin 9186, SYD). **E**, *G. redacta*, flower X4 (E.Shaw 757, SYD). **F**, *G. janamba*, habit X0.5 (R.Carolin 6817, SYD). Drawn by D.Mackay.

138. **Goodenia cirrifica** F.Muell., *Austral. J. Pharm.* 1: 81 (1886)

T: Alligator R., N.T., *M.Holtze*; lecto: MEL, *fide* R.C.Carolin, *Telopea* 3: 552 (1990); isolecto: DNA, K.

Ascending, divaricate herb to 40 cm tall, viscid, with not appressed, simple hairs and coarse, glandular hairs; upper branches usually zig-zag. Basal leaves ephemeral, obovate to oblanceolate, with lamina to 3 cm long, 8 mm wide; cauline leaves linear, incurved, entire, with lamina 10–50 mm long. Flowers in racemes to 30 cm long, or subumbels; bracts leaf-like; peduncle 5–20 mm long; bracteoles linear, to 10 mm long; pedicel to 10 mm long, articulate. Sepals narrowly elliptic to lanceolate, 2–3.5 mm long. Corolla 7–10 mm long, villous basally inside, without enations, auriculate; abaxial lobes 2–3 mm long; wings c. 1 mm wide, narrower above auricle. Indusium obdeltoid, 1.5 mm long. Ovules 4–6. Fruit ovoid, 5 mm long, splitting to midmark; valves entire, recurving. Seeds elliptic, 4 mm long, aculeate, brownish; wing 0.1–0.2 mm wide, white.

Occurs in Arnhem Land, N.T., in sclerophyll forest and woodland. Flowers Mar.–Oct. Map 336.

N.T.: Arnhem Land, *H.Basedow 114* (K); Oenpelli, *R.L.Specht 1117* (CANB, MEL); Pine Creek–Oenpelli, c. 6 km E of Mary R. crossing, *N.Byrnes NB787* (DNA); c. 55 km NNE of Pine Creek, 13 Mar. 1965, *M.Lazarides & L.G.Adams* (CANB 160557); c. 5 km S of Cahills Crossing, E Alligator R., Kakadu Natl Park, *H.S.Thompson 296* (CBG).

Corolla yellow; wings of adaxial corolla lobes narrower above auricle. A species of uncertain affinities. The seeds and hair type relate it to the *G. sepalosa* group but the presence of bracteoles and the entire capsule valves indicate the relationship is not particularly close. The peculiar 'zig-zag' habit combined with the glandular hairs distinguish it from other species.

139. **Goodenia armitiana** F.Muell., *Fragm.* 10: 110 (1877)

T: Einasleigh R., W.A., *W.E.de M.Armit 466*; lecto: MEL, *fide* R.C.Carolin, *Telopea* 3: 552 (1990).

G. linifolia W.Fitzg. ex K.Krause, *Pflanzenr.* 54: 84 (1912). T: Goody Goody, W.A., 4 Apr. 1905, *W.V.Fitzgerald*; holo: BM.

G. linifolia W.Fitzg., *J. & Proc. Roy. Soc. W. Australia* 3: 213 (1918), *nom. illeg., non* K.Krause (1912). T: Inglis Gap, W.A., *W.V.Fitzgerald*; lecto: K, *fide* R.C.Carolin, *Telopea* 3: 553 (1990).

G. armitiana var. *multicaulis* Blakely, *Austral. Naturalist* 11: 11 (1941). T: Connor Well, N.T., *B.A.Dale*, Oct. 1939; holo: NSW.

Erect herb to 40 cm tall, ±viscid or varnished, with glandular and sometimes simple hairs. Basal leaves linear-terete, involute; lamina 30–60 mm long, 0.5–1.5 mm wide. Flowers in racemes to 20 cm long, or subumbels; bracts leaf-like; pedicel 20–40 mm long, erect or ascending in fruit, articulate; bracteoles absent. Sepals lanceolate, c. 3 mm long. Corolla 8–10 mm long, almost glabrous inside, without enations, auriculate; abaxial lobes 2–3.5 mm long; wings 0.3–0.8 mm wide, narrower above auricle. Indusium depressed-ovate, 1–1.5 mm long. Ovules 30–35. Fruit ovoid-globular, usually ±compressed, 4–6 mm diam. Seeds orbicular, 3.5–4 mm wide, colliculate, grey-brown to black; wing c. 1 mm wide, whitish.

Occurs in northern Australia from the Kimberley, W.A., through N.T. to north-eastern Qld, in open communities in sandy soil. Flowers most of the year. Map 337.

W.A.: c. 100 km from De Grey R. on Broome road, *R.C.Carolin 1571* (SYD); Rudall R., *A.S.George 10813* (PERTH, SYD). N.T.: Gibson Ck, *J.Must 208* (DNA); c. 25 km W of Wollogorang, *R.C.Carolin 9232* (SYD). Qld: Chillagoe, *C.E.Hubbard & C.W.Winders 6744* (BRI).

Corolla yellow with a brownish throat; wings of adaxial corolla lobes narrower above auricle. The specimens with a conspicuous simple pubescence all tend to occur in the Fitzroy R. region of W.A. Those specimens from the eastern extremities of the range in Qld have a few scattered, simple hairs on the leaves but are neither glandular-hairy nor viscid. The glabrous forms are similar to *G. virgata* which has equal wings on the adaxial corolla lobes and divergent or reflexed fruiting pedicels.

Similar to *G. microptera* which, however, has flat leaves and a wider indusium.

140. **Goodenia virgata** Carolin, *Telopea* 2: 71 (1980)

T: 73 miles [c. 115 km] from Yuendumu on Alice Springs road, N.T., 24 Aug. 1970, *R.C.Carolin 7937*; holo: NSW; iso: SYD.

Erect to ascending herb, virgate, to 40 cm tall, with sparse, coarse hairs. Basal leaves flat or incurved towards top, linear to oblanceolate, dentate or entire, ±thick, with lamina 1–6 cm long, 1–4 mm wide; cauline leaves smaller, involute, usually linear. Flowers in racemes to 18 cm long, or subumbels; bracts leaf-like; pedicel 15–25 mm long, articulate; bracteoles absent. Sepals linear-lanceolate, 1.5–2 mm long. Corolla 8–12 mm long, with few hairs towards base inside, without enations, auriculate; abaxial lobes 4–5 mm long; wings c. 1.5 mm wide. Indusium broadly oblong, 1.5 mm long. Ovules 6–12. Fruit globular, 7 mm diam. Seeds orbicular, c. 2.5 mm diam., colliculate, grey; wing c. 0.8 mm wide, greyish.

Occurs in the southern Tanami Desert, W.A. and N.T., in sandy soil. Flowers May–Nov. Map 338.

W.A.: c. 6 km W of Dovers Hills, *A.S.George 9005* (PERTH, SYD). N.T.: c. 6 km S of Rabbit Flat, *R.C.Carolin 7917* (SYD); c. 55 km S of Napperby Stn, *M.Lazarides 6077* (AD, CANB, DNA, MEL).

Corolla yellow, with glandular and simple hairs outside. The pedicels have a prominent pulvinus at base and are divergent or reflexed in fruit. Related to *G. nuda* which, however, has smaller flowers, broader grey wing to the seed and coarser simple hairs on the stem and leaf; and to *G. armitiana* which however, has glandular-pubescent and viscid leaves, unequal wings on the adaxial corolla lobes and erect or slightly spreading fruiting pedicels.

141. **Goodenia triodiophila** Carolin, *Telopea* 2: 72 (1980)

T: 74 miles [c. 118 km] from Tom Price on Yampire Gorge road, W.A., 9 Aug. 1970, *R.C.Carolin 7773*; holo: NSW; iso: SYD.

Much-branched, stiff, ascending herb, to 40 cm tall, glabrous or with sparse, simple hairs; stems reddish or brownish, often hairy at base, green and glabrous above. Cauline leaves terete or linear, involute; lamina 4–10 cm long, c. 1 mm wide. Flowers in racemes to 15 cm long, or subumbels; bracts leaf-like; pedicel 8–30 mm long, articulate; bracteoles absent. Sepals lanceolate to narrowly elliptic, 1.5–3 mm long. Corolla 10–12 mm long, hairy outside and inside, without enations, auriculate; abaxial lobes 4–5 mm long; wings 0.5–1.8 mm wide, narrower above auricle. Indusium depressed-obovate, 1.5–1.8 mm long. Ovules 20–25. Fruit globular to ovoid, ±compressed, c. 6 mm diam. Seeds orbicular, 4–6 mm diam., transverse, colliculate, yellow-brown; wing 1–1.5 mm wide, yellowish.

Occurs in arid regions of inland Australia, in W.A., N.T. and western Qld, particularly in *Triodia* communities. Flowers Apr.–Oct. Map 339.

W.A.: c. 72 km from Marble Bar on Nullagine road, *R.C.Carolin 7687* (SYD); Sir Frederick Ra., *A.S.George 8317* (PERTH, SYD). N.T.: Livingstone Pass, Petermann Ranges, *R.C.Carolin 6265* (SYD); c. 110 km from Mt Doreen on The Granites road, *R.C.Carolin 7958* (SYD). Qld: No. 3 Tailings dam, Mount Isa,

P.Ollerenshaw & D.Kratzing 1145 (CBG, SYD).

Corolla yellow with a brownish throat, with glandular and simple hairs outside, sparsely hairy inside; sepals adnate to ovary only in lower 1/2. The corolla has a prominent pocket, and the wings of the adaxial lobes are narrower above the auricle. Differs from *G. armitiana* and *G. microptera*, which have short stocks not elongated basal stems, hairy leaves and lack a prominent pocket on the corolla.

142. **Goodenia microptera** F.Muell., *Fragm.* 3: 34 (1862)

T: near Nickol Bay, W.A., *P.Walcott*; holo: MEL.

G. heterochila var. *racemosa* Benth., *Fl. Austral.* 4: 71 (1868); *G. pritzelii* Domin, *Biblioth. Bot.* 22: 642 (1929). T: Camden Harbour, N.W. Australia, W.A., *herb. F.Mueller*; *n.v.*

Erect to ascending herb to 40 cm tall, villous, with long, dense multicellular, simple, and glandular hairs. Basal leaves narrowly oblong to oblanceolate, entire or dentate; lamina 4–10 cm long, 5–12 mm wide. Flowers in racemes to 20 cm long, or subumbels; bracts leaf-like; pedicel 5–40 mm long, articulate; bracteoles absent. Sepals narrowly elliptic, 2–2.5 mm long. Corolla 10–12 mm long, pubescent in throat, without enations, auriculate; abaxial lobes 3.5–4.5 mm long; wings 0.5–0.8 mm wide, narrower above auricle. Indusium depressed-ovate, 1.5 mm long, with a villous beard beneath. Ovules 6–12. Fruit compressed-globular, 7–8 mm diam. Seeds orbicular, 4–4.5 mm diam., reticulate, black; wing c. 1 mm wide, whitish.

Occurs in the Pilbara region, W.A., in rocky soil. Flowers chiefly Mar.–Aug. Map 340.

W.A.: Alpha Is., Monte Bello Islands, *F.L.Hill 599* (K); c. 70 km from Port Hedland on Broome road, *R.C.Carolin 7605* (SYD); c. 203 km from Onslow on Roeburne road, *R.C.Carolin 7850* (SYD); c.115 km from Roy Hill on Wittenoom road, *R.C.Carolin 7697* (SYD); 179 km E of Wittenoom, *D.E.Symon 10019* (AD, SYD).

Corolla yellow with a brownish throat; wings of adaxial lobes narrower above auricle. Related to *G. armitiana* which has glandular and some simple hairs but is not villous, involute leaves, and a wider wing on the lower margin of the adaxial corolla lobe.

143. **Goodenia nuda** E.Pritzel in F.L.E.Diels & E.Pritzel, *Bot. Jahrb. Syst.* 35: 562 (1905)

T: near Spring Stn, 55 km S of Roebourne, W.A., Apr. 1901, *F.L.E.Diels 2792*; holo: ?B (destroyed) *n.v.*; 96 miles [c. 153 km] from Onslow on Mt Stuart Road, W.A., *R.C.Carolin 7788*; neo: NSW, *fide* R.C.Carolin, *Telopea* 3: 553 (1990); isoneo: AD, SYD.

Illustration: K.Krause, *Pflanzenr.* 54: 87, fig. 16D–J (1912).

Erect to ascending herb to 50 cm tall, pale green to glaucous, glabrous or with few simple and glandular hairs. Basal leaves prominently 3-veined towards base, oblanceolate to narrowly elliptic, entire or with few, narrow teeth; lamina 4–10 cm long, 5–10 mm wide. Flowers in racemes to 25 cm long, or subumbels; bracts leaf-like; pedicel 20–50 mm long, articulate; bracteoles absent. Sepals narrowly elliptic to lanceolate, 3–3.5 mm long. Corolla 14–16 mm long, pubescent inside, without enations, auriculate; abaxial lobes 6–7 mm long; wings c. 2 mm wide. Indusium depressed-ovate, 2 mm long, bearded below. Ovules 20–30. Fruit ovoid to globular, 6–7 mm diam. Seeds elliptic, 3.5 mm long, colliculate, greyish; wing 0.5 mm wide, whitish.

Occurs in the Pilbara region, W.A. Flowers chiefly Apr.–Aug. Map 341.

W.A.: c. 30 km from Roy Hill on Wittenoom road, *R.C.Carolin 7709* (SYD); near Chichester Ra. Natl Park, *A.C.Beauglehole 48849* (SYD).

Corolla yellow. Similar to *G. microptera*, differing in the wider corolla wings, equal wings on the adaxial corolla lobes, and being glabrous or with few hairs.

144. **Goodenia pallida** Carolin, *Telopea* 3: 553 (1990)

T: 127 miles [c. 203 km] from Onslow on Roebourne road, W.A., 11 Aug. 1970, *R.C.Carolin 7845*; holo: NSW; iso: SYD.

Erect herb to 50 cm tall, glaucous, glabrescent, with short, simple and brown-headed, glandular hairs at 90°. Leaves mostly cauline, narrowly elliptic, dentate or entire, thick; lamina 5–7 cm long, 3–8 mm wide. Flowers in racemes to c. 15 cm long, or subumbels; bracts leaf-like; pedicel 3–4 cm long, articulate; bracteoles absent. Sepals linear-deltoid, 1.5–2 mm long. Corolla 14–16 mm long, with hairs in throat, without enations, auriculate; abaxial lobes 6–7 mm long; wings c. 2 mm wide. Indusium broadly ovate, 1.2 mm long. Ovules 15–20. Fruit subglobular, 6–7 mm diam. Seeds orbicular, 4–4.5 mm diam., colliculate, dark brown; wing 1–1.2 mm wide, yellow-brown, membranous.

Known only from the type collection, W.A. Flowers c. Aug. Map 342.

The pale purple colour of the corolla and the indumentum distinguish this from *G. nuda*.

145. **Goodenia prostrata** Carolin, *Telopea* 3: 554 (1990)

T: 22 miles [c. 35 km] from Roy Hill on Wittenoom road, W.A., 7 Aug. 1970, *R.C.Carolin 7702*; holo: NSW; iso: SYD.

Prostrate herb, glabrous except prominent axillary hair tufts, or with scattered, simple, glandular and/or multicellular hairs; stems to 30 cm long. Basal leaves oblanceolate, dentate, thick; lamina 2–4 cm long, 5–10 mm wide. Flowers in racemes to 30 cm long, or subumbels; bracts leaf-like; pedicel 10–25 mm long, articulate; bracteoles absent. Sepals narrowly elliptic, c. 2.5 mm long. Corolla 12–15 mm long, pubescent inside below throat, without enations, auriculate; abaxial lobes 4.5–5 mm long; wings c. 2.5 mm wide. Indusium broadly ovate, 1 mm long, villous. Ovules 8–12. Fruit globular, 4–5 mm diam. Seeds elliptic, 3 mm long, brown, colliculate; wing 0.2 mm wide, brown. Fig. 84F.

Occurs in the Pilbara region and adjacent areas, W.A., in sandy soil. Flowers chiefly May–Sept. Map 343.

W.A.: c. 17 km N of Roy Hill, *A.C.Beauglehole 11401* (SYD); railway crossing of Tom Price–Dampier railway and road from Wittenoom to Tom Price, *R.C.Carolin 7753* (SYD); c. 28 km NE of Mt Newman, *A.C.Beauglehole 48957* (SYD).

Corolla yellow with a brownish throat. Similar to *G. muelleriana*, differing in the habit, glabrescent leaves, narrower wing on the seed, and prominent tufts of axillary hairs.

146. **Goodenia muelleriana** Carolin, *Telopea* 3: 555 (1990)

T: 40 miles [c. 65 km] from Tom Price on Wittenoom road, W.A., 9 Aug. 1970, *R.C.Carolin 7761*; holo: NSW; iso: SYD.

Ascending to erect herb, to 40 cm tall, with long, simple hairs and large, brown-headed, glandular hairs. Basal leaves elliptic to oblanceolate, dentate to entire; lamina 3–6 cm long, 8–20 mm wide. Flowers in racemes to 20 cm long, or subumbels; bracts leaf-like; pedicel 20–35 mm long, articulate; bracteoles absent. Sepals lanceolate, c. 2 mm long. Corolla 12–15 mm long, pubescent below throat, without enations, auriculate; abaxial lobes 3.5–4 mm long; wings 2.5–3 mm wide. Indusium broadly ovate, 1 mm long, bearded below. Ovules 12–16. Fruit globular, c. 6 mm diam. Seeds orbicular, 2.5–3 mm wide, colliculate, dark grey; wing 1 mm wide, dark grey.

Occurs in the Pilbara region and adjacent areas, W.A. Flowers chiefly May–Sept. Map 344.

W.A.: c. 90 km from Port Hedland on Broome road, *R.C.Carolin 7593* (AD, SYD); c. 75 km from Marble Bar on Nullagine road, *R.C.Carolin 7688* (SYD); c. 153 km from Onslow on Mt Stuart road, *R.C.Carolin 7787* (SYD); c. 118 km from Tom Price on Yampire Gorge road, *R.C.Carolin 7771* (SYD).

Corolla yellow with a brownish throat. Previously included under *G. forrestii* but distinguished by its much less dense indumentum.

147. **Goodenia forrestii** F.Muell., *Victorian Naturalist* 9: 58 (1892) as *G. forestii*

T: Yule-Fortescue and Sherlock Rivers, W.A., *J.Forrest*; lecto: MEL *p.p.*, *fide* R.C.Carolin, *Telopea* 3: 556 (1990); isolecto: K.

Ascending to decumbent herb, with soft, dense, mostly simple hairs and some pale-headed, glandular hairs; stems to 50 cm long. Leaves elliptic to oblanceolate, dentate; lamina 2–6 cm long, 5–20 mm wide. Flowers in racemes to 25 cm long, or subumbels; bracts leaf-like; pedicel 15–30 mm long, articulate; bracteoles absent. Sepals lanceolate, 2.5–3 mm long. Corolla 12–15 mm long, pubescent below throat, without enations, auriculate; abaxial lobes 2.5–3 mm long; wings 2–2.5 mm wide. Indusium broadly oblong, c. 1 mm long, bearded below. Ovules 10–15. Fruit globular, c. 6 mm diam. Seeds orbicular, 5–6 mm diam., colliculate, dark grey; wing 1–2 mm wide, dark grey.

Occurs in near-coastal areas of north-western W.A., in scrub and woodland, in red sandy soil. Flowers chiefly May–Sept. Map 345.

W.A.: c. 65 km from Dampier on Onslow road, *R.C.Carolin 7887* (SYD); c. 285 km from Onslow on Rochlea road, *R.C.Carolin 7801* (SYD); c. 15 km S of Exmouth, *A.S.George 10330* (PERTH).

Corolla yellow with a brownish throat. Similar to *G. muelleriana*, differing in the much denser, softer indumentum, the hairs being shorter, finer and more patent. Also similar to *G. cycloptera* especially in the indumentum, but that species has very thick seed wings and a much more prominent corolla spur. *Goodenia heterochila* is also similar but has a reduced wing on the lower margin of the adaxial corolla lobe and a thick seed wing.

148. **Goodenia tenuiloba** F.Muell., *South. Sci. Rec.* n. ser., 1: 25 (1885)

T: near Mt Hale, upper Murchison R., W.A., 1884, *C.Crossland*; holo: MEL; iso: K.

Erect to ascending herb to 50 cm tall, with glandular and simple hairs. Basal leaves linear to oblong in outline, deeply pinnatifid with prominent lobes to 20 mm long and 2 mm wide, lamina 5–10 cm long, to 40 mm wide; rachis 1–5 mm wide. Flower in racemes to 25 cm long, or subumbels; bracts leaf-like; pedicel 20–40 mm long, articulate; bracteoles absent. Sepals lanceolate to narrowly elliptic, 3–5 mm long. Corolla 25–35 mm long, pubescent below throat, without enations, auriculate; abaxial lobes 10–12 mm long; wings 2–3 mm wide. Indusium depressed-obovate, 1.5 mm long, bearded below. Ovules 12–16. Fruit subglobular, c. 5 mm diam. Seeds orbicular, 3 mm diam., colliculate, dark brown; wing 0.5 mm wide, white. $n = 8$, W.J.Peacock, *Proc. Linn. Soc. New South Wales* 88: 13 (1963). Fig. 84A–B.

Occurs from the Pilbara region almost to Leonora, W.A., in scrub, in red sandy soil. Flowers chiefly May–Sept. Map 346.

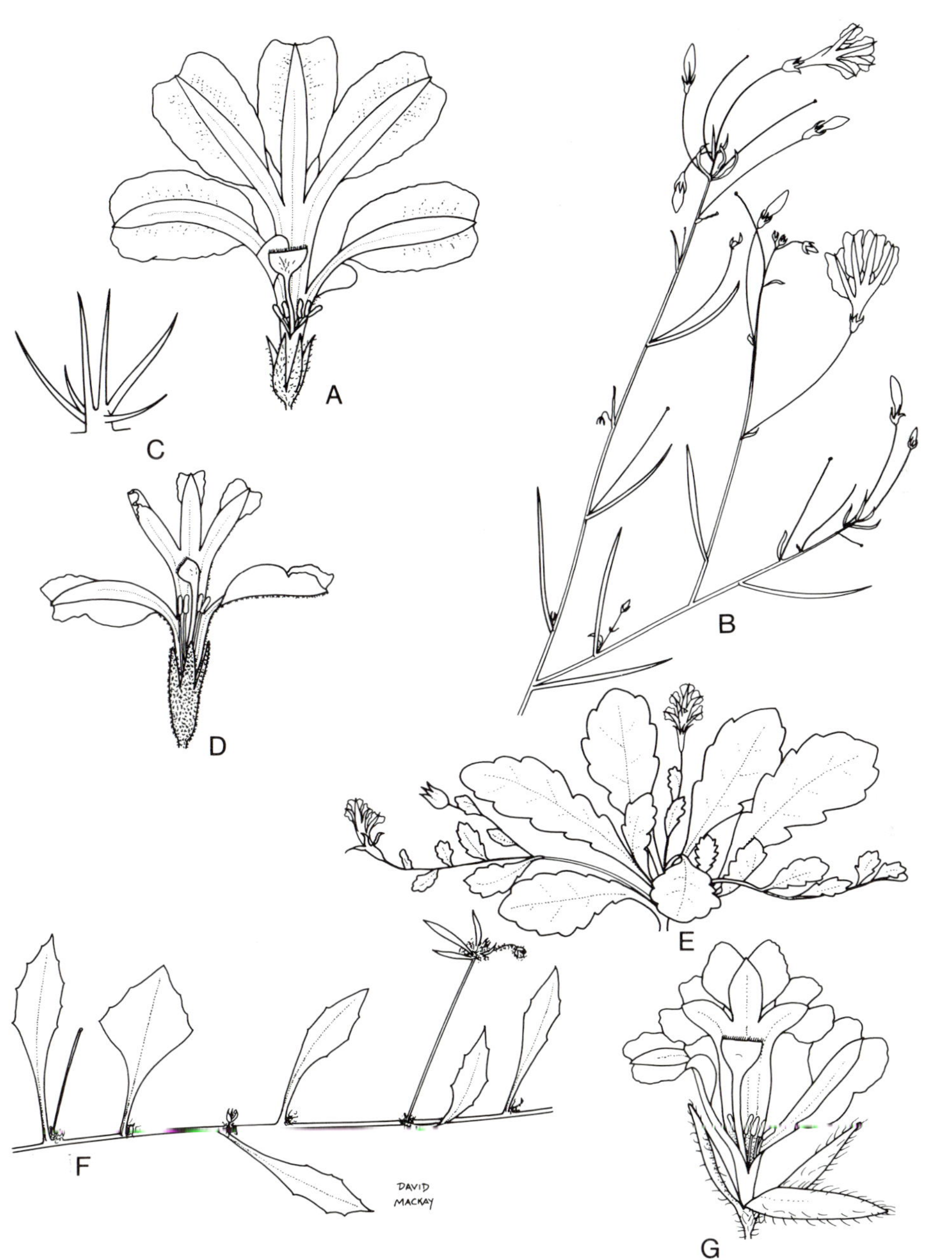

Figure 84. *Goodenia*. **A–B**, *G. tenuiloba*. **A**, flower X2; **B**, habit X0.5 (**A–B**, E.Errey 3051 & A.Beauglehole 59351, SYD). **C–E**, *G. stellata*. **C**, stellate hair X50; **D**, flower X2; **E**, habit X0.5 (**C–E**, R.Carolin 7770, SYD). **F**, *G. prostrata*, habit X1 (A.Ashby 4169, SYD). **G**, *G. sepalosa* var. *sepalosa*, flower X2 (K.Kenneally 8145, SYD). Drawn by D.Mackay.

W.A.: c. 25 km from Onslow on Ashburton Downs road, *R.C.Carolin 7843* (SYD); c. 9 km S of Exmouth, *A.S.George 6600* (PERTH, SYD); c. 28 km NE of Mt Newman Post Office, *A.C.Beauglehole 48965* (SYD); Mileura Stn, *W.J.Peacock 60861.2* (SYD); 56 km E of Meekatharra, *T.E.H.Aplin 2454* (PERTH, SYD).

Corolla yellow, with adaxial petals sometimes almost white. Buds constricted near base. The cytological voucher is *W.J.Peacock 60858.2* (SYD).

There are some recognisable variants in this species:

(i) Type form with deeply pinnatisect leaves, pubescent with simple and glandular hairs. Corolla possibly always bicoloured; wings making an acute angle at apex. Grows on rocky tablelands.

(ii) Leaves thick, deeply pinnate-lobed; glandular hairs very few, simple hairs long but not patent, scattered or absent. Corolla possibly always bicoloured, with a dense, short beard in the throat, which ends abruptly; wings making an acute angle at apex. Indusium densely pubescent below, (e.g. *Broadbent 1751*). Grows on saline flats.

(iii) Leaves thick, narrowly oblong to narrowly elliptic with short, narrow teeth, glabrous. Corolla not bicoloured, with short, simple hairs at 90° outside, with a beard of short hairs in the throat becoming less dense above but extending to the base of the lobes; wings making an obtuse angle at apex; abaxial lobes much shorter compared to connate part than in the 2 forms above. (e.g. *Broadbent 1779*).

(iv) Hairs almost all short and glandular. (Exmouth Gulf).

Differs from *G. microptera* and its allies which have narrower corolla wings. It also may be confused with *G. mimuloides* but in that species the glandular hairs are not easily visible and the leaves are much less dissected.

149. **Goodenia cusackiana** (F.Muell.) Carolin, *Telopea* 3: 556 (1990)

Velleia cusackiana F.Muell., *Victorian Naturalist* 12: 124 (1896). T: near the Fortescue R., W.A., *W.H.Cusack*; holo: MEL; iso: K.

Erect, scapose herb, woody at base, to 40 cm tall, densely silvery villous; stems leafy, to 15 cm long; hairs long, in axillary tufts. Leaves narrowly elliptic to lanceolate, entire; lamina 3–5 cm long, 3–5 mm wide. Flowers in racemes to 30 cm long; bracts linear to triangular, to 5 mm long; pedicel 1–4 mm long, articulate; bracteoles absent. Sepals lanceolate, 2–3 mm long. Corolla 8–10 mm long, pubescent in throat, without enations, auriculate; abaxial lobes 3–4 mm long; wings c. 1 mm wide. Indusium depressed-obovate, 1 mm long, thinly bearded below. Ovules 12–15. Fruit subglobular, 4–5 mm diam. Seeds orbicular, 3 mm diam., colliculate; wing 0.5 mm wide, thin, white.

Occurs in the Pilbara region, W.A., in rocky soil. Flowers July–Sept. Map 347.

W.A.: c. 40 km N of Cardabia homestead, *A.S.George 10216* (PERTH, SYD); c. 24 km W of Tambrey on Wittenoom–Roebourne road, *A.C.Beauglehole 11554* (SYD); Fig Tree Well, Yampire Gorge, *R.C.Carolin 7743* (SYD).

Corolla yellow, with simple and glandular hairs outside. A very distinctive species with a basal stem bearing leaves clothed with silvery hairs. Flowers are almost sessile, with very reduced bracts, and the auricles of the corolla are less differentiated than normally found in this section.

150. Goodenia vilmorinae F.Muell., *Fragm.* 3: 19 (1862)

T: between Bonny Creek and Mt Campbell, [N.T.], 1860–61, *J.Macdouall Stuart*; holo: MEL.

Ascending to erect, annual herb to 40 cm tall, villous and glandular-hairy. Leaves mostly basal, linear to lanceolate, entire to dentate, sometimes pinnate-lobed; lamina 4–10 cm long, 3–8 mm wide. Flowers in racemes to 10 cm long, or subumbels; bracts leaf-like; pedicel 10–35 mm long, articulate; bracteoles absent. Sepals lanceolate to elliptic, 4–5 mm long. Corolla 15–22 mm long, pubescent in throat, without enations, auriculate; abaxial lobes 6.5–8.5 mm long; wings c. 2 mm wide. Indusium depressed-obovate, 2–2.5 mm long. Ovules c. 30. Fruit subglobular, usually compressed, 7–9 mm diam. Seeds orbicular, 4–4.5 mm diam., colliculate, dark brown to nearly black; wing 1 mm wide, hyaline, thin. Fig. 65.

Occurs in arid regions of north-western and central Australia, in W.A., N.T., S.A. and Qld, in sandy soil. Flowers chiefly Apr.–Sept. Map 348.

W.A.: Mundabullangana Stn, *A.S.George 3461* (PERTH, SYD); c. 50 km from Roy Hill on Wittenoom road, *R.C.Carolin 7701* (SYD). N.T.: Winnecke goldfield, *T.S.Henshall 625* (DNA43768); Red Bank Gorge, *A.C.Beauglehole 45395* (SYD). Qld: c. 17 km S of Duchess, 6 July 1974, *P.Ollerenshaw & D.Kratzing* (CBG, SYD).

Corolla pale blue to lilac, with yellow in throat. The only other *Goodenia* lacking bracteoles and with a bluish corolla is *G. pallida*, which is not villous.

151. Goodenia hirsuta F.Muell., *Fragm.* 3: 35 (1862)

T: central Australia, N.T., *J.Macdouall Stuart*; holo: MEL.

G. mollissima F.Muell. ex Benth., *Fl. Austral.* 4: 73 (1868); ?*G. mitchellii* var. *mollissima* Domin, *Biblioth. Bot.* 89: 642 (1929). T: near Coopers Creek, Qld, *E.M.Bowman*; holo: K; iso: MEL.

Prostrate to decumbent herb, hirsute, with not appressed, simple hairs and short, glandular hairs; stems to 30 cm long. Basal leaves narrowly obovate, coarsely dentate; lamina 4–10 cm long, 15–35 mm wide. Flowers in racemes to 3 cm long, or subumbels; bracts leaf-like; pedicel 20–40 mm long, articulate; bracteoles usually absent. Sepals narrowly elliptic to lanceolate, 3.5–4.5 mm long. Corolla 14–15 mm long, hairy in throat, without enations, auriculate; abaxial lobes 4–5 mm long; wings 0.5–1 mm wide, sometimes narrower above auricle. Indusium broadly ovate, 1.5 mm long. Ovules 35–40. Fruit ovoid to ellipsoidal, 8 mm long. Seeds elliptic to orbicular, 3 mm long, glossy, black; wing scarcely delimited from body, c. 0.5 mm wide, paler than body, membranous. Fig. 45B.

Occurs in central arid regions of W.A., north-eastern N.T., S.A. and Qld, in a variety of soils. Flowers July–Oct. Map 349.

W.A.: NW end of Walter James Ra., *P.K.Latz 2363* (DNA). N.T.: Ingallana Ck, 29 July 1958, *G.Chippendale* (AD, CANB, DNA, MEL, NSW); c. 45 km from Yuendumu on Mt Doreen road, *R.C.Carolin 7950* (SYD). S.A.: Gosse Ra., 1886, *W.F.Schwarz* (MEL). Qld: Ayr, *W.A.Cowdry 103* (BRI).

Corolla yellow; adaxial corolla lobes with lower wings variable in width. Bracteoles are sometimes present and are linear, 1–5 mm long. In some specimens from central Australia the lower wing is very narrow but in others the 2 wings of these lobes are almost equal in width.

Previously *G. mollissima* and *G. hirsuta* have been placed in separate subgenera because of the reputed absence and presence of bracteoles respectively. However, the holotype of *G. mollissima* has linear bracteoles on at least some of the pedicels. In any case, the presence of bracteoles does not seem to be constant in the species. In some respects this species resembles *G. cycloptera* but that species has a prominent corolla spur, thick wing to the seed, and the indumentum is usually coarser. Similar also to *G. stellata* which, however, has stellate hairs.

152. **Goodenia stellata** Carolin, *Telopea* 2: 72 (1980)

T: 73 miles [c. 116 km] from Tom Price on Yampire Gorge road, W.A., 9 Aug. 1970, *R.C.Carolin 7770*; holo: NSW; iso: SYD.

Decumbent to prostrate herb, with stellate hairs; stems to 20 cm long. Basal leaves oblong-elliptic to ovate, crenate-dentate; lamina 4–8 cm long, 15–25 mm wide. Flowers in racemes to 5 cm long, or subumbels; bracts leaf-like; pedicel 10–20 mm long, articulate; bracteoles absent. Sepals lanceolate to narrowly oblong, c. 4 mm long. Corolla 12–15 mm long, hairy inside towards base, without enations, auriculate; abaxial lobes 5–6 mm long; wings 1 mm wide, narrower above auricle. Indusium broadly oblong, 1–1.5 mm long. Ovules 25–30. Fruit globular, 6–7 mm diam. Seeds orbicular, c. 4 mm diam., glossy, black; wing scarcely delimited from body, c. 0.5 mm wide, white or yellowish, membranous. Fig. 84C–E.

Occurs at scattered localities in the Pilbara and Gibson Desert, W.A., in red stony soil. Flowers July–Oct. Map 350.

W.A.: between Rawlinson and Alfred and Marie Ranges, 1876, *E.Giles* (MEL).

Corolla yellow; lower wing on adaxial corolla lobes slightly narrower than the rest. Distinguished from all other species by the combination of yellow corolla, ebracteolate pedicels and stellate hairs. The glossy, black seeds of this species and *G. hirsuta* are distinctive within the genus.

153. **Goodenia neogoodenia** (C.Gardner & A.S.George) Carolin, *Telopea* 3: 547 (1990)

Neogoodenia minutiflora C.Gardner & A.S.George, *J. Roy. Soc. W. Australia* 46: 138 (1963). T: 10 miles [c. 16 km] S of Mt Magnet, W.A., 20 Aug. 1960, *A.S.George 910*; holo: PERTH; iso: SYD.

Illustration: C.A.Gardner & A.S.George, *op. cit.* fig. 6.

Prostrate, annual herb, with scattered, short hairs, except on inflorescence; stems to 40 cm long. Leaves orbicular to rhomboidal, cordate, obtusely dentate; lamina to 8 mm diam. Flowers in racemes or spikes to 3 cm long; bracts linear to 3 mm long; pedicel to 2 mm long, not articulate; bracteoles absent. Sepals linear, c. 1 mm long. Corolla c. 1 mm long, glabrous inside, without enations, not auriculate; abaxial lobes less than 1 mm long; wings obsolete. Indusium subglobular, 0.5 mm diam. Ovule solitary. Fruit obovoid, compressed, 3–4 mm long, indehiscent. Seed elliptic, c. 3 mm wide, yellowish brown; wing with a narrow rim. Fig. 67.

Known only from the type locality and near Yalgoo, W.A. Grows in red loam in shallow depressions. Flowers c. August. Map 351.

W.A.: W of Yalgoo, *G.J.Keighery 5131* (PERTH).

The corolla is brownish. Distinguished from other members of the subsection by the leaves on slender petioles to 20 mm long, the minute flowers, solitary ovule and thin-walled indehiscent fruit.

Subsect. 3. Borealis

Goodenia subg. **Goodenia** sect. **Goodenia** subsect. **Borealis** Carolin, *Fl. Australia* 35: 331 (1992).

Type: *G. sepalosa* F.Muell. ex Benth.

Herbs. Leaves mostly cauline. Inflorescence a leafy raceme or head-like; pedicels ebracteolate, usually not articulate. Corolla yellow or mauve to pinkish, pubescent or glabrous inside, usually without enations inside, with 2 folds on connate part of abaxial lobes, auriculate; pouch not prominent. Fruit a capsule; valves 2, persistent, bifid, gaping. Seeds oblong to elliptic, with prominent rim; wing very narrow or obsolete, ±mucilaginous.

A subsection of 23 species found in northern Australia; 1 species extends to New Guinea, and 1 to the Philippines.

Ser. 1. Borealis

Goodenia subg. **Goodenia** sect. **Goodenia** subsect. **Borealis** ser. **Borealis** Carolin, *Fl. Australia* 35: 330 (1992).

Type: *G. sepalosa* F.Muell. ex Benth.

Style simple. Sepals to 2.5 mm wide.

The series contains 14 species in northern Australia, 1 extending to New Guinea.

154. Goodenia sepalosa F.Muell. ex Benth., *Fl. Austral.* 4: 72 (1868)

T: Camden Harbour, W.A., *Martin*; lecto: K, *fide* R.C.Carolin, *Telopea* 3: 556 (1990); isolecto: MEL.

Prostrate to ascending herb to 40 cm long, hirsute, sometimes with glandular hairs. Leaves mostly cauline, narrowly oblong to oblanceolate, often ±recurved, auriculate, dentate to entire; lamina 1–5 cm long, 3–8 mm wide. Inflorescence a raceme, often compact, rarely to 20 cm long; bracts leaf-like; pedicel 10–20 mm long, not articulate; bracteoles absent. Sepals narrowly elliptic to ovate, 7–12 mm long. Corolla 8–15 mm long, usually hairy inside down centre betwen 2 folds, without enations; abaxial lobes 3–5 mm long; wings 1–1.5 mm wide. Indusium very broadly obovate, 2 mm long. Ovules 8–12. Fruit globular, 3–4 mm diam. Seeds elliptic, 4 mm long, verrucose and granulose, yellowish brown.

Occurs from the north-west coast, W.A., to Arnhem Land, N.T.

Corolla yellow. Sepals wider (1.5–2 mm) than in other species of the section and at least 2/3rds as long as corolla. Similar to *G. byrnesii* which has much less dense hairs.

Two varieties are recognised.

Hairs mostly non-glandular **154a.** var. **sepalosa**

Hairs mostly glandular **154b.** var. **glandulosa**

154a. Goodenia sepalosa F.Muell. ex Benth. var. sepalosa

Hairs mostly non-glandular. Figs 26G, 45I, 84G.

Extends from the Kimberley, W.A., to Arnhem Land, N.T. Flowers most of the year. Map 352.

W.A.: Kununurra, *W.Leutert 30* (CANB); c. 6 km N of Broome–Derby road and John Price Point road junction, *R.C.Carolin 7527* (SYD); c. 51 km SW of Mary R. crossing, *D.E.Symon 5292* (AD); c. 48 km S of Port Hedland road junction with Broome–Derby road, 31 July 1970, *R.C.Carolin* (SYD). N.T.: Keep R., *A.S.Mitchell 339* (CANB, DNA, NSW, PERTH).

154b. **Goodenia sepalosa** var. **glandulosa** Carolin, *Telopea* 3: 556 (1990)

T: 9 miles [c. 15 km] S of Derby, 24 May 1967, W.A., *N.Byrnes*; holo: DNA.

Hairs mostly glandular.

Known only from the type collection, W.A. Flowers c. May. Map 353.

155. **Goodenia arachnoidea** Carolin, *Telopea* 3: 557 (1990)

T: 8 km SW of Theda Stn, Kimberley, W.A., 29 May 1971, *D.E.Symon 7101*; holo: AD; iso: PERTH, SYD.

Erect to ascending herb to 45 cm tall, with cobwebby and stiff, retrorse-arcuate hairs. Basal leaves obovate, dentate, with lamina 4–10 cm long, 1.5–3.5 cm wide; cauline leaves ±petiolate. Flowers in racemes to 30 cm long; bracts leaf-like; pedicel to 4 cm long, articulate; bracteoles absent. Sepals linear-deltoid, 2–5 mm long. Corolla 12–14 mm long, pubescent inside in a ring near base, without enations; abaxial lobes 4–5 mm long; wings c. 2 mm wide, narrower above auricle. Indusium broadly ovate, 1 mm long. Ovules 5–6. Fruit obovoid, 4–5 mm long. Seed elliptic, 4 mm long, obscurely verrucose, shining, brownish yellow. Fig. 85F.

Occurs in the north Kimberley, W.A., usually in open forest among sandstone outcrops. Flowers c. May. Map 354.

W.A.: Mitchell Plateau, NW Kimberley, *K.F.Kenneally 7136/A* (PERTH); King Edward R., c. 50 km NE of Mitchell R. homestead, *E.G.Errey & A.C.Beauglehole 2853/59153* (SYD); Vansittart Bay, *S.J.Forbes 2153* (MEL); Governor Is., NW Kimberley, *E.A.Chesterfield 254* (PERTH); Longini Landing, Kalumburu Mission, *D.E.Symon 7112* (AD).

Corolla yellow; lower wings of adaxial corolla lobes smaller. The dense cobwebby hairs distinguish this species from others of the series.

156. **Goodenia brachypoda** (F.Muell. ex Benth.) Carolin, *Telopea* 3: 558 (1990)

G. sepalosa var. *brachypoda* F.Muell. ex Benth., *Fl. Austral.* 4: 72 (1868). T: Victoria River, N.T, *F.Mueller*; holo: K; iso: ?MEL.

Decumbent to ascending herb, with soft, retrorse hairs at 90°; stems to 25 cm long. Leaves narrowly elliptic to oblong, auriculate, dentate; lamina 2–5 cm long, 4–13 mm wide. Flowers in racemes to 15 cm long; bracts leaf-like; pedicel 2–7 mm long, not articulate; bracteoles absent. Sepals linear 3.5–4 mm long, scarcely as long as corolla tube. Corolla 9–15 mm long, hairy in throat, without enations; abaxial lobes 3.5–4 mm long; wings c. 0.5 mm wide. Indusium depressed-obovate, c. 1 mm long, folded. Ovules c. 10. Fruit globular, c. 3.5 mm diam. Seeds elliptic, 2.5 mm long, verrucose, granulose, brownish to greyish.

Occurs in the north-eastern Kimberley, W.A., and occasionally in Arnhem Land, N.T. Flowers c. July–Sept. Map 355.

W.A.: W end of Cambridge Gulf, 1887, *A.J.Keiller* (MEL); near the Ord R., 1886, *H.J.O'Donnell* (MEL); Carlton Hill, Wyndham, *R.D.Royce 3318* (PERTH).

Corolla yellow. Similar to *G. sepalosa* in having broad sepals, but that species has coarser, longer hairs, with the wings of the corolla somewhat longer and wider, longer sepals and much denser pubescence in the corolla throat.

157. Goodenia leiosperma Carolin, *Telopea* 3: 558 (1990)

T: 39 miles [c. 62 km] S of Darwin, N.T., 18 Mar. 1961, *G.Chippendale*; holo: DNA.

Ascending to decumbent herb, with coarse hairs at ±90°; stems to 60 cm long. Basal leaves ephemeral; cauline leaves ovate to lanceolate, mostly auriculate, dentate, with lamina 5–10 cm long, 10–20 mm wide. Flowers in racemes to 40 cm long; bracts leaf-like; pedicel 4–10 cm long, not articulate; bracteoles absent. Sepals linear to linear-lanceolate, 5–6.5 mm long. Corolla 15–20 mm long, pubescent inside especially in central depression, wrinkled or with enations; abaxial lobes 6–7.5 mm long; wings c. 2 mm wide. Indusium depressed-obovate, 2 mm long. Ovules 10–15. Fruit globular, c. 8 mm diam. Seeds elliptic, 2.5–3.5 mm long, smooth, shining.

Occurs in the Victoria R. district, and Arnhem Land, N.T., in dry sclerophyll forest. Flowers chiefly Feb.–June. Map 356.

N.T.: c. 20 km S of Darwin, 25 May 1958, *G.Chippendale* (DNA); Fogg Dam area, 18 May 1959, *G.Chippendale* (DNA); c. 4 km W of Burrundie, 16 Mar. 1961, *G.Chippendale* (AD, DNA); McMinns Lagoon, 1896, *M.Holtze* (MEL); c. 8 km SW of Grove Hill, 17 Mar. 1961, *G.Chippendale* (DNA).

Corolla yellow. One specimen referred to this species (*N.Byrnes ND666* (DNA14382) from Litchfield Stn), has almost glabrous stems, glabrescent leaves and lacks the dense pubescence inside the corolla. Has been confused in the past with *G. hispida*, which has verrucose seeds, sparse pubescence on the adaxial corolla lobes where it is connate with the abaxial lobes, longer ovary and ovoid fruit.

158. Goodenia durackiana Carolin, *Telopea* 3: 559 (1990)

T: Kimberley Research Stn [near Kununurra], W.A., 6 Mar. 1963, *M.Lazarides 6743*; holo: PERTH; iso: CANB.

Erect to decumbent herb to 50 cm tall, with scattered hairs. Leaves mostly cauline, elliptic to oblong, ±auriculate, coarsely dentate, often glabrescent; lamina 3–6 cm long, 1–2.5 cm wide. Flowers in racemes to 40 cm long; bracts leaf-like; pedicel 2–5 cm long, not articulate; bracteoles absent. Sepals narrowly elliptic, c. 7 mm long. Corolla c. 15 mm long, pubescent in throat, without enations; abaxial lobes c. 5 mm long; wings 1–1.5 mm wide. Indusium depressed-obovate, 1 mm long. Ovules 20–30. Fruit globular, 8–10 mm diam. Seeds orbicular, 3–3.5 mm wide, smooth, dull, brown-yellow.

Occurs in the north-eastern Kimberley, W.A., in cracking clay soil, in grassland. Flowers c. Mar.–May. Map 357.

W.A.: Ord R., Apr.–May 1945, *K.M.Durack* (PERTH).

Corolla yellow. Pedicels divergent in fruit, with hairs mostly on the adaxial side. Differs from other species in the section in having smooth, dull seeds and the hairs on the pedicels almost restricted to the adaxial side.

159. Goodenia byrnesii Carolin, *Telopea* 3: 560 (1990)

T: 15 miles [c. 24 km] SW of Elliott, N.T., 12 Mar. 1969, *N.Byrnes 1433*; holo: NSW; iso: SYD.

Decumbent to prostrate herb, sparsely hirsute; stems to 30 cm long. Basal leaves ephemeral; cauline leaves ovate-oblong to narrowly oblong, ±auriculate, dentate, with lamina mostly 4–6 cm long, 5–10 mm wide. Flowers in racemes to 25 cm long; bracts leaf-like; pedicel 3–5 cm long, indistinctly articulate; bracteoles absent. Sepals narrowly elliptic, 5–6 mm long. Corolla 17–20 mm long, pubescent inside towards base, without enations; abaxial lobes 6–7 mm long; wings c. 1.5 mm wide. Indusium depressed-obovate,

2 mm long. Ovules c. 10. Fruit subglobular, 4–5 mm diam. Seeds elliptic, 4 mm long, verrucose and granulose, greyish yellow.

Occurs in northern Australia, in W.A., N.T. and north-western Qld, on black soil plains in grassland. Flowers Jan.–June. Map 358.

W.A.: Bloodwood Ck, Mt Anderson, *H.F.Broadbent 624* (PERTH). N.T.: Leila Lagoon, *R.C.Carolin 9294* (SYD); c. 67 km W of Wave Hill Police Stn, *R.A.Perry 2278* (DNA). Qld: Saxby R. crossing on Wondola road, *R.C.Carolin 8807* (SYD); c. 36 km from Nardoo on Burketown road, *R.C.Carolin 8908* (SYD).

Corolla yellow. The plant is unusually bright green. Similar to *G. sepalosa*, differing in less dense hairs, plant lighter green in colour, and the narrower leaves and sepals. It also shows similarities to *G. leiosperma* which has smooth seeds, larger fruit (c. 8 mm diam.), and broader leaves (10–20 mm wide); and to *G. hispida* which is more densely hairy and is darker green.

160. Goodenia campestris Carolin, *Telopea* 3: 561 (1990)

T: c. 35 miles [c. 56 km] S of Timber Creek, N.T., 10 May 1968, *R.C.Carolin 6667*; holo: NSW; iso: SYD.

Decumbent to ascending herb, with few, coarse hairs; stems to 50 cm long. Basal leaves ephemeral; cauline leaves ovate to lanceolate, ±auriculate, dentate, with lamina 1–7 cm long, 5–10 mm wide. Flowers in racemes to 30 cm long; bracts leaf-like; pedicel 1.5–7 cm long, not articulate; bracteoles absent. Sepals lanceolate to narrowly oblong, 1–2 mm long. Corolla 7–8 mm long, pubescent and wrinkled inside, without enations; abaxial lobes c. 2 mm long; wings 0.5–1 mm wide. Indusium broadly oblong, 1 mm long. Ovules 6–8. Fruit subglobular, 5 mm long. Seeds elliptic, 3.5 mm long, verrucose, granulose, yellowish.

Occurs in the Victoria R. district, N.T., on black soil plains in grassland. Flowers c. May. Map 359.

N.T.: Skull Ck, *R.C.Carolin 6659* (SYD); c. 8 km W of Victoria R. crossing, *N.Byrnes NB740* (DNA, NSW).

Corolla yellow; the abaxial corolla lobes shorter than connate part; ovary septum very short with few seeds. Close to *G. hispida*, which has broader leaves, longer sepals and larger flowers and fruit.

161. Goodenia malvina Carolin, *Telopea* 3: 562 (1990)

T: Kununurra, W.A., 7 Mar. 1963, *M.Lazarides 6780*; holo: CANB.

Decumbent to prostrate herb, sparsely hairy except stems; stems glabrous, to 50 cm long. Leaves cauline, sometimes decurrent, lanceolate to ovate, auriculate, dentate to narrowly pinnate-lobed; lamina 2.5–7 cm long, 5–13 mm wide. Flowers in racemes to 40 cm long; bracts leaf-like; pedicel 2–5 cm long, ±distinctly articulate; bracteoles absent. Sepals lanceolate to narrowly oblong, 2–4 mm long. Corolla 9–14 mm long, pubescent inside towards base, without enations; abaxial lobes 3–4 mm long; wings c. 1.5 mm wide, laciniate. Indusium broadly oblong, 1 mm long. Ovules 8–10. Fruit compressed-ovoid, 5–6 mm long. Seeds elliptic, c. 3.5 mm long, verrucose, granulose, yellowish.

Occurs in the north-eastern Kimberley, W.A., and Arnhem Land, N.T., in cracking clay soil. Flowers Mar.–May. Map 360.

W.A.: Ord R., Apr.–May 1945, *K.M.Durack* (PERTH). N.T.: c. 40 km E of Newry, 10 May 1959, *G.Chippendale* (DNA); c. 27 km NE of Beetaloo homestead, 10 Mar. 1959, *G.Chippendale* (DNA).

Corolla mauvish and yellowish towards top or pinkish, and has coarse, simple hairs and small, glandular hairs outside. Sepals are mostly 0.6–0.8 mm wide. The specimens from Arnhem Land show some very distinct differences from those collected further west, the corolla in particular is much smaller, the sepals shorter and the leaves are not decurrent. Similar to *G. hispida* which, however, has a yellow corolla, hispid stems, and hispid broader leaves (10–20 mm wide).

162. **Goodenia potamica** Carolin, *Telopea* 3: 564 (1990)

T: 24 miles [c. 38 km] W of Liverpool R. crossing, N.T., 27 June 1972, *J.Must 1069*; holo: DNA.

Prostrate to ascending herb, with arcuate hairs; stems to 20 cm long. Basal leaves ephemeral; cauline leaves elliptic to oblong, sometimes with a large, basal lobe on one side only, becoming very small towards the top, with lamina mostly 1–3 cm long, 5–8 mm wide. Flowers in racemes to 15 cm long; bracts leaf-like; pedicel 10–15 mm long, not articulate; bracteoles absent. Sepals oblong 1–1.5 mm long. Corolla 11–15 mm long, pubescent in throat, without enations; abaxial lobes 2.5–3 mm long; wings 1–1.5 mm wide. Indusium semi-orbicular, 4 mm diam. Ovules 8–10. Fruit ellipsoidal to globular, 3–5 mm long. Seeds elliptic-obovate, 2.5 mm long, verrucose, brown. Fig. 85A–B.

Occurs in the Liverpool R. area, Arnhem Land, N.T., in woodland. Flowers c. June. Map 361.

N.T.: Liverpool R., 1975, *F.Duncan* (SYD); 45 km upstream from mouth of the Liverpool R., *F.Duncan 803* (SYD).

Corolla yellow, with dark-headed, glandular hairs on calyx and pedicel which distinguish it from other members of the section. Of all species in this series, *G. potamica* approaches most closely to those of ser. *Calogyne*. In particular, its resemblance to *G. holtzeana* is striking but the trifid indusium of the latter and the broad, undivided, indusium of *G. potamica* distinguish them.

163. **Goodenia hispida** R.Br., *Prodr.* 577 (1810)

T: Carpentaria, Survey Point S [Point Blane, N.T.], 28 Jan. 1803, *R.Brown*; lecto: BM, *fide* R.C.Carolin, *Telopea* 3: 563 (1990); isolecto: K.

G. auriculata Benth., *Fl. Austral.* 4: 72 (1868). T: Depot Creek, upper Victoria R., [N.T.], 1855–56, *F.Mueller*; holo: K; iso: MEL.

Ascending herb to 60 cm tall, hispid or almost glabrous. Cauline leaves ovate to lanceolate, usually auriculate, dentate or entire; lamina 5–8 cm long, 1–2 cm wide. Flowers in racemes to 40 cm long; bracts leaf-like; pedicel 5–10 cm long, not articulate; bracteoles usually absent. Sepals lanceolate 4–7 mm long. Corolla 15–20 mm long, ±pubescent inside between ridges, without enations; abaxial lobes 7–8 mm long; wings c. 2 mm wide, often dentate. Indusium depressed-obovate, c. 2 mm long. Ovules 25–30. Fruit ovoid, 10–12 mm long. Seeds elliptic, 3.5–4 mm long, verrucose, granulose, pale brown.

Occurs in Arnhem Land and adjacent areas, N.T., in dry sclerophyll forest. Flowers chiefly Feb.–May. Map 362.

N.T.: Port Darwin, *M.Holtze 701* (MEL); c. 13 km NE of Mainoru, *R.C.Carolin 9409* (SYD); c. 5 km S of Mountain Valley homestead, *R.Swinbourne 630* (BRI, DNA); c. 8 km from Katherine on Wyndham road, *H.S.McKee 8531* (CANB, DNA, NSW); Timber Creek, *R.C.Carolin 6713* (SYD).

Corolla yellow, with simple and small, glandular hairs outside. Occasionally 2 linear bracteoles c. 2 mm long occur near the pedicel base. The species varies considerably but there appears to be almost complete intergradation between the very hispid forms in Arnhem Land (as in the type), and the less hispid forms with somewhat broader leaves from the Victoria R. district. Similar to *G. leiosperma* which has smooth seeds, a shorter ovary and globular fruit.

Figure 85 *Goodenia*. **A–B**, *G. potamica*. **A**, flower X2; **B**, habit X0.5 (**A–B**, J.Must 1069, SYD). **C**, *G. chthonocephalata*, habit X1 (P.Latz 7176, NT). **D–E**, *G. armstrongiana*. **D**, habit X0.5; **E**, flower X2 (**D–E**, D.Symon 7893, SYD). **F**, *G. arachnoidea*, habit X0.5 (K.Kenneally 7137, SYD). Drawn by D.Mackay.

164. Goodenia subauriculata C.White, *Proc. Roy. Soc. Queensland* 57: 33 (1946)

T: Pine Creek, N.T., *R.Tate*; holo: MEL.

Ascending to decumbent herb, hirsute; stems to 10 cm long. Leaves dentate, linear; lamina c. 8 cm long and 5 mm wide; upper leaves progressively broader and subauriculate. Inflorescence a spike, to 8 cm long; bracts leaf-like. Sepals lanceolate, c. 1.5 mm long. Corolla c. 4.5 mm long, with few hairs inside, without enations; abaxial lobes c. 1 mm long; wings c. 0.5 mm wide. Indusium broadly obovate, c. 0.5 mm long. Ovules few. Fruit ellipsoidal, 4.5 mm long, tardily separating. Seeds elliptic, 2.5–3 mm long, verrucose, granulose, pale yellow-brown.

Occurs in Arnhem Land and adjacent areas, N.T., and on Cape York Peninsula, Qld. Flowers c. Apr. Map 363.

Qld: Lockerbie, *L.J.Brass 18450* (CANB); Iron Ra., 13 Apr. 1944, *H.Flecker* (BRI).

Corolla brownish yellow. Sepals attached to top of ovary. Differs from most other members of this section in the almost sessile flowers, the sepals which are adnate to the ovary almost to the summit and thus attached to the very top of the fruit, and the small brownish yellow corolla. Closest to *G. chthonocephalata*.

165. Goodenia chthonocephalata Carolin, *Telopea* 3: 563 (1990)

T: Cox River Stn, N.T., 2 July 1977, *P.K.Latz 7176*; holo: DNA.

Cushion-like, annual herb, small, glabrescent. Leaves all basal, linear-lanceolate, entire; lamina to 8 cm long and 3 mm wide. Inflorescence with flowers sessile in leaf axils and thus in a head at ground level. Sepals linear, c. 0.5 mm long. Corolla 1–1.5 mm long, almost glabrous inside, without enations; lobes ±equal, to 1 mm long; wings obsolete. Indusium broadly elliptic, c. 0.2 mm wide. Ovules 30–40. Fruit ellipsoidal, 3 mm long. Seeds orbicular, c. 0.2 mm diam., verrucose, reddish brown with a paler border. Fig. 85C.

Known only from the type collection, N.T., in a damp situation. Flowers July. Map 364.

Corolla reddish when dry, with simple and multicellular hairs outside. Sepals attached to top of ovary. Differs from *G. subauriculata* which has larger, yellowish flowers, not carried in heads at ground level. Probably closest to *G. subauriculata*, differing in flower colour, and in flower heads carried at ground level.

166. Goodenia armstrongiana Vriese, *Natuurk. Verh. Holl. Maatsch. Wetensch. Haarlem* ser. 2, 10: 138 (1854)

T: Raffles Bay, [N.T.], Aug. 1846, *A.Armstrong*; holo: K.

Erect to decumbent herb, with coarse, antrorse hairs; stems to 60 cm long. Cauline leaves ±decurrent, ovate to narrowly elliptic, dentate to entire, with few hairs mostly on margins and on main vein below; lamina 1–4 cm long, 2–12 mm wide. Flowers in racemes to 40 cm long; bracts leaf-like; pedicel divergent, 10–30 mm long, not articulate; bracteoles absent. Sepals lanceolate, 1–2 mm long. Corolla 8–12 mm long, pubescent inside, without enations; abaxial lobes 3–4 mm long; wings 1.5–2 mm wide. Indusium broadly obovate, 1.5 mm long, with long hairs at base beneath. Ovules 10–20. Fruit ovoid, 4–6 mm long. Seeds elliptic, c. 3 mm long, without wing, verrucose, granulose, pale yellow-brown. Fig. 85D–E.

Occurs in Arnhem Land and the Victoria R. district, N.T., in northern Qld, and in New

Guinea, in dry sclerophyll woodland and savannah. Flowers chiefly Jan.–July. Map 365.

N.T.: c. 5 km S of Raffles Bay, 18 July 1961, *G.Chippendale* (CANB, DNA); c. 1.5 km N of Adelaide R., *R.C.Carolin 6417* (SYD); Fogg Dam area, 18 May 1959, *G.Chippendale* (DNA); Hemple Bay, Groote Eylandt, *R.L.Specht 313* (CANB, NSW). Qld: 6.5 km NNW of Beagle North Camp, c. 43 km NNE of Aurukun, *J.R.Clarkson 4355* (BRI, DNA, MO, NSW, PERTH, QRS).

Corolla usually yellow with a brownish throat, but sometimes white or very rarely 'pinkish'. Distinguished from other members of the section by its narrower leaves, more appressed hairs, attenuate base to the ovary and the tuft of long bristles at the base of the indusium.

167. **Goodenia argillacea** Carolin, *Telopea* 3: 563 (1990)

T: 8 miles [c. 13 km] NE of Mainoru R. crossing on Bulman road, N.T., 23 May 1974, *R.C.Carolin 9403*; holo: NSW; iso: SYD.

Herb with erect, main stem and weak, decumbent branches, glabrous; branches to 70 cm long. Cauline leaves oblong to linear, margins slightly recurved, sometimes slightly auriculate, obscurely dentate; lamina 2–4.5 cm long, 2.5–6 mm wide. Flowers in racemes to 60 cm long; bracts leaf-like; pedicel 2.5–4 cm long, slender, not articulate; bracteoles absent. Sepals lanceolate, 2.5 mm long. Corolla 10–12 mm long, with few hairs inside, without enations; abaxial lobes c. 2.5 mm long; wings c. 1 mm wide. Indusium depressed-obovate, 1 mm long. Ovules 6–8. Fruit obovoid, 3–3.5 mm long. Seeds elliptic, 3 mm long, minutely colliculate or smooth and dull.

Occurs in southern Arnhem Land, N.T., in *Melaleuca* scrub on calcareous clay. Flowers c. May. Map 366.

N.T.: c. 10 km beyond Mainoru R. on road to Bullman Stn, *R.Pullen 9371* (CANB, NSW).

Corolla brownish yellow, ±pubescent outside.

Ser. 2. Calogyne

Goodenia subg. **Goodenia** sect. **Goodenia** subsect. **Borealis** ser. **Calogyne** (R.Br.) Carolin, *Fl. Australia* 35: 330 (1992)

Calogyne R.Br., *Prodr.* 579 (1810). T: *G. pilosa* (R.Br.) Carolin

Style 3- or 4-fid. Sepals mostly to 0.4 mm wide.

The series contains 9 species; 8 endemic in northern Australia; 1 extends northwards to China and the Philippines.

168. **Goodenia porphyrea** (Carolin) Carolin, *Telopea* 3: 565 (1990)

Calogyne porphyrea Carolin, *Brunonia* 2: 4 (1979). T: Koolpinyah Stn, N.T., 14 Apr. 1966, *S.E.Pickering 98*; holo: DNA.

Decumbent to prostrate herb, with scattered, stiff, white hairs; stems to 50 cm long. Basal leaves obovate; cauline leaves narrowly oblong to lanceolate, dentate, with 2 large, basal teeth, with lamina 2.5–11 cm long, 2–5 mm wide. Flowers in racemes to 40 cm long; bracts leaf-like; pedicel to 1 cm long, not articulate; bracteoles absent. Sepals narrowly oblong to lanceolate, 4–5 mm long. Corolla 7–8 mm long, with scattered hairs inside, without enations; abaxial lobes 2–3 mm long; wings c. 1 mm wide. Style 3-fid. Median indusium truncate-obovate, 1.5 mm long. Ovules 4–8. Fruit globular, 5–6 mm diam.;

valves gaping. Seeds elliptic, 3–4 mm long, smooth, matt, yellowish brown.

Occurs in western Arnhem Land, N.T., in black soil. Flowers c. Mar.–Apr. Map 367.

N.T.: Humpty Doo, *C.S.Robinson R54* (DNA).

Corolla purplish, with stiff, white hairs outside; wings on abaxial corolla lobes scarcely 1/2 as long as lobes; ovary septum very short. Distinguished from other species in the series with purplish flowers by the narrower leaves which have stiff, white hairs.

169. **Goodenia pilosa** (R.Br.) Carolin, *Telopea* 3: 565 (1990)

Calogyne pilosa R.Br., *Prodr.* 579 (1810); *Goodenia dubia* Sprengel, *Syst. Veg.* 1: 271 (1824), *nom. illeg.* T: Carpentaria, island no. 12, [North Is., Sir Edward Pellew Group, N.T.], 16 Dec. 1802, *R.Brown*; lecto: BM, *fide* R.C.Carolin, *Brunonia* 2: 5 (1979) (as holo); isolecto: K, MEL.

Calogyne chinensis Benth., *J. Linn. Soc., Bot.* 5: 78 (1861). T: near Amoy [Xiamen], China, *Hance 1422*; holo: K.

Balingayum decumbens Blanco, *Fl. Filip.* 187 (1837). T: not designated.

Illustration: K.Krause, *Pflanzenr.* 54: 96, fig. 17A–B (1912) as *Calogyne pilosa.*

Prostrate to decumbent herb, with hairs not appressed or almost absent; stems to 50 cm long. Basal leaves narrowly oblong to elliptic, dentate, with lamina 4–8 cm long; cauline leaves smaller, ±auriculate. Flowers in racemes to 35 cm long; bracts leaf-like; pedicel 2–2.5 mm long, not articulate; bracteoles absent. Sepals lanceolate to narrowly oblong, 3–5 mm long. Corolla 8–15 mm long, hairy in throat, without enations; abaxial lobes 3–5 mm long; wings 1.5–2 mm wide, 2/3 as long as lobes. Style 3-fid. Median indusium ovate, 1.5 mm long. Ovules 10–15. Fruit subglobular, 2–3.5 mm diam.; valves gaping. Seeds pale brown, elliptic, 4 mm long, smooth or verrucose. Fig. 86G.

Occurs from Arnhem Land, N.T., to northern Qld, in sandy soil usually in damp situations. Also in Indonesia, China and the Philippines. Flowers May–Aug. Map 368.

N.T.: Yirrkala, *R.L.Specht 913* (AD, BRI); 16 km SW of Cape Arnhem, *D.E.Symon 7785* (AD); Goodparla Stn, *R.C.Carolin 6793* (AD, SYD). Qld: Camp Oven Waterhole, Carpentaria, *R.C.Carolin 9168* (SYD); Proserpine, *N.Michael 835* (BRI).

Corolla yellow, purplish in throat, hairy outside; ovary septum very short. Three forms are recognised by R.C.Carolin (*loc. cit.*) but given no formal status:

i. sprawling plant; leaves very hirsute; pedicels well developed; corolla hirsute outside; seeds glossy, smooth. Arnhem Land.

ii. compact plant; leaves with few hairs to almost glabrous; pedicels to 4 mm long; corolla with a few hairs outside; seeds verrucose. Arnhem Land.

iii. sprawling plant; leaves usually with a few hairs; pedicels usually more than 4 mm long; corolla hirsute outside; seeds verrucose, often almost smooth in the centre. Northern Qld.

170. **Goodenia neglecta** (Carolin) Carolin, *Telopea* 3: 565 (1990)

Calogyne neglecta Carolin, *Brunonia* 2: 7 (1979). T: Mudginberri Stn, N.T., 16 May 1968, *R.C.Carolin 6895*; holo: SYD.

Ascending herb to 30 cm tall, with glandular hairs and arcuate, simple hairs. Basal leaves obovate, dentate, glabrescent, with lamina 15–30 mm long, 6–12 mm wide; cauline leaves oblong to oblanceolate, smaller, often auriculate. Flowers in racemes to 25 cm long; bracts leaf-like; pedicel 2–3 mm long, slender, with mostly glandular hairs, not articulate; bracteoles absent. Sepals oblong to elliptic, 3 mm long. Corolla 10–15 mm long, hairy toward base inside, without enations; abaxial lobes 4–5 mm long; wings c. 1 mm wide. Style 3-fid. Median indusium ovate, 1 mm long, lateral ones broader. Ovules c. 6. Fruit globular, 3 mm diam.; valves gaping. Seeds elliptic, 3 mm long, colliculate, yellowish grey.

Occurs in Arnhem Land, N.T., in damp situations. Flowers Feb.–Apr. Map 369.

N.T.: Shoal Bay road, Darwin, *H.S.McKee 8396* (BRI); 12°31'S, 133°17'E, *M.Lazarides 7776* (CANB); c. 8 km NE of Jim Jim crossing, *R.C.Carolin 6908* (SYD).

Corolla yellow, with antrorse, ±appressed, mostly simple hairs outside, and wings 2/3 to 3/4 as long as corolla lobes. Ovary septum very short. The combination of yellow corolla and smooth seeds distinguish this species from all others in the series except some forms of G. *pilosa*, but these have narrower leaves.

171. **Goodenia heteroptera** (F.Muell.) B.D.Jackson, *Index Kewensis* 1: 1056 (1895)

Calogyne heteroptera F.Muell., *Fragm.* 10: 43 (1876). T. Newcastle Ra., Qld, *W.E.de M.Armit 377*; holo: MEL.

Herb to 5 cm tall, with simple and glandular hairs. Leaves elliptic to obovate, tapering towards base, dentate, glabrescent except margins and midrib; lamina 1–2 cm long, 3–7 mm wide. Flowers in racemes to 2 cm long; bracts leaf-like; pedicel to 5 mm long, not articulate; bracteoles absent. Sepals elliptic-oblong, 2.5–3 mm long. Corolla c. 6 mm long, hairy toward base inside, without enations; lobes almost equal; wings c. 0.5 mm wide. Style 3-fid. Median indusium ovate, 0.5 mm long, lateral ones similar. Ovules not seen. Fruit subglobular, 3 mm diam.; valves gaping. Seeds elliptic-oblong, 2.5 mm long, verrucose-granular, pale brown.

Known only from the type collection, Qld. Flowering not known. Map 370.

Corolla yellow, reddish brown in throat, with glandular and a few appressed hairs outside; wing of adaxial corolla lobe separated from auricle by a sinus. Ovary septum not seen. The small flowers and the sinus between the auricle and wing on the adaxial corolla lobes distinguish this species from others in the series.

172. **Goodenia holtzeana** (Specht) Carolin, *Telopea* 3: 565 (1990)

Calogyne holtzeana Specht, *Rec. Amer.-Austral. Sci. Exped. Arnhem Land* 3: 309 (1958). T: Port Darwin, N.T., 1884, *M.Holtze 431*; holo: MEL.

Erect to ascending herb, ±viscid in upper parts; stems to 60 cm long. Basal leaves often persistent, obovate to oblanceolate, dentate; with lamina 4.5–10 cm long, 15–30 mm wide; cauline leaves sessile, ovate to elliptic, smaller, auriculate, with hairs mostly simple, glabrescent except margins and midrib. Flowers in racemes to 40 cm long; bracts leaf-like; pedicel 10–25 mm long, glandular-hairy, not articulate; bracteoles absent. Sepals lanceolate, 4–6 mm long. Corolla 14–20 mm long, hairy inside, without enations; abaxial lobes 3–4 mm long; wings 1.5–2 mm wide. Style 3-fid. Median indusium ovate, c. 1 mm long, lateral ones narrower and overtopping central one. Ovules 15–20. Fruit globular to ovoid, 4–5 mm diam.; valves gaping. Seeds elliptic, 2 mm long, verrucose, brown. Figs 25A, 86E–F.

Extends from the Kimberley, W.A., to Arnhem Land, N.T., in eucalypt forest, usually in drier situations than *G. pilosa*. Flowers Feb.–June. Map 371.

Figure 86. *Goodenia*. **A–B**, *G. integerrima*. **A**, corolla X3: **B**, habit X0.5 (**A–B**, A.George 7291, PERTH). **C**, *G. kakadu*, habit X1 (P.Fryxell 4133 & L.Craven, SYD). **D**, *G. pumilio*, habit X1 (R.Carolin 9048, SYD). **E–F**, *G. holtzeana*. **E**, habit X1 (R.Carolin 6962, SYD) **F**, flower X2 (R.Carolin 9402, SYD). **G**. *G. pilosa*, habit X1 (A.Beauglehole 55012, SYD). **H**, *Selliera radicans*, habit X0.5 (fresh material, Sydney, Mar. 1987). Drawn by D.Mackay.

W.A.: Mt Barnett homestead, c. 225 km ENE of Derby, *E.A.Shaw 782* (AD). N.T.: 102 km W of Giddy R. crossing, *D.E.Symon 7746 p.p.* (AD, SYD); c. 2.5 km SW of Cannon Hill, *P.Martensz AE687* (CANB, DNA); Thoraks Reserve, c. 19 km E of Darwin, *D.J.Nelson 1090* (DNA); c. 13 km NE of Mainoru R. crossing, *R.C.Carolin 9402* (SYD).

Corolla yellow with brownish markings, with mostly glandular hairs outside; wings of abaxial corolla lobes c. 1/2 as long as lobes. Ovary septum c. 1/3 length of ovary. The glandular hairs are dark-headed and, together with the yellow corolla, distinguish this species from others in the series.

173. Goodenia heppleana (W.Fitzg.) Carolin, *Telopea* 3: 565 (1990)

Calogyne heppleana W.Fitzg., *J. & Proc. Roy. Soc. W. Australia* 3: 214 (1918). T: Isdell R., near Graces Knob, W.A., *W.V.Fitzgerald*; lecto: NSW, *fide* R.C.Carolin, *Brunonia* 2: 11 (1979); isolecto: K.

Erect to prostrate herb, with soft, arcuate, simple hairs and glandular hairs; stems to 50 cm long. Basal leaves oblanceolate, dentate, with lamina c. 8 cm long, 10–15 mm wide; cauline leaves smaller, almost sessile, sometimes cordate, not auriculate. Flowers in racemes to 40 cm long; bracts leaf-like; pedicel 18–25 mm long, mostly glandular-hairy, not articulate; bracteoles absent. Sepals lanceolate to narrowly elliptic, 3.5 mm long. Corolla 10–12 mm long, hairy in throat, without enations; abaxial lobes 4–5 mm long; wings 0.5–1 mm wide. Style 3-fid. Median indusium depressed-ovate, 0.5 mm long, laterals asymmetrically oblanceolate. Ovules not seen. Fruit globular, 4 mm diam.; valves not separating to base, not gaping. Seeds elliptic, 3.5 mm long, verrucose, yellowish brown.

Occurs in the north-eastern Kimberley, W.A., and Arnhem Land, N.T., in open forest and woodland. Flowers Feb.–June. Map 372.

W.A.: Thomsons Spring, Argyle, Ord R., *C.A.Gardner 7429* (PERTH). N.T.: c. 16 km E of Frances Ck, *R.C.Carolin 6740* (SYD); c. 36 km NE of Pine Creek, *N.Byrnes NB1506* (DNA); c. 25 km E of Pine Creek, *D.J.Nelson 285* (DNA, MEL, BRI).

Corolla yellow, with fine, simple hairs and glandular hairs outside; wings of abaxial corolla lobes almost as long as lobes. Ovary septum very short. Similar to *G. holtzeana* which has darker glandular hairs and stem-clasping, cauline leaves.

174. Goodenia symonii (Carolin) Carolin, *Telopea* 3: 565 (1990)

Calogyne symonii Carolin, *Brunonia* 2: 12 (1979). T: 13 km W of Giddy R. crossing, N.T., 18 June 1972, *D.E.Symon 7750*; holo: AD.

Ascending herb, without well-defined rosette, with glandular and simple hairs; stems to 20 cm long. Leaves elliptic to oblanceolate, tapering towards base, dentate, with lamina 1.5–3.5 cm long, 8–13 mm wide; upper leaves narrower. Flowers in racemes to 15 cm long; bracts leaf-like; pedicel 8–15 mm long, glandular-hairy, not articulate; bracteoles absent. Sepals lanceolate to narrowly elliptic, 3.5 mm long. Corolla 8–12 mm long, with few hairs inside, without enations; abaxial lobes 2–3 mm long; wings c. 1 mm wide. Style 3-fid. Median indusium almost sessile, oblong, c. 0.5 mm long, lateral ones smaller. Ovules 8. Fruit ovoid, c. 4 mm diam.; valves not dividing to base, not gaping. Seeds elliptic, 2.5 mm long, colliculate, matt, greyish yellow.

Occurs in Arnhem Land, N.T., in eucalypt woodland. Flowers Feb.–June. Map 373.

N.T.: 12°01'S, 132°45'E, *L.A.Craven 2240* (CANB); Port Darwin, *M.Holtze 524* (MEL); Howard Springs, *G.Chippendale 6169* (BRI, DNA, MEL); old BHP air strip, Arnhem Land, *D.E.Symon 7811* (AD, SYD).

Corolla purple, strigose, with coarse, simple and sometimes some glandular hairs outside; wings of abaxial corolla lobes less than 1/2 as long as lobes. Ovary septum is less than 1/4 length of ovary. The smooth seeds distinguish this species from others with purplish corollas in the series.

175. **Goodenia purpurea** (F.Muell.) Carolin, *Telopea* 3: 565 (1990)

Calogyne purpurea F.Muell., *Fragm.* 8: 57 (1873). T: Port Darwin, N.T., *F.Schultze 290*; holo: MEL.

Calogyne hians O.Schwarz, *Repert. Spec. Nov. Regni Veg.* 24: 103 (1927). T: Koolpinyah, N.T., *Bleeser 418*; holo: ?B (destroyed) *n.v.*

Ascending herb to 35 cm tall, with antrorse, simple hairs and more glandular hairs towards the top. Basal leaves narrowly oblong to lanceolate, with lamina 4.5–12 cm long, 5–14 mm wide; cauline leaves sessile, smaller, often auriculate. Flowers in racemes to 28 cm long; bracts leaf-like; pedicel 10–20 mm long, not articulate; bracteoles absent. Sepals oblong to elliptic, 2.5–3 mm long. Corolla 8–12 mm long, hairy towards base inside, without enations; abaxial lobes 3–4 mm long; wings c. 1.5 mm wide. Style 3-fid. Median indusium sessile, square, 1 mm long, lateral ones smaller. Ovules 12–18. Fruit globular, 3–4 mm diam.; valves eventually gaping. Seeds elliptic, 2 mm long, verrucose, yellowish brown.

Occurs in Arnhem Land, N.T., in eucalypt forest and woodland and in drier *Melaleuca* scrub. Flowers Feb.–June. Map 374.

N.T.: Elizabeth R., *R.C.Carolin 6438* (SYD); Berry Springs, *R.C.Carolin 6936* (SYD); Humpty Doo, *N.Byrnes* (DNA, SYD).

Corolla purple, with glandular hairs outside; wings of abaxial corolla lobes almost as long as lobes. Ovary septum 1/4 length of ovary. Distinguished from other species in the series with purplish corollas, by the verrucose seeds and glandular hairs.

176. **Goodenia quadrifida** (Carolin) Carolin, *Telopea* 3: 565 (1990)

Calogyne quadrifida Carolin, *Brunonia* 2: 15 (1979). T: Hardys Creek plains, N.T., 5 May 1967, *N.Byrnes NB600*; holo: DNA; iso: SYD.

Ascending herb to 25 cm tall, glabrous or glabrescent except inflorescence. Basal leaves narrowly oblong to oblanceolate, almost entire, with lamina 2–8 cm long, 2–4 mm wide; cauline leaves sessile, smaller, often auriculate. Flowers in racemes to 20 cm long; bracts leaf-like; pedicel 10–22 mm long, divergent in fruit, not articulate; bracteoles absent. Sepals lanceolate, 1.5–2 mm long. Corolla 8–13 mm long, hairy in throat, without enations; abaxial lobes 2–4 mm long; wings 1–2 mm wide. Style 4-fid. Indusia oblong to obovate, c. 1 mm long. Ovules 8–12. Fruit subglobular, 2–3 mm diam.; valves gaping. Seeds elliptic, 2 mm long, verrucose, yellowish brown.

Occurs rarely in Arnhem Land, N.T., on black soil plains. Flowers May. Map 375.

N.T.: Marrakai crossing, Adelaide R., *R.C.Carolin 6922* (SYD).

Corolla purplish brown, with glandular and simple hairs outside; wings of abaxial corolla lobes 2/3 as long as lobes. Ovary septum c. 1/4 length of ovary. The 4-fid style distinguishes this species from all others in the series.

Sect. 4. Amphichila

Goodenia sect. **Amphichila** DC., Prodr. 5: 516 (1836).

T: *G. pumilio* R.Br.

Weak, short-lived herb, with basal leaves, or stoloniferous and rooting and producing clusters of leaves at nodes. Flowers in terminal leafy racemes or solitary in axils; pedicel ebracteolate, not articulate. Sepals adnate to ovary almost to top. Corolla slightly hairy or glabrous inside, enations absent, auricles absent, red-purple; corolla pouch inconspicuous. Ovary septum 1/2–2/3 as long as locule; ovules 30–40, scattered on placentas. Fruit a capsule; valves 2. Seeds convex, c. 1 mm diam., smooth; wing narrow or obsolete.

The section contains 2 species in northern Australia, 1 extending to New Guinea.

177. **Goodenia pumilio** R.Br., *Prodr.* 579 (1810)

T: Endeavour R., [Qld], 1770, *J.Banks & D.C.Solander*; holo: BM.

Prostrate herb, stoloniferous, with scattered, stellate hairs; stems to 10 cm long. Leaves basal and in rosettes on stolons, ±petiolate, obovate to oblanceolate, entire or obscurely toothed, sometimes glabrescent; lamina 8–25 mm long, 4–10 mm wide. Flowers in ±condensed terminal racemes to 5 cm long, or sometimes solitary in axils of basal leaves; bracts leaf-like; pedicel 7–17 mm long, not articulate; bracteoles absent. Sepals lanceolate, largest c. 1 mm long. Corolla 1.0–1.5 mm long, ±glabrous inside; lobes ±equal, 0.5–1 mm long; wings obsolete. Indusium hemispherical, c. 0.2 mm long. Ovules c. 40 in several rows. Fruit obovoid, c. 2 mm long; valves entire. Seeds compressed, almost orbicular, c. 0.2 mm diam., glossy, smooth, brown; wing almost undifferentiated. Fig. 86D.

Occurs in northern N.T. and Qld, in marshes and swamps often with various *Melaleuca* spp., also in New Guinea. Flowers chiefly Apr.–July. Map 376.

N.T.: Oenpelli, *R.L.Specht 1234* (AD, CANB, K, MEL, NSW); Koolpinyah, *C.Dunlop 3616* (DNA); c. 8 km NE of Jim Jim crossing, *R.C.Carolin 6910* (AD, SYD). Qld: Woodleigh Stn near Cairns, 28 June 1938, *H.Flecker* (BRI); Gorge Ck, Mareeba, *H.S.Mckee 9471* (CANB, K, NSW); c. 15 km E of Werandangi homestead, *R.C.Carolin 9048* (SYD).

Corolla dark reddish purple. The stellate indumentum combined with the reddish flowers distinguish this species from all other *Goodenia* species. Its nearest relative, *G. kakadu*, has simple hairs.

178. **Goodenia kakadu** Carolin, *Telopea* 3: 566 (1990)

T: Kakadu National Park, Site 80, N.T., 30 May 1980, *L.A.Craven 6176*; holo: CANB; iso: SYD.

Prostrate herb; stems to 20 cm long, stoloniferous; hairs soft. Leaves in rosettes on stolons, narrowly oblong, tapering basally, thick, glabrescent; lamina c. 3 mm wide. Flowers solitary in leaf axils; pedicel to 12 mm long, not articulate. Sepals lanceolate to ovate; adaxial one c. 1.5 mm long; others c. 0.5 mm long. Corolla to 2 mm long, with few hairs inside; lobes equal, ovate, c. 1 mm long; wings obsolete. Indusium globular, to 0.4 mm diam. Ovules c. 30. Fruit obovoid, attenuate; valves entire. Seeds not seen. Fig. 86C.

Occurs at scattered localities in the Kimberley, W.A., and Arnhem Land, N.T. Grows in open herbfields in damp situations. Flowers Apr.–May. Map 377.

W.A.: Camp Ck Gauging Stn, Mitchell Plateau, *G.J.Keighery 4774* (PERTH); 2 km N of Kalumburu Mission, *P.A.Fryxell & L.A.Craven 4133* (CANB, SYD).

Corolla red; ovary attenuate towards base. Similar to *G. pumilio* which has an obtuse ovary, petiolate leaves, and stellate hairs.

Unplaced names

Calogyne raphanoides O.Schwarz, *Repert. Spec. Nov. Regni Veg.* 24: 103 (1927)

T: Port Darwin, N.T., *Bleeser 99*; *n.v.*

The type was destroyed in B and it has not been possible to match the description with any species (*fide* R.C.Carolin, *Brunonia* 2: 15, 1979).

Goodenia lasiophylla K.Krause, *Pflanzenr.* 54: 89 (1912)

T: between Ashburton and De Grey Rivers, W.A., *Clement*; holo: K.

Insufficient material for an accurate identification.

Goodenia marginata Vriese, *Natuurk. Verh. Holl. Wetensch. Haarlem.* ser. 2, 10: 143 (1854)

T: Argyle–Paramatta, N.S.W., *K.A.F.Hügel*; holo: W.

Probably not Goodeniaceae. Insufficient material for an accurate identification.

Goodenia melanoptera F.Muell., *Fragm.* 1: 115 (1859).

T: upper Victoria R., [N.T.], 1855–56, *F.Mueller*; holo: MEL.

Insufficient material for an accurate identification.

Goodenia stolonifera Vriese, *Natuurk. Verh. Holl. Wetensch. Haarlem* ser. 2, 10: 135 (1854)

T: Australia, *J.Verreaux*; *n.v.*

Probably not Goodeniaceae. Insufficient material for an accurate identification.

8. SELLIERA

R.C.Carolin

Selliera Cav., *Anales Hist. Nat.* 1: 41 (1799); named for Natale Sellier who drafted illustrations for A.J.Cavanilles' work.

Goodenia sect. *Selliera* (Cav.) G.Don, *Gen. Hist.* 3: 725 (1834)

Type: *S. radicans* Cav.

Glabrous perennials; stems prostrate, woody, rooting at nodes. Leaves sessile, attenuate towards base. Inflorescences with flowers in very condensed axillary racemes or solitary in leaf axils, bracteolate; pedicels not articulate. Sepals adnate to ovary. Corolla tubular, completely slit adaxially, without pouch; lobes ±equal, without wings or auricles, reddish brown outside, usually whitish inside. Stamens free. Ovary inferior, 2-locular; indusium ±horizontal; ovules numerous, in 2 rows in each locule. Fruit fleshy, tardily dehiscent or indehiscent. Seeds numerous, biconvex, with swollen mucilaginous wing. Embryo spathulate.

Monotypic genus, very variable; occurs in Australia, New Zealand and South America.

The taxonomic position and, indeed, status of the genus is equivocal. In many respects it resembles certain groups within *Goodenia* sect. *Goodenia*, but research to date has not determined whether it should be included in the genus *Goodenia*. If further work shows that this is so, *Selliera* should be reduced to a synonym of *Goodenia*.

G.Bentham, *Selliera, Fl. Austral.* 4: 81–83 (1869); K.Krause, *Selliera, Pflanzenr.* 54: 112–114 (1912).

Selliera radicans Cav., *Anales Hist. Nat.* 1: 41, t. 5, fig. 2 (1799)

Goodenia radicans (Cav.) Pers., *Syn. Pl.* 1: 195 (1805). T: San Carlos and near Coquimbo, Chile, *L.Née*; syn: both *n.v.*

Goodenia repens Labill., *Nov. Holl. Pl.* 53 (1804); *Selliera repens* (Labill.) Vriese, *Natuurk. Verh. Holl. Maatsch. Wetensch. Haarlem* ser. 2, 10: 163 (1854). T: [Tasmania]; holo: ?FI *n.v.* A possible fragment of the holotype (at P, donated by P.B.Webb) was seen.

S. herpystica Schldl., *Linnaea* 20: 598 (1847). T: Gawler R., near Bethany, S.A., *H.Behr*; holo: HAL.

Illustrations: J.J.H.de Labillardière, *op. cit.* t. 76, as *Goodenia repens*; M.Stones, *Curtis's Bot. Mag.* n. ser. 173: t. 395 (1962); G.R.Cochrane *et al.*, *Fl. Pl. Victoria & Tasmania* 48, fig. 189 (1980).

Prostrate herb; stems often matted, to 50 cm long, glabrous. Leaves ±glossy, spathulate to obovate, entire; lamina usually 1–11 cm long, 2–35 mm wide. Inflorescence with peduncles to 5 cm long; bracteoles linear, mostly to 3 mm long; pedicels to 12 mm long. Sepals ovate to oblong, 4–5 mm long, adnate to ovary almost to top. Corolla 5–12 mm long. Ovary attenuate at base; indusium subglobular, with silky hairs at base, glabrous or nearly so on lips. Seeds orbicular to elliptic, c. 2 mm wide, brown; wing c. 0.5 mm wide, usually crumpled, very sticky when wet. $n = 8$, J.B.Hair & E.J.Beuzenberg, *New Zealand J. Sci.* 3: 434 (1960); D.Moore, *Madroño* 17: 52 (1963). *Swamp Weed.* Figs 75, 86H.

Native in S.A., N.S.W. north to the Sydney region, Vic., and Tas.; occurs particularly near the coast in salt marshes, but also in some periodically flooded areas bordering inland lakes and streams. Also in coastal and highland sites in New Zealand and in salt marshes in southern Chile. Flowers chiefly spring and summer. Map 378.

S.A.: Lashmar Lagoon, Dudley Peninsula, Kangaroo Is., *D.E.Symon 8455* (AD). N.S.W.: Tarago, Lake Bathurst, 29 Apr. 1966, *R.Bonner* (NSW). Vic.: Dwyers Ck, The Grampians, *A.C.Beauglehole & A.E.Orchard ACB304577* (MEL). Tas.: 1.5 km NE of Sea Elephant, King Is., 29 May 1984, *P.Brown* (HO); East Lagoon, Tolberry near Longford, *A.M.Buchanan* (HO).

The syntypes were not seen, but the figure, A.J.Cavanilles, *loc. cit.*, and Chilean, New Zealand and Australian material appear to be conspecific. The species undoubtedly encompasses a number of ecotypes, some of which may warrant taxonomic status.

9. VELLEIA

R.C.Carolin

Velleia Smith, *Trans. Linn. Soc. London, Bot.* 4: 217 (1798); named for Thomas Velley, J.Smith's friend, who was an algologist and Lieutenant-Colonel of the Oxford militia.

Velleya Schultes in J.J.Roemer & J.R.Schultes, *Syst. Veg.* 5: 33 (1819) *orth. var.*

Type: *V. lyrata* R.Br.

Euthales R.Br., *Prodr.* 579 (1810); *Euthale* Vriese in J.G.C.Lehmann, *Pl. Preiss.* 1: 414 (1854) *orth. var.* T: *E. trinervis* (Labill.) R.Br.

Menoceras (R.Br.) Lindley, *Veg. Kingdom* 695 (1846). T: *V. paradoxa* R.Br..

Antherostylis C.Gardner, *J. Roy. Soc. W. Australia* 19: 91 (1933). T: *A. calcarata* C.Gardner

Glabrous or hairy herbs; scapes erect to prostrate. Leaves basal or cauline; axillary hairs conspicuous. Inflorescence with flowers in axillary dichasia or solitary, bracteolate; pedicels not articulate. Sepals 5 or 3, usually adnate to ovary basally only; adaxial one often larger. Corolla usually free from ovary almost to base, auriculate, bilabiate, yellow, orange, pink, mauve, rarely white; lobes usually unequal; with or without an anterior spur. Stamens free from each other, adnate to base of ovary. Ovary incompletely 2-locular; style simple; indusium erect, lips bristled; ovules several. Fruit capsular, 2- or 4-valved. Seeds flat, usually winged or with thickened rim. Embryo spathulate. Chromosome number $x = 8$, W.J.Peacock, *Proc. Linn. Soc. New South Wales* 88: 2–21 (1963).

The genus contains 21 species; 20 endemic in Australia; 1 extending to New Guinea and the Louisiade Archipelago.

The term *enation* is used to describe small outgrowths in the throat of the corolla. See family description and Fig. 88B.

G.Bentham, *Velleia, Fl. Austral.* 4: 45–50 (1869); K.Krause, *Pflanzenr.* 54: 27–40 (1912); R.C.Carolin, The genus *Velleia* Smith (Goodeniaceae), *Proc. Linn. Soc. New South Wales* 92: 27–57 (1967).

KEY TO SECTIONS

1 Sepals 5

  2 Sepals connate into a distinct tube — sect. 1. **Euthales**

  2: Sepals free — sect. 2. **Menoceras**

1: Sepals 3 — sect. 3. **Velleia**

KEY TO SPECIES

1 Sepals 3

  2 Bracteoles connate — **20. V. perfoliata**

  2: Bracteoles free

    3 Corolla lobes ±equal; scapes usually shorter than leaves — **21. V. montana**

    3: Corolla lobes very unequal; scapes usually longer than leaves

      4 Scapes hairy; hairs nowhere restricted to a single line

        5 Scapes mostly prostrate; hairs on scapes retrorse, coarse — **17. V. spathulata**

        5: Scapes mostly ascending; hairs on scapes not retrorse, soft — **18. V. pubescens**

      4: Scapes glabrous or glabrescent (but note *V. macrocalyx*)

        6 Adaxial sepal cordate at base — **15. V. lyrata**

        6: Adaxial sepal not cordate at base

          7 Pedicel with a single line of hairs; scapes ascending to decumbent — **19. V. macrocalyx**

          7: Pedicel glabrous; scapes erect — **16. V. parvisepta**

1: Sepals 5

  8 Leaves entire, terete to spathulate — **4. V. exigua**

  8: Leaves dentate to lobed, flat

    9 Leaves cauline, arranged along an erect stem

10 Leaves broad at base, sessile, 4–7 cm long **2. V. foliosa**

10: Leaves narrowing towards base, petiolate, 5–14 cm long **3. V. macrophylla**

9: Leaves all basal

11 Bracteoles connate into a funnel or disc

12 Sepals almost completely free; adaxial sepal 15–18 mm long; corolla pubescent outside **6. V. panduriformis**

12: Sepals connate into tube 2–6 mm long; adaxial sepal 5–11 mm long; corolla usually glabrous outside

13 Adaxial sepal 9–11 mm long; seeds smooth **7. V. connata**

13: Adaxial sepal 5–6 mm long; seeds papillate **8. V. discophora**

11: Bracteoles free or nearly so

14 Corolla lilac, white or pink

15 Corolla to 7 mm long **10. V. cycnopotamica**

15: Corolla more than 10 mm long

16 Sepals connate; auricle attached to wing; seeds smooth **5. V. daviesii**

16: Sepals free; auricle with membranous appendage and separated from wing; seeds wrinkled **9. V. rosea**

14: Corolla yellow or orange

17 Corolla not spurred

18 Sepals free; corolla to 9 mm long, yellow **11. V. hispida**

18: Sepals connate; corolla more than 8 mm long, usually orange **1. V. trinervis**

17: Corolla spurred

19 Indusium length less than or equal to indusium breadth; scapes glabrous or nearly so **14. V. glabrata**

19: Indusium length greater than breadth

20 Corolla lobes ±equal in length; corolla wings 1–1.5 mm wide; scapes glabrous **12. V. arguta**

20: Corolla lobes very unequal in length; corolla wings more than 2 mm wide; scapes pubescent **13. V. paradoxa**

Sect. 1. Euthales

Velleia sect. **Euthales** (R.Br.) Carolin, *Proc. Linn. Soc. New South Wales* 92: 28 (1967)

Euthales R.Br., *Prodr.* 579 (1810).

Type: *V. trinervis* Labill.

Leaves basal or cauline. Sepals 5, united into an attenuate tube at least as long as shortest lobes. Capsule 4-valved. Seeds with very narrow, mucilaginous wing.

Four species, all from south-western W.A.

1. **Velleia trinervis** Labill., *Nov. Holl. Pl.* 1: 54 (1805)

Euthales trinervis (Labill.) R.Br., *Prodr.* 580 (1810). T: south-western [W.A.], 1792 *J.J.H.Labillardière;* lecto: P, *fide* R.C.Carolin, *loc. cit.*; isolecto: FI, P.

Goodenia tenella Andrews, *Bot. Rep.* 7: t. 466 (1807) *nom. superfl.*

Euthales pilosella Vriese in J.G.C.Lehmann, *Pl. Preiss.* 1: 414 (1845); *Velleia pilosella* (Vriese) Ostenf. & C.Chr., *Biol. Meddel. Kongel. Danske Vidensk. Selsk.* 3: 122 (1921). T: near Princess Royal Harbour, W.A., Dec. 1840, *L.Preiss 1438*; lecto: L, *fide* R.C.Carolin, *Fl. Australia* 35: 334 (1992); isolecto: G, L, P.

V. trinervis var. *villosa* Benth., *Fl. Austral.* 4: 47 (1868). T: south-western W.A., *J.Drummond 4*: *188*; lecto: K, *fide* R.C.Carolin, *Fl. Australia* 35: 334 (1992).

V. trinervis var. *lanuginosa* E.Pritzel in F.L.E.Diels & E.Pritzel, *Bot. Jahrb. Syst.* 35: 556 (1905). T: near Mogumber, W.A., Aug. 1901, *F.L.E.Diels 4038*; holo: ?B (destroyed) *n.v.*

Herb, glabrous to pubescent; stock short; scapes ascending, to 40 cm tall. Leaves basal, linear to spathulate, dentate to almost entire; lamina 5–20 cm long, 5–25 mm wide. Bracteoles free; lower ones oblong to linear-deltoid, to 19 mm long. Adaxial sepal deltoid to ovate, 2–2.5 mm long. Corolla 8–12 mm long, pubescent outside, without enations, yellow-orange, red-brown in throat; wings ±to base of adaxial lobes, c. 2 mm wide. Ovary with septum c. 1/2 as long as locule; ovules c. 6; indusium broadly ovate, c. 2 mm wide. Seeds punctate, with thick mucilaginous rim. $2n$ = 16, W.J.Peacock, *Proc. Linn. Soc. New South Wales* 88: 8 (1963).

Occurs in south-western W.A. from north of Perth to Esperance, often in damp situations. Flowers chiefly Aug.–Dec. Map 379.

W.A.: Dinner Hill, c. 48 km W of Watheroo, *R.C.Carolin 3392* (SYD); Pinjarra, *C.A.Gardner 901* (PERTH); Mt le Grand, *A.S.George 2216* (PERTH); Mt Hamilla, Stirling Ra., 9 Oct. 1968, *J.W.Wrigley* (CBG); Mt Dale, *M.G.Corrick 9350* (MEL).

Chromosome vouchers are *W.J.Peacock 60878.3, 60882.3, 60887.2* (SYD).

2. **Velleia foliosa** (Benth.) K.Krause, *Pflanzenr.* 54: 40 (1912)

V. macrophylla var. *foliosa* Benth., *Fl. Austral.* 4: 48 (1868). T: south-western Australia, *J.Drummond 192*; lecto: K, *fide* R.C.Carolin, *Proc. Linn. Soc. New South Wales* 92: 33 (1967); isolecto: MEL, NSW, P.

Perennial, glabrous; stems to 30 cm long. Leaves cauline, sessile, crowded, narrowly obovate, dentate; lamina 4–7 cm long, 10–15 mm wide. Scapes to 20 cm long. Bracteoles free, the lower ones linear to lanceolate, to 13 mm long. Adaxial sepal ovate, c. 5 mm long. Corolla c. 12 mm long, pubescent outside and inside, without enations, orange–yellow, reddish in throat; wings ±to base of adaxial lobes, c. 2 mm wide. Ovary with septum to 1/2 as long as locule; ovules c. 4; indusium obovate, with a bunch of hairs on lower surface. Seeds elliptic, c. 2 mm long, punctate.

Endemic in the Stirling Ra., W.A., in rocky sites. Flowers chiefly Oct.–Dec. Map 380.

W.A.: Ross Peak, 7 Dec. 1934, *C.A.Gardner* (PERTH); near top of Bluff Knoll, *A.S.George 3117* (PERTH); Talyuberlup, Stirling Ra. Natl Park, *J.S.Beard 7598* (NSW).

This and the next species differ from all other *Velleia* species in the stock being elongated into a distinct stem. It differs from *V. macrophylla* in its oblong leaves which do not narrow towards the base.

3. **Velleia macrophylla** (Lindley) Benth., *Fl. Austral.* 4: 47 (1868)

Euthales macrophylla Lindley, *Bot. Reg.* 26: misc. 54 (1840); *Goodenia macrophylla* (Lindley) F.Muell., *Fragm.* 6: 11 (1867). T: Western Australia, ex *Hort. Soc. Nat. London* 1840, grown from seed purchased of *J.Drummond*; lecto: CGE, *fide* R.C.Carolin, *Proc. Linn. Soc. New South Wales* 92: 34 (1967).

Perennial, glabrous; stems to 1 m tall; scapes to 50 cm long. Leaves cauline, not crowded, obovate to elliptic, dentate, ±petiolate; lamina 5–20 cm long, 0.5–8 cm wide. Bracteoles free, lower ones linear to lanceolate, to 13 mm long. Adaxial sepal oblong, to 7 mm long. Corolla similar to preceding species, c. 12 mm long. Ovary with septum to 1/2 as long as locule; ovules c. 4; indusium obovate, with a bunch of hairs on lower surface. Seeds elliptic, c. 4 mm long, punctate.

Occurs in the extreme south-west of W.A. Flowers chiefly Nov.–Jan. Map 381.

W.A.: Pemberton, *T.E.H.Aplin 1383* (PERTH); Scott R., *R.D.Royce 74* (PERTH); Bow R., Nov. 1912, *S.W.Jackson* (NSW); c. 48 km S of Nannup, *V.Mann 80 & A.S.George* (K, NSW); Shannon R., c. 22 km E of Northcliffe, *R.Pullen 9950* (CANB, NSW).

The tallest species in the genus.

4. **Velleia exigua** (F.Muell.) Carolin, *Fl. Australia* 35: 334 (1992)

Goodenia exigua F.Muell., *Fragm.* 3: 142 (1863); *Selliera exigua* (F.Muell.) Benth., *Fl. Austral.* 4: 82 (1868). T: Moir Inlet, W.A., *G.Maxwell*; lecto: MEL, *fide* R.C.Carolin, *Fl. Australia* 35: 334 (1992); isolecto: K.

Perennial, to 10 cm tall, glabrous, stoloniferous. Leaves cauline, clustered in irregular whorls, terete-spathulate, with yellowish aristate tip to 10 mm long; lamina to 3 mm wide. Inflorescence with flowers in few-flowered axillary cymes or solitary. Bracteoles linear, 1–2 mm long. Adaxial sepal deltoid, c. 1 mm long. Corolla c. 2 mm long, yellow, glabrous outside and inside, without enations; wings short, c. 0.5 mm wide. Ovary with septum very short; ovules 3 or 4; indusium obloid, c. 0.5 mm wide. Capsule subglobular. Seeds subglobular, c. 1 mm diam., punctate; wing obsolete.

Occurs in the Stirling Ra. and at Moirs Inlet, W.A., in saline clays. Map 382.

W.A.: just N of Quarderwardup Lake, Stirling Ra., *G.J.Keighery 9843* (PERTH); unnamed salt lake, SE of Ellen Peak, Stirling Ra., *G.J.Keighery 5525* (PERTH).

This species is quite unlike any other species of *Velleia* with its habit of terete leaves clustered in irregular whorls, its flowers solitary and almost sessile, its ovary adnate to sepals in the lower half, its capsule dehiscing irregularly, and its almost wingless seed.

Sect. 2. Menoceras

Velleia sect. **Menoceras** R.Br., *Prodr.* 580 (1810).

Menoceras (R.Br.) Lindley, *Veg. Kingdom* 695 (1847).

Type: *V. paradoxa* R.Br., lecto: *fide* R.C.Carolin, *Proc. Linn. Soc. New South Wales* 92: 34 (1967).

Velleia sect. *Aceratia* F.Muell., *Trans. & Proc. Philos. Soc. Victoria* 1: 17 (1855). T: *V. connata* F.Muell.

Velleia sect. *Pentasepala* K.Krause, *Pflanzenr.* 54: 28 (1912). T: *V. paradoxa* R.Br.; lecto: *fide* R.C.Carolin, *Proc. Linn. Soc. New South Wales* 92: 34 (1967).

Antherostylis C.Gardner, *J. Roy. Soc. W. Australia* 19: 91 (1933). T: *Antherostylis calcarata* C.Gardner

Leaves basal. Sepals 5, free or connate into a tube shorter than lobes. Capsule 2- or

4-valved. Seeds winged or with a thickened mucilaginous rim.

A group of 10 species; most restricted to W.A., but some extend over all mainland States.

5. **Velleia daviesii** F.Muell., *Fragm.* 10: 10 (1876)

T: Ularing, W.A., Oct. 1891, *J.Young*; holo: MEL.

Annual herb, pubescent, yellowish; stock short; scapes erect, 20–40 cm long. Leaves oblanceolate, ±lyrately pinnatifid, often glabrescent; lamina to 20 cm long. Bracteoles free or nearly so, leaf-like; lower ones to 40 mm long. Sepals connate into a short tube; adaxial one broadly ovate, to 10 mm long. Corolla to 20 mm long, pubescent outside and inside, with enations, lilac to white; wings more than 1/2 as long as adaxial lobes. Ovary with septum c. 1/2 as long as locule; ovules c. 20; indusium broadly ovate, c. 5 mm wide, villous. Capsule ovoid. Seeds orbicular, 4 mm diam., punctate; wing 1 mm wide.

Occurs in the eastern Goldfields region of W.A., in scrub and steppe grassland, in sandy soil. Flowers chiefly Aug.–Dec. Map 383.

W.A.: Fraser Ra., *C.A.Gardner 2912* (K, PERTH); Coolgardie Goldfields, *E.Pritzel 850* (AD, K, NSW, P); c. 41 km SW of Coolgardie on Great Eastern Hwy, *M.D.Tindale 41 & E.M.Bennett* (NSW).

Only 3 lilac-flowered species are pubescent, i.e., *V. daviesii*, *V. rosea* and *V. cycnopotamica*. *Velleia daviesii* is distinguished by its punctate, not wrinkled, seeds.

6. **Velleia panduriformis** A.Cunn. ex Benth., *Fl. Austral.* 4: 46 (1868)

T: Goodenough Bay and Point Cunningham, NW coast, [W.A.], *A.Cunningham*; holo: K; iso: BM, MEL.

Herb, glaucous; stock short; scapes erect, to 1 m tall. Leaves shortly petiolate, obovate, dentate, glabrous; lamina to 20 cm long, to 7 cm wide. Bracteoles connate into funnel-like disc, to 12 cm diam., often split on one side. Sepals ±free, dentate; adaxial one broadly elliptic, 15–18 mm long. Corolla 20–25 mm long, pubescent outside, bearded inside, with enations, deep or brownish yellow; wings much shorter than adaxial lobes, to 1 mm wide. Ovary with septum swollen, to 1/2 as long as locule; ovules c. 20; indusium transversely oblong, c. 5 mm wide, lips ±glabrous towards centre. Capsule ovoid. Seeds orbicular, 8–9 mm diam., punctate; wing c. 2 mm wide. $2n = 16$, W.J.Peacock, *Proc. Linn. Soc. New South Wales* 88: 8 (1963). *Pindan Poison.* Figs 73, 88A–B.

Occurs in the Kimberley, W.A. Grows in red sand in tall open shrubland and low open woodland. Flowers chiefly Apr.–Sept. Map 384.

W.A.: 20 km out of Derby on road to Mt House, *D.E.Symon 10132* (AD, SYD); Hardmans Ck W of Fitzroy Ck, *R.C.Carolin 7455* (SYD); c. 11 km S of James Price Point, *R.C.Carolin 7496* (SYD); Meda R., *C.A.Gardner 1624* (PERTH); c. 25 km NE of Karunjie Stn, *N.H.Speck 5011* (CANB).

The rosette of large leaves usually dies off before flowering begins. One branch of the dichasium grows more strongly than the other at each node of the scape, giving the appearance of a raceme of clusters. This also occurs in *V. connata* (very occasionally and to a much reduced degree) and sometimes in *V. macrocalyx*. Reputed but not proven to be toxic to stock, see C.A.Gardner & H.W.Bennetts, *Toxic Pl. W. Australia* 190–191 (1956).

7. **Velleia connata** F.Muell., *Trans. & Proc. Philos. Soc. Victoria* 1: 18 (1855)

T: Murray scrub [S.A. or Vic.?], *F.Mueller*; iso: K.

V. helmsii K.Krause, *Pflanzenr.* 54: 33 (1912). T: Victoria Desert Camp 53, W.A., *R.Helms*; holo: ?B (destroyed) *n.v.*; iso: AD, K, MEL, NSW.

Illustrations: K.Krause, *op. cit.* 34, fig. 9J–K; G.M.Cunningham *et al.*, *Pl. W. New South Wales* 638 (1981).

Herb, glabrous, glaucous; stock short; scapes ascending, to 60 cm long. Leaves obovate, dentate to lyrately pinnatifid; lamina 5–29 cm long, to 8 cm wide. Bracteoles connate into funnel, to 9 cm diam. Sepals connate into tube 4–6 mm long; adaxial one ovate, 9–11 mm long. Corolla 15–17 mm long, usually glabrous outside, bearded inside, with enations, yellow or brownish to white or pink; wings much shorter than adaxial lobes, usually less than 1 mm wide. Ovary with septum swollen, 1/2 as long as locule; ovules 15–20; indusium depressed-ovate, c. 4 mm wide; lips ±glabrous towards centre. Capsule compressed-ovoid. Seeds orbicular, 5–6 mm diam., punctate; wing to 2 mm wide.

Occurs in all States except Tas., in a variety of communities in drier areas. Flowers most of the year. Map 385.

W.A.: near Miss Gibson Hill, c. 26°50' S, 126°17'E, *A.S.George 4002* (PERTH). N.T.: c. 29 km NE of Barrow Creek, *M.Lazarides 5823* (BRI, CANB, DNA, K, MEL, NSW, PERTH). S.A.: Wilpena Pound, 30 Nov. 1933, *J.B.Cleland* (AD, K). Qld: Cunnamulla, 1 June 1942, *G.H.Allen* (CANB). N.S.W.: c. 93 km N of Ledknappa crossing, *P.L.Milthorpe & G.M.Cunningham 1746* (NSW).

Related to *V. panduriformis* and *V. discophora*. Rarely, the scapes branch in a manner similar to that of *V. panduriformis*.

8. Velleia discophora F.Muell., *Fragm.* 10: 10 (1876)

T: Ularing, W.A., 13–17 Oct. 1875, *J.Young*; holo: MEL.

Illustration: K.Krause, *Pflanzenr.* 54: 34, fig. 9a–c (1912); C.A.Gardner & H.W.Bennetts, *Poison. Pl. W. Australia* t. LII (1956).

Herb, glabrous, glaucous; stock short; scapes ascending, to 80 cm long. Leaves oblanceolate, dentate to lyrately pinnatifid; lamina to 20 cm long, to 50 mm wide. Bracteoles connate into funnel, to 5 cm wide. Sepals connate into short tube; adaxial one ovate, 5–6 mm long. Corolla 10–13 mm long, glabrous outside, villous inside, with enations, yellow; wings ±to base of adaxial lobes, 1–2 mm wide. Ovary with septum very short; ovules c. 12; indusium semi-orbicular, c. 2.5 mm wide, villous. Capsule ovoid. Seeds ±orbicular, 4 mm diam., papillate; wing c. 1 mm wide. $2n = 16$, W.J.Peacock, *Proc. Linn. Soc. New South Wales* 88: 8 (1963). *Cabbage Poison.*

Occurs in southern W.A. from the central wheatbelt to the edge of the Great Victoria Desert and N to Meekatharra; grows in sandy and gravelly soils usually in open communities, appearing often after fire. Flowers chiefly Sept.–Feb. Map 386.

W.A.: Salmon Gums, *R.D.Royce 4036* (PERTH); Merredin, *M.Koch 2837* (K, NSW); c. 21 km E of Meekatharra, *N.T.Burbidge 4718* (CANB, PERTH); c. 19 km N of Lake Grace, *P.R.Jeffries 641030* (PERTH); Borrikin Rock, *B.H.Smith 122* (MEL, NSW).

The papillate seeds separate this species from the others in this section. Chromosome voucher is *W.J.Peacock 6111.35.1* (SYD). Toxic to stock under some circumstances, see see C.A.Gardner & H.W.Bennetts, *Toxic Pl. W. Australia* 191–192 (1956).

9. Velleia rosea S.Moore, *J. Linn. Soc., Bot.* 34: 202 (1899)

T: between Wilson Pool [Creek] and Lake Darlot, W.A., *S.Moore*; holo: BM.

V. rosea var. *erecta* K.Krause, *Pflanzenr.* 54: 37 (1912). T: Murrin-Murrin, W.A., *W.J.George*; holo: ?B (destroyed) *n.v.*

Illustration: R.Erickson *et al.*, *Fl. Pl. W. Austral.* 155, fig. 492 (1973).

Annual herb, pubescent; stock short; scapes prostrate or ascending, to 15 cm long. Leaves oblanceolate, ±coarsely dentate to pinnatisect; lamina 3–7 cm long, 5–20 mm wide. Bracteoles usually free, to 15 mm long. Sepals free; adaxial one obovate to narrowly elliptic, c. 5 mm long. Corolla c. 13 mm long, pubescent outside, pubescent in throat, without enations, pink lilac or white; wings ±to base of adaxial lobes, 2–3 mm wide. Ovary with septum very short; ovules c. 4; indusium depressed-ovate, c. 2 mm wide, sprinkled with hairs. Capsule compressed. Seeds orbicular, 4–5 mm diam, wrinkled; wing 1 mm wide. $2n$ = 16, W.J.Peacock, *Proc. Linn. Soc. New South Wales* 88: 9 (1963) as *V. cycnopotamica*. Figs 76, 87C.

Occurs in W.A. from the Kalgoorlie area N to Meekatharra, and W to Cowcowing and the Geraldton area, in sandy or loamy soil. Flowers chiefly May–Oct. Map 387.

W.A.: Cowcowing, *M.Koch 1324* (AD, MEL); c. 32 km S of Menzies, *R.C.Carolin 3036* (SYD); Malcolm, *C.A.Gardner 2480* (PERTH); c. 45 km N of Paynes Find, *A.S.George 685* (PERTH); c. 30 km N of Agnew on road to Wiluna, *T.E.H.Aplin 2365* (PERTH).

The wrinkled seeds and the membranous appendage on the petal wing below the auricle distinguish this species from all others in the section except *V. cycnopotamica* which has much smaller flowers and an inflated capsule. Chromosome vouchers are *W.J.Peacock 60854.1, 60856.1, 60872.2, 6082.2* (SYD).

10. Velleia cycnopotamica F.Muell., *Fragm.* 6: 7 (1867)

T: Swan River, W.A., *J.Drummond 410*; holo: MEL; iso: G, P.

Illustration: K.Krause, *Pflanzenr.* 54: 36, fig. 10D–F (1912).

Annual herb, ±pubescent; stock short; scapes ascending, to 25 cm long. Leaves oblong to oblanceolate, dentate to lyrate; lamina 2–6 cm long, 3–10 mm wide. Bracteoles free, to 15 mm long. Sepals free or nearly so; adaxial one oblong to narrowly elliptic, c. 4 mm long. Corolla 5–6 mm long, pubescent to almost glabrous outside, ±glabrous inside, without enations, pink or lilac to white; wings ±to base of adaxial lobes, to 1 mm wide. Ovary with septum very short; ovules c. 4; indusium semi-orbicular, c. 1 mm wide, with few hairs. Capsule inflated-globular. Seeds orbicular, 3–4 mm diam, wrinkled; wing c. 1 mm wide. $2n$ = 16, W.J.Peacock, *Proc. Linn. Soc. New South Wales* 88: 8 (1963).

Occurs from Geraldton to Albany area, W.A., and on the Eyre Peninsula, S.A., in open habitats and often in cultivated land. Flowers chiefly Aug.–Oct. Map 388.

W.A.: Wongan Hills, *A.Morrison 120* (K, PERTH); Corrigin, *R.C.Carolin 3142* (SYD); c. 15 km E of Calingiri, *T.E.H.Aplin 151* (PERTH); S.A.: Wudinna, *C.W.Johns 14* (AD).

The wings of the petals have membranous appendages below the auricles. See also *V. rosea*. Chromosome vouchers are *W.J.Peacock 60812.1, 60817.1, 60881.3, 6097.3* (SYD).

11. Velleia hispida W.Fitzg., *W. Austral. Nat. Hist. Soc.* 1: 25 (1904)

T: Nannine, W.A., Sept. 1903, *W.V.Fitzgerald*; holo: NSW.

Herb, ±pubescent; stock short; scapes ascending, to 16 cm long. Leaves narrowly obovate to narrowly elliptic, dentate or lyrate; lamina 5–8 cm long, 12–20 mm wide. Bracteoles free or nearly so, to 2 cm long. Sepals free; adaxial one ovate, 5 mm long. Corolla to 9 mm long, pubescent outside, almost glabrous inside, without enations, yellow; wings c. 1/2 as long as adaxial lobes, 0.5 mm wide. Ovary with septum less than 1/2 as long as locule; ovules 6–8; indusium broadly ovate, 0.5 mm wide, with few hairs. Capsule globular. Seeds orbicular, c. 2.5 mm diam., punctate; wing 0.5 mm wide. Fig. 87D.

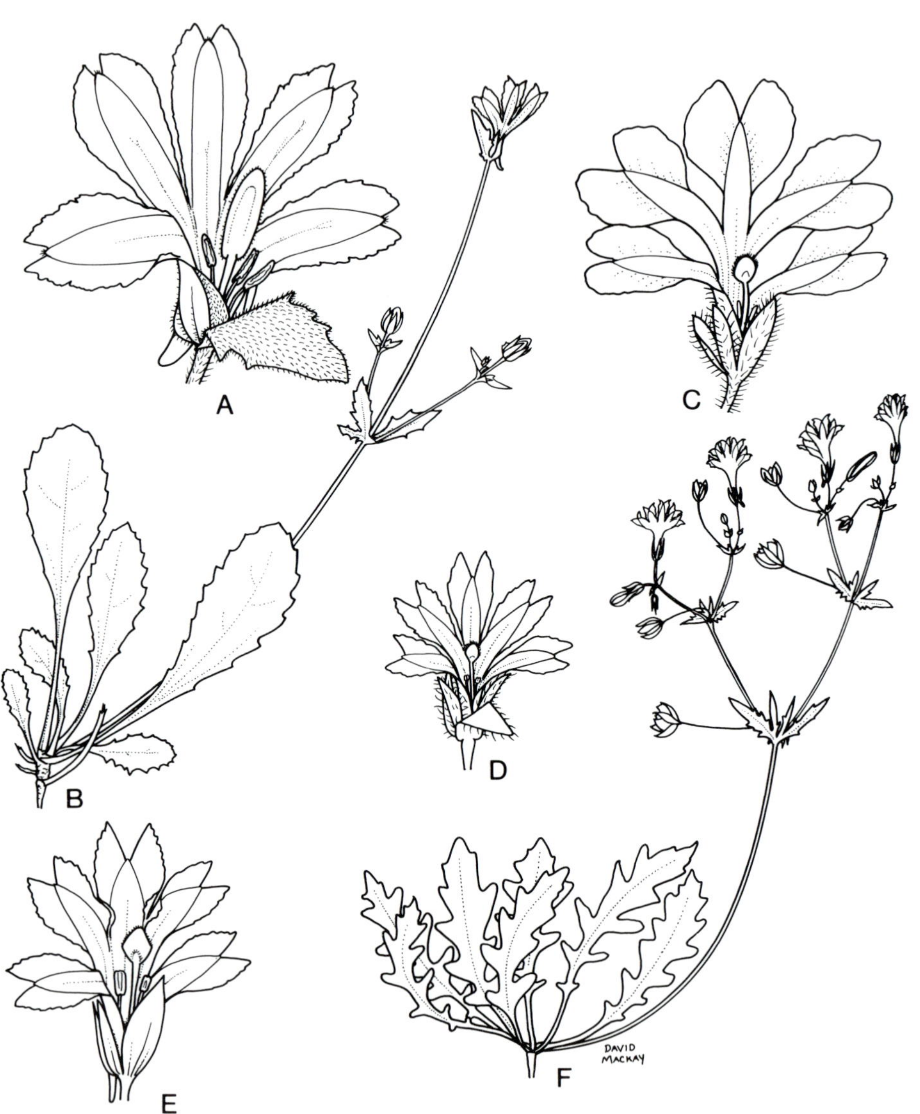

Figure 87. *Velleia*. **A–B**, *V. arguta*. **A**, flower X2; **B**, habit X0.5 (**A–B**, D.Symon 12178, SYD). **C**, *V. rosea*, flower X2 (R.Carolin 3036, SYD). **D**, *V. hispida*, flower X4 (A.Beauglehole 49154, SYD). **E–F**, *V. glabrata*. **E**, flower X2; **F**, habit X0.5 (**E–F**, H.Eichler 17473, SYD). Drawn by D.Mackay.

Extends from the Laverton area NW to Meekatharra and NE into the Great Victoria Desert, W.A. Flowers chiefly July–Nov. Map 389.

W.A.: c. 40 km from Meekatharra towards Wiluna, *W.J.Peacock 60866.3* (SYD); c. 38 km N of Sandstone, *A.S.George 2656* (PERTH); Tuckanarra, *C.A.Gardner 2278* (PERTH); 171 km by road SW of Warburton Mission on Laverton road, *A.C.Beauglehole 60114* (SYD); near Old Minnie Creek homestead, E of Laverton, *A.S.George 4656* (PERTH).

This species is sometimes confused with *V. cycnopotamica*, but its yellow flowers and punctate seeds distinguish it clearly. The corolla is adnate to the ovary almost to its top.

12. **Velleia arguta** R.Br., *Prodr.* 580 (1810)

T: base of mountains near inlet no. XII [Spencer Gulf, S.A.], Mar. 1802, *R.Brown*; ?holo: BM.

Antherostylis calcarata C.Gardner, *J. Roy. Soc. W. Australia* 19: 92 (1933). T: Junana Rocks, W.A., 24 Oct. 1931, *C.A.Gardner 2909*; holo: PERTH; iso: K.

Illustration: G.R.Cochrane *et al.*, *Fl. Pl. Victoria & Tasmania* 58, fig. 247 (1980) as *Velleia paradoxa*.

Perennial herb, ±glabrous; stock short; scapes ascending, to 40 cm long. Leaves oblanceolate to narrowly elliptic, ±dentate; lamina 4–12 cm long, to 15 mm wide. Bracteoles free, lanceolate to ovate, to 20 mm long. Sepals free; adaxial one oblong to narrowly ovate, 10–12 mm long. Corolla spurred, 12–20 mm long, pubescent outside, ±glabrous inside, with few wrinkles near base, deep yellow, brownish towards throat; wings ±to base of adaxial lobes, 1–1.5 mm wide. Ovary with septum less than 1/2 as long as locule; ovules 10–16; indusium oblong, 5–6 mm long, c. twice as long as broad, pubescent. Capsule ovoid. Seeds elliptic, c. 4 mm wide, smooth; wing 2 mm wide. Fig. 87A–B.

Occurs in the drier areas of southern Australia from south-eastern W.A. through S.A. to western N.S.W. and Vic.; in grassland or mallee, in sand or sometimes rocky areas. Flowers chiefly July–Jan. Map 390.

W.A.: c. 1.5 km west of Cocklebiddy, 29 Aug. 1955, *A.R.Main* (PERTH); Eucla to Madura, 27 Aug. 1963, *R.C.Carolin* (SYD). S.A.: Wonoka Hill, *K.Hill 2133 & L.A.S.Johnson* (NSW). N.S.W.: Broken Hill, Dec. 1917, *E.C.Andrews* (NSW). Vic.: Dimboola, 1901, *H.E.D'Alton* (MEL).

The elongated indusium and almost equal corolla lobes with narrow wings distinguish this species from others in the section. The bracteoles sometimes have ciliate margins, and the corolla pouch has a spur 3–6 mm long.

13. **Velleia paradoxa** R.Br., *Prodr.* 580 (1810)

T: Cow Pasture Plains, N.S.W., Oct. 1803, *R.Brown*; lecto: BM, *fide* R.C.Carolin, *Proc. Linn. Soc. New South Wales* 92: 45 (1967); isolecto: BRI, K, NSW.

V. paradoxa var. *humilis* DC., *Prodr.* 7(2): 518 (1839). T: none given; deCandolle cited '*V. paradoxa* Lindl. bot. reg. t. 971. Gart. mag. 1828. t. 7.'

V. paradoxa var. *stenoptera* F.Muell. ex Benth., *Fl. Austral.* 4: 48 (1868). T: Bentham referred 'the Queensland and New England and Richmond river specimens' to this name.

Illustrations: J.Lindley, *Bot. Reg.* 12: t. 971 (1826); K.Krause, *Pflanzenr.* 54: 36, fig. 10A–C (1912).

Perennial herb, softly pubescent; stock short; scapes ascending or decumbent, to 40 cm long. Leaves obovate to elliptic, ±dentate, often glabrescent; lamina 7–25 cm long, 15–35 mm wide. Bracteoles free, oblong to ovate, to 4 cm long. Sepals free; adaxial one ovate to oblong, 4–9 mm long. Corolla spurred, 10–20 mm long, pubescent outside, glabrous or sparsely hairy inside, without enations, yellow; wings ±to base of adaxial lobes, to 3 mm wide. Ovary with septum c. 1/4 as long as locule; ovules 8–14; indusium obovate, 3–6 mm wide, orifice very curved, pubescent. Capsule ovoid, ±compressed. Seeds orbicular, 3–5 mm diam., smooth; wing to 1 mm wide. $2n = 16$, W.J.Peacock, *Proc. Linn. Soc. New South Wales* 88: 8 (1963). Fig. 74.

Extends from the Eyre Peninsula, S.A., through Vic., Tas. and N.S.W. to southern Qld, in open forest and grassland, but not in the driest areas. Flowers chiefly Aug.–Feb. Map 391.

S.A.: Naracoorte, 26 Oct. 1933, *E.H.Ising* (AD). Qld: Darling Downs between Cambooya and Clifton, *C.T.White 12663* (BRI). N.S.W.: North of Ulan, *W.J.Peacock 6111.8.1* (SYD); Jenolan Caves, Dec. 1899, *W.F.Blakely* (NSW). Vic.: Mansfield, *R.A.Black 587.000(2)* (MEL). Tas.: Launceston, Dec. 1915, *F.A.Rodway* (NSW).

The corolla pocket has a spur to 8 mm long. Chromosome vouchers are *W.J.Peacock 6011.2.1, 6111.8.1* (SYD).

14. **Velleia glabrata** Carolin, *Proc. Linn. Soc. New South Wales* 92: 46 (1967)

T: Urumburi, S of Thargomindah, Qld, 16 Aug. 1964, *R.C.Carolin 4080*; holo: NSW.

Illustration: G.M.Cunningham *et al.*, *Pl. W. New South Wales* 639 (1981).

Annual herb, glabrous or with a few soft hairs; stock short; scapes ascending, to 38 cm long. Leaves oblong to oblanceolate, dentate to lyrate; lamina 4–8 cm long, to 20 mm wide. Bracteoles connate only at base, to 2 cm long. Sepals free; adaxial one ovate to broadly elliptic, 5–6 mm long. Corolla 12–14 mm long with short spur, pubescent outside, ±villous inside at base, without enations, yellow; wings ±to base of adaxial lobes, c. 2 mm wide. Ovary with septum very short; ovules 6–12; indusium depressed-ovate, c. 2 mm wide, orifice only slightly curved, slightly pubescent. Capsule subglobular. Seeds orbicular, 4–5 mm diam., punctate; wing c. 1 mm wide. Fig. 87E–F.

Occurs in drier parts of W.A., N.T., S.A., Qld, and N.S.W.; mostly south of Tropic of Capricorn, in a variety of communities. Flowers chiefly July–Feb. Map 392.

W.A.: c. 11 km S of Meekatharra, *C.A.Gardner 2302* (PERTH). S.A.: Wildcat Bore, Everard Park homestead, *H.Eichler 17473* (AD, SYD); Churina Well on Cook–Vokes road, *D.E.Symon 12757* (AD, SYD). Qld: Charleville, Dec. 1916, *E.W.Bick* (BRI). N.S.W.: 'Cooneybar', Byrock, *G.M.Cunningham 873* (NSW).

The corolla spur reaches 3 mm long. This species is sometimes difficult to distinguish from *V. paradoxa*. In *V. glabrata* the length of the indusium is almost equal to its breadth, and the adaxial lobes of the corolla are about as long as the connate part; in *V. paradoxa* the indusium is distinctly longer than broad, and the adaxial lobes of the corolla are as long as, or longer than, the connate part.

Sect. 3. Velleia

Velleia Smith sect. **Velleia**

Velleia sect. *Trisepala* K.Krause, *Pflanzenr.* 54: 28 (1912). T: *V. lyrata* R.Br.; lecto: *fide* R.C.Carolin, *Proc. Linn. Soc. New South Wales* 92: 47 (1967).

Leaves basal. Sepals 3, free or nearly so. Capsule 4-valved. Seeds with thickened rim but scarcely winged.

Seven species, occurring in eastern Australia, 1 extending to New Guinea and the Louisiade Archipelago.

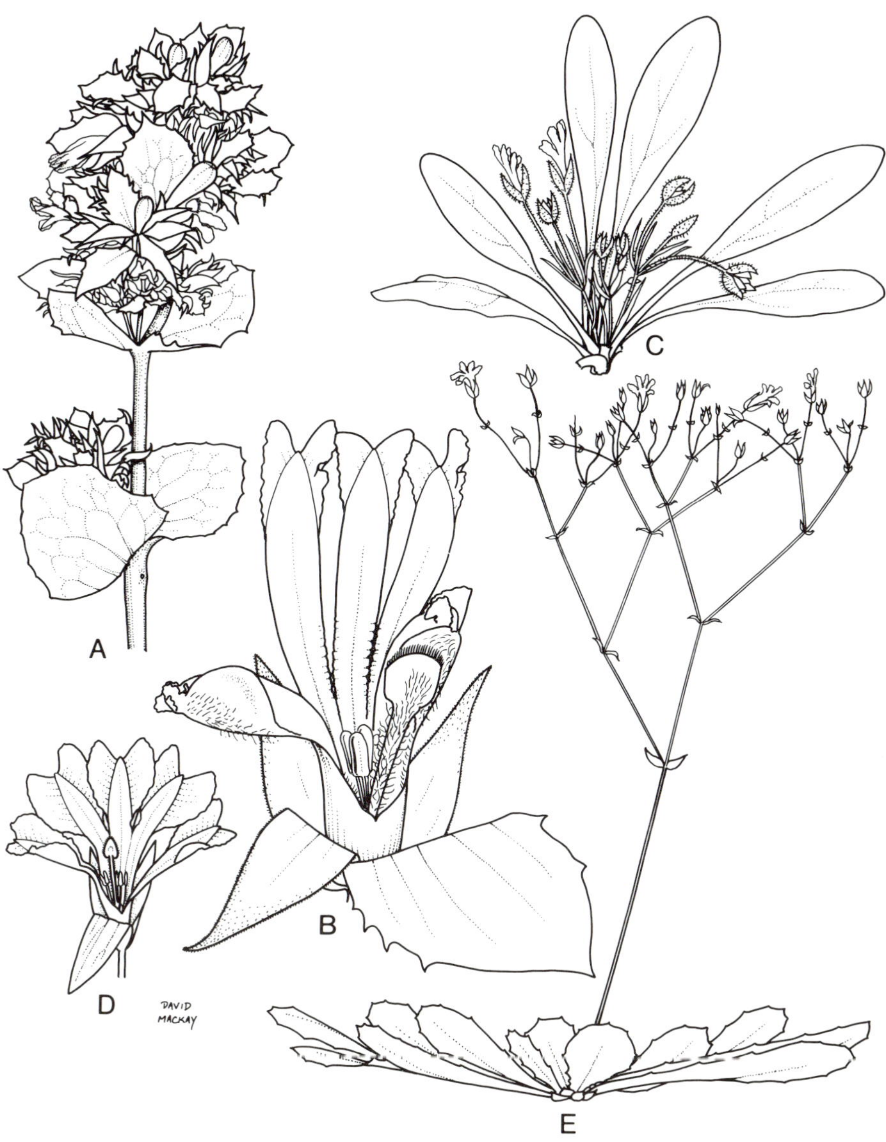

Figure 88. *Velleia*. **A–B**, *V. panduriformis*. **A**, inflorescence X0.5; **B**, flower X2 (R.Carolin 7538, SYD). **C**, *V. montana*, habit X1 (J.McLuckie & A.Petrie, Jan 1925, SYD). **D–E**, *V. spathulata*. **D**, flower X2; **E**, habit X0.5 (**D–E**, R.Carolin 562, SYD). Drawn by D.Mackay.

15. Velleia lyrata R.Br., *Prodr.* 580 (1810)

T: South Head of Port Jackson, N.S.W., 1803, *R.Brown*; holo: BM; iso: P.

Illustrations: J.B.Ker-Gawler, *Bot. Reg.* 7: t. 551 (1821); K.Krause, *Pflanzenr.* 54: 30, fig. 7a–d (1912).

Perennial herb, ±glabrous; stock short; scapes ascending or erect, to 50 cm long. Leaves obovate to oblanceolate, dentate to lyrate; lamina 3–18 cm long, 1–4 cm wide. Bracteoles free, to 15 mm long. Adaxial sepal ovate, cordate at base, 4–8 mm long. Corolla 10–15 mm long, pubescent outside and inside, with small enations, yellow; wings almost to base of adaxial lobes, 1–2 mm wide. Ovary with septum 1/2 as long as locule; ovules 4–10; indusium broadly ovate, 1–2 mm wide. Capsule subglobular. Seeds orbicular to elliptic, 1–2.5 mm wide, punctate. $2n = 16$, W.J.Peacock, *Proc. Linn. Soc. New South Wales* 88: 9 (1963).

Grows in damp situations near the Qld/N.S.W. border and in the Sydney region. Flowers chiefly Aug.–Apr. Map 393.

Qld: Gurulmundi–Wollebee road, *K.A.W.Williams 75046* (BRI, SYD). N.S.W.: Glenbrook, *H.S.McKee 6751* (SYD); Waterfall, June 1919, *A.A.Hamilton* (NSW); Wondabyne, Sept. 1923, *W.F.Blakely & D.W.C.Shiress* (NSW); Debenham road, N of Somersby Falls road, 30 Mar. 1979, *H.Falding & H.Bryant* (NSW).

The scapes are glabrous. The cordate base to the adaxial sepal distinguishes this species from the others in this section, except *V. spathulata* which is prostrate or decumbent. Chromosome voucher is *W.J.Peacock 5811* (SYD).

16. Velleia parvisepta Carolin, *Proc. Linn. Soc. New South Wales* 92: 49 (1967)

T: Dubbo, N.S.W., 8 Nov. 1960, *W.J.Peacock*; holo: NSW.

Perennial herb, glabrous; stock short; scapes erect or ascending, to 40 cm long. Leaves oblanceolate, dentate or lobed; lamina 8–15 cm long, 8–20 mm wide. Bracteoles free, to 10 mm long. Adaxial sepal elliptic, not cordate, 7–8 mm long. Corolla 8–12 mm long, pubescent outside, almost glabrous inside, with enations, yellow; wings ±to base of adaxial lobes, c. 1 mm wide. Ovary with septum almost obsolete; ovules 6–8; indusium depressed-ovate, c. 1 mm wide. Capsule subglobular. Seeds orbicular, 2 mm diam., punctate.

Only known from the Goonoo State Forest, N.S.W., in damp situations in open forest. Flowers Sept.–Jan. Map 394.

N.S.W.: Goonoo Forest, 26 Sept. 1951, *G.M.Chippendale & E.F.Constable* (NSW); Goonoo State Forest, 25 km ENE of Mogriguy on Dubbo–Mendooran road, *R.Coveny 10071* (NSW).

The bracteoles are linear to oblanceolate. Differs from *V. lyrata* in the shorter septum in the ovary, and in the non-cordate base to the adaxial sepal.

17. Velleia spathulata R.Br., *Prodr.* 580 (1810)

T: Kingstown, Newcastle, N.S.W., Oct.–Nov. 1804, *R.Brown*; lecto: BM, *fide* R.C.Carolin, *Proc. Linn. Soc. New South Wales* 92: 51 (1967); isolecto: MEL (*p.p.*).

Illustration: K.Krause, *Pflanzenr.* 54: 30, fig. 7E–F (1912).

Perennial herb, variously hairy; stock short; scapes prostrate or decumbent, to 25 cm long. Leaves oblanceolate, entire or dentate; lamina to 10 cm long, to 25 mm wide. Bracteoles free, to 12 mm long. Adaxial sepal lanceolate to oblong, ±cordate, 4–6 mm long. Corolla 6–11 mm long, pubescent outside, glabrous inside, with enations, yellow with purplish markings; wings short, c. 1 mm wide. Ovary with septum 1/2 as long as locule; ovules

18–25; indusium semi-orbicular, c. 1 mm wide. Capsule ovoid. Seeds orbicular, 1.5 mm diam., punctate. $2n = 16$, W.J.Peacock, *Proc. Linn. Soc. New South Wales* 88: 8 (1963). Fig. 88D–E.

Occurs along the east coast from Cape York, Qld, southwards to Sydney, N.S.W., in damp sandy situations. Also in western New Guinea and the Louisiade Archipelago. Flowers most of the year. Map 395.

Qld: Elimbah, *H.S.McKee 9734* (CANB, NSW); Coolum Beach, 4 Apr. 1945, *M.S.Clemens* (K). N.S.W.: Narrabeen Swamps, Apr. 1901, *A.A.Hamilton* (NSW); Barcoonger State Forest, *W.J.Peacock 611.22.3* (SYD); Coffs Harbour, May 1909, *J.L.Boorman* (NSW).

This species has coarse, retrorse hairs on the scapes, appressed villous hairs on the pedicels, and narrow sepals. It is thus distinguished from *V. lyrata* (glabrous pedicels and ovate sepals), and *V. pubescens* (pedicels with spreading hairs). Chromosome vouchers are *W.J.Peacock 611.22.3, 611.24.2, 6012.2.4* (SYD).

18. **Velleia pubescens** R.Br., *Prodr.* 581 (1810)

T: Shoalwater Bay, and Thirsty Sound, [Qld], *R.Brown 87*; lecto: BM, *fide* R.C.Carolin, *Proc. Linn. Soc. New South Wales* 92: 53 (1963); isolecto: MEL, P.

Illustration: K.Krause, *Pflanzenr.* 54: 30, fig. 7G–H (1912)

Perennial herb, pubescent; stock short; scapes ascending to decumbent, to 15 cm long. Leaves oblanceolate, dentate; lamina 5–12 cm long, to 25 mm wide. Bracteoles free, to 3 cm long. Adaxial sepal ovate-elliptic, sometimes cordate, 6–11 mm long. Corolla c. 14 mm long, pubescent outside and inside, with enations, yellow; wings ±to base of adaxial lobes, c. 1 mm wide. Ovary with septum c. 1/2 as long as locule; ovules c. 20; indusium depressed-obovate, c. 1 mm wide. Capsule subglobular. Seeds orbicular to elliptic, 2 mm diam., punctate.

Occurs near Shoalwater Bay and around Herberton, Qld. Rarely collected and possibly not well conserved. Flowering unknown. Map 396.

Qld: Broad Sound, 1871, *E.Bowman* (MEL); Bay of Inlets, 1770, *J.Banks & D.Solander* (BM); Herberton, *M.Michael 389* (BRI).

The pedicels have spreading hairs. Similar to *V. lyrata,* differing in the dense indumentum.

19. **Velleia macrocalyx** Vriese in T.L.Mitchell, *J. Exped. Trop. Australia* 258 (1848)

T: Belyando River, [Qld], *T.L.Mitchell*; holo: L.

V. prostrata Ewart & L.Kerr, *Proc. Roy. Soc. Victoria* n.s. 39: 7 (1926). T: Wycliffe Well, N.T., June 1924, *A.J.Ewart*; holo: MEL.

Illustration: W.H.de Vriese, *Natuurk. Verh. Holl. Maatsch. Wetensch. Haarlem* ser. 2, 10: t. 34 (1854).

Perennial herb, almost glabrous; stock short; scapes ascending or decumbent, to 35 cm long. Leaves elliptic to oblanceolate, entire or dentate; lamina 5–15 cm long, 2–4 cm wide. Bracteoles free, to 4 cm long. Adaxial sepal narrowly elliptic to ovate, to 12 mm long. Corolla 10–12 mm long, pubescent outside, ±villous inside, without enations, yellow; wings narrow, ±to base of adaxial lobes. Ovary septum c. 1/2 as long as locule; ovules 14–16; indusium transversely oblong, 2 mm wide. Capsule ovoid. Seeds broadly elliptic, 1.5 mm wide, punctate.

Occurs in N.T. and Qld, from Barkly Tableland and Sandover R. to Burdekin R., in woodland and grassland. Flowers chiefly June–Sept. Map 397.

N.T.: c. 9 km from Borroloola towards Wollogorang, *C.H.Gittins 2482* (NSW); c. 56 km NE of Alexandria Stn, *R.A.Perry 1510* (BRI, CANB, DNA, K, MEL); Argadargada homestead paddock, 19 Sept. 1954, *G.Chippendale* (DNA). Qld: Massacre Inlet, *L.J.Brass 190* (BRI, CANB).

Differs from all other members of this section in the branching of the dichasia. At each node of the scape, one dichotomy sometimes grows more strongly than the other, thus giving a paniculate or racemose appearance. In addition, the bracteoles are linear to elliptic and the pedicels have a single line of hairs arising between lateral sepals.

20. **Velleia perfoliata** R.Br., *Prodr.* 581 (1810)

T: Blue Mountains, N.S.W., 1803, *A.Gordon*; holo: BM.

Perennial herb, glabrous; stock short; scapes erect, to 50 cm long. Leaves elliptic to oblanceolate, dentate; lamina 10–25 cm long, 2.5–6 cm wide. Bracteoles connate into open funnel to 8 cm diam. Adaxial sepal broadly elliptic, 8 mm long. Corolla 10–12 mm long, pubescent outside, almost glabrous inside, without enations, yellow; wings ±to base of adaxial lobes, to 2 mm wide. Ovary with septum c. 1/4 as long as locule; ovules c. 10; indusium semi-orbicular, c. 3 mm wide. Capsule globular. Seeds orbicular, c. 3 mm wide, punctate.

Occurs from near Wisemans Ferry and the Colo R. to the Goulburn R. valley, N.S.W., in open forest. Flowers probably Sept.–Dec. Map 398.

N.S.W.: Windsor–Singleton road, Colo R., *R.C.Carolin 5448* (SYD); c. 15 km W of 'Mandalay', Colo R. on Windsor–Singleton road, *J.O'Hara & R.Coveny 3455* (NSW); 5 road miles [c. 8 km] E of Putty road, c. 19 km W of St Albans, *J.Campbell & J.Pickard 1644* (NSW); near 'The Boree Track', 19.3 km SW of Wollombi, *J.Campbell & J.Pickard 1201* (NSW); Dingo Ck, Goulburn R. valley, Hunter Valley, *T.Tame 1146* (NSW).

Differs from all others in the section in the connate bracteoles which form a disc-like, dentate funnel to 8 cm diam. Plants have stout stocks. A rare species which is unsatisfactorily conserved.

21. **Velleia montana** J.D.Hook., *Hooker's London J. Bot.* 6: 265 (1847)

T: Hampshire Hills, Tas., *R.C.Gunn 227*; lecto: K, *fide* R.C.Carolin, *Proc. Linn. Soc. New South Wales* 92: 56 (1967).

Illustration: J.D.Hooker, *Fl. Tasman.* t. 68B (1854).

Perennial herb, glabrous or pubescent; stock short; scapes decumbent or ascending, to 10 cm long. Leaves oblanceolate to obovate, dentate or entire; lamina 1.5–8 cm long, 6–30 mm wide. Bracteoles free, to 5 mm long. Adaxial sepals ovate to oblong, 5–6 mm long. Corolla 7–10 mm long, pubescent outside, pubescent inside, without enations, yellow; wings ±to base of adaxial lobes, c. 0.5 mm wide. Ovary with septum c. 1/2 as long as locule; ovules c. 20; indusium depressed-ovate, c. 1 mm wide. Capsule subglobular. Seeds orbicular to elliptic, c. 1.5 mm diam., punctate. $2n = 16$, W.J.Peacock, *Proc. Linn. Soc. New South Wales* 88: 9 (1963). Fig. 88C.

Occurs in the eastern highlands of N.S.W. and Vic. southwards from New England, and in Tas. Flowers chiefly Nov.–Feb. Map 399.

N.S.W.: Dainers Gap, *M.Gray 0093* (CANB); Clarence–Wolgan, Nov. 1906, *J.H.Maiden* (NSW); Barrington Tops, 7 Jan. 1934, *L.R.Fraser, J.W.Vickery & N.H.Burgess* (SYD). Vic.: Mt Buffalo, *T.B.Muir 643* (MEL). Tas.: Great Lake, 26 Dec. 1937, *E.P.Rodway* (NSW); Lake Leake, *R.Melville 2536* (NSW).

The scapes are usually shorter than the leaves, and the bracteoles are ±linear.

Excluded name

Euthales filiformis Vriese in J.G.C.Lehmann, *Pl. Preiss.* 1: 414 (1845) as *Euthale*

T: near York, W.A., 30 Mar. 1840, *L.Preiss 1889*; holo: L, *fide* J.H.Kern, *Blumea* 13: 116 (1965).

This is *Stellaria filiformis* (Benth.) Mattf. (Caryophyllaceae), *fide* J.H.Kern, *loc. cit.*

10. PENTAPTILON

R.C.Carolin

Pentaptilon E.Pritzel in F.L.E.Diels & E.Pritzel, *Bot. Jahrb. Syst.* 35: 564 (1905); from the Greek *penta* (five) and *ptilon* (wing) referring to the five-winged fruit.

Type: *P. careyi* (F.Muell.) E.Pritzel

Erect herbs, scapose, from a basal stock, glandular-hairy on inflorescence. Leaves mostly basal, tomentose with branched hairs. Inflorescence a terminal thyrse, bracteolate; pedicels not articulate. Sepals adnate to ovary to top. Corolla bilabiate, indistinctly auriculate; lobes unequal; corolla pocket short. Stamens free, epigynous. Ovary winged, 2-locular almost to the top; style simple; indusium horizontal, with white bristles on lips; ovules 2–6 per locule. Fruit indehiscent, soft, with 3 large and 2 smaller ±bladdery wings alternating with persistent sepals. Seeds compressed, with a hard aculeate testa; wing narrow.

A monotypic genus, endemic in W.A. characterised especially by the indehiscent winged fruit.

Pentaptilon careyi (F.Muell.) E.Pritzel in F.L.E.Diels & E.Pritzel, *Bot. Jahrb. Syst.* 35: 564 (1905)

Catospermum careyi F.Muell., *Melb. Chem. & Drugg.* 6: 96 (1884). T: between Northampton and Shark Bay, W.A., *H.Stuart Carey*; holo: MEL.

Illustration: K.Krause, *Pflanzenr.* 54: 115, fig 23 (1912).

Herb to 40 cm tall. Leaves obovate to elliptic, attenuate towards base, entire, tomentose; lamina 3–7 cm long, 1–3.5 cm wide. Inflorescence bracts linear-oblong, to 25 mm long; peduncles to 4 cm long; flowers almost sessile between bracteoles; bracteoles linear, to 10 mm long. Sepals oblong to elliptic, c. 1 mm long. Corolla 7–9 mm long, with glandular hairs outside and a dense tuft of white hairs in throat, yellow on wings, brownish red in throat. Indusium oblong, c. 1 mm long. Fruit 7–8 mm long and wide, including wings. Figs 24E, 25I, 89C–D.

Occurs from Northampton to mouth of Murchison R., W.A., in sandy soil in heath and mallee communities. Flowers Aug.–Oct. Map 400.

W.A.: c. 42 km N of Murchison R., *W.R.Barker 2184* (AD, SYD); c. 16 km N of Northampton, *R.C.Carolin 3325* (SYD).

Easily recognised by the swollen wings on the ovary and fruit.

11. VERREAUXIA

R.C.Carolin

Verreauxia Benth., *Fl. Austral.* 4: 105 (1868); named for the French naturalist, Jules Verreaux, who collected plants in Australia during the late 1840s.

Type: *V. verreauxii* (Vriese) Carolin

Herbs or shrubs with 1–few erect stems. Leaves tapering towards base, entire, densely tomentose or villous. Inflorescence loose or spike-like, often a branched thyrse on a terminal scape, bracteolate; pedicel articulate. Sepals adnate to top of ovary. Corolla yellow, sometimes brownish outside; lobes unequal; wings ±equal. Stamens free, epigynous. Ovary 1-locular, with reddish or golden multicellular hairs between 3 of the sepaline ribs; ovule solitary; style simple; indusium horizontal, oblong, ±grooved, with short white bristles on lips. Fruit a compressed, hairy nut surmounted by sepals. Seeds flat, ±distinctly rimmed, with hard testa; wing obsolete. Embryo spathulate.

A genus of 3 species endemic in south-western Australia, characterised especially by the indehiscent very hairy nut-like fruit. Related to *Pentaptilon* which has the same type of branched hairs, but in which the ovary and fruit are winged.

1 Leaves all basal; flowers pedicellate in loose thyrses **1. V. verreauxii**

1: Leaves cauline; flowers sessile in spike-like thyrses

2 Leaves felted, usually greyish **2. V. reinwardtii**

2: Leaves villous, usually yellowish **3. V. villosa**

1. Verreauxia verreauxii (Vriese) Carolin, *Telopea* 2: 75 (1980)

Dampiera verreauxii Vriese, *Natuurk. Verh. Holl. Maatsch. Wetensch. Haarlem* ser. 2, 10: 118 (1854). T: without locality, Australia, per *J.Verreaux*; holo: P.

V. paniculata Benth., *Fl. Austral.* 4: 105 (1868), *nom. superfl.* as *Dampiera verreauxii* listed in synonymy.

Illustrations: W.H.de Vriese, *op. cit.*, t. 20, as *Dampiera verreauxii*; K.Krause, *Pflanzenr.* 54: 170, fig. 30D–H (1912) as *V. paniculata.*

Perennial, densely villous-tomentose; stock short, thick; scapes to 50 cm tall. Leaves basal, elliptic to obovate; lamina to 60 mm long, to 30 mm wide. Inflorescence a loose, terminal thyrse, with scattered glandular hairs; peduncles to 50 mm long; pedicels c. 16 mm long. Bract and bracteoles linear, mostly to 2 mm long. Corolla c. 10 mm long, sparsely villous and with glandular hairs outside, ±glabrous inside; scarcely auriculate; abaxial lobes elliptic-oblong, 4–5 mm long; wings ciliate. Ovary surface visible beneath hairs.

Now only known to occur near Beverley, W.A., in sandy soil in low open woodland. Flowers Dec.–Mar. Map 401.

W.A.: WSW of Beverley, *S.van Leeuwen 242* (PERTH); W of Beverley, *S.Patrick 411* (PERTH).

Differs from the other 2 species in the habit and the less densely hairy flowers. The type was probably collected by J.Drummond (*Drummond 4: 186* cited by Krause, loc. cit.). Thought to be extinct until recently when 2 populations were rediscovered. There has been only sporadic collecting during the summer in the area where this species occurs and it is possible that there are more populations.

Figure 89. **A–B**, *Verreauxia reinwardtii*. **A**, habit X0.5 (NSW 52411, NSW); **B**, fruit X7.5 (A.Travers 17, NSW). **C–D**, *Pentaptilon careyi*. **C**, fruit X4; **D**, habit X0.5 (**C–D**, W.Barker 2184, AD). Drawn by D.Mackay.

2. **Verreauxia reinwardtii** (Vriese) Benth., *Fl. Austral.* 4: 105 (1868)

Scaevola reinwardtii Vriese in J.G.C.Lehmann, *Pl. Preiss.* 1: 409 (1845); *Dampiera reinwardtii* (Vriese) Vriese, *Natuurk. Verh. Holl. Maatsch. Wetensch. Haarlem* ser. 2, 10: 97 (1854). T: Quangen, Victoria [District], W.A., 20 Mar. 1840, *L.Preiss 1454*; lecto: LD, *fide* R.C.Carolin, *Fl. Australia* 35: 334 (1992).

Illustrations: W.H.de Vriese, *op. cit.* t. 5, as *Dampiera reinwardtii*; K.Krause, *Pflanzenr.* 54: 170, fig. 30A–C (1912).

Erect shrub, branched, to 1 m tall, greyish tomentose. Leaves cauline, obovate to orbicular; lamina mostly 50–70 mm long, 10–30 mm wide. Inflorescence an interrupted spike, the clusters distant from each other; scape to 40 cm long, usually branched at base; peduncles and pedicels very short; bracts and bracteoles narrowly deltoid, mostly to 2 mm long. Corolla 8–10 mm long, tomentose outside and in throat, ±auriculate; abaxial lobes oblong, to 4 mm long; wings entire, glabrous. Ovary surface obscured by short, white tomentum, and usually with long reddish or golden jointed hairs. Figs 24C, 89A–B.

Occurs from Geraldton S to Perth and inland to Mullewa and Narrogin, W.A., in sandy soil in heath and open woodland. Flowers chiefly Oct.–Feb. Map 402.

W.A.: Brand Hwy, 31 km N of junction with road between Wanneroo and Gingin, *P.Weston 267* (SYD); 16 km S of Badgingarra, *P.G.Wilson 3872* (PERTH); c. 43 km from Dongara towards Eneabba, 22 Sept. 1968, *M.E.Phillips* (CBG, SYD); Wongamine Nature Reserve, *G.J.Keighery 6925* (PERTH); 3 km S of Wannamal, *G.J.Keighery 2009* (PERTH).

3. **Verreauxia villosa** E.Pritzel in F.L.E.Diels & E.Pritzel, *Bot. Jahrb. Syst.* 35: 573 (1904)

T: near Menzies, Oct. 1901, *F.L.E.Diels 5191*; holo: ?B (destroyed) *n.v.*

V. dyeri E.Pritzel ex Hemsley in W.J.Hooker, *Icon. Pl.* t. 2782 (1905). T: railway between Cunderdin and Dedari, W.A., *G.H.Thistleton-Dyer*; lecto: K, *fide* R.C.Carolin, *Fl. Australia* 35: 334 (1992).

Very similar to the preceding species. Shrub yellowish, silky above the felted tomentum. Flower clusters more condensed on scape. Corolla 5–8 mm long. Fig. 77.

Occurs between Cunderdin, Hyden and Menzies, W.A., in sandy soil in open woodland. Flowers chiefly Oct.–Jan. Map 403.

W.A.: W of Boorabbin, *A.M.Ashby 3175* (AD, SYD); Bendering, 16 Nov. 1923, *C.A.Gardner* (PERTH); near Narembeen S of Merredin, Sept. 1929, *W.E.Blackall* (PERTH); 4.3 km NW along Brennand road from Emu Fence road S of Southern Cross, *A.S.George 16452* (PERTH).

MAPS

Number in brackets refers to the page on which the taxon is described.

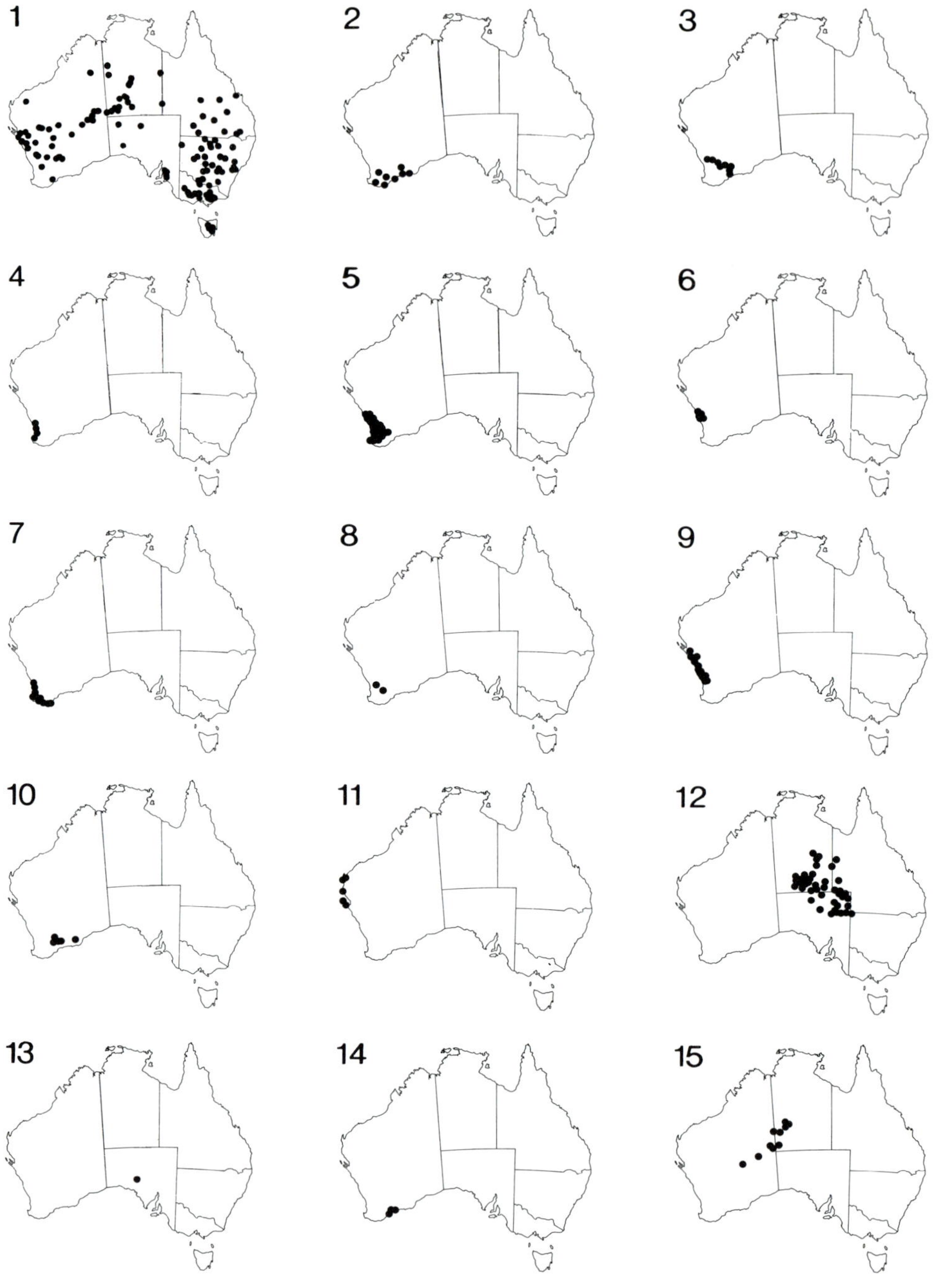

1. Brunonia australis (3)
2 Anthotium humile (15)
3. Anthotium rubriflorum (16)
4. Anthotium junciforme (16)
5. Lechenaultia biloba (19)
6. Lechenaultia stenosepala (20)
7. Lechenaultia expansa (20)
8. Lechenaultia pulvinaris (21)
9. Lechenaultia floribunda (22)
10. Lechenaultia papillata (22)
11. Lechenaultia subcymosa (23)
12. Lechenaultia divaricata (24)
13. Lechenaultia aphylla (24)
14. Lechenaultia heteromera (24)
15. Lechenaultia lutescens (26)

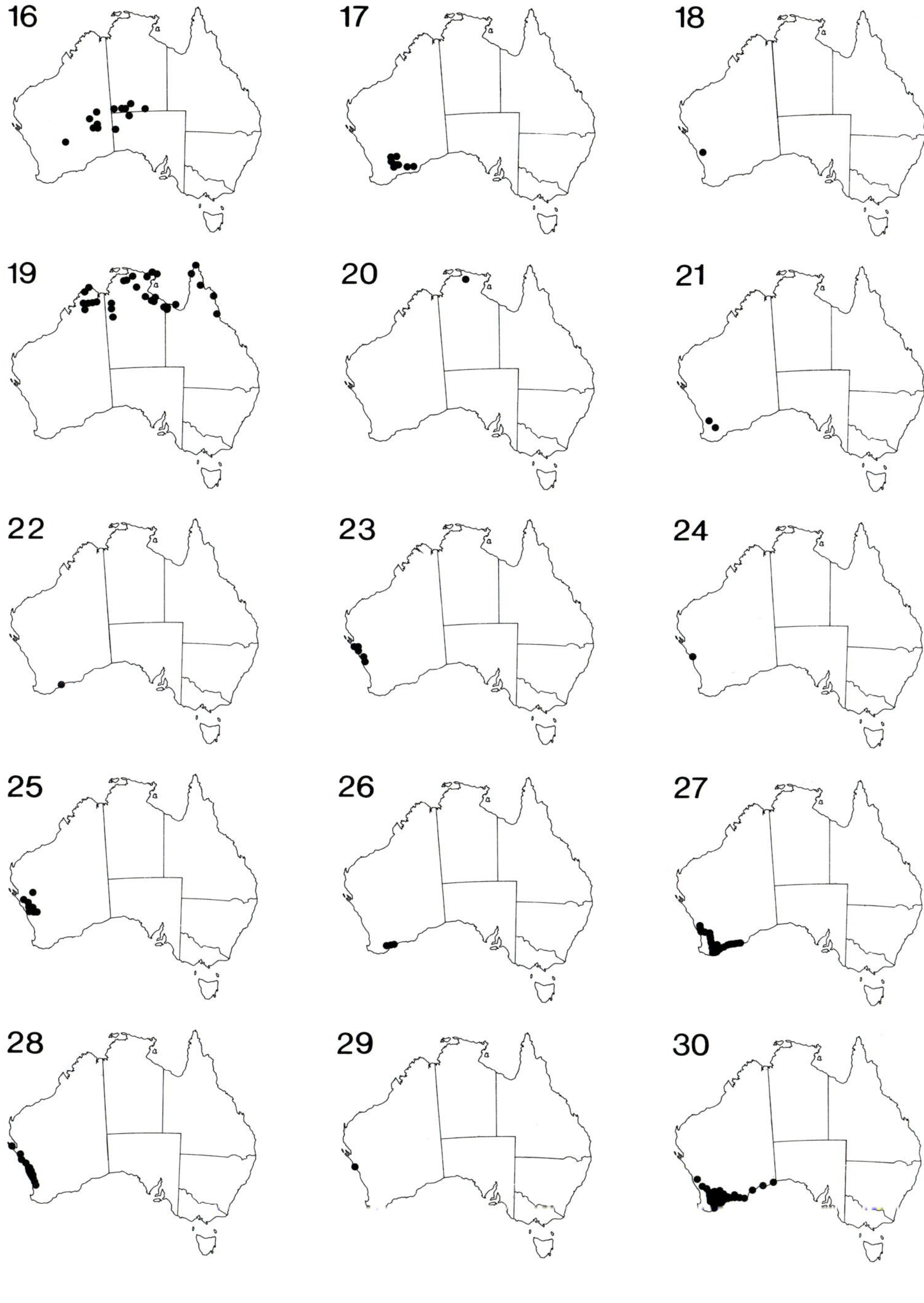

16. Lechenaultia striata (26)
17. Lechenaultia brevifolia (27)
18. Lechenaultia juncea (27)
19. Lechenaultia filiformis (28)
20. Lechenaultia ovata (28)
21. Lechenaultia laricina (29)
22. Lechenaultia superba (30)
23. Lechenaultia hirsuta (30)
24. Lechenaultia longiloba (31)
25. Lechenaultia macrantha (31)
26. Lechenaultia acutiloba (31)
27. Lechenaultia tubiflora (32)
28. Lechenaultia linarioides (33)
29. Lechenaultia chlorantha (33)
30. Lechenaultia formosa (34)

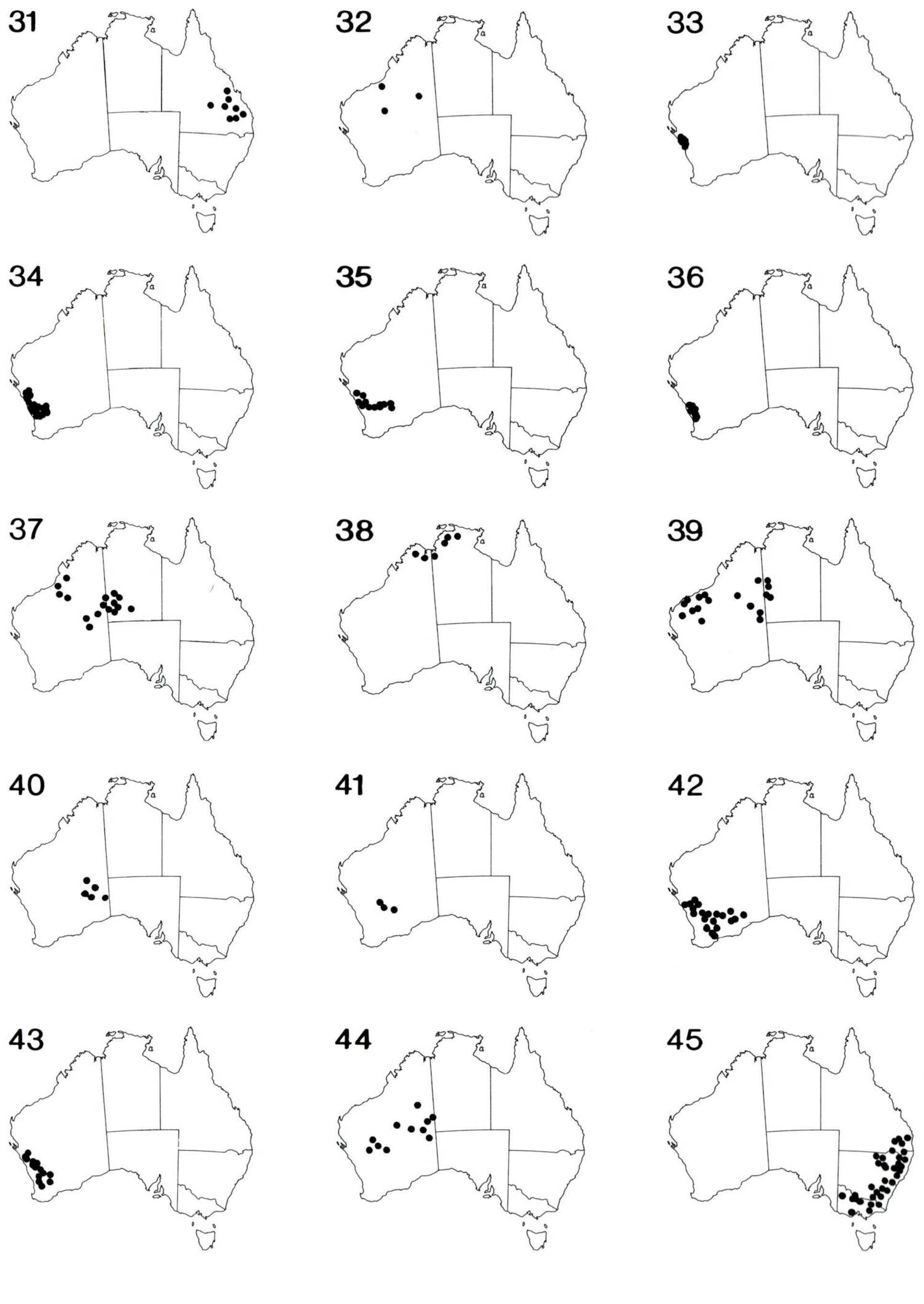

31. Dampiera discolor (41)
32. Dampiera atriplicina (41)
33. Dampiera krauseana (43)
34. Dampiera spicigera (43)
35. Dampiera stenostachya (44)
36. Dampiera teres (44)
37. Dampiera cinerea (44)
38. Dampiera conospermoides (45)
39. Dampiera candicans (45)
40. Dampiera ramosa (47)
41. Dampiera plumosa (47)
42. Dampiera eriocephala (47)
43. Dampiera wellsiana (48)
44. Dampiera dentata (48)
45. Dampiera purpurea (49)

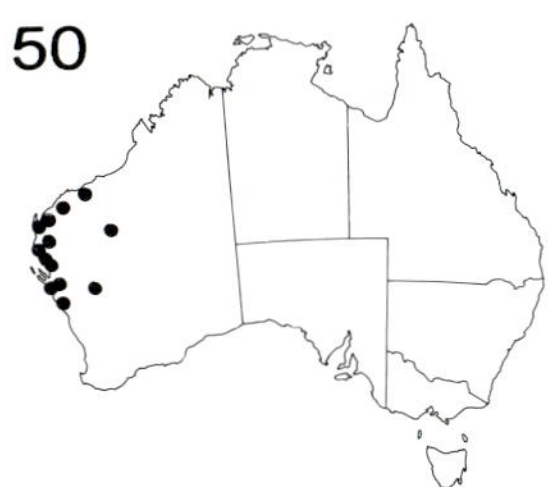

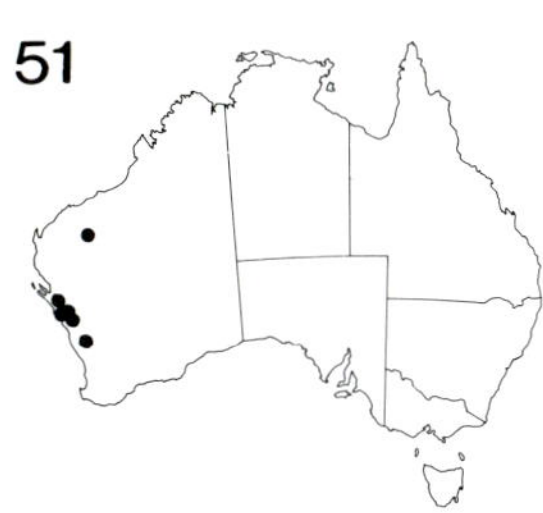

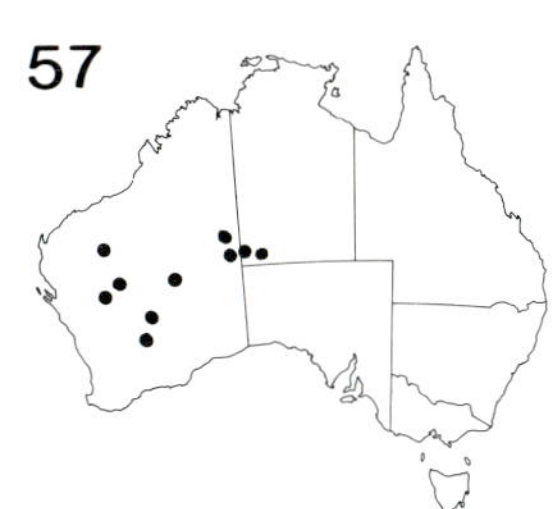

46. Dampiera ferruginea (50)

47. Dampiera altissima (50)

48. Dampiera salahae (51)

49. Dampiera tephrea (51)

50. Dampiera incana var. incana (53)

51. Dampiera incana var. fuscescens (53)

52. Dampiera orchardii (53)

53. Dampiera scaevolina (54)

54. Dampiera tenuicaulis var. tenuicaulis (54)

55. Dampiera tenuicaulis var. curvula (55)

56. Dampiera diversifolia (55)

57. Dampiera roycei (55)

58. Dampiera marifolia (56)

59. Dampiera haematotricha subsp. haematotricha (57)

60. Dampiera haematotricha subsp. dura (57)

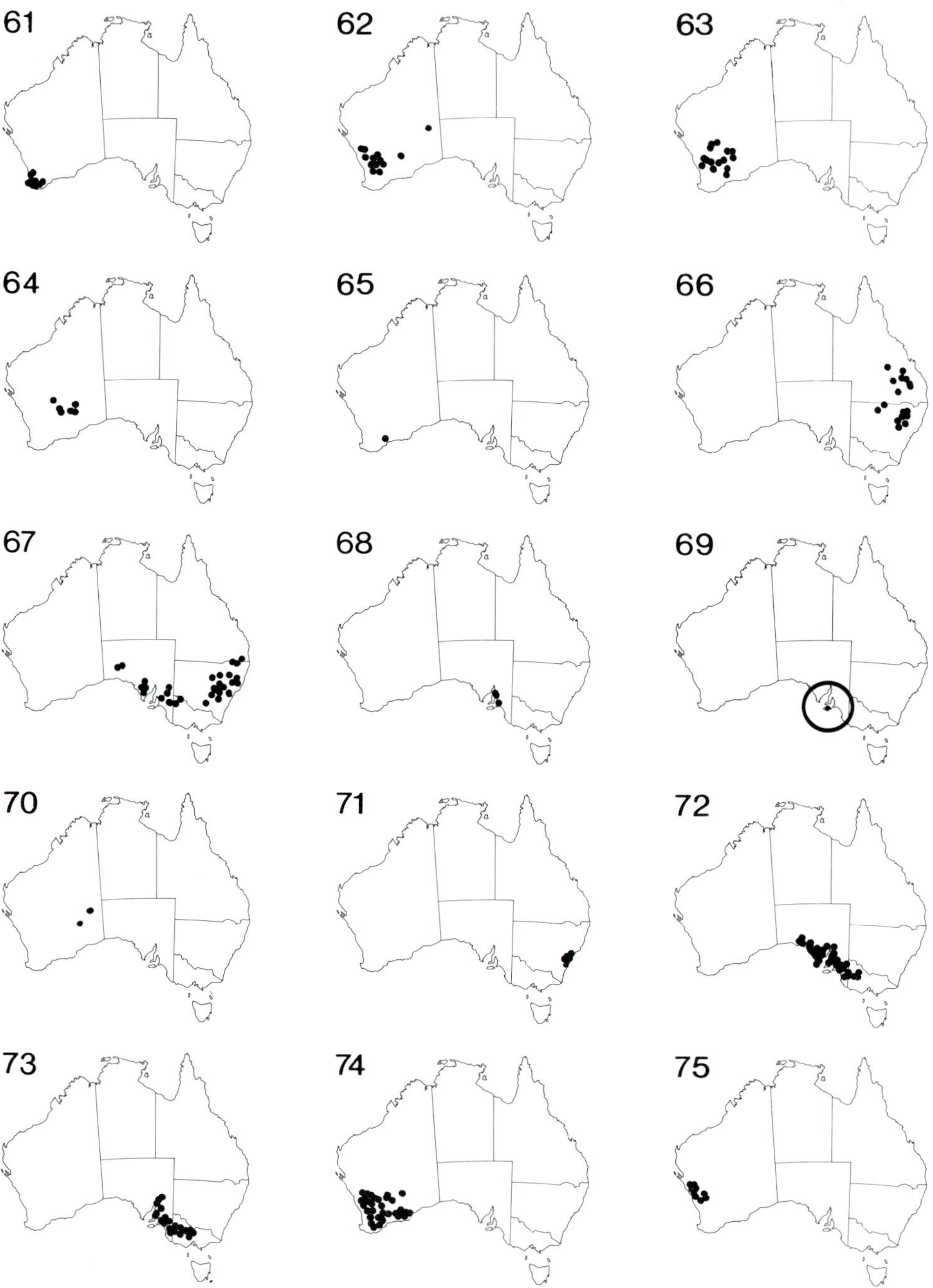

61. Dampiera hederacea (57)
62. Dampiera tomentosa (59)
63. Dampiera luteiflora (59)
64. Dampiera stenophylla (60)
65. Dampiera fitzgeraldensis (60)
66. Dampiera adpressa (60)
67. Dampiera lanceolata var. lanceolata (61)
68. Dampiera lanceolata var. intermedia (62)
69. Dampiera lanceolata var. insularis (62)
70. Dampiera eriantha (62)
71. Dampiera scottiana (62)
72. Dampiera rosmarinifolia (63)
73. Dampiera dysantha (63)
74. Dampiera lavandulacea (64)
75. Dampiera oligophylla (65)

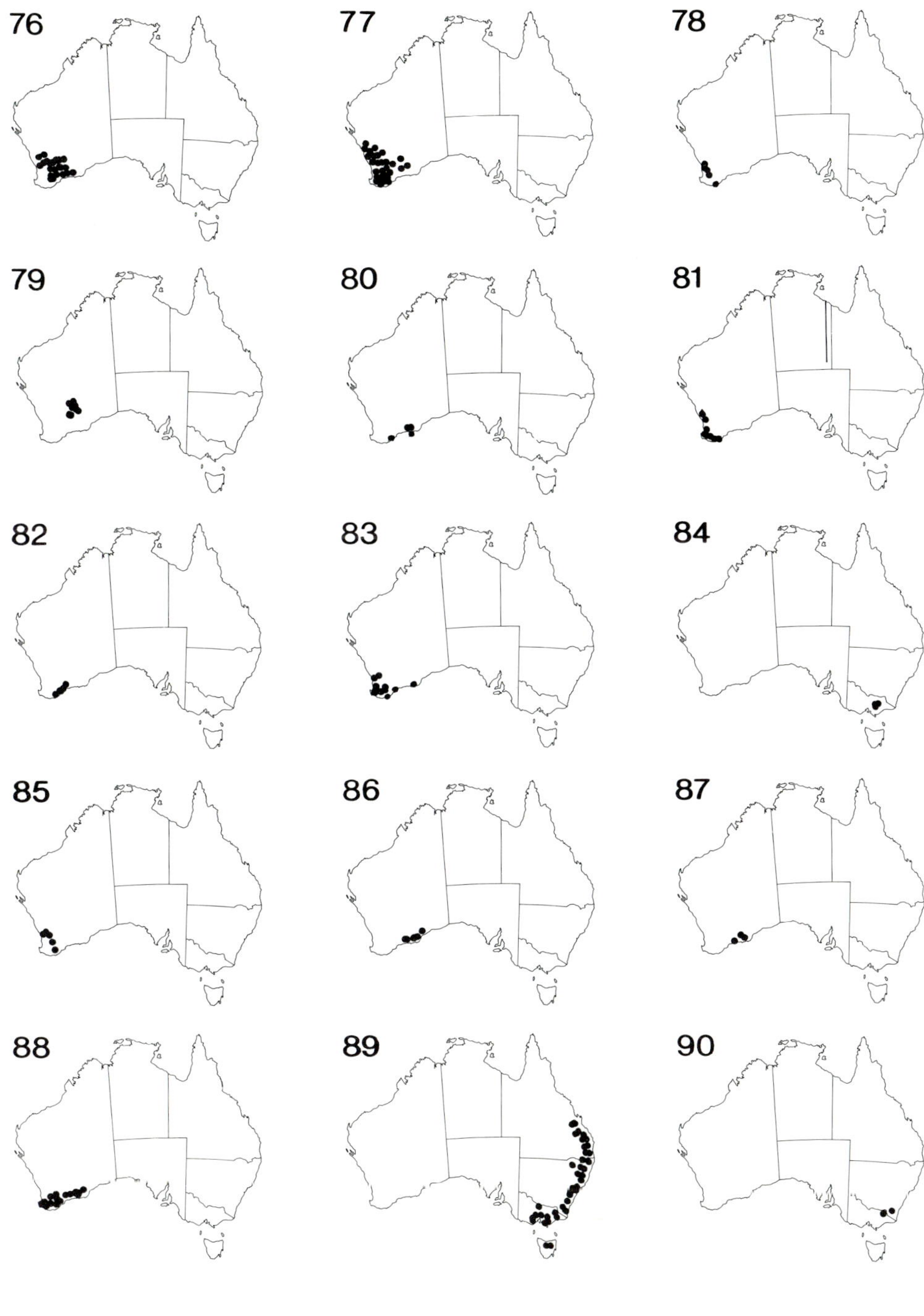

76. Dampiera juncea (65)
77. Dampiera linearis (65)
78. Dampiera pedunculata (66)
79. Dampiera latealata (67)
80. Dampiera decurrens (68)
81. Dampiera trigona (68)
82. Dampiera loranthifolia (69)
83. Dampiera leptoclada (69)
84. Dampiera galbraithiana (70)
85. Dampiera triloba (70)
86. Dampiera parvifolia (72)
87. Dampiera sericantha (72)
88. Dampiera fasciculata (72)
89. Dampiera stricta (73)
90. Dampiera fusca (73)

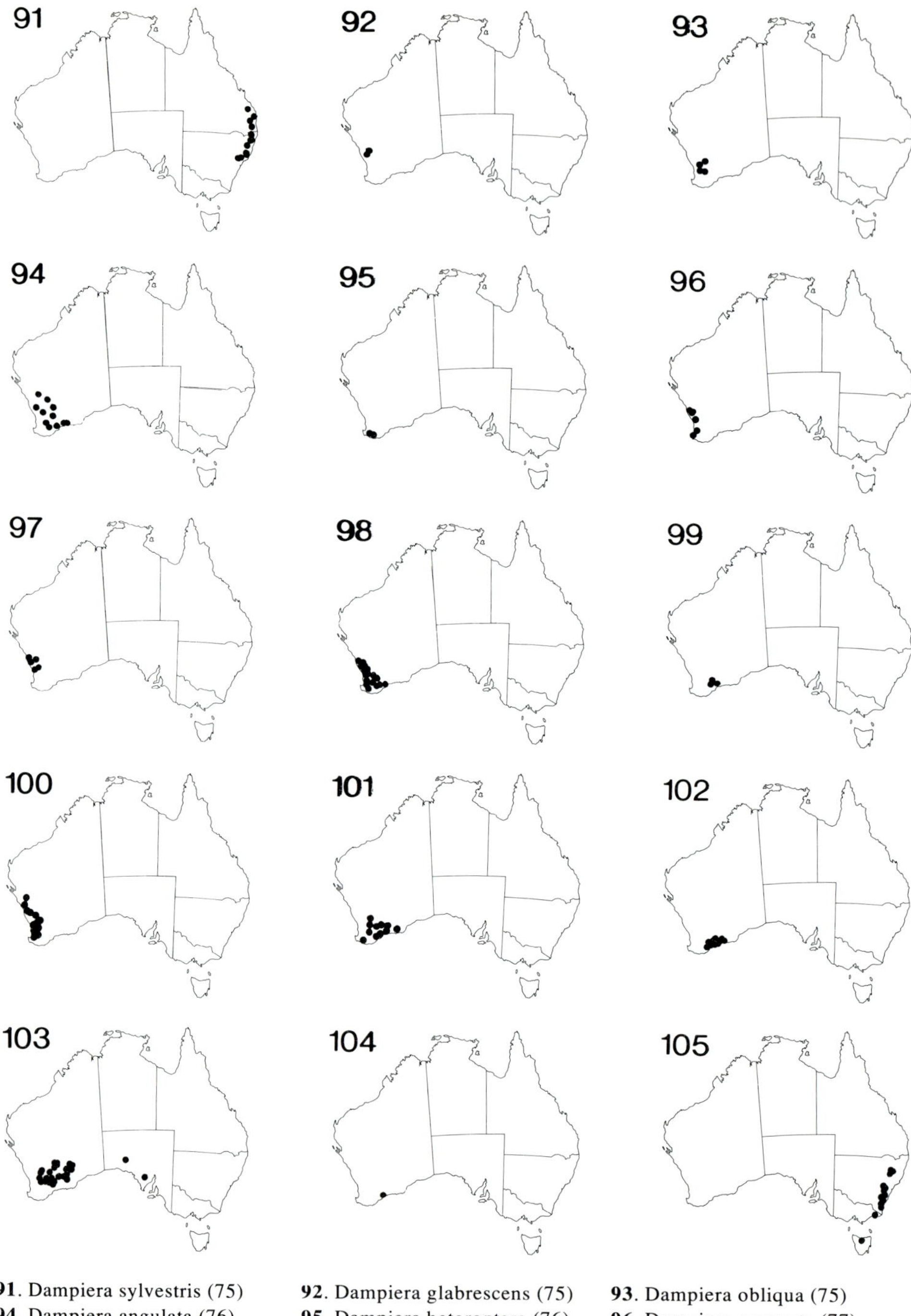

91. Dampiera sylvestris (75)
94. Dampiera angulata (76)
97. Dampiera carinata (77)
100. Dampiera lindleyi (78)

103. Coopernookia strophiolata (81)

92. Dampiera glabrescens (75)
95. Dampiera heteroptera (76)
98. Dampiera alata (77)
101. Dampiera sacculata (79)

104. Coopernookia georgei (83)

93. Dampiera obliqua (75)
96. Dampiera coronata (77)
99. Dampiera deltoidea (78)
102. Coopernookia polygalacea (81)
105. Coopernookia barbata (83)

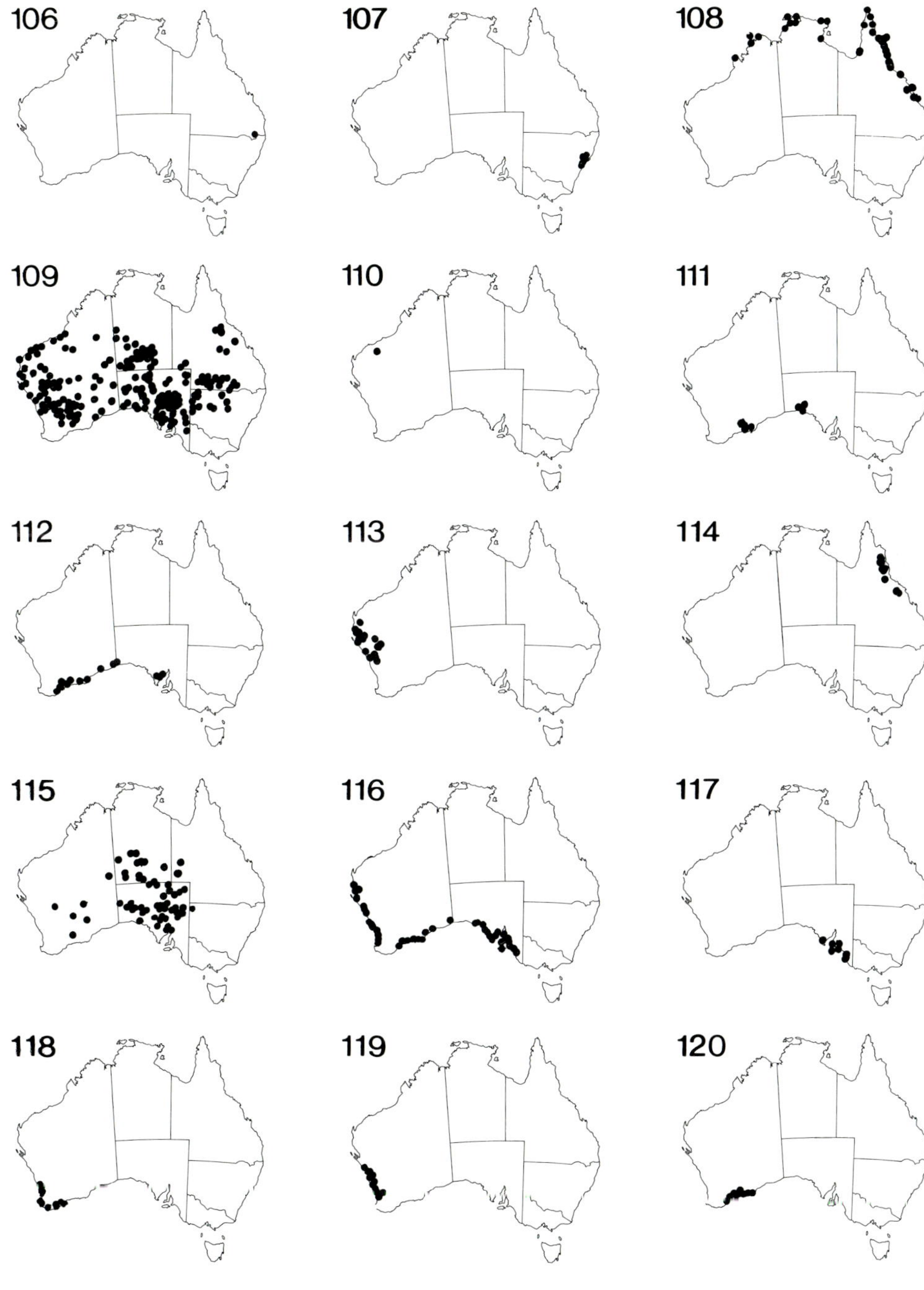

106. Coopernookia scabridiuscula (84)
107. Coopernookia chisholmii (84)
108. Scaevola taccada (97)
109. Scaevola spinescens (97)
110. Scaevola acacioides (99)
111. Scaevola bursariifolia (99)
112. Scaevola myrtifolia (100)
113. Scaevola tomentosa (100)
114. Scaevola enantophylla (101)
115. Scaevola collaris (103)
116. Scaevola crassifolia (103)
117. Scaevola angustata (104)
118. Scaevola nitida (104)
119. Scaevola thesioides subsp. thesioides (105)
120. Scaevola thesioides subsp. filifolia (106)

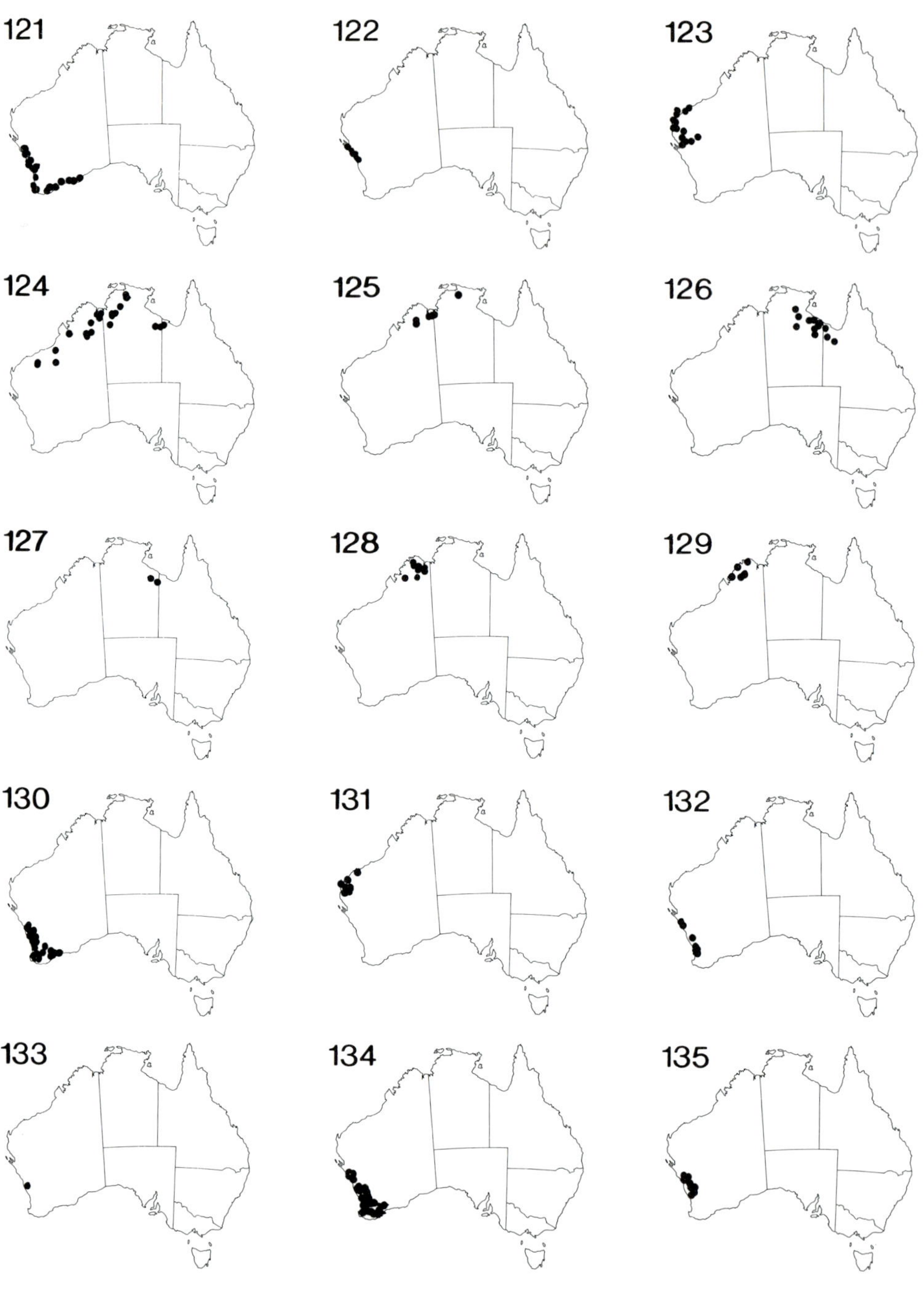

121. Scaevola globulifera (106)
122. Scaevola porocarya (107)
123. Scaevola cunninghamii (107)
124. Scaevola browniana subsp. browniana (108)
125. Scaevola browniana subsp. grandior (109)
126. Scaevola revoluta subsp. revoluta var. revoluta (110)
127. Scaevola revoluta subsp. revoluta var. viscida (110)
128. Scaevola revoluta subsp. stenostachya (110)
129. Scaevola macrostachya (111)
130. Scaevola glandulifera (111)
131. Scaevola pulchella (113)
132. Scaevola anchusifolia (113)
133. Scaevola eneabba (114)
134. Scaevola lanceolata (114)
135. Scaevola virgata (115)

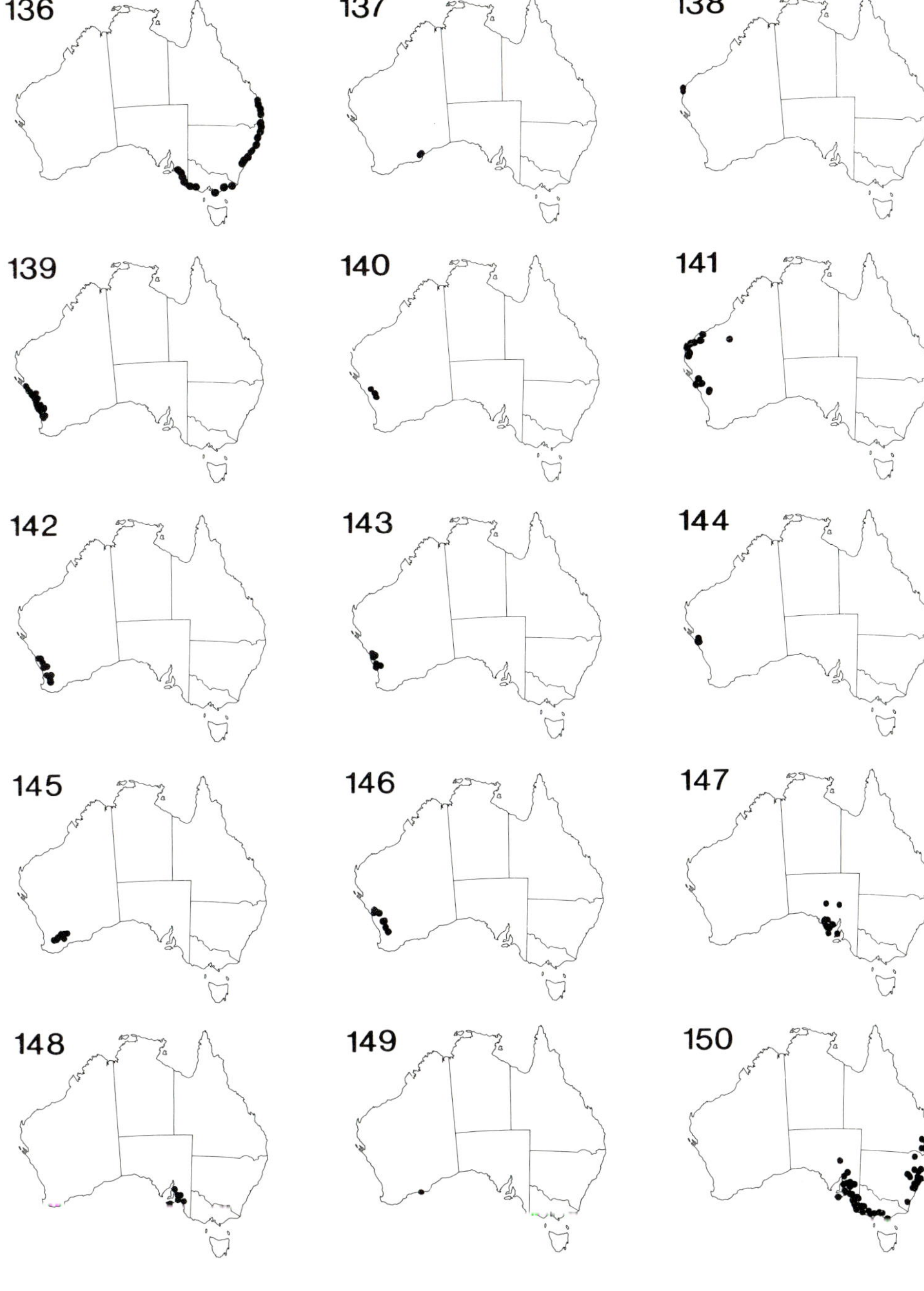

136. Scaevola calendulacea (115)
137. Scaevola brookeana (116)
138. Scaevola spicigera (116)
139. Scaevola canescens (117)
140. Scaevola globosa (117)
141. Scaevola sericophylla (117)
142. Scaevola repens var. repens (118)
143. Scaevola repens var. angustifolia (119)
144. Scaevola oldfieldii (119)
145. Scaevola pulvinaris (119)
146. Scaevola humifusa (120)
147. Scaevola linearis subsp. linearis (121)
148. Scaevola linearis subsp. confertifolia (121)
149. Scaevola paludosa (121)
150. Scaevola albida (122)

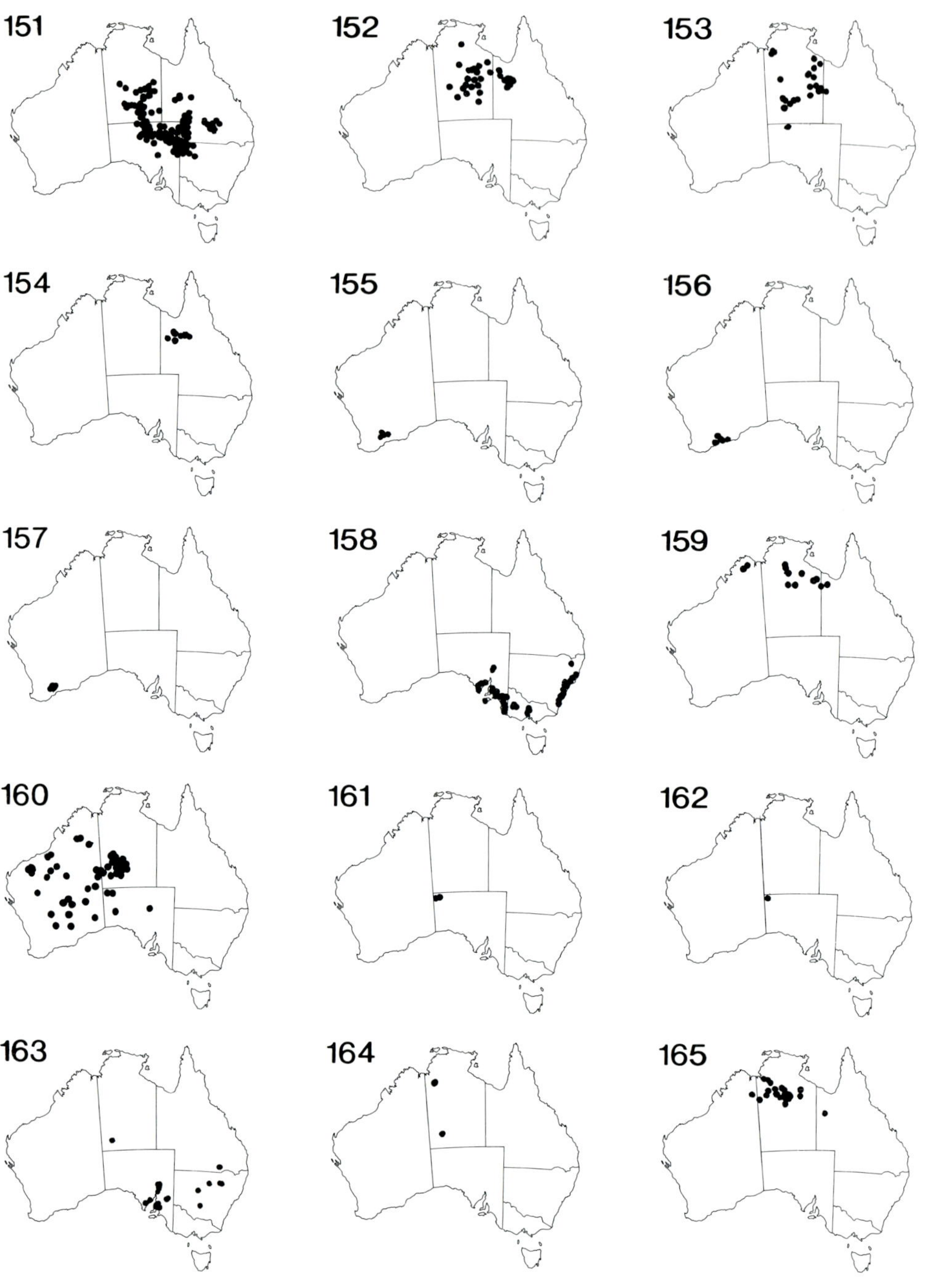

151. Scaevola parvibarbata (123)
152. Scaevola ovalifolia (123)
153. Scaevola glabrata (124)
154. Scaevola glutinosa (124)
155. Scaevola densifolia (126)
156. Scaevola cuneiformis (126)
157. Scaevola argentea (127)
158. Scaevola aemula (127)
159. Scaevola amblyanthera var. amblyanthera (128)
160. Scaevola amblyanthera var. centralis (128)
161. Scaevola collina (129)
162. Scaevola obovata (129)
163. Scaevola humilis (130)
164. Scaevola graminea (130)
165. Scaevola laciniata (132)

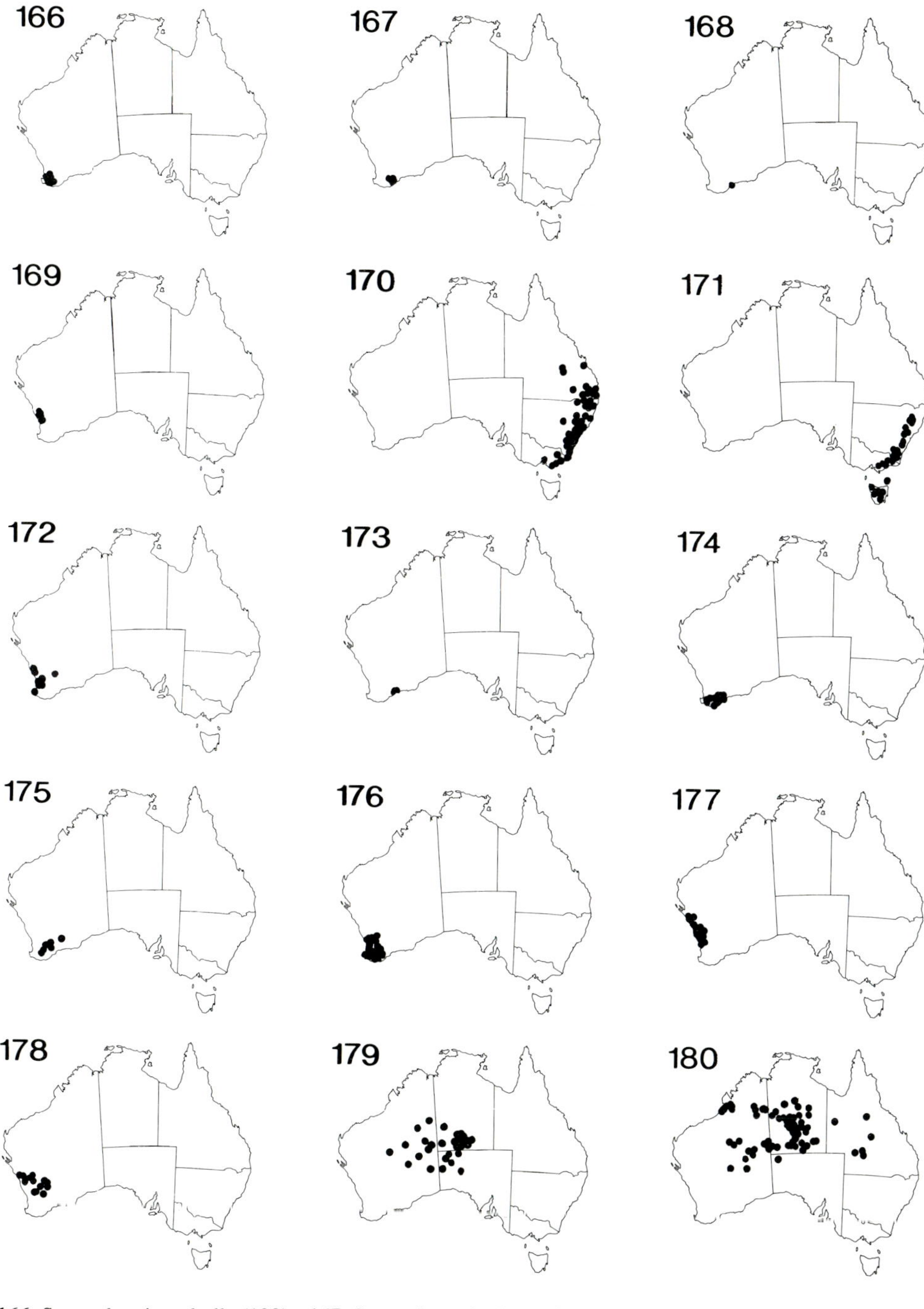

166. Scaevola microphylla (132)
167. Scaevola auriculata (132)
168. Scaevola macrophylla (133)
169. Scaevola platyphylla (133)
170. Scaevola ramosissima (134)
171. Scaevola hookeri (135)
172. Scaevola pilosa (135)
173. Scaevola tenuifolia (136)
174. Scaevola striata var. striata (137)
175. Scaevola striata var. arenaria (137)
176. Scaevola calliptera (137)
177. Scaevola phlebopetala (138)
178. Scaevola hamiltonii (139)
179. Scaevola basedowii (139)
180. Scaevola parvifolia subsp. parvifolia (141)

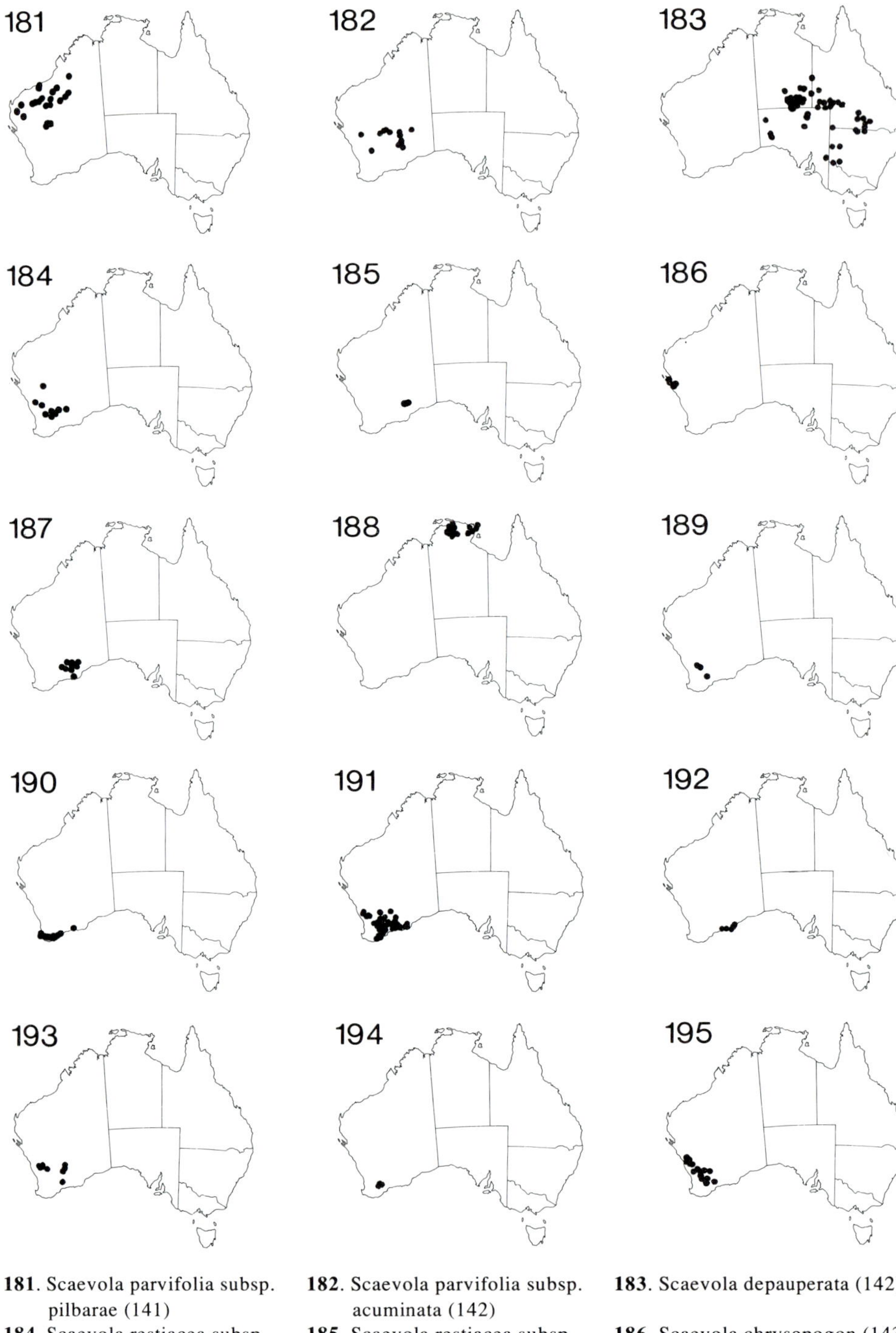

181. Scaevola parvifolia subsp. pilbarae (141)

182. Scaevola parvifolia subsp. acuminata (142)

183. Scaevola depauperata (142)

184. Scaevola restiacea subsp. restiacea (143)

185. Scaevola restiacea subsp. divaricata (143)

186. Scaevola chrysopogon (143)

187. Scaevola oxyclona (144)

188. Scaevola angulata (144)

189. Scaevola tortuosa (145)

190. Diaspasis filifolia (146)

191. Goodenia scapigera (167)

192. Goodenia decursiva (167)

193. Goodenia watsonii subsp. watsonii (168)

194. Goodenia watsonii subsp. glandulosa (168)

195. Goodenia pinifolia (169)

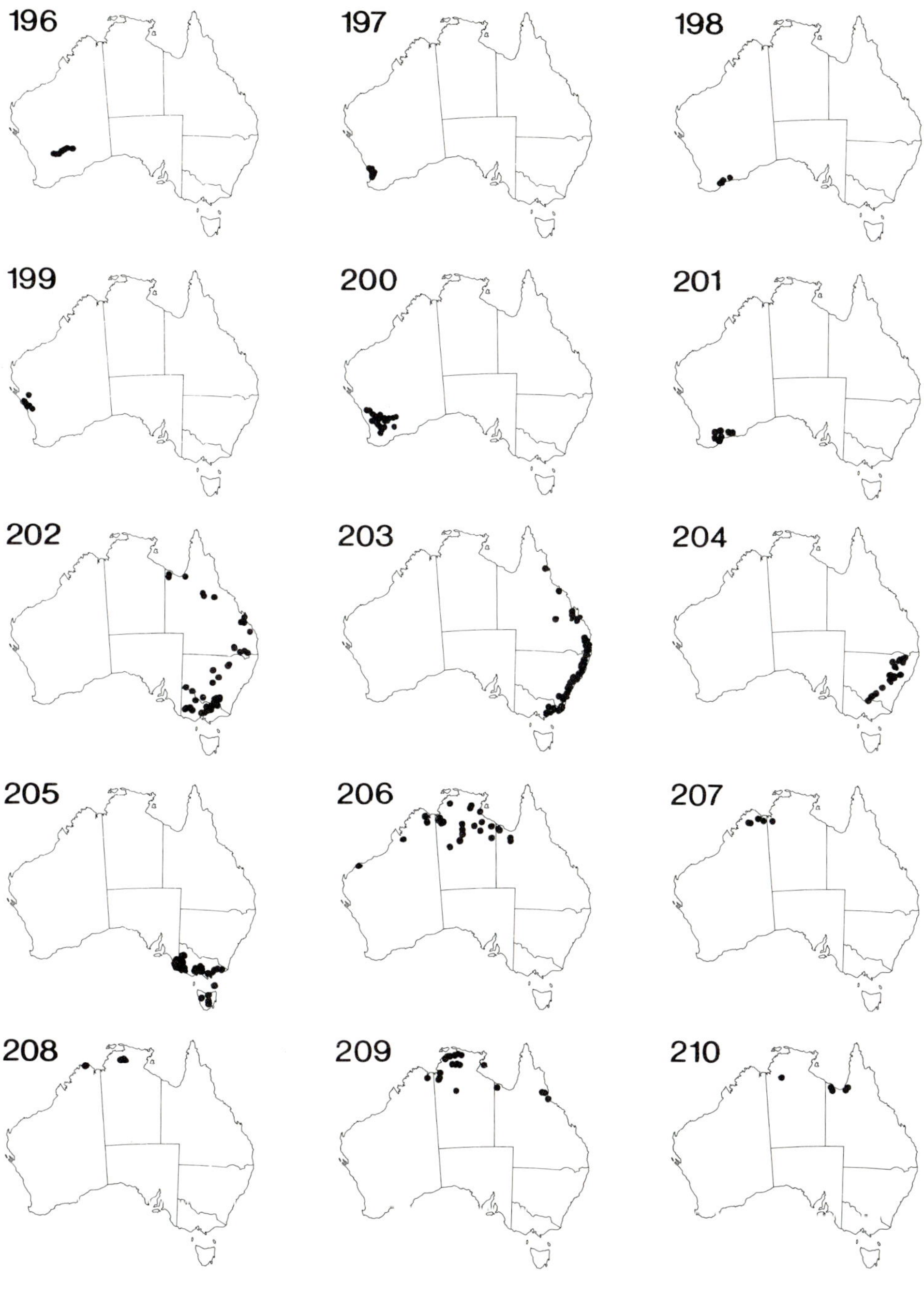

196. Goodenia elderi (169)
197. Goodenia fasciculata (169)
198. Goodenia stenophylla (170)
199. Goodenia drummondii (170)
200. Goodenia helmsii (171)
201. Goodenia viscida (171)
202. Goodenia gracilis (173)
203. Goodenia paniculata (174)
204. Goodenia macbarronii (174)
205. Goodenia humilis (175)
206. Goodenia lamprosperma (176)
207. Goodenia bicolor (176)
208. Goodenia gloeophylla (177)
209. Goodenia purpurascens (177)
210. Goodenia minutiflora (177)

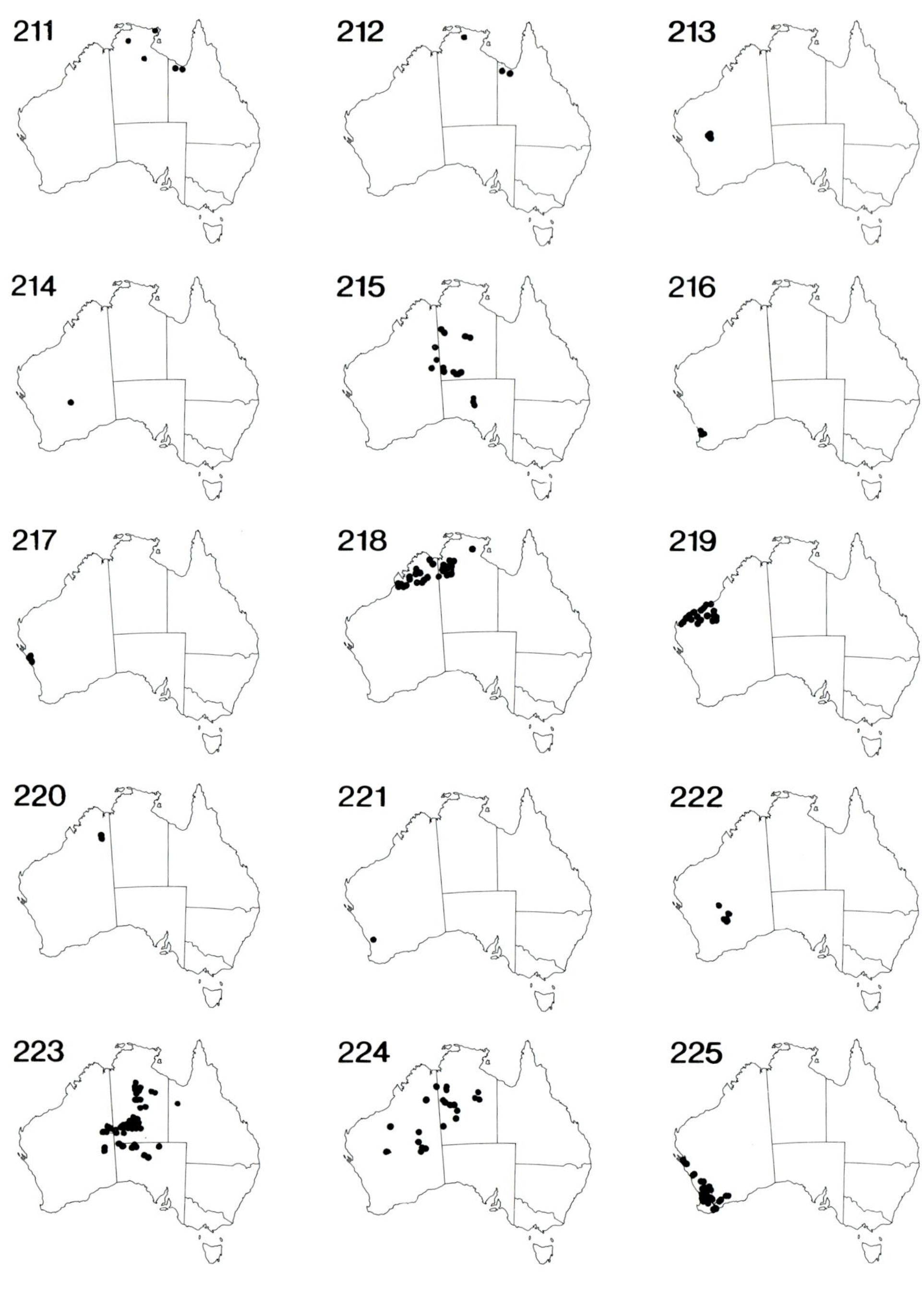

211. Goodenia viscidula (178)
212. Goodenia paludicola (178)
213. Goodenia berringbinensis (179)
214. Goodenia lyrata (179)
215. Goodenia modesta (181)
216. Goodenia claytoniacea (181)
217. Goodenia sericostachya (182)
218. Goodenia scaevolina (183)
219. Goodenia stobbsiana (183)
220. Goodenia suffrutescens (184)
221. Goodenia arthrotricha (184)
222. Goodenia eremophila (184)
223. Goodenia ramelii (185)
224. Goodenia azurea (185)
225. Goodenia caerulea (186)

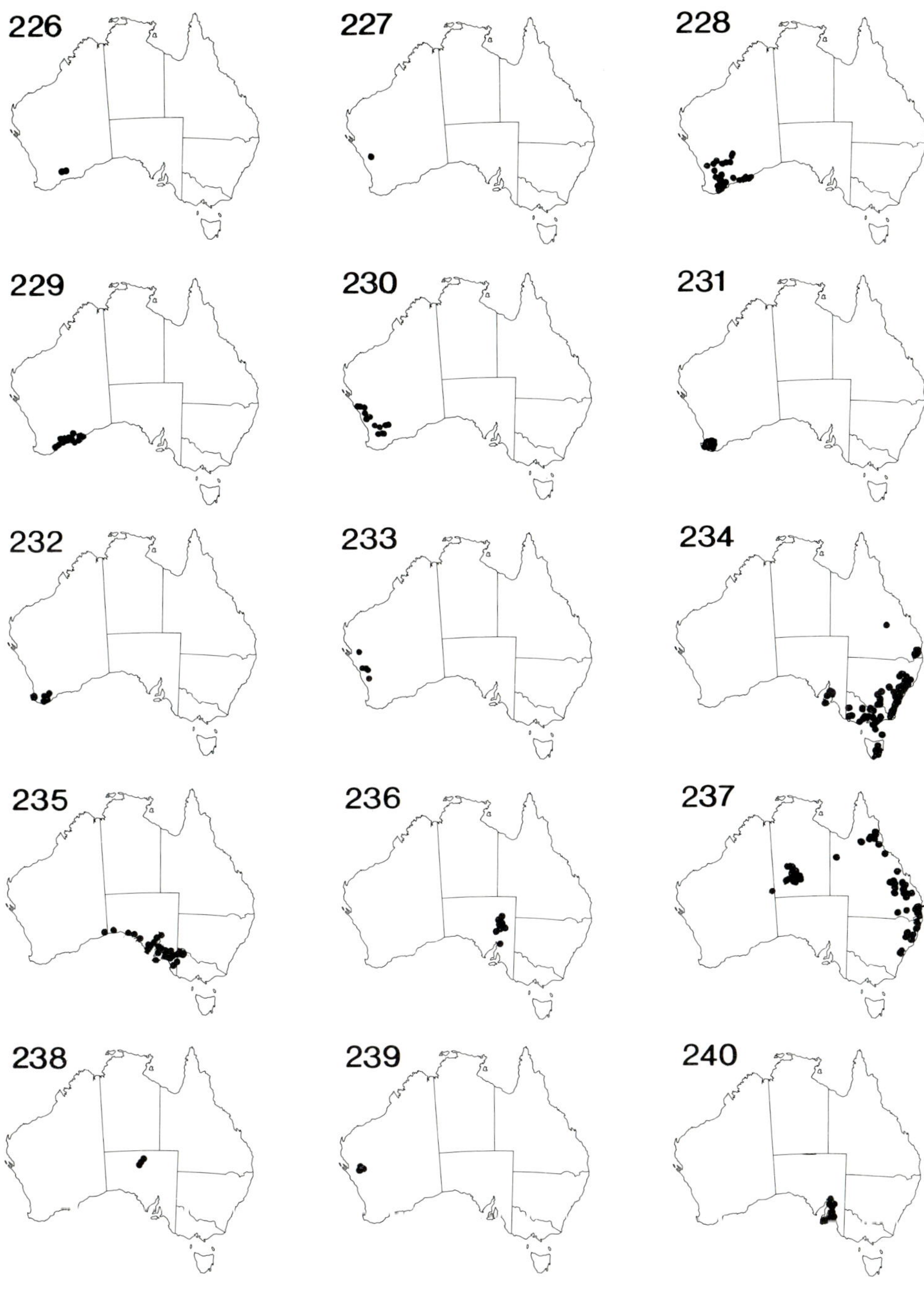

226. Goodenia trichophylla (187)
227. Goodenia perryi (187)
228. Goodenia incana (188)
229. Goodenia pterigosperma (188)
230. Goodenia glareicola (190)
231. Goodenia eatoniana (190)
232. Goodenia leptoclada (190)
233. Goodenia hassallii (191)
234. Goodenia ovata (192)
235. Goodenia varia (193)
236. Goodenia vernicosa (193)
237. Goodenia grandiflora (194)
238. Goodenia chambersii (195)
239. Goodenia kingiana (195)
240. Goodenia amplexans (196)

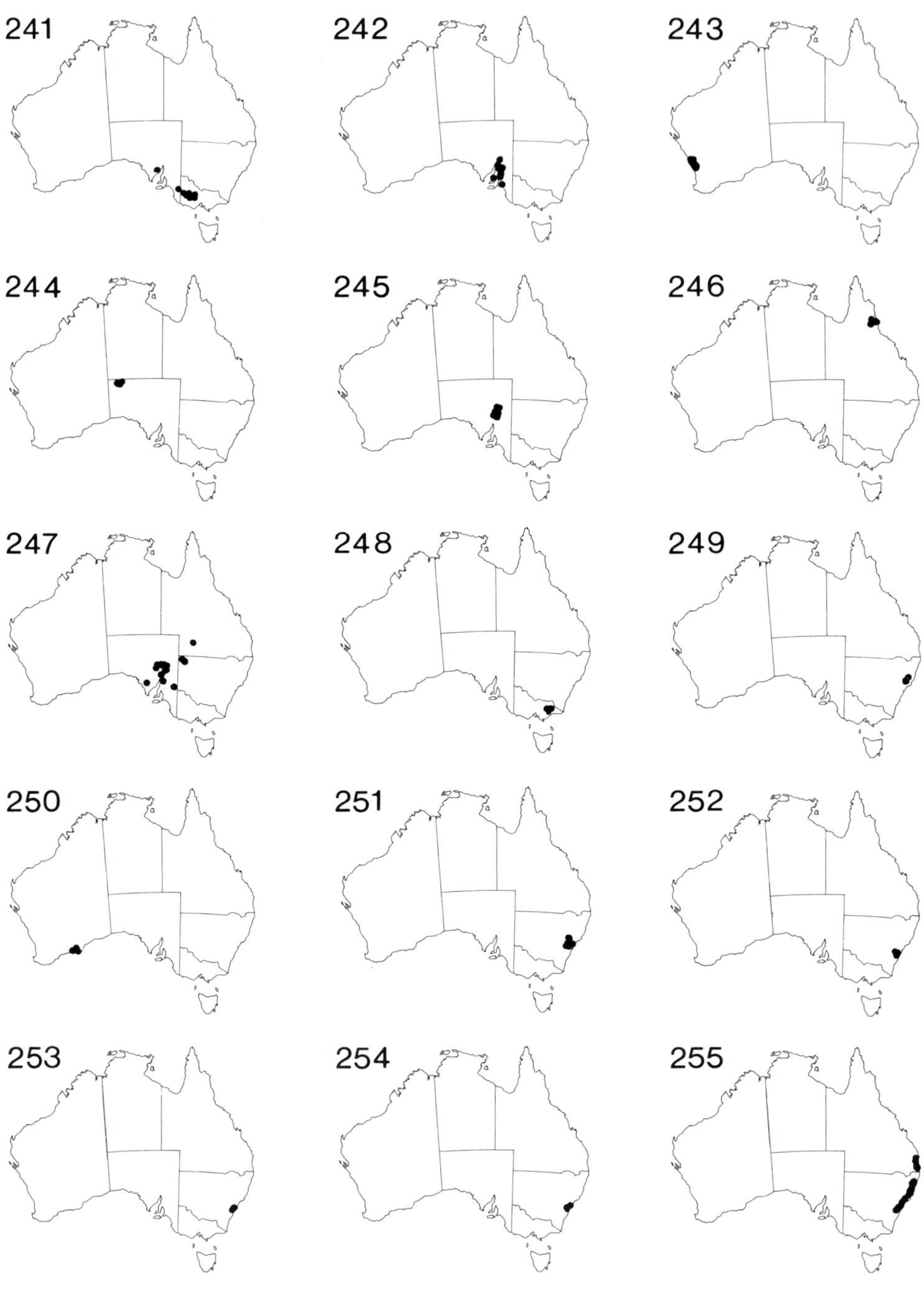

241. Goodenia benthamiana (196) **242**. Goodenia albiflora (197) **243**. Goodenia xanthotricha (197)
244. Goodenia brunnea (198) **245**. Goodenia saccata (198) **246**. Goodenia stirlingii (198)
247. Goodenia calcarata (199) **248**. Goodenia macmillanii (199) **249**. Goodenia fordiana (200)
250. Goodenia quadrilocularis (200) **251**. Goodenia decurrens (202) **252**. Goodenia rostrivalvis (202)
253. Goodenia dimorpha var. dimorpha (203) **254**. Goodenia dimorpha var. angustifolia (203) **255**. Goodenia stelligera (203)

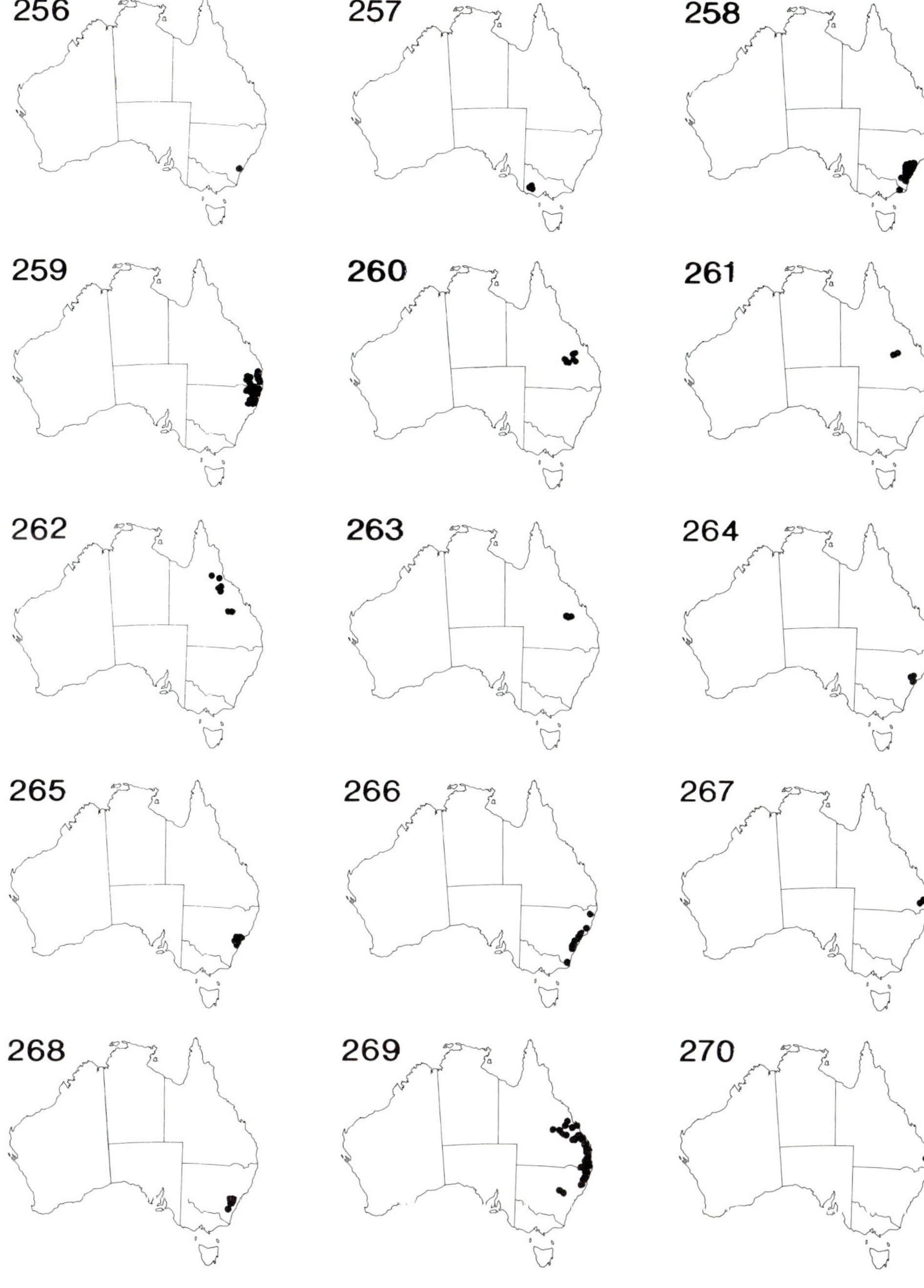

256. Goodenia glomerata (204)

257. Goodenia lineata (205)

258. Goodenia bellidifolia subsp. bellidifolia (206)

259. Goodenia bellidifolia subsp. argentea (206)

260. Goodenia racemosa var. racemosa (207)

261. Goodenia racemosa var. latifolia (207)

262. Goodenia disperma (207)

263. Goodenia viridula (207)

264. Goodenia stephensonii (208)

265. Goodenia heterophylla subsp. heterophylla (209)

266. Goodenia heterophylla subsp. eglandulosa (209)

267. Goodenia heterophylla subsp. teucriifolia (209)

268. Goodenia heterophylla subsp. montana (210)

269. Goodenia rotundifolia (210)

270. Goodenia arenicola (211)

271. Goodenia delicata (211)
272. Goodenia hederacea subsp. hederacea (212)
273. Goodenia hederacea subsp. alpestris (212)
274. Goodenia xanthosperma (213)
275. Goodenia centralis (221)
276. Goodenia goodeniacea (221)
277. Goodenia rupestris (223)
278. Goodenia blackiana (223)
279. Goodenia geniculata (224)
280. Goodenia lanata (224)
281. Goodenia affinis (226)
282. Goodenia convexa (226)
283. Goodenia tripartita (227)
284. Goodenia willisiana (227)
285. Goodenia robusta (228)

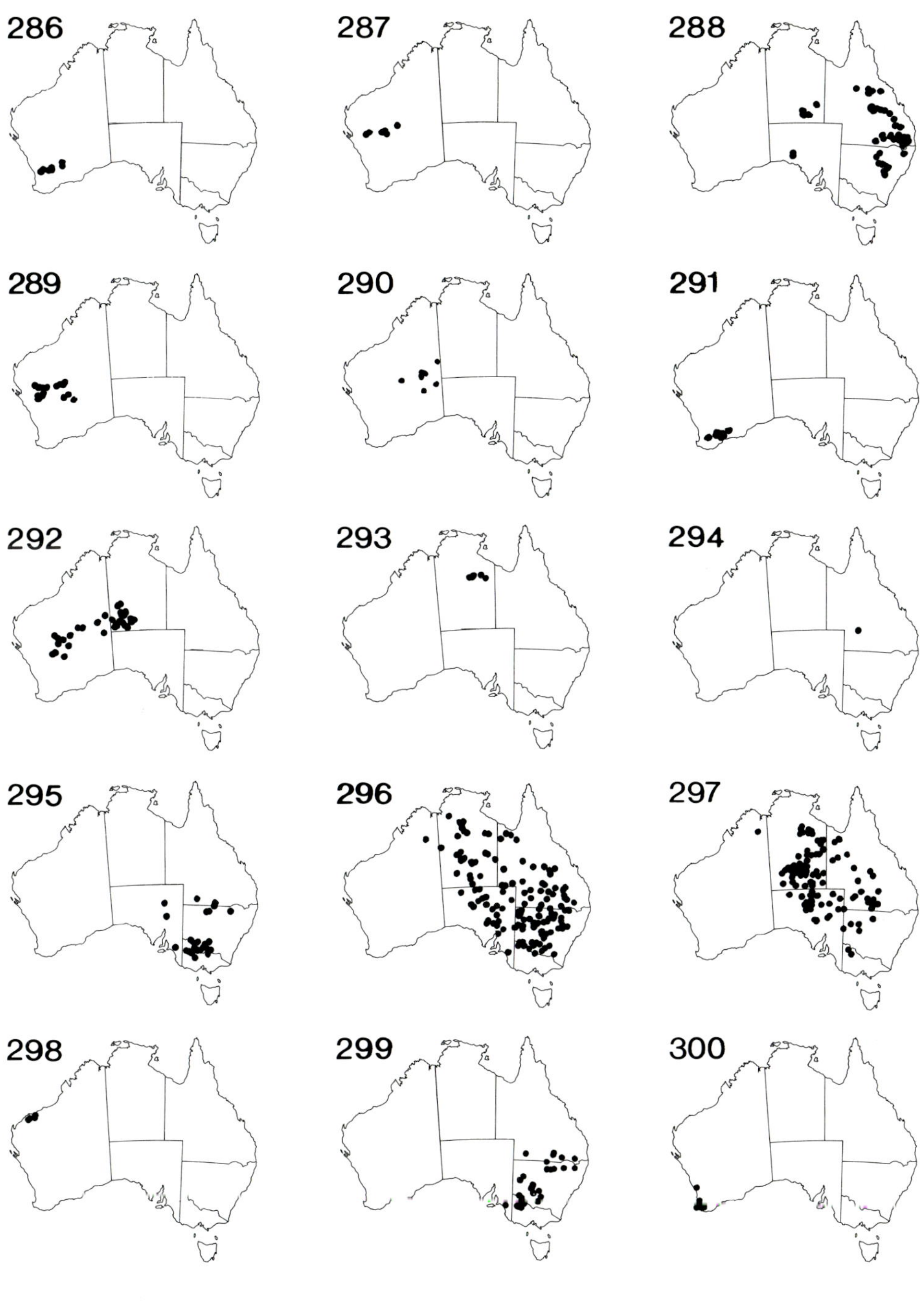

286. Goodenia dyeri (228)
287. Goodenia wilunensis (228)
288. Goodenia glabra (229)
289. Goodenia peacockiana (230)
290. Goodenia schwerinensis (230)
291. Goodenia laevis (231)
292. Goodenia mueckeana (231)
293. Goodenia nigrescens (232)
294. Goodenia angustifolia (232)
295. Goodenia glauca (233)
296. Goodenia fascicularis (233)
297. Goodenia lunata (234)
298. Goodenia pascua (234)
299. Goodenia heteromera (235)
300. Goodenia pusilla (235)

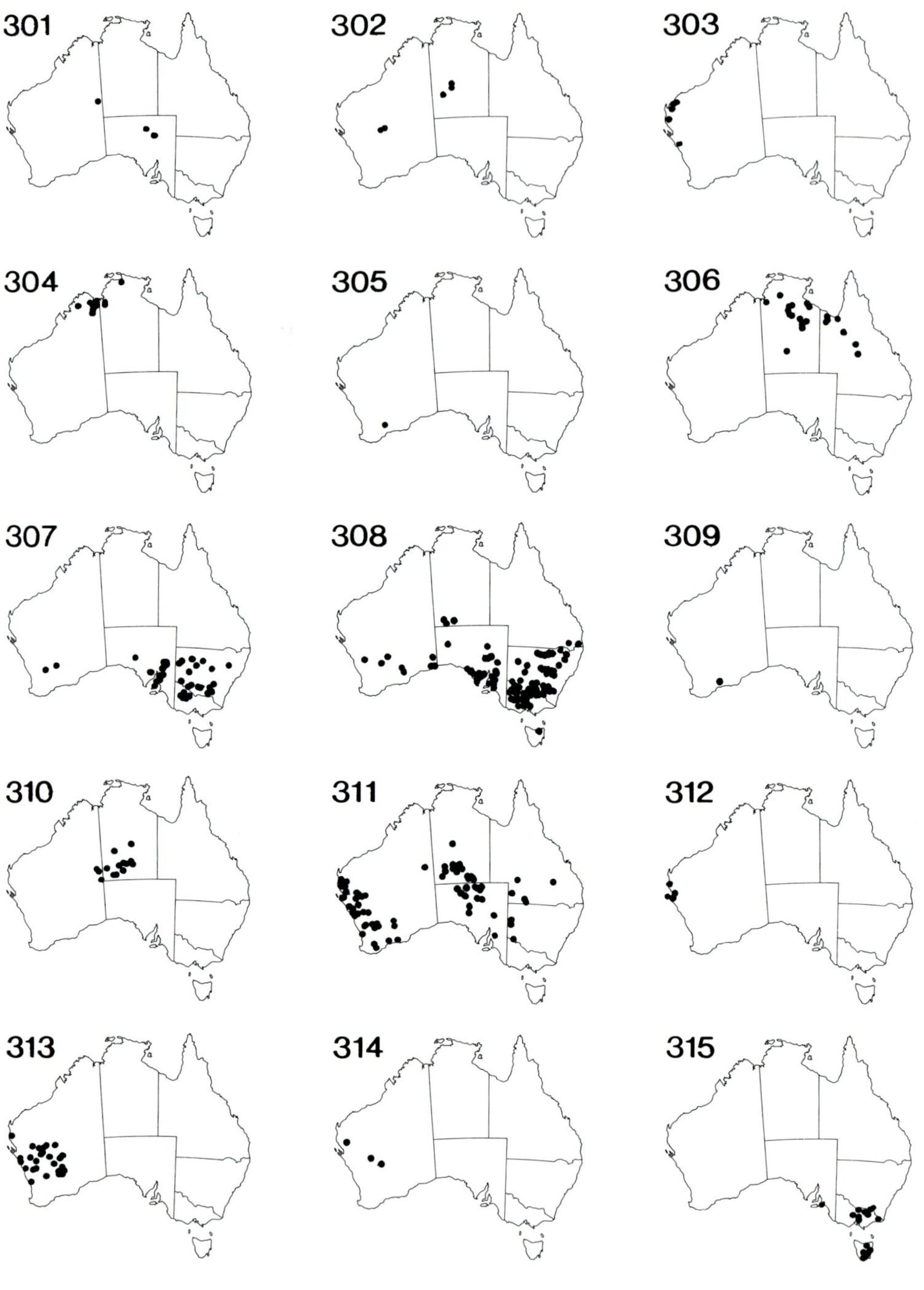

301. Goodenia anfracta (236)
302. Goodenia maideniana (236)
303. Goodenia corynocarpa (237)
304. Goodenia coronopifolia (237)
305. Goodenia integerrima (238)
306. Goodenia strangfordii (238)
307. Goodenia pusilliflora (239)
308. Goodenia pinnatifida (239)
309. Goodenia phillipsiae (240)
310. Goodenia gibbosa (240)
311. Goodenia berardiana (241)
312. Goodenia ochracea (243)
313. Goodenia mimuloides (243)
314. Goodenia macroplectra (245)
315. Goodenia elongata (245)

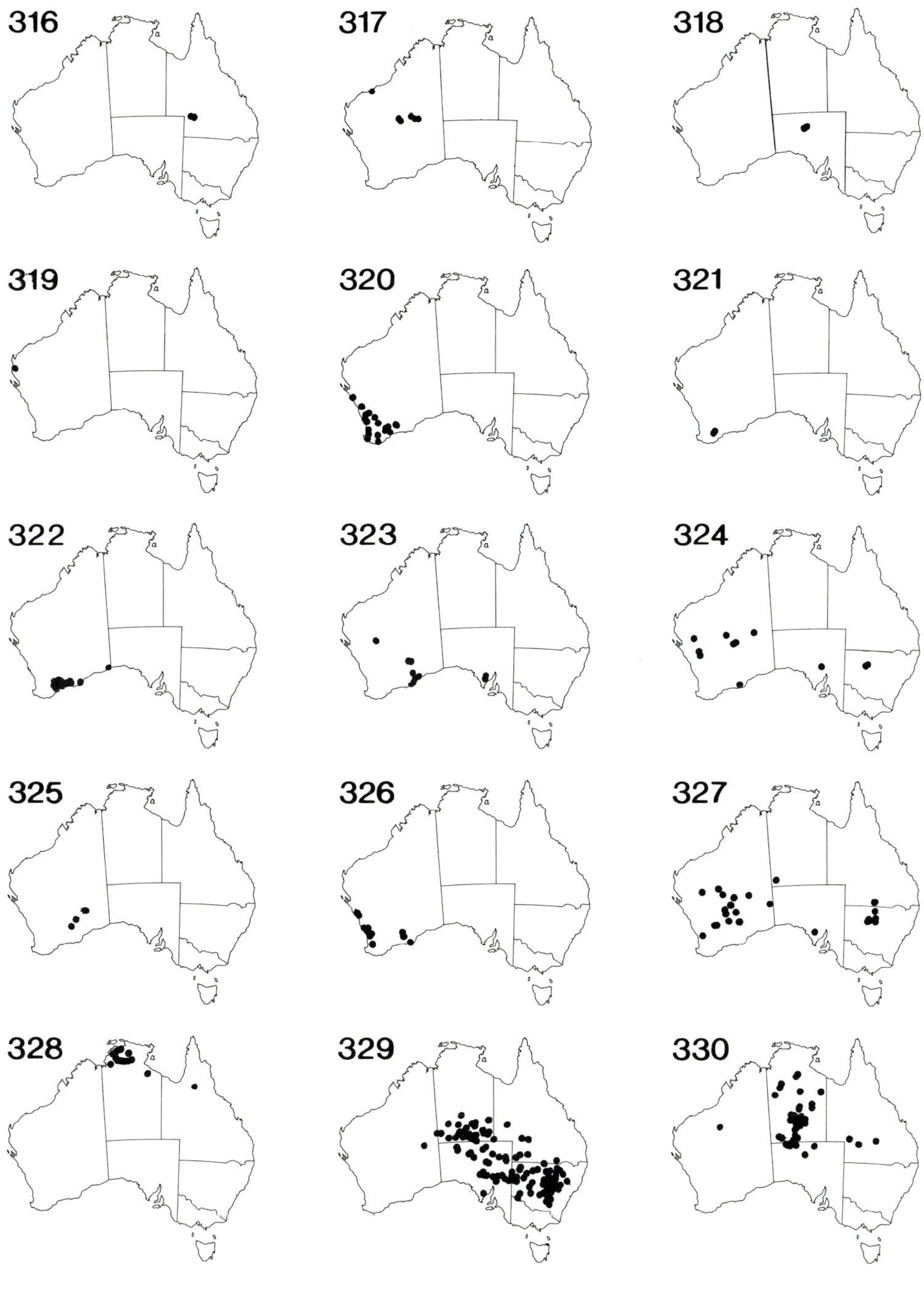

316. Goodenia megasepala (246) **317**. Goodenia iyouta (246) **318**. Goodenia lobata (248)
319. Goodenia salmoniana (248) **320**. Goodenia pulchella (248) **321**. Goodenia filiformis (250)
322. Goodenia concinna (250) **323**. Goodenia quasilibera (250) **324**. Goodenia occidentalis (251)
325. Goodenia krauseana (251) **326**. Goodenia micrantha (252) **327**. Goodenia havilandii (252)
328. Goodenia janamba (253) **329**. Goodenia cycloptera (253) **330**. Goodenia heterochila (254)

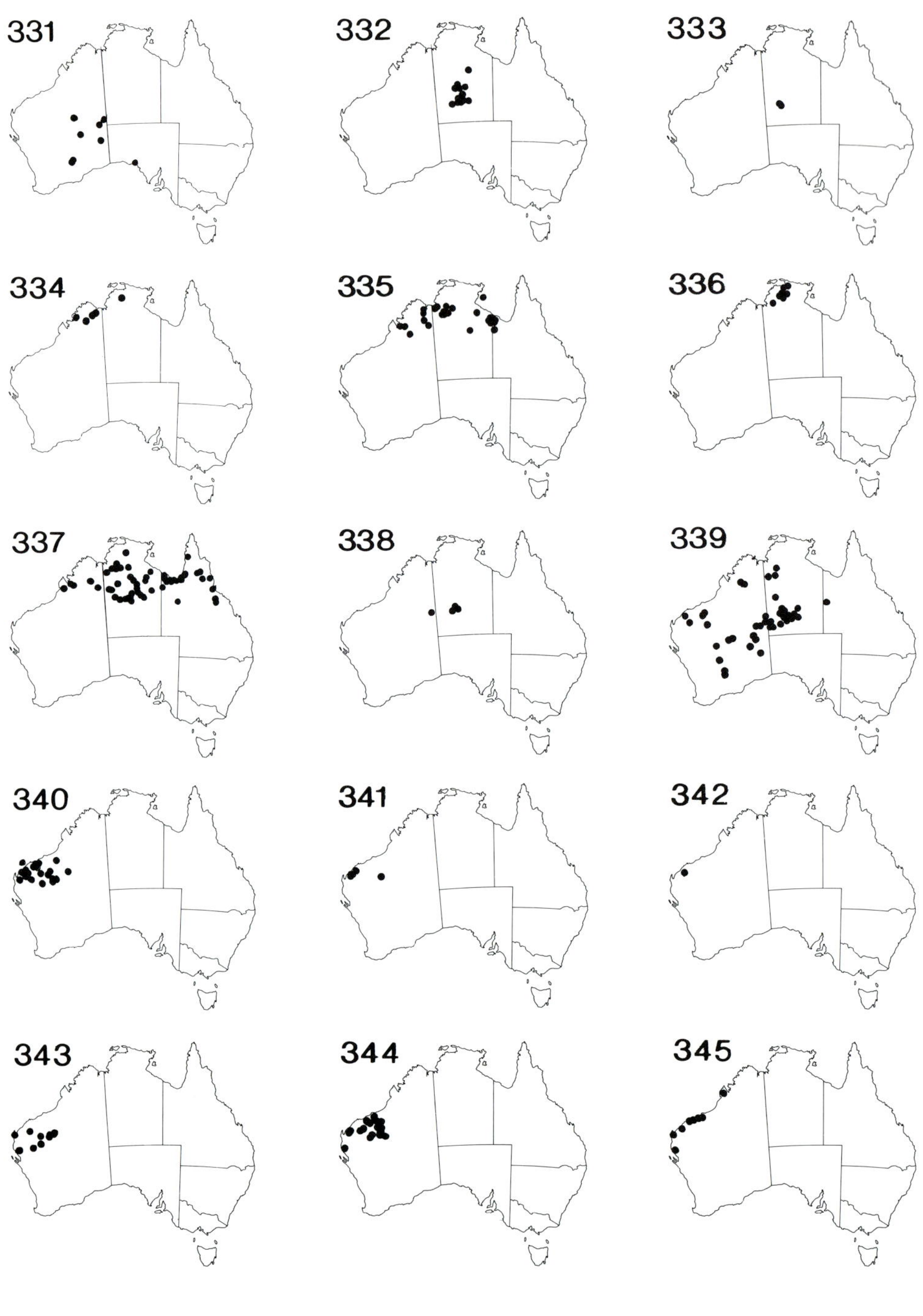

331. Goodenia glandulosa (254)
332. Goodenia larapinta (255)
333. Goodenia faucium (255)
334. Goodenia redacta (255)
335. Goodenia odonnellii (256)
336. Goodenia cirrifica (258)
337. Goodenia armitiana (258)
338. Goodenia virgata (259)
339. Goodenia triodiophila (259)
340. Goodenia microptera (260)
341. Goodenia nuda (260)
342. Goodenia pallida (261)
343. Goodenia prostrata (261)
344. Goodenia muelleriana (261)
345. Goodenia forrestii (262)

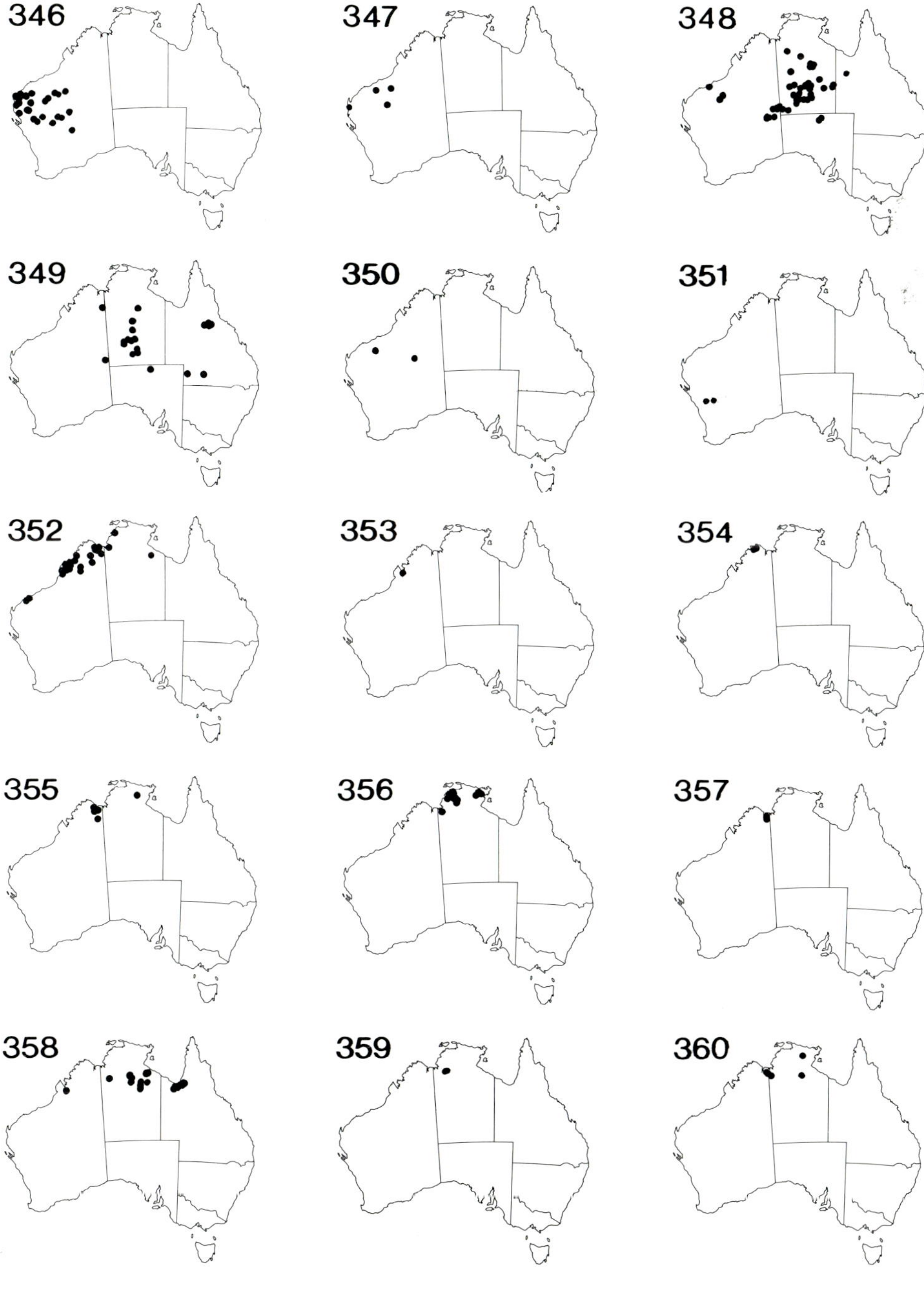

346. Goodenia tenuiloba (262)
347. Goodenia cusackiana (264)
348. Goodenia vilmorinae (265)
349. Goodenia hirsuta (265)
350. Goodenia stellata (266)
351. Goodenia neogoodenia (266)
352. Goodenia sepalosa var. sepalosa (267)
353. Goodenia sepalosa var. glandulosa (268)
354. Goodenia arachnoidea (268)
355. Goodenia brachypoda (268)
356. Goodenia leiosperma (269)
357. Goodenia durackiana (269)
358. Goodenia byrnesii (269)
359. Goodenia campestris (270)
360. Goodenia malvina (270)

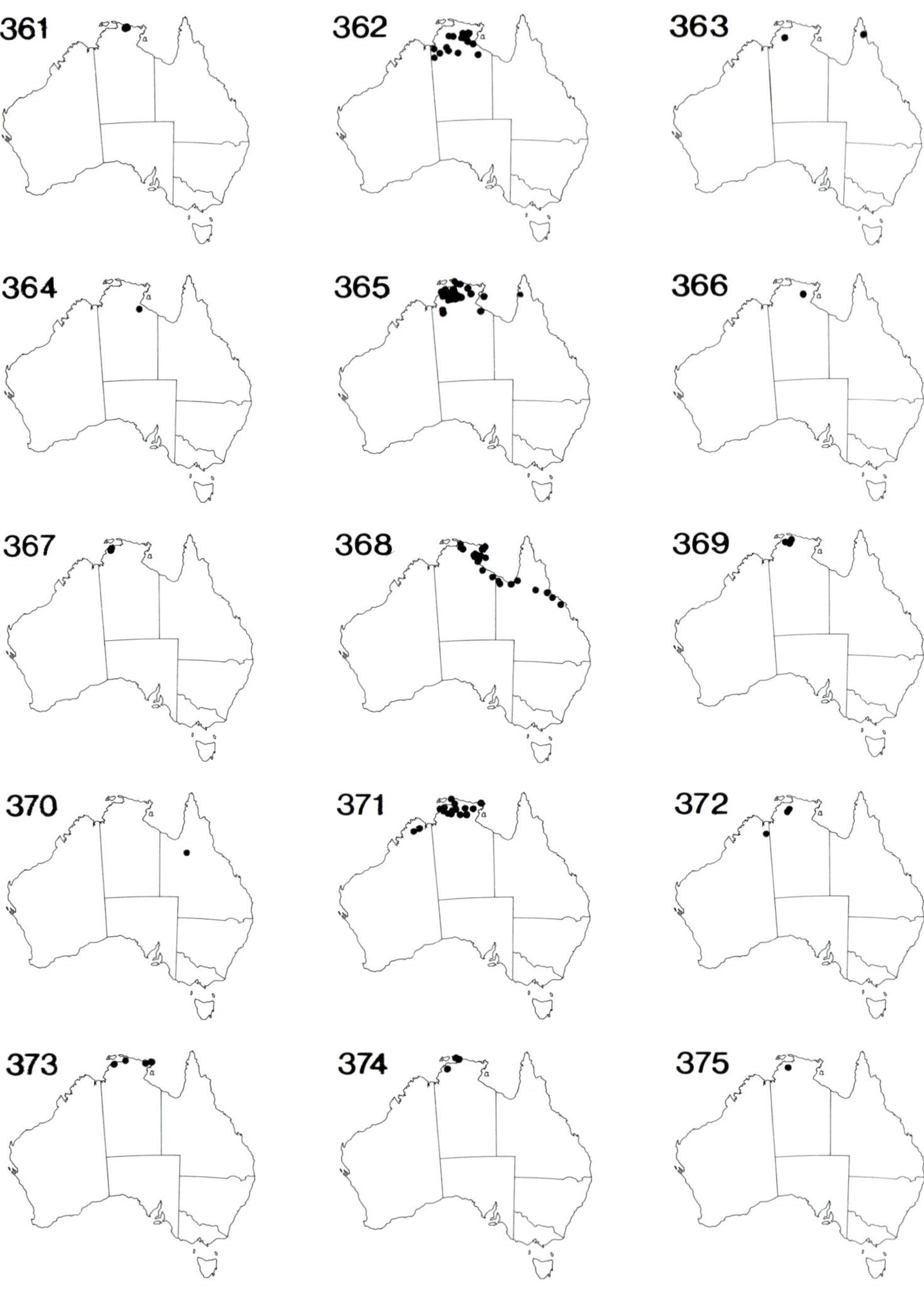

361. Goodenia potamica (271)
362. Goodenia hispida (271)
363. Goodenia subauriculata (273)
364. Goodenia chthonocephalata (273)
365. Goodenia armstrongiana (273)
366. Goodenia argillacea (274)
367. Goodenia porphyrea (274)
368. Goodenia pilosa (275)
369. Goodenia neglecta (275)
370. Goodenia heteroptera (276)
371. Goodenia holtzeana (276)
372. Goodenia heppleana (278)
373. Goodenia symonii (278)
374. Goodenia purpurea (279)
375. Goodenia quadrifida (279)

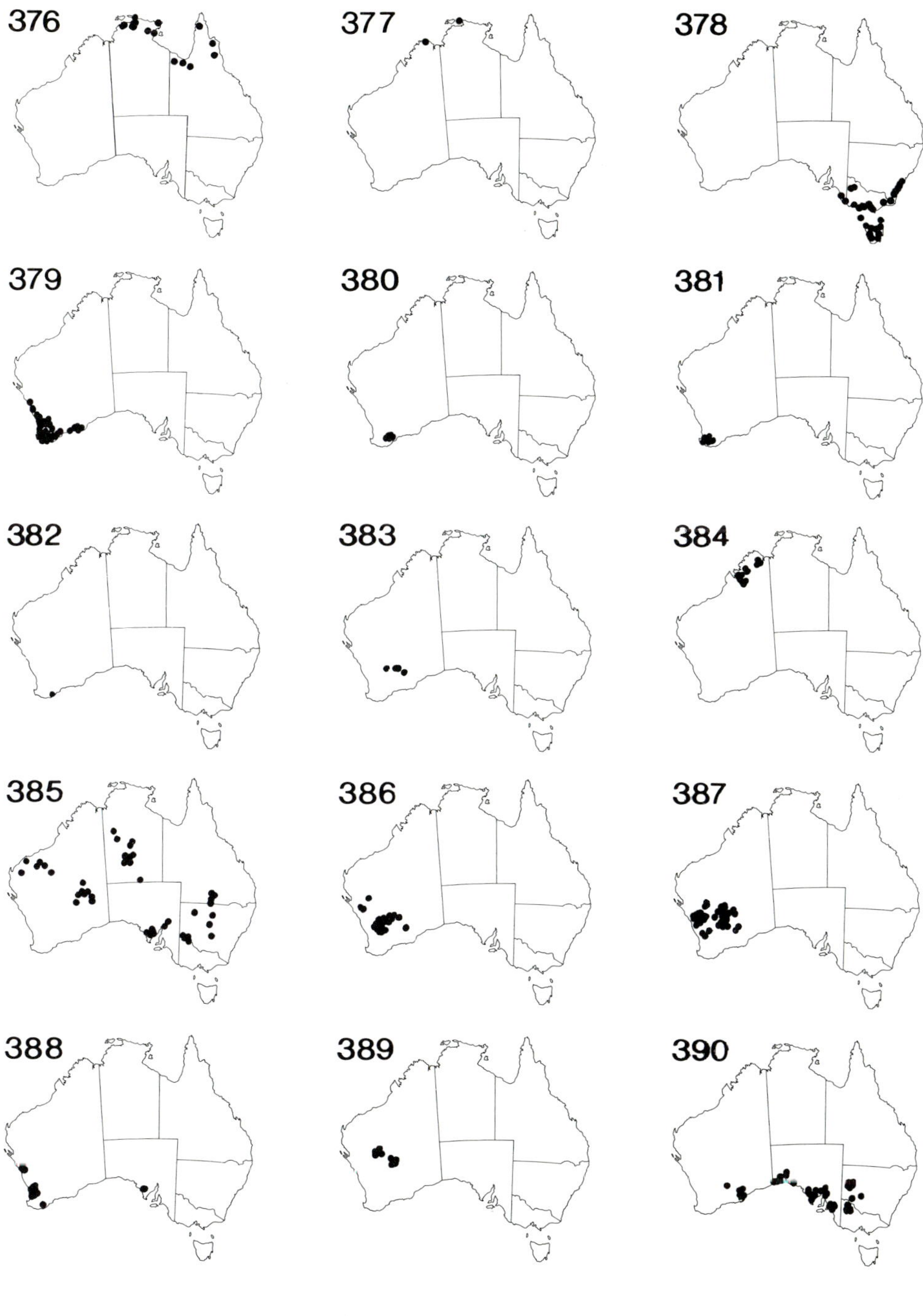

376. Goodenia pumilio (280)
377. Goodenia kakadu (280)
378. Selliera radicans (282)
379. Velleia trinervis (285)
380. Velleia foliosa (285)
381. Velleia macrophylla (286)
382. Velleia exigua (286)
383. Velleia daviesii (287)
384. Velleia panduriformis (287)
385. Velleia connata (287)
386. Velleia discophora (288)
387. Velleia rosea (288)
388. Velleia cycnopotamica (289)
389. Velleia hispida (289)
390. Velleia arguta (291)

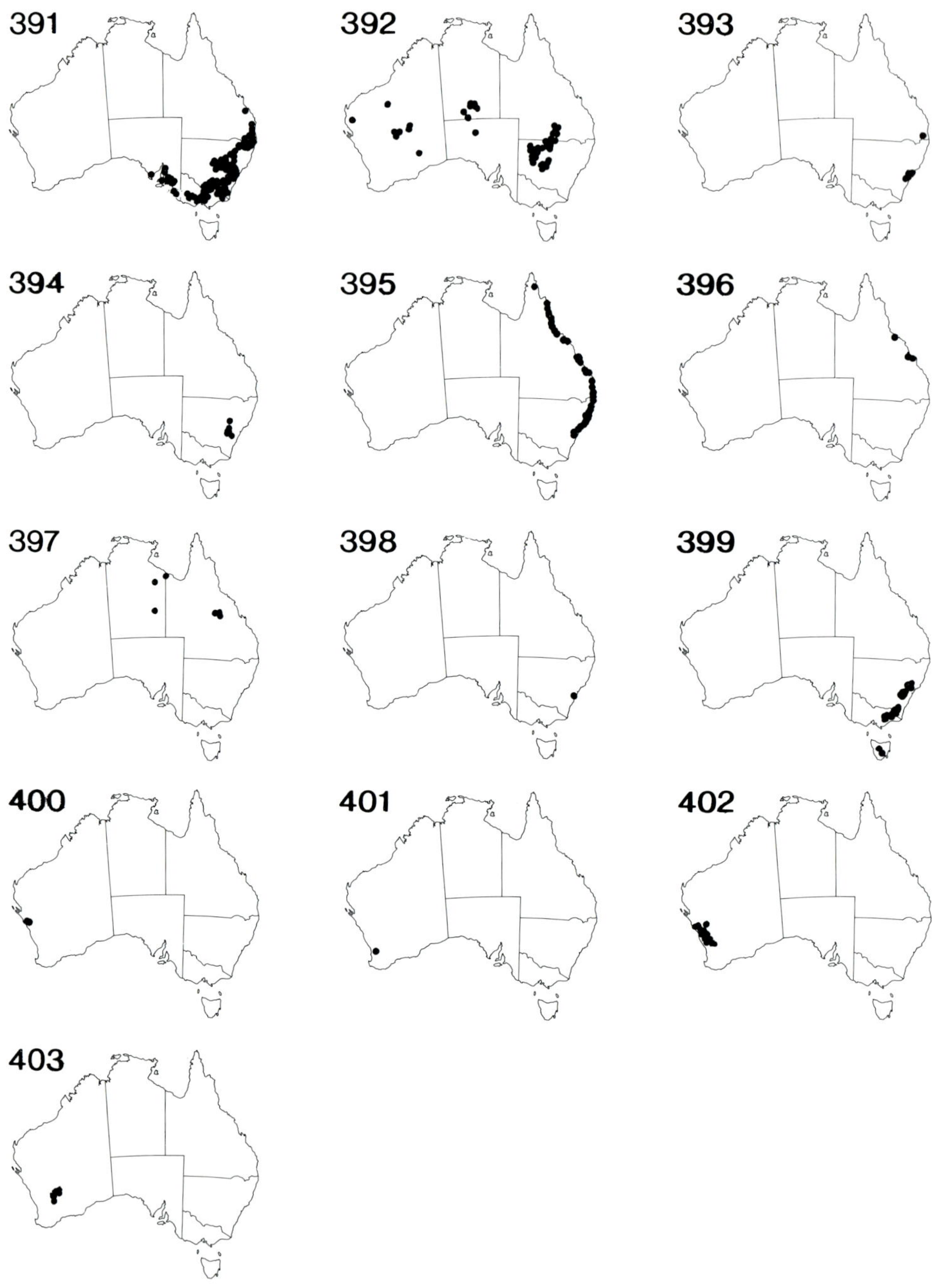

391. Velleia paradoxa (291)
392. Velleia glabrata (292)
393. Velleia lyrata (294)
394. Velleia parvisepta (294)
395. Velleia spathulata (294)
396. Velleia pubescens (295)
397. Velleia macrocalyx (295)
398. Velleia perfoliata (296)
399. Velleia montana (296)
400. Pentaptilon careyi (297)
401. Verreauxia verreauxii (298)
402. Verreauxia reinwardtii (300)
403. Verreauxia villosa (300)

APPENDIX

New taxa, combinations and lectotypifications

New taxa, combinations and lectotypifications occurring in this Volume of the *Flora of Australia* are formally published here. Names are arranged alphabetically except infrageneric taxa which are arranged alphabetically in descending order of rank. For economy, the entries are brief; the treatment in the main text is more comprehensive. The date of publication of this Volume will be given in Volumes 50 and 54.

For the lectotypes here chosen by R.C.Carolin, and by M.T.M.Rajput and R.C.Carolin, the following approach has been adopted. For taxa described by Robert Brown, Brown's own collections at BM have been selected unless there is a strong reason for not doing so. In all cases below, the BM specimens were adequate for defining and applying the names.

L.Preiss's specimens at LD are the reference collection for taxa described in Lehmann's *Plantae Preissianae* (see M.D.Crisp, Plantae Preissianae Types at Lund, *Austral. Syst. Bot. Soc. Newsletter* 36: 4–6, 1983). Where possible these have been selected as lectotypes.

Types for G.Bentham's taxa are selected from material at K annotated by him. Likewise, types for F.Mueller's taxa are selected from material at MEL annotated by him, unless there is good reason for doing otherwise.

In some cases where these guidelines cannot be followed the most complete specimen corresponding to the protologue has been selected.

For generic and infrageneric names, the lectotypes are names that are considered representative of the circumscription in the protologue.

GOODENIACEAE

DAMPIERA

M.T.M. Rajput & R.C.Carolin

Dampiera ser. **Angulares** Rajput & Carolin, ser. nov.

Caules in sectione transversali triangulares.

Type: *Dampiera stricta* (Smith) R.Br.

A formal diagnosis is published here in order to validate the name, used illegitimately under Article 22 of the *International Code of Botanical Nomenclature* in *Telopea* 3: 194 (1988). The epithet refers to the stems.

Dampiera carinata Benth., *Fl. Austral.* 4: 111 (1868)

T: south-western W.A., *J.Drummond 2: 397*; lecto (here chosen): MEL; isolecto: BM.

Dampiera cuneata R.Br., *Prodr.* 588 (1810)

T: King George Sound, [W.A.], Dec. 1801, *R.Brown*; lecto (here chosen): BM.

Dampiera glabrescens Benth., *Fl. Austral.* 4: 119 (1868)

T: south-western W.A., *J.Drummond 4: 194*; lecto (here chosen): K; isolecto: BM, MEL.

Dampiera haematotricha subsp. **dura** (Benth.) Rajput & Carolin, comb. nov.

D. altissima var. *dura* Benth., *Fl. Austral.* 4: 113 (1868). T: south-western W.A., *J.Drummond 5: 71*; holo: K; iso: BM, MEL, NSW.

GOODENIA

R.C.Carolin

Goodenia Smith, *Spec. Bot. New Holl.* 15 (1793)

Type: *G. ovata* Smith; lecto (here chosen).

Goodenia subg. **Monochila** (G.Don) Carolin, stat. nov.

Goodenia sect. *Monochila* G.Don, *Gen. Hist.* 3: 725 (1834). T: *G. scapigera* R.Br.; lecto (here chosen).

Goodenia sect. **Caeruleae** (Benth.) Carolin, stat. nov.

Goodenia sect. *Eugoodenia* ser. *Caeruleae* Benth., *Fl. Austral.* 4: 53, 65 (1868). T: *G. caerulea* R.Br.; lecto (here chosen).

Goodenia sect. *Ochrosanthus* G.Don, *Gen. Hist.* 3: 724 (1834)

Type: *G. ovata* Smith; lecto (here chosen).

Goodenia sect. **Porphyranthus** G.Don, *Gen. Hist.* 3: 725 (1834)

T: *G. purpurascens* R.Br.; lecto (here chosen).

Goodenia subsect. **Borealis** Carolin, subsect. nov.

Pedicelli ebracteolati. Valvae capsularum persistentes, hiantes. Semina elliptica, ala minima vel obsoleta.

Type: *G. sepalosa* F.Muell. ex. Benth.

A subsection of 23 species in Australia, 1 extending also to New Guinea and 1 to the Philippines. The epithet refers to the northern distribution of the included species.

Goodenia subsect. **Caeruleae** (Benth.) Carolin, stat. nov.

Goodenia sect. *Eugoodenia* ser. *Caeruleae* Benth., *Fl. Austral.* 4: 53, 65 (1868). T: *G. caerulea* R.Br.; lecto (here chosen).

Goodenia subsect. **Ebracteolatae** K.Krause, *Pflanzenr*. 54: 46 (1912)

Goodenia ser. *Pedicellosae* Benth., *Fl. Austral.* 4: 54, 73 (1868). T: *G. cycloptera* R.Br.; lecto (here chosen).

Goodenia subsect. **Scaevolina** Carolin, subsect. nov.

Suffrutices. Pedicelli bracteolati. Corolla caerulea. Semina colliculata alis parvis.

Type: *G. scaevolina* F.Muell.

A subsection of 8 species, mostly in northern Australia but 2 in the south-west. The subsectional epithet is taken from the name of the type species.

Goodenia ser. **Borealis** Carolin, ser. nov.

Sepala ad 2.5 mm lata. Stylus simplex.

Type: *G. sepalosa* F.Muell. ex Benth.

A series of 14 species in Australia, 1 extending to New Guinea. The epithet refers to the northern distribution of the included species.

Goodenia ser. *Bracteolatae* Benth., *Fl. Austral.* 4: 52, 59 (1868)

Type: *G. ovata* Smith; lecto (here chosen).

Goodenia ser. **Calogyne** (R.Br.) Carolin, comb. et stat. nov.

Calogyne R.Br., *Prodr.* 579 (1810). T: *G. pilosa* (R.Br.) Carolin, *fide* R.C.Carolin, *Brunonia* 2: 3 (1979).

Goodenia ser. *Foliosae* Benth., *Fl. Austral.* 4: 53, 69 (1868)

Type: *G. strangfordii* F.Muell.; lecto (here chosen).

Goodenia ser. *Racemosae* Benth., *Fl. Austral.* 4: 51, 57 (1868)

Type: *G. decurrens* R.Br.; lecto (here chosen).

Goodenia ser. *Rosulatae* K.Krause, *Pflanzenr.* 54: 46, 52 (1912)

Type: *G. geniculata* R.Br.; lecto (here chosen).

Goodenia amplexans var. *angustifolia* K.Krause, *Pflanzenr.* 54: 62 (1912)

T: Western R., Kangaroo Is., S.A., Sept. 1908, *Rogers*, lecto (here chosen): NSW; isolecto: B (destroyed).

Goodenia arthrotricha F.Muell. ex Benth., *Fl. Austral.* 4: 62 (1868)

T: south-western W.A., *J.Drummond 4: 190*; lecto (here chosen): K; isolecto: BM, MEL.

Goodenia boormannii K.Krause, *Pflanzenr.* 54: 56 (1912)

T: Dubbo, N.S.W., Jan. 1898, *J.Boorman*; lecto (here chosen): W; isolecto: ?B (destroyed) *n.v.*, NSW.

Krause cited two sheets, at B and W, and the former was apparently destroyed in 1942.

Goodenia grandiflora var. *macmillanii* (F.Muell.) K.Krause, *Pflanzenr.* 54: 75 (1912)

T: near Macalister R., Gippsland, Vic., *F.Mueller*; lecto (here chosen): MEL; isolecto: ?BM, K, MEL.

Goodenia scapigera var. *parviflora* Benth., *Fl. Austral.* 4: 57 (1868)

T: Phillips Ra. [i.e. The Barrens], W.A., *G.Maxwell*; lecto (here chosen): K; isolecto: MEL.

Goodenia semiteres Domin, *Biblioth. Bot.* 89: 644 (1929)

T: near Pentland, Qld, Mar. 1910, *K.Domin 8788*; lecto (here chosen): PR; isolecto: PR.

The lectotype is the better of the two sheets at PR seen and annotated by Domin.

Lobelia longiscapa Vriese in J.G.C.Lehmann, *Pl. Preiss.* 1: 398 (1845)

T: interior of south-western W.A., 1 Nov. 1839, *J.A.L.Preiss 1435*; lecto (here chosen): LD; isolecto: G.

Stekhovia Vriese, *Natuurk. Verh. Holl. Maatsch. Wetensch. Haarlem* ser. 2, 10: 166 (1854)

Type: *S. scapigera* (R.Br.) Vriese; lecto (here chosen).

LECHENAULTIA

D.A.Morrison

Lechenaultia aphylla D.Morrison, sp. nov.

Herbae vel fruticuli divaricati, caulibus insigniter porcatis. Folia sparsa, triangularia vel anguste triangularia, 0.5–1 mm longa, 0.5–1 mm lata, caduca. Inflorescentia monochasialis laxa vel flores interdum solitarii terminales, bracteis et bracteolis foliiformibus. Pseudocapsula 4-valvibus, 15 mm longa, non constricta, articulis circa 6–9 per seriebus, lateraliter complanatis, 1.5 mm longis, 1–1.5 mm diam., avellaneis ad atropurpureis, sulcis verticalibus exiguis.

T: 11.0 km NE of Mt Finke next to Dentons Track, S.A., 30°53'S, 134°07'W, 5 Oct. 1987, *D.E.Symon NPWS 1093*; holo: SYD; iso: AD.

Only collected once, the specimens being without flowers. The apparently leafless habit and the various fruit characteristics are distinctive within *Lechenaultia*. No other species of the genus is known from the region.

The specific epithet (Latin, *aphyllus*, leafless) refers to the almost leafless habit.

SCAEVOLA

R.C.Carolin

Scaevola amblyanthera var. **centralis** Carolin, var. nov.

Suffrutex parvus pilis glandulosis numerosis simplicibisque paucioribus. Folia sessilia anguste elliptica vel anguste obovata, plerumque dentata. Indusium ad basin pilis erectis albidis vel purpurascentibus.

T: 3 km NW of Kunoth Well, Hamilton Downs Stn, N.T., 11 Nov. 1974, *D.J.Nelson 2383*; holo: DNA.

The varietal epithet refers to the geographical location of the taxon in Central Australia.

Scaevola argentea Carolin, sp. nov.

Suffrutex prostratus ramosus ad 15 cm altus pilis confertis argenteis. Folia sessilia, elliptica ad obovata, 8–20 mm longa, recurvata. Flores spicis vel thyrsis compactis dispositi, bracteis imbricatis. Indusium ad basin pilis erectis purpureis.

T: 3 miles [c. 5 km] S of Lake Cobham, W.A., 6 Jan. 1972, *K.Newbey 3467*; holo: PERTH.

The specific epithet (Latin *argenteus*, silver) refers to the indumentum.

Scaevola attenuata R.Br., *Prodr.* 583 (1810)

T: unlabelled [probably from King George Sound, W.A., Dec. 1801, *R.Brown*]; lecto (here chosen): BM; isolecto: K.

Neither the lectotype nor the isolectotype bears a collecting label, but Brown's field notes indicate that they were collected at King George Sound.

Scaevola enantophylla F.Muell., *Fragm.* 8: 58 (1873)

Based on *S. oppositifolia sensu* F.Mueller, *Fragm.* 6: 225 (1868), non Roxb. T: Rockingham Bay, Qld, *J.Dallachy*; lecto (here chosen): MEL; isolecto: K.

Scaevola ovalifolia var. *glabra* R.Br., Prodr. 584 (1810).

T: Carpentaria, [N.T.], *R.Brown*; lecto (here chosen): BM; isolecto: MEL.

Scaevola hookeri (Vriese) F.Muell. ex J.D.Hook., *Fl. Tasman.* 1: 231 (1856)

Merkusia hookeri Vriese, *Ned. Kruidk. Arch.* 2: 159 (1851). T: Hampshire Hills, Tas., Feb. 1837, *R.C.Gunn 848*; lecto (here chosen): K.

Scaevola obovata Carolin, sp. nov.

Suffrutex ad 50 cm altus pilis simplicibus patentibus glandulosisque. Folia sessilia, obovata ad oblongo-obovata. Sepala crista sinuata deminuta. Corolla caerulea, 16–18 mm longa, barbulis simplicibus. Indusium ad basin pilis erectis purpurascentibus.

T: Dulgunia Hill, Tomkinson Ra., S.A., 4 Sept. 1978, *J.Z.Weber 5398*; holo: AD

The specific epithet (Latin *obovatus*, obovate) refers to the shape of the leaves.

Scaevola parvifolia var. *brevifolia* F.Muell., *J. Bot.* 15: 304 (1877).

T: [Queen] Victoria Spring, W.A., *Birch*; lecto (here chosen): MEL.

Scaevola restiacea subsp. **divaricata** Carolin, subsp. nov.

Suffrutex divaricatus ramis lateralis numerosis spiniformibus. Folia caulina parva. Indusium pilis albis paucis.

T: 15 miles [c. 24 km] E of Zanthus, W.A., 3 Oct. 1956, *R.D.Royce 5576*; holo: PERTH.

The subspecific epithet (Latin *divaricatus*, divaricate, widely spreading) refers to the branching habit.

Scaevola thesioides subsp. **filifolia** (E.Pritzel) Carolin, stat. nov.

S. thesioides var. *filifolia* E.Pritzel in F.L.E.Diels & E.Pritzel, *Bot. Jahrb.* 35: 571 (1905). T: near Esperance, W.A., Nov. 1901, *F.L.E.Diels 5937*; holo: ?B (destroyed); neo: 4 miles (6.4 km) S of Trulove [Truslove], W.A., 15 Oct. 1931, *W.E.Blackall 1041*; PERTH, *fide* R.C.Carolin, *Telopea* 3: 497 (1990).

VELLEIA

R.C.Carolin

Euthales pilosella Vriese in J.G.C.Lehmann, *Pl. Preiss.* 1: 414 (1845)

T: near Princess Royal Harbour, W.A., Dec. 1840, *J.A.L.Preiss 1438*; lecto: L (here chosen); isolecto: G, L, P.

There is no sheet of this collection at LD.

Velleia exigua (F.Muell.) Carolin, comb. nov.

Goodenia exigua F.Muell., *Fragm.* 3: 142 (1863). T: Moir Inlet, W.A., *G.Maxwell*; lecto (here chosen): MEL; isolecto: K.

Velleia trinervis var. *villosa* Benth., *Fl. Austral.* 4: 47 (1868)

T: south-western W.A., *J.Drummond 4: 188*; lecto (here chosen): K.

VERREAUXIA

R.C.Carolin

Verreauxia dyeri E.Pritzel ex Hemsley in W.J.Hooker, *Icon. Pl.* t. 2782 (1905)

T: railway between Cunderdin and Dedari, W.A., *G.H.Thistleton-Dyer*; lecto (here chosen): K.

Verreauxia reinwardtii (Vriese) Benth., *Fl. Austral.* 4: 105 (1868)

Scaevola reinwardtii Vriese in J.G.C.Lehmann, *Pl. Preiss.* 1: 409 (1845). T: Quangen, Victoria [district], W.A., 20 Mar. 1840, *J.A.L.Preiss 1454*; lecto (here chosen): LD.

SUPPLEMENTARY GLOSSARY

barbulae: in *Scaevola*, outgrowths on the margin of the wings or in the throat of the corolla; they may be simple or have apical hairs or papillae.

calli: in *Dampiera* small outgrowths in the throat of the corolla (acting as tactile guides for pollinators)

elaiosome: an appendage of a seed, usually rich in oil, not essential for the viability of the seed but attractive to fauna (especially ants) as a food for larvae etc. and hence an aid to dispersal by such fauna.

Abbreviations and Contractions

Literature

Author abbreviations follow the *Draft Index of Author Abbreviations compiled at the Herbarium, Royal Botanic Gardens Herbarium, Kew* (HMSO, London, 1980).

Journal titles are abbreviated in accordance with G.H.M.Lawrence *et al.*, *Botanico-Periodicum-Huntianum* (Hunt Botanical Library, Pittsburgh, 1968).

Other literature is abbreviated in accordance with F.A.Stafleu & R.S.Cowan, *Taxonomic Literature*, 2nd edn (Bohn, Scheltema & Holkema, Utrecht, 1976–1987), except that upper case initial letters are used for proper names and significant words. The *Flora of Australia* is abbreviated to *Fl. Australia.*

Herbaria

Abbreviations of herbaria are in accordance with P.K.Holmgren, N.H.Holmgren & L.C.Barnett, *Index Herbariorum* Part I, 8th edn (New York Botanical Garden, 1990). Those most commonly cited in the *Flora* are:

AD	State Herbarium of South Australia, Adelaide
BM	Natural History Museum, London
BRI	Queensland Herbarium, Brisbane
CANB	Australian National Herbarium, Canberra
CBG	Australian National Botanic Gardens Herbarium, Canberra
DNA	Northern Territory Herbarium, Darwin
HO	Tasmanian Herbarium, Hobart
K	Royal Botanic Gardens, Kew
MEL	National Herbarium of Victoria, Melbourne
NSW	National Herbarium of New South Wales, Sydney
PERTH	Western Australian Herbarium, Perth
QRS	Australian National Herbarium, Atherton

States, Territories, nearby countries

Abbreviations of Australian States and Territories and nearby countries as used in statements of distribution and citation of collections.

A.C.T.	Australian Capital Territory
N.Caled.	New Caledonia
N.S.W.	New South Wales
N.T.	Northern Territory
N.Z.	New Zealand
P.N.G.	Papua New Guinea
Qld	Queensland
S.A.	South Australia
Tas.	Tasmania
Vic.	Victoria
W.A.	Western Australia

Abbreviations and Contractions

General abbreviations

alt.	altitude
app.	appendix
auct.	*auctoris*/*auctorum* (of an author or authors)
auct. mult.	*auctoris multorum* (of many authors)
c.	*circa* (about)
Ck	Creek
cm	centimetre
col.	colour
coll.	collector
colln	collection
comb.	*combinatio*/combination
cons.	conservation
cult.	cultivated
Dept	Department
diam.	diameter
E	east
ed.	editor
edn	edition
et al.	*et alii*/and others
eds	editors
fam.	*familia*/family
f.	*forma*/form
fig./figs	figure/figures (in other works)
Fig.	Figure (referring to a Figure in this Volume of the *Flora*)
gen.	*genus*/genus
holo	holotype
hort.	*hortus* (garden) or *hortensis* (of a garden)
Hwy	Highway
in litt.	*in litteris* (in correspondence)
Is.	Island
iso	isotype
km	kilometre
lat.	latitude
lecto	lectotype
loc. cit.	*loco citato* (in the same work and page as just cited)
loc. id.	*loco idem* (in the same place as just cited)
long.	longitude
L.S.	longitudinal section
m	metre
mm	millimetre
Mt	Mount
Mtn	Mountain
Mtns	Mountains
N	north
n	haploid chromosome number
2n	diploid chromosome number
Natl	National
nom. cons.	*nomen conservandum* (conserved name)
nom. illeg.	*nomen illegitimum* (illegitimate name)

nom. inval.	*nomen invalidum* (name not validly published)
nom. nud.	*nomen nudum*
nom. prov.	*nomen provisorium* (provisional name)
nom. rej.	*nomen rejiciendum* (rejected name)
nom. superfl.	*nomen superfluum* (superfluous name)
nov.	*novus*/new
n. ser.	new series
n.v.	*non vidi* (not seen)
op. cit.	*opere citato* (in the work cited above)
orth.	orthography, orthographic
p./pp.	page/pages
pers. comm.	by personal communication
p.p.	*pro parte* (in part)
R.	River
Ra.	Range
Rd	Road
S	south
sect.	*sectio*/section
SEM	Scanning Electron Micrograph
ser.	*series*/series
s. lat.	*sensu lato* (in a wide sense)
s.n.	*sine numero* (without number)
sp./spp.	species (singular/plural)
s. str.	*sensu stricto* (in a narrow sense)
St	Street
stat.	*status*/status
Stn	(pastoral) Station
subg.	subgenus
subsp.	subspecies
suppl.	supplement
syn	syntype
synon.	synonym
T	Type (collection)
t.	*tabula* (plate)
trib.	*tribus*/tribe
trig.	trigonometric station
T.S.	transverse section
typ. cons.	*typus conservandus* (conserved type)
var.	*varietas*/variety
W	west
x	basic chromosome number

Symbols

*	naturalised taxon
[ ]	misapplied name or *nomen invalidum*; also, in localities, denotes a place name later than that originally cited or on the herbarium sheet
(f)	female
(m)	male

Publication date of previous volumes

Volume 1	22 August 1981
Volume 3	24 April 1989
Volume 4	12 November 1984
Volume 8	9 December 1982
Volume 18	8 June 1990
Volume 19	27 June 1988
Volume 22	17 May 1984
Volume 25	25 December 1985
Volume 29	27 July 1982
Volume 45	15 May 1987
Volume 46	2 May 1986

For the publication date of Volume 35, see Volumes 50 & 54.

INDEX

Accepted names are in roman, synonyms and doubtful names in *italic*.

Principal page references are in **bold**, figures in *italic*.

***Flora of Australia* — Index to families of flowering plants, current at January 1992.**

Bold figures denote published volumes.

	Volume		Volume		Volume
Acanthaceae	33	Cactaceae	**4**	Elaeagnaceae	16
Aceraceae	**25**	Caesalpiniaceae	12	Elaeocarpaceae	7
Actinidiaceae	6	Callitrichaceae	32	Elatinaceae	6
Agavaceae	**46**	Campanulaceae	34	Epacridaceae	9
Aizoaceae	**4**	Cannabaceae	**3**	Ericaceae	9
Akaniaceae	**25**	Cannaceae	**45**	Eriocaulaceae	40
Alangiaceae	**22**	Capparaceae	**8**	Erythroxylaceae	24
Alismataceae	39	Caprifoliaceae	36	Eucryphiaceae	10
Aloeaceae	**46**	Cardiopteridaceae	**22**	Euphorbiaceae	23
Alseuosmiaceae	10	Caryophyllaceae	5	Eupomatiaceae	2
Amaranthaceae	5	Casuarinaceae	**3**	Fabaceae	13,14,15
Anacardiaceae	**25**	Celastraceae	**22**	Fagaceae	**3**
Annonaceae	2	Centrolepidaceae	40	Flacourtiaceae	**8**
Apiaceae	27	Cephaloteceae	10	Flagellariaceae	40
Apocynaceae	28	Ceratophyllaceae	2	Frankeniaceae	**8**
Aponogetonaceae	39	Chenopodiaceae	**4**	Fumariaceae	2
Aquifoliaceae	**22**	Chrysobalanaceae	10	Gentianaceae	28
Araceae	39	Cistaceae	**8**	Geraniaceae	27
Araliaceae	27	Clusiaceae	6	Gesneriaceae	33
Arecaceae	39	Combretaceae	**18**	Globulariaceae	32
Aristolochiaceae	2	Commelinaceae	40	Goodeniaceae	**35**
Asclepiadaceae	28	Connaraceae	10	Grossulariaceae	10
Asteraceae	37,38	Convolvulaceae	30	Gunneraceae	**18**
Austrobaileyaceae	2	Corsiaceae	47	Gyrostemonaceae	**8**
Balanopaceae	**3**	Corynocarpaceae	**22**	Haemodoraceae	**45**
Balanophoraceae	**22**	Costaceae	**45**	Haloragaceae	**18**
Basellaceae	5	Crassulaceae	10	Hammelidaceae	**3**
Bataceae	**8**	Cucurbitaceae	**8**	Hanguanaceae	**46**
Berberidaceae	2	Cunoniaceae	10	Hernandiaceae	2
Betulaceae	**3**	Cuscutaceae	30	Himantandraceae	2
Bignoniaceae	33	Cymodoceaceae	39	Hippocrateaceae	**22**
Bixaceae	**8**	Cyperaceae	41,42	Hydatellaceae	**45**
Bombacaceae	7	Datiscaceae	**8**	Hydrocharitaceae	39
Boraginaceae	30	Davidsoniaceae	10	Hydrophyllaceae	30
Brassicaceae	**8**	Dichapetalaceae	**22**	Icacinaceae	**22**
Bromeliaceae	**45**	Dilleniaceae	6	Idiospermaceae	2
Brunoniaceae	**35**	Dioscoreaceae	**46**	Iridaceae	**46**
Burmanniaceae	47	Dipsacaceae	36	Juncaceae	40
Burseraceae	**25**	Donatiaceae	34	Juncaginaceae	39
Byblidaceae	10	Droseraceae	**8**	Lamiaceae	31
Cabombaceae	2	Ebenaceae	10	Lauraceae	2

BOTG-1340

QK 431 F567 V.35 1992